ACE YOUR COURSE WITH CIRCUIT WORKS!

FREE FOR DOWNLOAD

The Circuit Works simulator will help you learn the principles and relationships that underlie basic first- and second-order circuits. Confused by resistors, capacitors, inductors, independent sources, and transformers? With Circuit Works you can:

- Choose a circuit from a library of 100 active and passive circuits, with resistors, capacitors, inductors, independent sources, controlled current and voltage sources, transformers, and op-amps.
- Select an input signal generator from a library of 20 periodic and 15 aperiodic parameterized signals.

- Display the frequency-domain magnitude and phase (Bode) plots.
- Assign 2 sets of values to components and see in color-differentiated format how the change affects the frequency- and time-domain responses of the circuit.
- And much more…Circuit Works takes the work out of circuits!

Circuit Works is free to purchasers of **Basic Engineering Circuit Analysis, 7e,** by J. David Irwin. To download Circuit Works for free, simply go to **www.wiley.com/college/irwin** and register with the unique password included on the inside cover of this text.

IBM-PC compatible with Windows 95, 98, NT, 2000 and ME.

THE PROBLEM-SOLVING COMPANION– SOLUTIONS FOR YOUR SUCCESS!

The Problem-Solving Companion features additional problems for every chapter, with complete, worked-out solutions to walk you through the problem-solving process. Every step is shown in great detail, so you'll no longer have to wonder – now how did they get that? Includes a list of the most important formulas for solving circuits problems. Go to **www.wiley.com/college/irwin** to check out the free demo or purchase the Problem-Solving Companion.

WILEY

The Range of Current

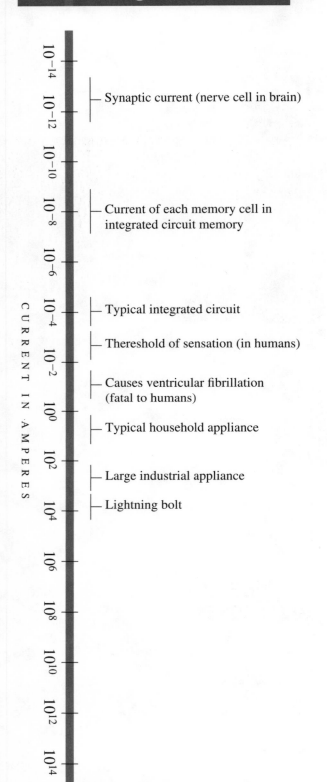

CURRENT IN AMPERES

- 10^{-14}
- 10^{-12} — Synaptic current (nerve cell in brain)
- 10^{-10}
- 10^{-8} — Current of each memory cell in integrated circuit memory
- 10^{-6}
- 10^{-4} — Typical integrated circuit
- 10^{-2} — Thereshold of sensation (in humans)
- 10^{0} — Causes ventricular fibrillation (fatal to humans)
- 10^{2} — Typical household appliance
- 10^{4} — Large industrial appliance
- Lightning bolt
- 10^{6}
- 10^{8}
- 10^{10}
- 10^{12}
- 10^{14}

The Range of Voltage

VOLTAGE IN VOLTS

- 10^{-8} — Antenna of radio receiver
- 10^{-6} — Electroencephalography (EEG)— Voltage between 2 points on human scalp
- 10^{-4} — Electrocardiograms (EKGs)— Voltage produced by heart across human chest
- 10^{-2}
- 10^{0} — Flashlight battery (1.5 volts)
- Voltage on integrated circuits
- Car battery (12 volts)
- 10^{2} — House wiring—U.S. (110V)
- House wiring—Europe (220V)
- Large industrial motors
- Voltage on TV picture tube
- 10^{4}
- 10^{6} — Cross-country (high-tension) transmission lines
- 10^{8} — Lightning

Basic
Engineering
Circuit Analysis

Basic
Engineering
Circuit Analysis

Seventh Edition
A Wiley First Edition

J. David Irwin

AUBURN UNIVERSITY

John Wiley & Sons, Inc.
New York   Chichester   Weinheim   Brisbane   Toronto   Singapore

ACQUISITIONS EDITOR	Bill Zobrist
ASSISTANT EDITOR	Jenny Welter
MARKETING MANAGER	Katherine Hepburn
SENIOR PRODUCTION EDITOR	Christine Cervoni
SENIOR DESIGNER	Karin Kincheloe
ILLUSTRATION COORDINATOR	Gene Aiello
TEXT DESIGNER	Jerry Wilke
COVER DESIGNER	Carol Grobe
COVER PHOTOS	Circuit board © Alan Carruthers/Photo Researchers Stocktreck © The Stock Market

This book was set in 10/12 Times by Preparè and printed and bound by Von Hoffmann Press, Inc. The cover was printed by Von Hoffmann Press, Inc.

This book is printed on acid free paper.

The Quality Assurance logo is meant to signify the author's and Wiley's commitment to produce the highest-quality text possible, with a minimum of technical and typographical errors. While it is difficult to claim that this text is absolutely free of these distractions, all text material and solutions have been meticulously prepared and proofed. The author, as well as independent reviewers have made every effort to find every technical or typographical error. It is our sincerest hope that by undertaking these additional efforts to ensure the quality of this text, the reader will be unfettered in their pursuit of learning. We welcome your comments on how the text can be further improved. Please write to: Electrical Engineering Editor, Higher Education Division, John Wiley & Sons, Inc., 605 Third Avenue, NY, NY 10158.

Library of Congress Cataloging-in-Publication Data:

Irwin, J. David,
    Basic engineering circuit analysis / J. David Irwin.—7th ed.
        p. cm.
    ISBN 0-471-40740-2 (cloth : alk. paper)
        1. Electric circuit analysis. I. Title.

TK454 .I78 2001
621.319'2—dc21   2001023737

Printed in the United States of America

10 9 8 7 6 5 4 3

To my loving family:
Edie
Geri, Bruno, Andrew and Ryan
John, Julie, John David and Abi
Laura

Brief Contents

Contents

7...AC Steady-State Analysis 258

8...Magnetically Coupled Networks 317

9...Steady-State Power Analysis 347

13...Application of the Laplace Transform to Circuit Analysis *518*

14...Fourier Analysis Techniques *561*

15...Two-Port Networks *609*

16...Basic Semiconductor Electronic Circuits *631*

Appendix Complex Numbers *655*

Index *661*

Answers to Selected
Problems...*www.wiley.com/college/irwin*

Preface

The Circuit Analysis Course in the Engineering Curriculum

Circuit analysis is not only fundamental to the areas of electrical and computer engineering—the concepts studied also have tentacles that extend to areas far beyond the boundaries of these topics. Today, this subject may get less time in the curriculum than in the past because of constraints that range from attempts to reduce the total number of hours required for graduation to the size or even growth of the university's core requirement. Therefore, a viable book on this topic must be designed to enhance the learning process on every possible front; that is, it must (1) be easily readable, (2) contain numerous learning aids so that the material can be quickly grasped, and (3) provide ample opportunity for readers to become proficient by testing their understanding. Thus, success is dependent not only on learning the material, but also on learning it quickly. I have designed this book to match those needs.

How This Text Meets the Needs of Instructor and Student

Seven editions of *Basic Engineering Circuit Analysis* have provided readers with a thorough understanding of the fundamental concepts of circuit analysis and their applications to real-world problems. Over the almost 20-year life cycle of this text, instructors and students alike have found the book's most helpful features to be:

1. the clear and concise explanations
2. a variety of effective learning aids
3. numerous problems with varying degrees of difficulty
4. a variety of real-world examples that demonstrate the usefulness of the material under discussion.

Like the previous editions, the seventh edition is designed as a core text for students majoring in electrical and computer engineering, as well as a basic introduction to electric circuits for students in other engineering disciplines. The goal is to provide an effective and efficient environment for students to obtain a thorough understanding of the analysis of electric circuits and an introduction to their design. The material is presented at a level suitable for students who are taking or have completed the introductory courses in college calculus and physics. I have taken great care to provide complete and clear discussions, as well as an abundant supply of new learning aids.

ACCURACY The text and problems/solutions have been thoroughly checked for clarity and accuracy not only by the author, but also by more than a dozen academics and one honor student. This group consists of 11 reviewers who are well-known professors of ECE, and they are:

Robert Krueger, University of Wisconsin
Ashok Goel, Michigan Technological University

Clifford Pollock, Cornell University
Peyton Peebles, University of Florida
Marty Kaliski, Cal Poly, San Luis Obispo
Darryl Morrell, Arizona State University
Peddapullaiah Sannuti, Rutgers University
Paul Greiling, UCLA
Jorge Aravena, LSU
James Rowland, University of Kansas
Thomas H. Shumpert, Auburn University

In addition, the following individuals served as accuracy checkers for every aspect of the text:

Professor Scott Smith, Boise State University
Professor Val Tareski, North Dakota State University
Mr. Jung Young Lee, UC Berkeley

FLEXIBILITY While the organization and presentation of the material have been designed to enable students to understand and apply the fundamentals as quickly as possible, the organization also provides instructors maximum flexibility to use the text in a variety of different course settings and learning styles. Some sections or chapters can be skipped completely and still provide a coherent presentation in a natural progression from start to finish. For example, those instructors who do not wish to cover the CAD tools, that is, PSPICE, MATLAB, and Microsoft EXCEL, can skip them with no loss of continuity. On the other hand, one could make computer-generated solutions a major focus of the course. In addition, the design material permits instructors to introduce this important topic early in the curriculum.

Pedagogical Structure Designed to Reinforce Learning

Students don't all learn in the same way. Some are visual learners, while others are more kinesthetic. A "learning styles survey" appears after the preface to help each student determine his or her particular learning style and gives guidance on how to tailor his or her study habits. Pedagogical features are included to fit different learning styles.

- *Learning Goals*, listed at the beginning of each chapter, provide an overview of the topics within the chapter and the skills and knowledge students should achieve.

- *Learning Hints* that appear on many pages of the text help shorten the learning curve. These comments in the margin provide guidance for understanding different facets of the presentation and problems of all types. Coupled with myriad examples, Learning Hints provide readers with a companion tutor. Additionally, they aid the instructor and the student by conveying some of the subtleties that are typically implicit in lecture or in traditional presentation.

- *Learning by Example* sections, more often than any other component, provide students with the means for acquiring and evaluating new knowledge. The numerous worked-out examples in the text are the hallmark feature.

- An expanded number of real-world examples, labeled *Learning by Application*, appear in many sections of the text, and at the end of every applicable chapter, answer the question "Why do we study circuit theory?" Applications frequently deal with design issues ranging from very simple matters, such as finding the value of some specific component, to modeling the collapse of the Tacoma Narrows Bridge.

- *Learning by Doing* and *Learning Extensions* are assessment tools coordinated within the text. The Learning by Doing exercises are quick, simple reinforcements of the principles and provide a check of the reader's understanding of the material. Learning Extensions provide practice for the reader in applying the basic concepts, as well as guidance in understanding the techniques needed to solve the end-of-chapter problems.

▶ *Problem-Solving Strategies* are placed to assist the student in selecting the proper solution technique, or combination of techniques applicable in a particular situation. This assistance not only helps the student understand the subtle differences among various techniques in their application to a particular problem, but also helps eliminate the psychological barrier that sometimes exists in determining a suitable method of attack.

▶ *Computer-aided Design (CAD) Tools* allow students, like all modern engineers, to apply the power of the computer to solve a variety of problems. Special icons are employed within the book to indicate sections where the CAD tools are used. The very latest version of PSPICE by Cadence is used, and this version coupled with the use of both MATLAB and Microsoft EXCEL are integrated within the text and coordinated with the Student Study Guide (discussed later) where Electronics Workbench is also introduced.

▶ *Learning by Design* sections appear at the end of each applicable chapter. This feature provides the reader with an understanding of how to apply what they have learned to the design of circuits. The use of engineering design in a curriculum is a major component of the ABET criteria. The inclusion of this material permits its introduction to the student at an early stage in the curriculum.

▶ *Learning Check* includes both the Summary and Problems, and appears at the end of every chapter. The important topics are reviewed concisely in the Summary as a quick reminder for readers. The problems are segmented by chapter subdivision and graduated in difficulty to permit users to test their understanding of the material and hone their skills in solving different types of problems. *The problem sets also include some problems specifically designed to mimic those that appear on the Fundamentals of Engineering (FE) Exam taken by students in preparation for becoming a Registered Professional Engineer.*

New to This Edition

This edition generally follows the presentation outlined in previous editions. However, I have made some important changes that are designed to meet the current needs of the course and that respond to the experience and comments of colleagues who have taught from the book. As is my custom, the problem sets are new, the presentation has been streamlined to include important new material, and a number of new and exciting real-world applications are presented throughout. Other specific improvements are outlined on a chapter-by-chapter basis as follows:

Chapter 2—Standard resistor values are presented, and their tolerances and power ratings are discussed using examples. The delta/wye and wye/delta transformations have also been inserted here.

Chapter 3—MATLAB is introduced here as a CAD tool for solving a set of linearly independent simultaneous equations for both nodal and loop analysis. The presentation of op-amps is updated to reflect the use of modern components.

Chapter 4—Microsoft EXCEL is introduced and the new version of PSPICE by Cadence is introduced and employed in the solution of dc circuits.

Chapter 5—Standard values for capacitors and inductors are given, and their tolerances and specific component ratings are discussed using examples.

Chapter 6—Both first- and second-order transient circuits are combined in this chapter. MATLAB plotting routines are used to obtain transient response plots, and PSPICE is used in the solution of transient problems.

Chapter 7—Both MATLAB and PSPICE are employed to obtain solutions to ac steady-state circuit analysis problems. PSPICE can also be used to sweep the frequency to obtain plots of magnitude and phase.

Chapter 8—The Magnetically Coupled Networks chapter has been moved forward to this location.

Chapter 11—Variable frequency network performance has moved to this chapter, and MATLAB is employed to obtain the magnitude and phase plots from a transfer function. The Operational Transconductance Amplifier (OTA) is also introduced as a viable component for wireless applications, which include high-frequency active filters and low-voltage circuitry.

Chapter 12—MATLAB is used to obtain the inverse Laplace transform and perform a convolution.

Chapter 14—Fourier analysis techniques have been moved to this location. PSPICE is used to obtain the Fourier series components of a waveform.

Chapter 15—Two-Port Networks have been moved from an appendix to a full chapter.

Chapter 16—This new chapter provides an introduction to semiconductor electronics. Diodes and transistors and some of their applications are examined.

Companion Web Site

Among other items, this site contains *Answers to Selected Problems.*

Supplements

The ***Student Study Guide*** for the seventh edition contains additional detailed examples that track the chapter presentation to aid and check the student's understanding of the problem-solving process. Many of these examples involve computer simulations with PSPICE, MATLAB, Microsoft EXCEL, and Electronics Workbench. A CD bound into the study guide includes circuit simulations and five easy-to-use video segments for demonstrating PSPICE solutions.

EGrade Anonymous Quizzing is also available to students using this text. Students are encouraged to visit our web site at *www.wiley.com/college/circuitsextra* and register to begin taking practice quizzes on eGrade to increase their circuits problem-solving skills. EGrade questions are organized by topic and are automatically scored to provide immediate feedback, so the student can either drill specifically in problem areas (focusing on topics he or she needs more work in) or just do general practice drills to prepare for a test.

Problem-Solving Companion is available for download from the text web site. This *Companion* contains over 70 additional problems with extremely detailed worked-out solutions to walk you through the problem-solving process. It also includes techniques for solving linearly independent simultaneous equations. Visit the web site at *www.wiley.com/college/ elec/irwin407402* for an example from this resource.

Circuits Extra—Check out the latest offerings for users of Wiley circuits texts.

EGrade On-line Assessment is also available for this text. EGrade is a tool that allows instructors to automate the process of assigning, delivering, grading, and routing all kinds of homework, quizzes, and tests, while providing students with immediate scoring and feedback on their work. Electric circuits test banks in eGrade format are available for instructors who would like to include a web component in their course in the form of on-line homework and quizzing. Questions are arranged by topic and are in a variety of formats, including fill-in-the-blank, multiple choice, true/false, and more. For more information, and to see a demo of eGrade, visit *www.wiley.com/college/egrade*.

The ***Solutions Manual***, containing solutions to all learning extensions and end-of-chapter problems, and *PowerPoint Slides* for this text are available only to instructors who have adopted the text for classroom use. The solutions manual and PowerPoint slides are available on the web site at *www.wiley.com/college/irwin*, under the *Instructor's Companion Site*. You must first register for a password on-line and supply your course information for confirmation before you will receive access to these resources.

Circuit Works is a simulator based on a set library of 100 circuits, with adjustable parameters and 35 parameterized signal generators. This tool helps you to succeed by learning the principles and relationships that underlie basic first- and second-order circuits having resistors, capacitors, inductors, op-amps, dependent and independent sources, and tranformers. *Circuit Works* is free to purchasers of this text. To download *Circuit Works* for free, simply go to the web site at *http://www.wiley.com/college/circuitworks* and register with the unique password included on the inside cover of this text.

Acknowledgements

Over the approximate 20-year period that this text has been employed, it is estimated that more than one thousand instructors have used my book to teach circuit analysis to hundreds of thousands of students. I am most grateful for the confidence that has been demonstrated in the educational soundness of the text. In addition, I have received numerous evaluations and suggestions from professors and their students over the years, and their feedback has helped me continuously improve the presentation. For this seventh edition, I would especially like to express my appreciation to a number of individuals and groups. I owe Bill Dillard at Auburn University a special debt of gratitude for the numerous contributions that he made to the development of new material in the text, as well as some of the supporting material. I am also most appreciative of the suggestions for improvement that were made by Professor John Choma of the University of Southern California and Professor Mark Nelms of Auburn University.

I was fortunate to have an outstanding group of reviewers for the book. They are:

Robert Krueger, University of Wisconsin
Ashok Goel, Michigan Technological University
Clifford Pollock, Cornell University
Peyton Peebles, University of Florida
Marty Kaliski, Cal Poly, San Luis Obispo
Darryl Morrell, Arizona State University
Peddapullaiah Sannuti, Rutgers University
Paul Greiling, UCLA
Jorge Aravena, Louisiana State University
James Rowland, University of Kansas
Tom Shumpert, Auburn University

The book was carefully checked for accuracy by three very competent individuals: Professor Scott F. Smith, Boise State University, Professor Val Tareski, North Dakota State University, and Jung Young Lee, a honor student at UC Berkeley. I sincerely appreciate the superior job that was done by these groups. In addition, I would like to thank two professors and their students for the special group discussions that took place in Professor James Rowland's class at the University of Kansas and Professor John Durkin's class at the University of Akron.

While this is the seventh edition of this book, it is the very first edition for John Wiley & Sons, Inc. The preparation of this book and the numerous ancillary documents that support it have been handled with both enthusiasm and great care. The combined wisdom and leadership of Bill Zobrist, my Executive Editor, has resulted in a tremendous team effort that has addressed every aspect of the presentation. This team included the following individuals:

Marketing Manager, Katherine Hepburn
Senior Production Editor, Christine Cervoni
Senior Designer, Karin Kincheloe

Illustration Coordinator, Gene Aiello
Assistant Editor, Jenny Welter
Senior Market Research Analyst, Carl Kulo
Online Marketing Coordinator, Bonnie Kubat
Developmental Editor, Johnna Barto
Marketing Assistant, Cecily Iddings
Editorial Assistant, Susannah Barr

Each member of this team played a vital role in preparing the package that is the *Seventh Edition of Basic Engineering Circuit Analysis*, and I am most appreciative of their many contributions.

As in the past, I am pleased to acknowledge the support that has been provided by numerous individuals to earlier editions of this book. The Auburn personnel who have helped are:

Paula R. Marino
Thomas A. Baginski
Charles A. Gross
James L. Lowry
Zhi Ding
David C. Hill
Henry Cobb
Les Simonton
Betty Kelley
E. R. Graf
L. L. Grigsby
M. A. Honnell
R. C. Jaeger

M. S. Morse
Sung-Won Park
C. L. Rogers
Travis Blalock
Kevin Driscoll
Keith Jones
George Lindsey
David Mack
John Parr
Monty Rickles
James Trivltayakhun
Susan Williamson
Jacinda Woodward

Many of my colleagues throughout the United States, some of whom are now retired, have also made numerous suggestions for improving the book:

M. E. Shafeei,
  Penn State University at Harrisburg
Leonard J. Tung, Florida A&M University/
  Florida State University
Darrell Vines, Texas Tech University
David Anderson, University of Iowa
Richard Baker, UCLA
James L. Dodd, Mississippi State University
Earl D. Eyman, University of Iowa
Arvin Grabel, Northeastern University
Paul Gray, University of Wisconsin,
  Platteville
Mohammad Habli,
  University of New Orleans
John Hadjilogiou,
  Florida Institute of Technology
Ralph Kinney, LSU
K. S. P. Kumar, University of Minnesota
James Luster, Snow College

Ian McCausland, University of Toronto
Arthur C. Moeller, Marquette University
M. Paul Murray, Mississippi State University
Burks Oakley, II, University of
  Illinois at Champaign-Urbana
John O'Malley, University of Florida
William R. Parkhurst, Wichita State
  University
George Prans, Manhattan College
James Rowland, University of Kansas
Robert N. Sackett, Normandale
  Community College
Richard Sanford, Clarkson University
Ronald Schultz, Cleveland State University
Karen M. St. Germaine, University of
  Nebraska
Janusz Strazyk, Ohio University
Saad Tabet, Florida State University
Seth Wolpert, University of Maine

Finally, I wish to express my deep appreciation to my wife, Edie, who has been most supportive of my efforts with this book.

J. David Irwin

How Do I Learn Best?

This questionnaire aims to find out something about your preferences for the way you work with information. You will have a preferred learning style, and one part of that learning style is your preference for the intake and the output of ideas and information.

Circle the letter of the answer that best explains your preference. Circle more than one if a single answer does not match your perception. Leave blank any question that does not apply.

1. You are about to give directions to a person who is standing with you. She is staying in a hotel in town and wants to visit your house later. She has a rental car. Would you
   a. draw a map on paper?
   b. tell her the directions?
   c. write down the directions (without a map)?
   d. pick her up at the hotel in your car?

2. You are not sure whether a word should be spelled "dependent" or "dependant." Do you
   c. look it up in the dictionary?
   a. see the word in your mind and choose by the way it looks?
   b. sound it out in your mind?
   d. write both versions down on paper and choose one?

3. You have just received a copy of your itinerary for a world trip. This is of interest to a friend. Would you
   b. call her immediately and tell her about it?
   c. send her a copy of the printed itinerary?
   a. show her on a map of the world?
   d. share what you plan to do at each place you visit?

4. You are going to cook something as a special treat for your family. Do you
   d. cook something familiar without the need for instructions?
   a. thumb through the cookbook looking for ideas from the pictures?
   c. refer to a specific cookbook where there is a good recipe?

5. A group of tourists has been assigned to you to find out about wildlife reserves or parks. Would you
   d. drive them to a wildlife reserve or park?
   a. show them slides and photographs?
   c. give them pamphlets or a book on wildlife reserves or parks?
   b. give them a talk on wildlife reserves or parks?

6. You are about to purchase a new CD player. Other than price, what would most influence your decision?
   b. The salesperson telling you what you want to know.
   c. Reading the details about it.
   d. Playing with the controls and listening to it.
   a. Its fashionable and upscale appearance.

7. Recall a time in your life when you learned how to do something like playing a new board game. Try to avoid choosing a very physical skill, e.g., riding a bike. How did you learn best? By
   a. visual clues—pictures, diagrams, charts?
   c. written instructions?
   b. listening to somebody explaining it?
   d. doing it or trying it?

8. You have an eye problem. Would you prefer that the doctor
   b. tell you what is wrong?
   a. show you a diagram of what is wrong?
   d. use a model to show what is wrong?

9. You are about to learn to use a new program on a computer. Would you
   d. sit down at the keyboard and begin to experiment with the program's features?
   c. read the manual that comes with the program?
   b. call a friend and ask questions about it?

10. You are staying in a hotel and have a rental car. You would like to visit friends whose address/location you do not know. Would you like them to
   a. draw you a map on paper?
   b. tell you the directions?
   c. write down the directions (without a map)?
   d. pick you up at the hotel in their car?

11. Apart from price, what would most influence your decision to buy a particular book?
   d. You have used a copy before.
   b. A friend talking about it.
   c. Quickly reading parts of it.
   a. The appealing way it looks.

12. A new movie has arrived in town. What would most influence your decision to go (or not go)?
   b. You heard a radio review about it.
   c. You read a review about it.
   a. You saw a preview of it.

13. Do you prefer a lecturer or teacher who likes to use
   c. a textbook, handouts, readings?
   a. flow diagrams, charts, graphs?
   d. field trips, labs, practical sessions?
   b. discussion, guest speakers?

Count Your Choices:	a.	b.	c.	d.
	☐	☐	☐	☐
	V	A	R	K

Now match the letter or letters you have recorded most to the same letter or letters in the learning styles chart. You may have more than one learning style preference—many people do. Next to each letter in the chart are suggestions that will refer you to different learning aids throughout this text.

Learning Styles Chart

Visual

INTAKE: TO TAKE IN THE INFORMATION	TO MAKE A STUDY PACKAGE	TEXT FEATURES THAT MAY HELP YOU THE MOST	OUTPUT: TO DO WELL ON EXAMS
• Pay close attention to charts, drawings, and handouts your instructor uses. • Underline. • Use different colors. • Use symbols, flow charts, graphs, different arrangements on the page, white space.	Convert your lecture notes into "page pictures." To do this: • Use the "Intake" strategies. • Reconstruct images in different ways. • Redraw pages from memory. • Replace words with symbols and initials. • Look at your pages.	**Learning Goals** **Charts of real-world values** **Endpaper charts** **Photos of circuit components** **Learning by Example** **Learning by Application** **Problem-Solving Strategies** **Learning by Design** **Learning Check (Problems)** **Tools for Learning (MATLAB and PSPICE)** *Answers to Selected Problems* *Problem-Solving Companion* *Student Study Guide*	• Recall your "page pictures." • Draw diagrams where appropriate. • Practice turning your visuals back into words.

Aural

INTAKE: TO TAKE IN THE INFORMATION	TO MAKE A STUDY PACKAGE	TEXT FEATURES THAT MAY HELP YOU THE MOST	OUTPUT: TO DO WELL ON EXAMS
• Attend lectures and tutorials. • Discuss topics with students and instructors. • Explain new ideas to other people. • Use a tape recorder. • Leave spaces in your lecture notes for later recall. • Describe overheads, pictures, and visuals to somebody who was not in class.	You may take poor notes because you prefer to listen. Therefore: • Expand your notes by talking with others and with information from your textbook. • Tape record summarized notes and listen. • Read summarized notes out loud. • Explain your notes to another "aural" person.	**Learning Goals** **Learning Hints** **Learning by Example** **Problem-Solving Strategies** **Learning by Application** **Learning Check (Summary & Problems)**	• Talk with the instructor. • Spend time in quiet places recalling the ideas. • Practice writing answers to old exam questions. • Say your answers out loud.

Reading/Writing

INTAKE: TO TAKE IN THE INFORMATION	TO MAKE A STUDY PACKAGE	TEXT FEATURES THAT MAY HELP YOU THE MOST	OUTPUT: TO DO WELL ON EXAMS
• Use lists and headings. • Use dictionaries, glossaries, and definitions. • Read handouts, textbooks, and supplementary library readings. • Use lecture notes.	• Write out words again and again. • Reread notes silently. • Rewrite ideas and principles into other words. • Turn charts, diagrams, and other illustrations into statements.	**Learning Goals** **Learning Hints** **Endpaper charts** **Learning by Example** **Learning by Application** **Problem-Solving Strategies** **Learning by Design** **Learning Check (Summary & Problems)** **Tools for Learning (MATLAB and PSPICE)** *Answers to Selected Problems* *Problem-Solving Companion* *Student Study Guide*	• Write exam answers. • Practice with multiple-choice questions. • Write paragraphs, beginnings and endings. • Write your lists in outline form. • Arrange your words into hierarchies and points.

Kinesthetic

INTAKE: TO TAKE IN THE INFORMATION	TO MAKE A STUDY PACKAGE	TEXT FEATURES THAT MAY HELP YOU THE MOST	OUTPUT: TO DO WELL ON EXAMS
• Use all your senses. • Go to labs, take field trips. • Listen to real-life examples. • Pay attention to applications. • Use hands-on approaches. • Use trial-and-error methods.	You may take poor notes because topics do not seem concrete or relevant. Therefore: • Put examples in your summaries. • Use case studies and applications to help with principles and abstract concepts. • Talk about your notes with another "kinesthetic" person. • Use pictures and photographs that illustrate an idea.	**Learning by Doing** **Learning Hints** **Learning by Example** **Learning Applications** **Extending Your Learning** **Learning by Design** **Learning Check (Problems)** **Tools for Learning (MATLAB and PSPICE)** *Answers to Selected Problems* *Problem-Solving Companion* *Student Study Guide*	• Write practice answers. • Role-play the exam situation.

Source: *Adapted from VARK pack. © Copyright Version 2.0 (2000) held by Neil D. Fleming, Christchurch, New Zealand, and Charles C. Bonwell, Green Mountain Falls, Colorado 80819 (719) 684-9261. This material may be used for faculty or student development if attribution is given. It may not be published in either paper or electronic form without consent of the authors. There is a VARK web site at www.active-learning-site.com.*

Basic Concepts | 1

Today we live in a predominantly electrical world. Although this statement may sound strange at first, a moment's reflection will indicate its inherent truth. The two primary areas of electrotechnology that permeate essentially every aspect of our lives are power and information. Without them, life as we know it would undergo stupendous changes. We have learned to generate, convert, transmit, and utilize these technologies for the enhancement of the whole human race.

Electrotechnology is a driving force in the changes that are occurring in every engineering discipline. For example, surveying is now done with lasers and electronic range finders, and automobiles employ electronic dashboards and electronic ignition systems. Industrial processes that range from chemical refineries and metal foundries to wastewater treatment plants use (1) electronic sensors to obtain information about the process, (2) instrumentation systems to gather the information, and (3) computer control systems to process the information and generate electronic commands to actuators, which correct and control the process.

Fundamental to electrotechnology is the area of circuit analysis. A thorough knowledge of this subject provides an understanding of such things as cause and effect, amplification and attenuation, feedback and control, and stability and oscillation. Of critical importance is the fact that the same principles applied to engineering systems can also be applied to economic and social systems. Thus, the ramifications of circuit analysis are immense, and a solid understanding of this subject is well worth the effort expended to obtain it.

In this chapter we will introduce some of the basic quantities that will be used throughout the text. Specifically, we will define electric current, voltage, power and energy, as well as the difference between direct current and alternating current. In addition, we will classify electric elements as either passive or active, the latter of which can be further subdivided into both independent and dependent. This basic introduction will lay the groundwork for our further study of a wide variety of electric circuits.

1.1 System of Units

The system of units we employ is the international system of units, the Système International des Unités, which is normally referred to as the SI standard system. This system, which is composed of the basic units meter (m), kilogram (kg), second (s), ampere (A), degree kelvin (°K), and candela (cd), is defined in all modern physics texts and therefore will not be defined here. However, we will discuss the units in some detail as we encounter them in our subsequent analyses.

The standard prefixes that are employed in SI are shown in Fig. 1.1. Note the decimal relationship between these prefixes. These standard prefixes are employed throughout our study of electric circuits.

Figure 1.1
Standard SI prefixes.

| 10^{-12} | 10^{-9} | 10^{-6} | 10^{-3} | 1 | 10^3 | 10^6 | 10^9 | 10^{12} |

pico (p) nano (n) micro (μ) milli (m)   kilo (k) mega (M) giga (G) tera (T)

Circuit technology has changed drastically over the years. For example, in the early 1960s the space on a circuit board occupied by the base of a single vacuum tube was about the size of a quarter (25-cent coin). Today that same space could be occupied by an Intel Pentium integrated circuit chip containing 3.1 million transistors. These chips are the engine for a host of electronic equipment.

1.2 Basic Quantities

Before we begin our analysis of electronic circuits, we must define terms that we will employ. However, in this chapter and throughout the book our definitions and explanations will be as simple as possible to foster an understanding of the use of the material. No attempt will be made to give complete definitions of many of the quantities because such definitions are not only unnecessary at this level but are often confusing. Although most of us have an intuitive concept of what is meant by a circuit, we will simply refer to an *electric circuit* as an interconnection of electrical components, each of which we will describe with a mathematical model.

The most elementary quantity in an analysis of electric circuits is the electric *charge*. Our interest in electric charge is centered around its motion, since charge in motion results in an energy transfer. Of particular interest to us are those situations in which the motion is confined to a definite closed path.

An electric circuit is essentially a pipeline that facilitates the transfer of charge from one point to another. The time rate of change of charge constitutes an electric *current*. Mathematically, the relationship is expressed as

$$i(t) = \frac{dq(t)}{dt} \quad \text{or} \quad q(t) = \int_{-\infty}^{t} i(x)\, dx \qquad \textbf{1.1}$$

where i and q represent current and charge, respectively (lowercase letters represent time dependency and capital letters are reserved for constant quantities). The basic unit of current is the ampere (A) and 1 ampere is 1 coulomb per second.

Although we know that current flow in metallic conductors results from electron motion, the conventional current flow, which is universally adopted, represents the movement of positive charges. It is important that the reader think of current flow as the movement of positive charge regardless of the physical phenomena that take place. The symbolism that will be used to represent current flow is shown in Fig. 1.2. $I_1 = 2$ A in Fig. 1.2a indicates that at any point in the wire shown,

$I_1 = 2$ A

Circuit 1

(a)

$I_2 = -3$ A

Circuit 2

(b)

Figure 1.2
Conventional current flow: (a) positive current flow; (b) negative current flow.

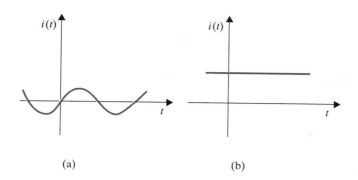

Figure 1.3
Two common types of current: (a) alternating current (ac); (b) direct current (dc).

2 C of charge pass from left to right each second. $I_2 = -3$ A in Fig. 1.2b indicates that at any point in the wire shown, 3 C of charge pass from right to left each second. Therefore, it is important to specify not only the magnitude of the variable representing the current, but also its direction.

There are two types of current that we encounter often in our daily lives, alternating current (ac) and direct current (dc), which are shown as a function of time in Fig. 1.3. *Alternating current* is the common current found in every household and is used to run the refrigerator, stove, washing machine, and so on. Batteries, which are used in automobiles or flashlights, are one source of *direct current*. In addition to these two types of currents, which have a wide variety of uses, we can generate many other types of currents. We will examine some of these other types later in the book. In the meantime, it is interesting to note that the magnitude of currents in elements familiar to us ranges from soup to nuts, as shown in Fig. 1.4.

We have indicated that charges in motion yield an energy transfer. Now we define the *voltage* (also called the *electromotive force* or *potential*) between two points in a circuit as the difference in energy level of a unit charge located at each of the two points. Work or energy, $w(t)$ or W, is measured in joules (J); 1 joule is 1 newton meter (N · m). Hence, voltage [$v(t)$ or V] is measured in volts (V) and 1 volt is 1 joule per coulomb; that is, 1 volt = 1 joule per coulomb = 1 newton meter per coulomb.

If a unit positive charge is moved between two points, the energy required to move it is the difference in energy level between the two points and is the defined voltage. It is extremely important that the variables that are used to represent voltage between two points be defined in such a way that the solution will let us interpret which point is at the higher potential with respect to the other.

LEARNING by Doing

D 1.1   Determine the amount of time required for 100 C of charge to pass through the circuit in Fig. 1.2a.

ANSWER   *50 s*

LEARNING Hint

$$V = \frac{W}{q}$$

Current in amperes (A)

10^6	
10^4	Lightning bolt
10^2	Large industrial motor current
10^0	Typical household appliance current
10^{-2}	Causes ventricular fibrillation in humans
10^{-4}	Human threshold of sensation
10^{-6}	
10^{-8}	Integrated circuit (IC) memory cell current
10^{-10}	
10^{-12}	
10^{-14}	Synaptic current (brain cell)

Figure 1.4
Typical current magnitudes.

Figure 1.5
Voltage representations.

(a)   (b)   (c)

In Fig. 1.5a the variable that represents the voltage between points A and B has been defined as V_1, and it is assumed that point A is at a higher potential than point B, as indicated by the + and − signs associated with the variable and defined in the figure. The + and − signs define a reference direction for V_1. If $V_1 = 2$ V, then the difference in potential of points A and B is 2 V and point A is at the higher potential. If a unit positive charge is moved from point A through the circuit to point B, it will give up energy to the circuit and have 2 J less energy when it reaches point B. If a unit positive charge is moved from point B to point A, extra energy must be added to the charge by the circuit, and hence the charge will end up with 2 J more energy at point A than it started with at point B.

For the circuit in Fig. 1.5b, $V_2 = -5$ V means that the potential between points A and B is 5 V and point B is at the higher potential. The voltage in Fig. 1.5b can be expressed as shown in Fig. 1.5c. In this equivalent case, the difference in potential between points A and B is $V_2 = 5$ V, and point B is at the higher potential.

Note that it is important to define a variable with a reference direction so that the answer can be interpreted to give the physical condition in the circuit. We will find that it is not possible in many cases to define the variable so that the answer is positive, and we will also find that it is not necessary to do so. A negative number for a given variable gives exactly the same information as a positive number for a new variable that is the same as the old variable, except that it has an opposite reference direction. Hence, when we define either current or voltage, it is absolutely necessary that we specify both magnitude and direction. Therefore, it is incomplete to say that the voltage between two points is 10 V or the current in a line is 2 A, since only the magnitude and not the direction for the variables has been defined.

The range of magnitudes for voltage, equivalent to that for currents in Fig. 1.4, is shown in Fig. 1.6. Once again, note that this range spans many orders of magnitude.

Voltage in volts (V)	
10^8	Lightning bolt
10^6	High-voltage transmission lines Voltage on a TV picture tube
10^4	Large industrial motors AC outlet plug in U.S. households
10^2	
10^0	Car battery Voltage on integrated circuits Flashlight battery
10^{-2}	
10^{-4}	Voltage across human chest produced by the heart (EKG)
10^{-6}	Voltage between two points on human scalp (EEG) Antenna of a radio receiver
10^{-8}	
10^{-10}	

Figure 1.6
Typical voltage magnitudes.

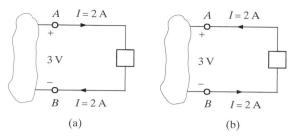

Figure 1.7 *Voltage–current relationships for (a) energy absorbed and (b) energy supplied.*

At this point we have presented the conventions that we employ in our discussions of current and voltage. *Energy* is yet another important term of basic significance. Figure 1.7 illustrates the voltage–current relationships for energy transfer. In this figure, the block representing a circuit element has been extracted from a larger circuit for examination. In Fig. 1.7a, energy is being supplied *to* the element by whatever is attached to the terminals. Note that 2 A, that is, 2 C, of charge are moving from point *A* to point *B* through the element each second. Each coulomb loses 3 J of energy as it passes through the element from point *A* to point *B*. Therefore, the element is absorbing 6 J of energy per second. Note that when the element is *absorbing* energy, a positive current enters the positive terminal. In Fig. 1.7b energy is being supplied *by* the element to whatever is connected to terminals *A-B*. In this case, note that when the element is *supplying* energy, a positive current enters the negative terminal and leaves via the positive terminal. In this convention a negative current in one direction is equivalent to a positive current in the opposite direction, and vice versa. Similarly, a negative voltage in one direction is equivalent to a positive voltage in the opposite direction.

LEARNING Example 1.1

Suppose that your car will not start. To determine whether the battery is faulty, you turn on the light switch and find that the lights are very dim, indicating a weak battery. You borrow a friend's car and a set of jumper cables. However, how do you connect his car's battery to yours? What do you want his battery to do?

SOLUTION   Essentially, his car's battery must supply energy to yours, and therefore it should be connected in the manner shown in Fig. 1.8. Note that the positive current leaves the positive terminal of the good battery (supplying energy) and enters the positive terminal of the weak battery (absorbing energy). Note that the same connections are used when charging a battery.

In practical applications there are often considerations other than simply the electrical relations (e.g., safety). Such is the case with jump-starting an automobile. Automobile batteries produce explosive gases that can be ignited accidentally, causing severe physical injury. Be safe—follow the procedure described in your auto owner's manual.

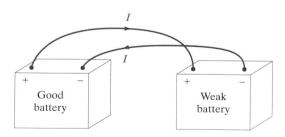

Figure 1.8 *Diagram for Example 1.1.*

We have defined voltage in joules per coulomb as the energy required to move a positive charge of 1 C through an element. If we assume that we are dealing with a differential amount of charge and energy, then

$$v = \frac{dw}{dq} \qquad \textbf{1.2}$$

Multiplying this quantity by the current in the element yields

$$vi = \frac{dw}{dq}\left(\frac{dq}{dt}\right) = \frac{dw}{dt} = p \qquad \textbf{1.3}$$

which is the time rate of change of energy or power measured in joules per second, or watts (W). Since, in general, both v and i are functions of time, p is also a time-varying quantity. Therefore, the change in energy from time t_1 to time t_2 can be found by integrating Eq. (1.3); that is,

$$\Delta w = \int_{t_1}^{t_2} p \, dt = \int_{t_1}^{t_2} vi \, dt \qquad \textbf{1.4}$$

Figure 1.9
Sign convention for power.

At this point, let us summarize our sign convention for power. To determine the sign of any of the quantities involved, the variables for the current and voltage should be arranged as shown in Fig. 1.9. The variable for the voltage $v(t)$ is defined as the voltage across the element with the positive reference at the same terminal that the current variable $i(t)$ is entering. This convention is called the *passive sign convention* and will be so noted in the remainder of this book. The product of v and i, with their attendant signs, will determine the magnitude and sign of the power. If the sign of the power is positive, power is being absorbed by the element; if the sign is negative, power is being supplied by the element.

LEARNING Example 1.2

Given the two diagrams shown in Fig. 1.10, determine whether the element is absorbing or supplying power and how much.

SOLUTION In Fig. 1.10a the power is $P = (2\,\text{V})(-4\,\text{A}) = -8\,\text{W}$. Therefore, the element is supplying power. In Fig. 1.10b, the power is $P = (2\,\text{V})(2\,\text{A}) = 4\text{W}$. Therefore, the element is absorbing power.

Figure 1.10
Elements for Example 1.2.

(a)   (b)

E1.1 Determine the amount of power absorbed or supplied by the elements in Fig. E1.1.

ANSWER (a) $P = -48$ W; (b) $P = 8$ W.

Figure E1.1   (a)   (b)

LEARNING Example 1.3

We wish to determine the unknown voltage or current in Fig. 1.11.

(a)   (b)

Figure 1.11   *Elements for Example 1.3.*

SOLUTION   In Fig. 1.11a, a power of −20 W indicates that the element is delivering power. Therefore, the current enters the negative terminal (terminal *A*), and from Eq. (1.3) the voltage is 4 V. Thus *B* is the positive terminal, *A* is the negative terminal, and the voltage between them is 4 V.

In Fig 1.11b, a power of +40 W indicates that the element is absorbing power and, therefore, the current should enter the positive terminal *B*. The current thus has a value of −8 A, as shown in the figure.

E1.2   Determine the unknown variables in Fig. E1.2.

Figure E1.2   (a)   (b)

ANSWER   (a) $V_1 = -20$ V; (b) $I = -5$ A.

Finally, it is important to note that these electrical networks satisfy the principle of conservation of energy. For our present purposes this means that the power supplied in a network is exactly equal to the power absorbed.

1.3 Circuit Elements

Thus far we have defined voltage, current, and power. In the remainder of this chapter we will define both independent and dependent current and voltage sources. Although we will assume ideal elements, we will try to indicate the shortcomings of these assumptions as we proceed with the discussion.

In general, the elements we will define are terminal devices that are completely characterized by the current through the element and/or the voltage across it. These elements, which we will employ in constructing electric circuits, will be broadly classified as being either active or passive. The distinction between these two classifications depends essentially on one thing—whether they supply or absorb energy. As the words themselves imply, an *active* element is capable of generating energy and a *passive* element cannot generate energy.

However, we will show later that some passive elements are capable of storing energy. Typical active elements are batteries, generators, and transistor models. The three common passive elements are resistors, capacitors, and inductors.

In Chapter 2 we will launch an examination of passive elements by discussing the resistor in detail. However, before proceeding with that element, we first present some very important active elements.

1. Independent voltage source

2. Independent current source

3. Two dependent voltage sources

4. Two dependent current sources

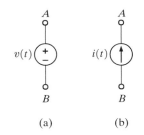

Figure 1.12
Symbols for (a) independent voltage source, (b) independent current source.

INDEPENDENT SOURCES An *independent voltage source* is a two-terminal element that maintains a specified voltage between its terminals *regardless of the current through it*. The general symbol for an independent source, a circle, is shown in Fig. 1.12a. As the figure indicates, terminal A is $v(t)$ volts positive with respect to terminal B.

In contrast to the independent voltage source, the *independent current source* is a two-terminal element that maintains a specified current *regardless of the voltage across its terminals*. The general symbol for an independent current source is shown in Fig. 1.12b, where $i(t)$ is the specified current and the arrow indicates the positive direction of current flow.

In their normal mode of operation, independent sources supply power to the remainder of the circuit. However, they may also be connected into a circuit in such a way that they absorb power. A simple example of this latter case is a battery-charging circuit such as that shown in Example 1.1.

It is important that we pause here to interject a comment concerning a shortcoming of the models. In general, mathematical models approximate actual physical systems only under a certain range of conditions. Rarely does a model accurately represent a physical system under every set of conditions. To illustrate this point, consider the model for the voltage source in Fig. 1.12a. We assume that the voltage source delivers v volts regardless of what is connected to its terminals. Theoretically, we could adjust the external circuit so that an infinite amount of current would flow, and therefore the voltage source would deliver an infinite amount of power. This is, of course, physically impossible. A similar argument could be made for the independent current source. Hence, the reader is cautioned to keep in mind that models have limitations and thus are valid representations of physical systems only under certain conditions.

LEARNING by Doing

D 1.4 Determine the power supplied by the 12-V source.

ANSWER *24 W*

LEARNING Example 1.4

Determine the power absorbed or supplied by the elements in the network in Fig. 1.13.

Figure 1.13
Network for Example 1.4.

SOLUTION The current flow is out of the positive terminal of the 24-V source, and therefore this element is supplying $(2)(24) = 48$ W of power. The current is into the positive terminals of elements 1 and 2, and therefore elements 1 and 2 are absorbing $(2)(6) = 12$ W and $(2)(18) = 36$ W, respectively. Note that the power supplied is equal to the power absorbed.

LEARNING Hint

Elements that are connected in series have the same current.

LEARNING EXTENSION

E1.3   Find the power that is absorbed or supplied by the elements in Fig. E1.3.

Figure E1.3

DEPENDENT SOURCES   In contrast to the independent sources, which produce a particular voltage or current completely unaffected by what is happening in the remainder of the circuit, dependent sources generate a voltage or current that is determined by a voltage or current at a specified location in the circuit. These sources are very important because they are an integral part of the mathematical models used to describe the behavior of many electronic circuit elements.

For example, metal-oxide-semiconductor field-effect transistors (MOSFETs) and bipolar transistors, both of which are commonly found in a host of electronic equipment, are modeled with dependent sources, and therefore the analysis of electronic circuits involves the use of these controlled elements.

In contrast to the circle used to represent independent sources, a diamond is used to represent a dependent or controlled source. Figure 1.14 illustrates the four types of dependent sources. The input terminals on the left represent the voltage or current that controls the dependent source, and the output terminals on the right represent the output current or voltage of the controlled source. Note that in Figs. 1.14a and d the quantities μ and β are dimensionless constants because we are transforming voltage to voltage and current to current. This is not the case in Figs. 1.14b and c; hence when we employ these elements a short time later, we must describe the units of the factors r and g.

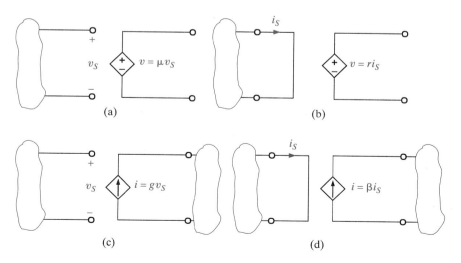

Figure 1.14
Four different types of dependent sources.

LEARNING Example 1.5

Given the two networks shown in Fig. 1.15, we wish to determine the outputs.

SOLUTION In Fig. 1.15a the output voltage is $V_o = \mu V_S$ or $V_o = 20 V_S = (20)(2\ V) = 40\ V$. Note that the output voltage has been amplified from 2 V at the input terminals to 40 V at the output terminals; that is, the circuit is a voltage amplifier with an amplification factor of 20.

In Fig. 1.15b, the output current is $I_o = \beta I_S = (50)(1\ mA) = 50\ mA$; that is, the circuit has a current gain of 50, meaning that the output current is 50 times greater than the input current.

(a)   (b)

Figure 1.15
Circuits for Example 1.5.

LEARNING EXTENSION

E1.4 Determine the power supplied by the dependent sources in Fig. E1.4.

ANSWER (a) Power supplied = 80 W; (b) power supplied = 160 W.

(a)   (b)

Figure E1.4

LEARNING Example 1.6

Let us find the current I_o in the network in Fig. 1.16.

Figure 1.16
Circuit used in
Example 1.6.

SOLUTION First, we must determine the power absorbed or supplied by each element in the network. Using the sign convention for power, we find

$$P_{2A} = (6)(-2) = -12\ W$$
$$P_1 = (6)(I_o) = 6I_o\ W$$
$$P_2 = (12)(-9) = -108\ W$$
$$P_3 = (10)(-3) = -30\ W$$
$$P_{4V} = (4)(-8) = -32\ W$$
$$P_{DS} = (8I_x)(11) = (16)(11) = 176\ W$$

Since energy must be conserved,

$$-12 + 6I_o - 108 - 30 - 32 + 176 = 0$$

or

$$6I_o + 176 = 12 + 108 + 30 + 32$$

Hence,

$$I_o = 1A$$

LEARNING EXTENSION

E1.5 Find the power that is absorbed or supplied by the circuit elements in the network in Fig. E1.5.

Figure E1.5

ANSWER $P_{24\,V} = 96$ W supplied, $P_1 = 32$ W absorbed, $P_{4I_x} = 64$ W absorbed.

LEARNING Check

Summary

▶ **The standard prefixes employed**

$$p = 10^{-12} \qquad k = 10^3$$
$$n = 10^{-9} \qquad M = 10^6$$
$$\mu = 10^{-6} \qquad G = 10^9$$
$$m = 10^{-3} \qquad T = 10^{12}$$

▶ **The relationships between current and charge**

$$i(t) = \frac{dq(t)}{dt} \quad \text{or} \quad q(t) = \int_{-\infty}^{t} i(x)\, dx$$

▶ **The relationships among power, energy, current, and voltage**

$$p = \frac{dw}{dt} = vi$$

$$\Delta w = \int_{t_1}^{t_2} p\, dt = \int_{t_1}^{t_2} vi\, dt$$

▶ **The passive sign convention**   The passive sign convention states that if the voltage and current associated with an element are as shown in Fig. 1.9, the product of v and i, with their attendant signs, determines the magnitude and sign of the power. If the sign is positive, power is being absorbed by the element, and if the sign is negative, the element is supplying power.

▶ **Independent and dependent sources**   An ideal independent voltage (current) source is a two-terminal element that maintains a specified voltage (current) between its terminals regardless of the current (voltage) through (across) the element. Dependent or controlled sources generate a voltage or current that is determined by a voltage or current at a specified location in the circuit.

▶ **Conservation of energy**   The electric circuits under investigation satisfy the conservation of energy.

Problems

SECTION 1.2

1.1 If 60 C of charge pass through an electric conductor in 30 seconds, determine the current in the conductor.

1.2 If the current in an electric conductor is 2.4 A, how many coulombs of charge pass any point in a 30-second interval?

1.3 Determine the time interval required for a 12-A battery charger to deliver 4800 C.

1.4 A lightning bolt carrying 30,000 A lasts for 50 microseconds. If the lightning strikes an airplane flying at 20,000 feet, what is the charge deposited on the plane?

1.5 Determine the energy required to move 240 C through 6 V.

1.6 Determine the amount of power absorbed or supplied by the element in Fig. P1.6 if (a) $V_1 = 4$ V, $I = 2$ A and (b) $V_1 = -4$ V, $I = -2$ A.

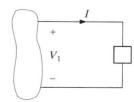

Figure P1.6

1.7 Repeat Problem 1.6 if (a) $V_1 = -6$ V, $I = 3$ A and (b) $V_1 = 6$ V, $I = -3$ A.

1.8 Determine the missing quantity in the circuits in Fig. P1.8.

(a)　　　　　　(b)

Figure P1.8

1.9 Repeat Problem 1.8 for the circuits in Fig. P1.9.

(a)　　　　　　(b)

Figure P1.9

1.10 Repeat Problem 1.8 for the circuits in Fig. P1.10.

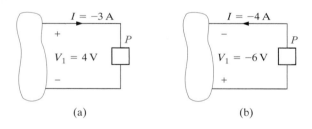

(a)　　　　　　(b)

Figure P1.10

1.11 Determine the power supplied to the elements in Fig. P1.11.

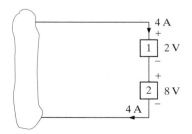

Figure P1.11

1.12 Determine the power supplied to the elements in Fig. P1.12.

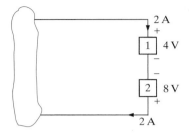

Figure P1.12

1.13 Two elements are connected in series, as shown in Fig. P1.13. Element 1 supplies 24 W of power. Is element 2 absorbing or supplying power, and how much?

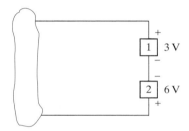

Figure P1.13

1.14 Two elements are connected in series, as shown in Fig. P1.14. Element 1 absorbs 36 W of power. Is element 2 absorbing or supplying power, and how much?

Figure P1.14

SECTION 1.3

Important note: The values used in the problems of Section 1.3 are not arbitrary. They have been selected to satisfy the basic laws of circuit analysis that will be studied in the following chapters.

1.15 Determine the power that is absorbed or supplied by the circuit elements in Fig. P1.15.

(a)   (b)

Figure P1.15

1.16 Find the power that is absorbed or supplied by the circuit elements in Fig. P1.16.

(a)   (b)

Figure P1.16

1.17 Compute the power that is absorbed or supplied by the elements in the network in Fig. P1.17.

Figure P1.17

1.18 Find I_x in the network in Fig. P1.18.

Figure P1.18

1.19 Is the source V_S in the network in Fig. P1.19 absorbing or supplying power, and how much?

Figure P1.19

1.20 Find V_x in the network in Fig. P1.20.

Figure P1.20

1.21 Find I_o in the network in Fig. P1.21.

Figure P1.21

2 ─│ *Resistive Circuits*

LEARNING Goals

In this chapter we introduce the basic concepts and laws that are fundamental to circuit analysis. These laws are Ohm's law, Kirchhoff's current law (KCL), and Kirchhoff's voltage law (KVL). We cannot overemphasize the importance of these three laws because they will be used extensively throughout our entire study of circuit analysis. The reader who masters their use quickly will not only find the material in this text easy to learn, but will be well positioned to grasp subsequent topics in the field of electrical engineering.

As a general rule, most of our activities will be confined to analysis; that is, to the determination of a specific voltage, current, or power somewhere in a network. The techniques we introduce have wide application in circuit analysis, even though we will discuss them within the framework of simple networks.

Our approach here is to begin with the simplest passive element, the resistor, and the mathematical relationship that exists between the voltage across it and the current through it, as specified by Ohm's law. As we build our confidence and proficiency by successfully analyzing some elementary circuits, we will introduce other techniques, such as voltage division and current division, that will accelerate our work.

In this chapter we introduce circuits containing dependent sources, which are used to model active devices such as transistors. Thus, our study of circuit analysis provides a natural introduction to many topics in the area of electronics.

Finally, we present a real-world application to indicate the usefulness of circuit analysis, and then we briefly introduce the topic of circuit design in an elementary fashion. In future chapters these topics will be revisited often to present some fascinating examples that describe problems we encounter in our everyday lives.

2.1 Ohm's Law

Ohm's law is named for the German physicist Georg Simon Ohm, who is credited with establishing the voltage–current relationship for resistance. As a result of his pioneering work, the unit of resistance bears his name.

Ohm's law states that the voltage across a resistance is directly proportional to the current flowing through it. The resistance, measured in ohms, is the constant of proportionality between the voltage and current.

A circuit element whose electrical characteristic is primarily resistive is called a resistor and is represented by the symbol shown in Fig. 2.1a. A resistor is a physical device that can be purchased in certain standard values in an electronic parts store. These resistors, which find use in a variety of electrical applications, are normally carbon composition or wirewound. In addition, resistors can be fabricated using thick oxide or thin metal films for use in hybrid circuits, or they can be diffused in semiconductor integrated circuits. Some typical discrete resistors are shown in Fig. 2.1b. An example of a resistor constructed with integrated circuit processing is shown in Chapter 16.

The mathematical relationship of Ohm's law is illustrated by the equation

$$v(t) = R \times i(t), \text{ where } R \geqq 0 \qquad\qquad \textbf{2.1}$$

LEARNING Hint

The passive sign convention will be employed in conjunction with Ohm's law.

Figure 2.1 (a) Symbol for a resistor; (b) some practical devices. (1), (2), and (3) are high-power resistors. (4) and (5) are high-wattage fixed resistors. (6) is a high-precision resistor. (7)–(12) are fixed resistors with different power ratings.

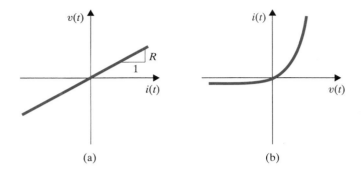

Figure 2.2
Graphical representation of the voltage–current relationship for (a) a linear resistor and (b) a diode.

D 2.1 Determine $v(t)$ in the circuits.

(a)

(b)

ANSWER *(a)* 10 V
(b) −10 V

or equivalently, by the voltage–current characteristic shown in Fig. 2.2a. Note carefully the relationship between the polarity of the voltage and the direction of the current. In addition, note that we have tacitly assumed that the resistor has a constant value and therefore that the voltage–current characteristic is linear.

The symbol Ω is used to represent ohms, and therefore,

$$1\ \Omega = 1\ \text{V/A}$$

Although in our analysis we will always assume that the resistors are *linear* and are thus described by a straight-line characteristic that passes through the origin, it is important that readers realize that some very useful and practical elements do exist that exhibit a *nonlinear* resistance characteristic; that is, the voltage–current relationship is not a straight line. Diodes and transistors, which are briefly described in Chapter 16, are examples of elements that exhibit nonlinear characteristics, and are used extensively in electric circuits. A typical characteristic for a diode is shown in Fig. 2.2b.

Since a resistor is a passive element, the proper current–voltage relationship is illustrated in Fig. 2.1a. The power supplied to the terminals is absorbed by the resistor. Note that the charge moves from the higher to the lower potential as it passes through the resistor and the energy absorbed is dissipated by the resistor in the form of heat. As indicated in Chapter 1, the rate of energy dissipation is the instantaneous power, and therefore

$$p(t) = v(t)i(t) \tag{2.2}$$

which, using Eq. (2.1), can be written as

$$p(t) = Ri^2(t) = \frac{v^2(t)}{R} \tag{2.3}$$

This equation illustrates that the power is a nonlinear function of either current or voltage and that it is always a positive quantity.

Conductance, represented by the symbol G, is another quantity with wide application in circuit analysis. By definition, conductance is the reciprocal of resistance; that is,

$$G = \frac{1}{R} \tag{2.4}$$

The unit of conductance is the siemens, and the relationship between units is

$$1\ \text{S} = 1\ \text{A/V}$$

Using Eq. (2.4), we can write two additional expressions,

$$i(t) = Gv(t) \tag{2.5}$$

and

$$p(t) = \frac{i^2(t)}{G} = Gv^2(t) \qquad \qquad \textbf{2.6}$$

Equation (2.5) is another expression of Ohm's law.

Two specific values of resistance, and therefore conductance, are very important: $R = 0$ and $R = \infty$,

In examining the two cases, consider the network in Fig. 2.3a. The variable resistance symbol is used to describe a resistor such as the volume control on a radio or television set. As the resistance is decreased and becomes smaller and smaller, we finally reach a point where the resistance is zero and the circuit is reduced to that shown in Fig. 2.3b; that is, the resistance can be replaced by a short circuit. On the other hand, if the resistance is increased and becomes larger and larger, we finally reach a point where it is essentially infinite and the resistance can be replaced by an open circuit, as shown in Fig. 2.3c. Note that in the case of a short circuit where $R = 0$,

$$v(t) = Ri(t)$$
$$= 0$$

(a)

(b)

(c)

Figure 2.3
Short-circuit and open-circuit descriptions.

Therefore, $v(t) = 0$, although the current could theoretically be any value. In the open-circuit case where $R = \infty$,

$$i(t) = v(t)/R$$
$$= 0$$

Therefore, the current is zero regardless of the value of the voltage across the open terminals.

LEARNING Example 2.1

In the circuit in Fig. 2.4a, determine the current and the power absorbed by the resistor.

(a)

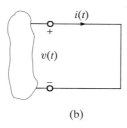

(b)

Figure 2.4
Circuits for Examples 2.1 and 2.2.

SOLUTION Using Eq. (2.1), we find the current to be

$$I = V/R = 12/2\text{k} = 6 \text{ mA}$$

Note that because many of the resistors employed in our analysis are in kΩ, we will use k in the equations in place of 1000. The power absorbed by the resistor is given by Eq. (2.2) or (2.3) as

$$P = VI = (12)(6 \times 10^{-3}) = 0.072 \text{ W}$$
$$= I^2R = (6 \times 10^{-3})^2(2\text{k}) = 0.072 \text{ W}$$
$$= V^2/R = (12)^2/2\text{k} = 0.072 \text{ W}$$

LEARNING Example 2.2

Given the network in Fig. 2.4b, we wish to find R and V_S.

SOLUTION Using the power relationship, we find that

$$R = P/I^2 = (80 \times 10^{-3})/(4 \times 10^{-3})^2 = 5\,\text{k}\Omega$$

The voltage can now be derived using Ohm's law as

$$V_S = IR = (4 \times 10^{-3})(5\text{k}) = 20\ \text{V}$$

The voltage could also be obtained from the remaining power relationships in Eqs. (2.2) and (2.3).

Before leaving this initial discussion of circuits containing sources and a single resistor, it is important to note a phenomenon that we will find to be true in circuits containing many sources and resistors. The presence of a voltage source between a pair of terminals tells us precisely what the voltage is between the two terminals regardless of what is happening in the balance of the network. What we do not know is the current in the voltage source. We must apply circuit analysis to the entire network to determine this current. Likewise, the presence of a current source connected between two terminals specifies the exact value of the current through the source between the terminals. What we do not know is the value of the voltage across the current source. This value must be calculated by applying circuit analysis to the entire network. Furthermore, it is worth emphasizing that when applying Ohm's law, the relationship $V = IR$ specifies a relationship between the voltage *directly across* a resistor R and the current that is *present* in this resistor. Ohm's law does not apply when the voltage is present in one part of the network and the current exists in another. This is a common mistake made by students who try to apply $V = IR$ to a resistor R in the middle of the network while using a V at some other location in the network.

LEARNING EXTENSIONS

E2.1 Given the circuits in Fig. E2.1, find (a) the current I and the power absorbed by the resistor in Fig. E2.1a, and (b) the voltage across the current source and the power supplied by the source in Fig. E2.1b.

ANSWER (a) $I = 0.3$ mA, $P = 3.6$ mW, (b) $V_S = 3.6$ V, $P = 2.16$ mW.

Figure E2.1   (a)   (b)

E2.2 Given the circuits in Fig. E2.2, find (a) R and V_S in the circuit in Fig. E2.2a, and (b) find I and R in the circuit in Fig. E2.2b.

ANSWER (a) $R = 10$ kΩ, $V_S = 4$ V, (b) $I = 20.8$ mA, $R = 576$ Ω.

Figure E2.2   (a)   (b)

2.2 Kirchhoff's Laws

The previous circuits that we have considered have all contained a single resistor and were analyzed using Ohm's law. At this point we begin to expand our capabilities to handle more complicated networks that result from an interconnection of two or more of these simple elements. We will assume that the interconnection is performed by electrical conductors (wires) that have zero resistance; that is, perfect conductors. Because the wires have zero resistance, the energy in the circuit is in essence lumped in each element, and we employ the term *lumped-parameter circuit* to describe the network.

To aid us in our discussion, we will define a number of terms that will be employed throughout our analysis. As will be our approach throughout this text, we will use examples to illustrate the concepts and define the appropriate terms. For example, the circuit shown in Fig. 2.5 will be used to describe the terms *node, loop,* and *branch*. A *node* is simply a point of connection of two or more circuit elements. The reader is cautioned to note that although one node can be spread out with perfect conductors, it is still only one node. For example, node 5 consists of the entire bottom connector of the circuit. In other words, if we start at some point in the circuit and move along perfect conductors in any direction until we encounter a circuit element, the total path we cover represents a single node. Therefore, we can assume that a node is one end of a circuit element together with all the perfect conductors that are attached to it. Examining the circuit, we note that there are numerous paths through it. A *loop* is simply any *closed path* through the circuit in which no node is encountered more than once. For example, starting from node 1, one loop would contain the elements R_1, v_2, R_4, and i_1; another loop would contain R_2, v_1, v_2, R_4, and i_1; and so on. However, the path R_1, v_1, R_5, v_2, R_3, and i_1 is not a loop because we have encountered node 3 twice. Finally, a *branch* is a portion of a circuit containing only a single element and the nodes at each end of the element. The circuit in Fig. 2.5 contains eight branches.

Figure 2.5
Circuit used to illustrate KCL.

Given the previous definitions, we are now in a position to consider Kirchhoff's laws, named after German scientist Gustav Robert Kirchhoff. These two laws are quite simple but extremely important. We will not attempt to prove them because the proofs are beyond our current level of understanding. However, we will demonstrate their usefulness and attempt to make the reader proficient in their use. The first law is *Kirchhoff's current law* (KCL), which states that *the algebraic sum of the currents entering any node is zero.* In mathematical form the law appears as

LEARNING Hint

KCL is an extremely important and useful law.

$$\sum_{j=1}^{N} i_j(t) = 0 \qquad\qquad 2.7$$

where $i_j(t)$ is the jth current entering the node through branch j and N is the number of branches connected to the node. To understand the use of this law, consider node 3 shown in Fig. 2.5. Applying Kirchhoff's current law to this node yields

$$i_2(t) - i_4(t) + i_5(t) - i_7(t) = 0$$

LEARNING by Doing

D 2.3 Write the KCL equation for the following node:

ANSWER
$$i_1(t) - i_2(t) - i_3(t) - i_4(t) + i_5(t) = 0$$

We have assumed that the algebraic signs of the currents entering the node are positive and, therefore, that the signs of the currents leaving the node are negative.

If we multiply the foregoing equation by -1, we obtain the expression

$$-i_2(t) + i_4(t) - i_5(t) + i_7(t) = 0$$

which simply states that *the algebraic sum of the currents leaving a node is zero.* Alternatively, we can write the equation as

$$i_2(t) + i_5(t) = i_4(t) + i_7(t)$$

which states that *the sum of the currents entering a node is equal to the sum of the currents leaving the node.* Both of these italicized expressions are alternative forms of Kirchhoff's current law.

Once again it must be emphasized that the latter statement means that the sum of the *variables* that have been defined entering the node is equal to the sum of the *variables* that have been defined leaving the node, not the actual currents. For example, $i_j(t)$ may be defined entering the node, but if its actual value is negative, there will be positive charge leaving the node.

Note carefully that Kirchhoff's current law states that the *algebraic* sum of the currents either entering or leaving a node must be zero. We now begin to see why we stated in Chapter 1 that it is critically important to specify both the magnitude and the direction of a current.

LEARNING Example 2.3

Let us write KCL for every node in the network in Fig. 2.5 assuming that the currents leaving the node are positive.

SOLUTION   The KCL equations for nodes 1 through 5 are

$$-i_1(t) + i_2(t) + i_3(t) = 0$$
$$i_1(t) - i_4(t) + i_6(t) = 0$$
$$-i_2(t) + i_4(t) - i_5(t) + i_7(t) = 0$$
$$-i_3(t) + i_5(t) - i_8(t) = 0$$
$$-i_6(t) - i_7(t) + i_8(t) = 0$$

Note carefully that if we add the first four equations we obtain the fifth equation. What does this tell us? Recall that this means that this set of equations is not linearly independent. We can show that the first four equations are, however, linearly independent. Store this idea in memory because it will become very important when we learn how to write the equations necessary to solve for all the currents and voltages in a network in the following chapter.

LEARNING Example 2.4

The network in Fig. 2.5 is represented by the topological diagram shown in Fig. 2.6. We wish to find the unknown currents in the network.

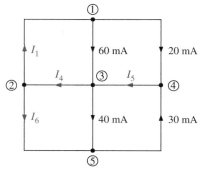

Figure 2.6
Topological diagram for the circuit in Fig. 2.5.

SOLUTION   Assuming the currents leaving the node are positive, the KCL equations for nodes 1 through 4 are

$$-I_1 + 0.06 + 0.02 = 0$$
$$I_1 - I_4 + I_6 = 0$$
$$-0.06 + I_4 - I_5 + 0.04 = 0$$
$$-0.02 + I_5 - 0.03 = 0$$

The first equation yields I_1 and the last equation yields I_5. Knowing I_5 we can immediately obtain I_4 from the third equation. Then the values of I_1 and I_4 yield the value of I_6 from the second equation. The results are $I_1 = 80$ mA, $I_4 = 70$ mA, $I_5 = 50$ mA, and $I_6 = -10$ mA.

As indicated earlier, dependent or controlled sources are very important because we encounter them when analyzing circuits containing active elements such as transistors. The following example presents a circuit containing a current-controlled current source.

LEARNING Example 2.5

Let us write the KCL equations for the circuit shown in Fig. 2.7.

Figure 2.7
Circuit containing a dependent current source.

SOLUTION The KCL equations for nodes 1 through 4 follow.

$$i_1(t) + i_2(t) - i_5(t) = 0$$

$$-i_2(t) + i_3(t) - 50i_2(t) = 0$$

$$-i_1(t) + 50i_2(t) + i_4(t) = 0$$

$$i_5(t) - i_3(t) - i_4(t) = 0$$

If we added the first three equations, we would obtain the negative of the fourth. What does this tell us about the set of equations?

Finally, it is possible to generalize Kirchhoff's current law to include a closed surface. By a closed surface we mean some set of elements completely contained within the surface that are interconnected. Since the current entering each element within the surface is equal to that leaving the element (i.e., the element stores no net charge), it follows that the current entering an interconnection of elements is equal to that leaving the interconnection. Therefore, Kirchhoff's current law can also be stated as follows: *The algebraic sum of the currents entering any closed surface is zero.*

LEARNING Example 2.6

Let us find I_4 and I_1 in the network represented by the topological diagram in Fig. 2.6.

Figure 2.8
Diagram used to demonstrate KCL for a surface.

SOLUTION This diagram is redrawn in Fig. 2.8; node 1 is enclosed in surface 1 and nodes 3 and 4 are enclosed in surface 2. A quick review of the previous example indicates that we derived a value for I_4 from the value of I_5. However, I_5 is now completely enclosed in surface 2. If we apply KCL to surface 2, assuming the currents out of the surface are positive, we obtain

$$I_4 - 0.06 - 0.02 - 0.03 + 0.04 = 0$$

or

$$I_4 = 70 \text{ mA}$$

which we obtained without any knowledge of I_5. Likewise for surface 1, what goes in must come out and, therefore, $I_1 = 80$ mA. The reader is encouraged to cut the network in Fig. 2.6 into two pieces in any fashion and show that KCL is always satisfied at the boundaries.

LEARNING EXTENSIONS

E2.3 Given the networks in Fig. E2.3, find (a) I_1 in Fig. E2.3a and (b) I_T in Fig. E2.3b.

ANSWER (a) $I_1 = -50$ mA, (b) $I_T = 70$ mA.

Figure E2.3   (a)   (b)

E2.4 Find (a) I_1 in the network in Fig. E2.4a and (b) I_1 and I_2 in the circuit in Fig. E2.4b.

ANSWER (a) $I_1 = 6$ mA,
(b) $I_1 = 8$ mA and
$I_2 = 5$ mA.

Figure E2.4   (a)   (b)

E2.5 Find the current i_x in the circuits in Fig. E2.5.

ANSWER (a) $i_x = 4$ mA,
(b) $i_x = 12$ mA.

Figure E2.5   (a)   (b)

LEARNING by Doing

D 2.4 Write the KVL equation for the following loop, traveling clockwise:

ANSWER
$-V_S + V_{R_1} + V_{R_2} + V_{R_3} = 0$

Kirchhoff's second law, called *Kirchhoff's voltage law* (KVL), states that the *algebraic sum of the voltages around any loop is zero*. As was the case with Kirchhoff's current law, we will defer the proof of this law and concentrate on understanding how to apply it. Once again the reader is cautioned to remember that we are dealing only with lumped-parameter circuits. These circuits are conservative, meaning that the work required to move a unit charge around any loop is zero.

Recall that in Kirchhoff's current law, the algebraic sign was required to keep track of whether the currents were entering or leaving a node. In Kirchhoff's voltage law the algebraic sign is used to keep track of the voltage polarity. In other words, as we traverse the circuit, it is necessary to sum to zero the increases and decreases in energy level. Therefore, it is important we keep track of whether the energy level is increasing or decreasing as we go through each element.

Finally, we employ the convention V_{ab} to indicate the voltage of point a with respect to point b: that is, the variable for the voltage between point a and point b, with point a considered positive relative to point b. Since the potential is measured between two points, it is convenient to use an arrow

LEARNING Example 2.7

Consider the circuit shown in Fig. 2.9. If V_{R_1} and V_{R_2} are known quantities, let us find V_{R_3}.

Figure 2.9
Circuit used to illustrate KVL.

network is a single loop, we have only one closed path. We will adopt a policy of considering an increase in energy level as negative and a decrease in energy level as positive. Using this policy and starting at point a in the network and traversing it in a clockwise direction, we obtain the equation

$$+V_{R_1} - 5 + V_{R_2} - 15 + V_{R_3} - 30 = 0$$

which can be written as

$$+V_{R_1} + V_{R_2} + V_{R_3} = 5 + 15 + 30$$

$$= 50$$

Now suppose that V_{R_1} and V_{R_2} are known to be 18 V and 12 V, respectively. Then $V_{R_3} = 20$ V.

SOLUTION In applying KVL, we must traverse the circuit and sum to zero the increases and decreases in energy level. Since the

LEARNING Example 2.8

Consider the network in Fig. 2.10.

Figure 2.10
Circuit used to explain KVL.

Let us demonstrate that only two of the three possible loop equations are linearly independent.

SOLUTION Note that this network has three closed paths: the left loop, right loop, and outer loop. Applying our policy for writing KVL equations and traversing the left loop starting at point *a*, we obtain

$$V_{R_1} + V_{R_4} - 16 - 24 = 0$$

The corresponding equation for the right loop starting at point *b* is

$$V_{R_2} + V_{R_3} + 8 + 16 - V_{R_4} = 0$$

The equation for the outer loop starting at point *a* is

$$V_{R_1} + V_{R_2} + V_{R_3} + 8 - 24 = 0$$

Note that if we add the first two equations, we obtain the third equation. Therefore, as we indicated in Example 2.3, the three equations are not linearly independent. Once again, we will address this issue in the next chapter and demonstrate that we need only the first two equations to solve for the voltages in the circuit.

between the two points with the head of the arrow located at the positive node. Note that the double-subscript notation, the + and − notation, and the single-headed arrow notation are all the same if the head of the arrow is pointing toward the positive terminal and the first subscript in the double-subscript notation. All of these equivalent forms for labeling voltages are shown in Fig. 2.11. The usefulness of the arrow notation stems from the fact that we may want to label the voltage between two points that are far apart in a network. In this case, the other notations are often confusing.

(a)

(b)

(c)

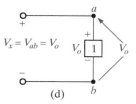

(d)

Figure 2.11
Equivalent forms for labeling voltage.

LEARNING Example 2.9

Consider the network in Fig. 2.12a. Let us apply KVL to determine the voltage between two points. Specifically, in terms of the double-subscript notation, let us find V_{ae} and V_{ec}.

SOLUTION The circuit is redrawn in Fig. 2.12b. Since points *a* and *e* as well as *e* and *c* are not physically close, the arrow notation is very useful. Our approach to determining the unknown voltage is

(a)

(b)

Figure 2.12
Network used in Example 2.9.

to apply KVL with the unknown voltage in the closed path. Therefore, to determine V_{ae} we can use the path *aefa* or *abcdea*. The equations for the two paths in which V_{ae} is the only unknown are

$$V_{ae} + 10 - 24 = 0$$

and

$$16 - 12 + 4 + 6 - V_{ae} = 0$$

Note that both equations yield $V_{ae} = 14$ V. Even before calculating V_{ae}, we could calculate V_{ec} using the path *cdec* or

cefabc. However, since V_{ae} is now known, we can also use the path *ceabc*. KVL for each of these paths is

$$4 + 6 + V_{ec} = 0$$

$$-V_{ec} + 10 - 24 + 16 - 12 = 0$$

and

$$-V_{ec} - V_{ae} + 16 - 12 = 0$$

Each of these equations yields $V_{ec} = -10$ V.

LEARNING Hint

KVL is an extremely important and useful law.

In general, the mathematical representation of Kirchhoff's voltage law is

$$\sum_{j=1}^{N} v_j(t) = 0 \qquad \textbf{2.8}$$

where $v_j(t)$ is the voltage across the jth branch (with the proper reference direction) in a loop containing N voltages. This expression is analogous to Eq. (2.7) for Kirchhoff's current law.

LEARNING Example 2.10

Given the network in Fig. 2.13 containing a dependent source, let us write the KVL equations for the two closed paths *abda* and *bcdb*.

SOLUTION The two KVL equations are

$$V_{R_1} + V_{R_2} - V_S = 0$$

$$20V_{R_1} + V_{R_3} - V_{R_2} = 0$$

Figure 2.13
Network containing a dependent source.

E2.6 Find V_{ad} and V_{eb} in the network in Fig. E2.6.

Figure E2.6

ANSWER $V_{ad} = 26$ V, $V_{eb} = 10$ V.

E2.7 Find V_{bd} in the circuit in Fig. E2.7.

Figure E2.7

ANSWER $V_{bd} = 11$ V.

Figure 2.14

Circuits used to explain Ohm's law.

Before proceeding with the analysis of simple circuits, it is extremely important that we emphasize a subtle but very critical point. Ohm's law as defined by the equation $V = IR$ refers to the relationship between the voltage and current as defined in Fig. 2.14a. If the direction of either the current or the voltage, but not both, is reversed, the relationship between the current and the voltage would be $V = -IR$. In a similar manner, given the circuit in Fig. 2.14b, if the polarity of the voltage between the terminals A and B is specified as shown, then the direction of the current I is from point B through R to point A. Likewise, in Fig. 2.14c, if the direction of the current is specified as shown, then the polarity of the voltage must be such that point D is at a higher potential than point C and, therefore, the arrow representing the voltage V is from point C to point D.

2.3 Single-Loop Circuits

VOLTAGE DIVISION   At this point we can begin to apply the laws we have presented earlier to the analysis of simple circuits. To begin, we examine what is perhaps the simplest circuit—a single closed path, or loop, of elements. The elements of a single loop carry the same current and, therefore, are said to be in series. However, we will apply Kirchhoff's voltage law and Ohm's law to the circuit to determine various quantities in the circuit.

Our approach will be to begin with a simple circuit and then generalize the analysis to more complicated ones. The circuit shown in Fig. 2.15 will serve as a basis for discussion. This circuit consists of an independent voltage source that is in series with two resistors. We have assumed that the current flows in a clockwise direction. If this assumption is correct, the solution of the equations that yields the current will produce a positive value. If the current is actually flowing in the opposite direction, the value of the current variable will simply be negative, indicating that the current is flowing in a direction opposite to that assumed. We have also

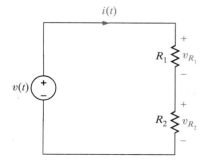

Figure 2.15

Single-loop circuit.

made voltage polarity assignments for v_{R_1} and v_{R_2}. These assignments have been made using the convention employed in our discussion of Ohm's law and our choice for the direction of $i(t)$; that is, the convention shown in Fig. 2.14a.

Applying Kirchhoff's voltage law to this circuit yields

$$-v(t) + v_{R_1} + v_{R_2} = 0$$

or

$$v(t) = v_{R_1} + v_{R_2}$$

However, from Ohm's law we know that

$$v_{R_1} = R_1 i(t)$$

$$v_{R_2} = R_2 i(t)$$

Therefore,

$$v(t) = R_1 i(t) + R_2 i(t)$$

Solving the equation for $i(t)$ yields

$$i(t) = \frac{v(t)}{R_1 + R_2} \qquad \textbf{2.9}$$

Knowing the current, we can now apply Ohm's law to determine the voltage across each resistor:

$$
\begin{aligned}
v_{R_1} &= R_1 i(t) \\
&= R_1 \left[\frac{v(t)}{R_1 + R_2} \right] \qquad \textbf{2.10}\\
&= \frac{R_1}{R_1 + R_2} v(t)
\end{aligned}
$$

LEARNING Hint

The manner in which voltage divides between two series resistors

Similarly,

$$v_{R_2} = \frac{R_2}{R_1 + R_2} v(t) \qquad \textbf{2.11}$$

Although simple, Eqs. (2.10) and (2.11) are very important because they describe the operation of what is called a *voltage divider*. In other words, the source voltage $v(t)$ *is divided between the resistors R_1 and R_2 in direct proportion to their resistances.*

In essence, if we are interested in the voltage across the resistor R_1, we bypass the calculation of the current $i(t)$ and simply multiply the input voltage $v(t)$ by the ratio

$$\frac{R_1}{R_1 + R_2}$$

As illustrated in Eq. (2.10), we are using the current in the calculation, but not explicitly.

Note that the equations satisfy Kirchhoff's voltage law, since

$$-v(t) + \frac{R_1}{R_1 + R_2} v(t) + \frac{R_2}{R_1 + R_2} v(t) = 0$$

LEARNING Example 2.11

Consider the circuit shown in Fig. 2.16. The circuit is identical to Fig. 2.15 except that R_1 is a variable resistor such as the volume control for a radio or television set. Suppose that $V_S = 9$ V, $R_1 = 90$ kΩ, and $R_2 = 30$ kΩ.

Figure 2.16
Voltage-divider circuit.

Let us examine the change in both the voltage across R_2 and the power absorbed in this resistor as R_1 is changed from 90 kΩ to 15 kΩ.

SOLUTION   Since this is a voltage-divider circuit, the voltage V_2 can be obtained directly as

$$V_2 = \left[\frac{R_2}{R_1 + R_2} \right] V_S$$

$$= \left[\frac{30k}{90k + 30k} \right] (9)$$

$$= 2.25 \text{ V}$$

Now suppose that the variable resistor is changed from 90 kΩ to 15 kΩ. Then

$$V_2 = \left[\frac{30k}{30k + 15k} \right] 9$$

$$= 6 \text{ V}$$

The direct voltage-divider calculation is equivalent to determining the current I and then using Ohm's law to find V_2. Note that the larger voltage is across the larger resistance. This voltage-divider concept and the simple circuit we have employed to describe it are very useful because, as will be shown later, more complicated circuits can be reduced to this form.

Finally, let us determine the instantaneous power absorbed by the resistor R_2 under the two conditions $R_1 = 90$ kΩ and $R_1 = 15$ kΩ. For the case $R_1 = 90$ kΩ, the power absorbed by R_2 is

$$P_2 = I^2 R_2 = \left(\frac{9}{120k} \right)^2 (30k)$$

$$= 0.169 \text{ mW}$$

In the second case

$$P_2 = \left(\frac{9}{45k} \right)^2 (30k)$$

$$= 1.2 \text{ mW}$$

The current in the first case is 75 μA, and in the second case it is 200 μA. Since the power absorbed is a function of the square of the current, the power absorbed in the two cases is quite different.

Let us now demonstrate the practical utility of this simple voltage-divider network.

LEARNING Example 2.12

Consider the circuit in Fig. 2.17a, which is an approximation of a high-voltage dc transmission facility. We have assumed that the bottom portion of the transmission line is a perfect conductor and will justify this assumption in the next chapter. The load can be represented by a resistor of value 183.5 Ω. Therefore, the equivalent circuit of this network is shown in Fig. 2.17b.

Let us determine both the power delivered to the load and the power losses in the line.

SOLUTION   Using voltage division, the load voltage is

$$V_{load} = \left[\frac{183.5}{183.5 + 16.5} \right] 400k$$

$$= 367 \text{ kV}$$

Line resistance is 0.04125 Ω/mile

Figure 2.17
A high-voltage dc transmission facility.

400-mile transmission line

(a)

(b)

(continued)

The input power is 800 MW and the power transmitted to the load is

$$P_{load} = I^2 R_{load}$$
$$= 734 \text{ MW}$$

Therefore, the power loss in the transmission line is

$$P_{line} = P_{in} - P_{load} = I^2 R_{line}$$
$$= 66 \text{ MW}$$

Since $P = VI$, suppose now that the utility company supplied power at 200 kV and 4 kA. What effect would this have on our transmission network? Without making a single calculation, we know that because power is proportional to the square of the current, there would be a large increase in the power loss in the line and, therefore, the efficiency of the facility would decrease substantially. That is why, in general, we transmit power at high voltage and low current.

MULTIPLE SOURCE/RESISTOR NETWORKS   At this point we wish to extend our analysis to include a multiplicity of voltage sources and resistors. For example, consider the circuit shown in Fig. 2.18a. Here we have assumed that the current flows in a clockwise direction, and we have defined the variable $i(t)$ accordingly. This may or may not be the case, depending on the value of the various voltage sources. Kirchhoff's voltage law for this circuit is

$$+v_{R_1} + v_2(t) - v_3(t) + v_{R_2} + v_4(t) + v_5(t) - v_1(t) = 0$$

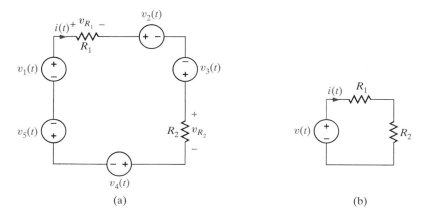

Figure 2.18
Equivalent circuits with multiple sources.

(a)   (b)

or, using Ohm's law,

$$\left(R_1 + R_2\right)i(t) = v_1(t) - v_2(t) + v_3(t) - v_4(t) - v_5(t)$$

which can be written as

$$\left(R_1 + R_2\right)i(t) = v(t)$$

where

$$v(t) = v_1(t) + v_3(t) - \left[v_2(t) + v_4(t) + v_5(t)\right]$$

so that under the preceding definitions, Fig. 2.18a is equivalent to Fig. 2.18b. In other words, the sum of several voltage sources in series can be replaced by one source whose value is the algebraic sum of the individual sources. This analysis can, of course, be generalized to a circuit with N series sources.

Now consider the circuit with N resistors in series, as shown in Fig. 2.19a. Applying Kirchhoff's voltage law to this circuit yields

$$v(t) = v_{R_1} + v_{R_2} + \cdots + v_{R_N}$$
$$= R_1 i(t) + R_2 i(t) + \cdots + R_N i(t)$$

and therefore,

$$v(t) = R_S i(t) \qquad\qquad \textbf{2.12}$$

(a)   (b)

where

$$R_S = R_1 + R_2 + \cdots + R_N \qquad \textbf{2.13}$$

and hence,

$$i(t) = \frac{v(t)}{R_S} \qquad \textbf{2.14}$$

Note also that for any resistor R_i in the circuit, the voltage across R_i is given by the expression

$$v_{R_i} = \frac{R_i}{R_S} v(t) \qquad \textbf{2.15}$$

which is the voltage-division property for multiple resistors in series.

Equation (2.13) illustrates that *the equivalent resistance of N resistors in series is simply the sum of the individual resistances*. Thus, using Eq. (2.13), we can draw the circuit in Fig. 2.19b as an equivalent circuit for the one in Fig. 2.19a.

LEARNING **Example 2.13**

Given the circuit in Fig. 2.20a, let us find I, V_{bd}, and the power absorbed by the 30-kΩ resistor. Finally, let us use voltage division to find V_{bc}.

Figure 2.20   *Circuit used in Example 2.13.*

SOLUTION   KVL for the network yields the equation

$$10kI + 20kI + 12 + 30kI - 6 = 0$$
$$60kI = -6$$
$$I = -0.1 \text{ mA}$$

Therefore, the magnitude of the current is 0.1 mA, but its direction is opposite to that assumed.

The voltage V_{bd} can be calculated using either of the closed paths *abdea* or *bcdb*. The equations for both cases are

$$10kI + V_{bd} + 30kI - 6 = 0$$

and

$$20kI + 12 - V_{bd} = 0$$

Using $I = -0.1$ mA in either equation yields $V_{bd} = 10$ V. Finally, the power absorbed by the 30-kΩ resistor is

$$P = I^2 R = 0.3 \text{ mW}$$

Now from the standpoint of determining the voltage V_{bc}, we can simply add the sources since they are in series, add the remaining resistors since they are in series, and reduce the network to that shown in Fig. 2.20b. Then

$$V_{bc} = \frac{20k}{20k + 40k}(-6)$$
$$= -2 \text{ V}$$

E2.8 Find I and V_{bd} in the circuit in Fig. E2.8.

Figure E2.8

ANSWER $I = -0.05$ mA and $V_{bd} = 10$ V

E2.9 In the network in Fig. E2.9, if V_{ad} is 3 V, find V_S.

Figure E2.9

ANSWER $V_S = 9$ V.

2.4 Single-Node-Pair Circuits

CURRENT DIVISION An important circuit is the single-node-pair circuit. In this case the elements have the same voltage across them and, therefore, are in *parallel*. We will, however, apply Kirchhoff's current law and Ohm's law to determine various unknown quantities in the circuit.

Following our approach with the single-loop circuit, we will begin with the simplest case and then generalize our analysis. Consider the circuit shown in Fig. 2.21. Here we have an independent current source in parallel with two resistors.

Since all of the circuit elements are in parallel, the voltage $v(t)$ appears across each of them. Furthermore, an examination of the circuit indicates that the current $i(t)$ is into the upper node of the circuit and the currents $i_1(t)$ and $i_2(t)$ are out of the node. Since KCL essentially states that what goes in must come out, the question we must answer is how $i_1(t)$ and $i_2(t)$ divide the input current $i(t)$.

Applying Kirchhoff's current law to the upper node, we obtain

$$i(t) = i_1(t) + i_2(t)$$

Figure 2.21
Simple parallel circuit.

and, employing Ohm's law, we have

$$i(t) = \frac{v(t)}{R_1} + \frac{v(t)}{R_2}$$

$$= \left(\frac{1}{R_1} + \frac{1}{R_2} \right) v(t)$$

$$= \frac{v(t)}{R_p}$$

where

$$\frac{1}{R_p} = \frac{1}{R_1} + \frac{1}{R_2} \qquad 2.16$$

LEARNING Hint

The parallel resistance equation

$$R_p = \frac{R_1 R_2}{R_1 + R_2} \qquad 2.17$$

Therefore, the equivalent resistance of two resistors connected in parallel is equal to the product of their resistances divided by their sum. Note also that this equivalent resistance R_p is always less than either R_1 or R_2. Hence, by connecting resistors in parallel we reduce the overall resistance. In the special case when $R_1 = R_2$, the equivalent resistance is equal to half of the value of the individual resistors.

The manner in which the current $i(t)$ from the source divides between the two branches is called *current division* and can be found from the preceding expressions. For example,

$$v(t) = R_p i(t)$$

$$= \frac{R_1 R_2}{R_1 + R_2} i(t) \qquad 2.18$$

and

LEARNING Hint

The manner in which current divides between two parallel resistors

$$i_1(t) = \frac{v(t)}{R_1}$$

$$i_1(t) = \frac{R_2}{R_1 + R_2} i(t) \qquad 2.19$$

LEARNING by Doing

D 2.5 Find I_1 and I_2 in the following circuit:

ANSWER $I_1 = 1$ mA
$I_2 = 4$ mA

and

$$i_2(t) = \frac{v(t)}{R_2}$$

$$= \frac{R_1}{R_1 + R_2} i(t) \qquad 2.20$$

Equations (2.19) and (2.20) are mathematical statements of the current-division rule.

LEARNING Example 2.14

Given the network in Fig. 2.22a, let us find I_1, I_2, and V_o.

SOLUTION First, it is important to recognize that the current source feeds two parallel paths. To emphasize this point, the circuit is redrawn as shown in Fig. 2.22b. Applying current division, we obtain

$$I_1 = \left[\frac{40k + 80k}{60k + (40k + 80k)} \right] (0.9 \times 10^{-3})$$

$$= 0.6 \text{ mA}$$

and

$$I_2 = \left[\frac{60k}{60k + (40k + 80k)} \right] (0.9 \times 10^{-3})$$

$$= 0.3 \text{ mA}$$

Note that the larger current flows through the smaller resistor, and vice versa. In addition, note that if the resistances of the two paths are equal, the current will divide equally between them. KCL is satisfied since $I_1 + I_2 = 0.9$ mA.

The voltage V_o can be derived using Ohm's law as

$$V_o = 80k I_2$$

$$= 24 \text{ V}$$

The problem can also be approached in the following manner. The total resistance seen by the current source is 40 kΩ, i.e., 60 kΩ in parallel with the series combination of 40 kΩ and 80 kΩ as shown in Fig. 2.23c. The voltage across the current source is then

$$V_1 = (0.9 \times 10^{-3})40k$$

$$= 36 \text{ V}$$

Now that V_1 is known, we can apply voltage division to find V_o.

$$V_o = \left(\frac{80k}{80k + 40k} \right) V_1$$

$$= \left(\frac{80k}{120k} \right) 36$$

$$= 24 \text{ V}$$

Figure 2.22
Circuit used in Example 2.14.

(a) (b) (c)

LEARNING Example 2.15

A typical car stereo consists of a 2-W audio amplifier and two speakers represented by the diagram shown in Fig. 2.23a. The output circuit of the audio amplifier is in essence a 430-mA current source and the speakers each have a resistance of 4 Ω. Let us determine the power absorbed by the speakers.

SOLUTION The audio system can be modeled as shown in

Fig. 2.23b. Since the speakers are both 4-Ω devices, the current will split evenly between them and the power absorbed by each speaker is

$$P = I^2 R$$

$$= (215 \times 10^{-3})^2 (4)$$

$$= 184.9 \text{ mW}$$

Figure 2.23
Circuits used in Example 2.15.

(a) (b)

E2.10   Find the currents I_1 and I_2 and the power absorbed by the 40-kΩ resistor in the network in Fig. E2.10.

ANSWER $I_1 = 12$ mA, $I_2 = -4$ mA, and $P_{40\ k\Omega} = 5.76$ W.

Figure E2.10

MULTIPLE SOURCE/RESISTOR NETWORKS   Let us now extend our analy-
sis to include a multiplicity of current sources and resistors in parallel. For example, consider
the circuit shown in Fig. 2.24a. We have assumed that the upper node is $v(t)$ volts positive
with respect to the lower node. Applying Kirchhoff's current law to the upper node yields

$$i_1(t) - i_2(t) - i_3(t) + i_4(t) - i_5(t) - i_6(t) = 0$$

or

$$i_1(t) - i_3(t) + i_4(t) - i_6(t) = i_2(t) + i_5(t)$$

(a)   (b)

Figure 2.24
Equivalent circuits.

The terms on the left side of the equation all represent sources that can be combined algebraically into a single source; that is,

$$i_o(t) = i_1(t) - i_3(t) + i_4(t) - i_6(t)$$

which effectively reduces the circuit in Fig. 2.24a to that in Fig. 2.24b. We could, of course, generalize this analysis to a circuit with N current sources. Using Ohm's law, the currents on the right
side of the equation can be expressed in terms of the voltage and individual resistances so that the
KCL equation reduces to

$$i_o(t) = \left(\frac{1}{R_1} + \frac{1}{R_2}\right) v(t)$$

Now consider the circuit with N resistors in parallel, as shown in Fig. 2.25a. Applying
Kirchhoff's current law to the upper node yields

$$i_o(t) = i_1(t) + i_2(t) + \cdots + i_N(t)$$

$$= \left(\frac{1}{R_1} + \frac{1}{R_2} + \cdots + \frac{1}{R_N}\right) v(t) \qquad \textbf{2.21}$$

(a)   (b)

Figure 2.25
Equivalent circuits.

or

$$i_o(t) = \frac{v(t)}{R_p} \qquad \textbf{2.22}$$

where

$$\frac{1}{R_p} = \sum_{i=1}^{N} \frac{1}{R_i} \qquad \textbf{2.23}$$

so that as far as the source is concerned, Fig. 2.25a can be reduced to an equivalent circuit, as shown in Fig. 2.25b.

The current division for any branch can be calculated using Ohm's law and the preceding equations. For example, for the *j*th branch in the network of Fig. 2.25a,

$$i_j(t) = \frac{v(t)}{R_j}$$

Using Eq. (2.22), we obtain

$$i_j(t) = \frac{R_p}{R_j} i_o(t) \qquad \textbf{2.24}$$

which defines the current-division rule for the general case.

LEARNING Example 2.16

Given the circuit in Fig. 2.26a, we wish to find the current in the 12-kΩ load resistor.

SOLUTION To simplify the network in Fig. 2.26a, we add the current sources algebraically and combine the parallel resistors in the following manner:

$$\frac{1}{R_p} = \frac{1}{18k} + \frac{1}{9k} + \frac{1}{12k}$$

$$R_p = 4 \text{ k}\Omega$$

Using these values we can reduce the circuit in Fig. 2.26a to that in Fig. 2.26b. Now, applying current division, we obtain

$$I_L = -\left[\frac{4k}{4k + 12k}\right](1 \times 10^{-3})$$

$$= -0.25 \text{ mA}$$

Figure 2.26
Circuits used in
Example 2.16.

(a) (b)

E2.11 Find the power absorbed by the 6-kΩ resistor in the network in Fig. E2.11.

ANSWER $P = 2.67$ mW.

Figure E2.11

2.5 Series and Parallel Resistor Combinations

We have shown in our earlier developments that the equivalent resistance of N resistors in series is

$$R_S = R_1 + R_2 + \cdots + R_N \qquad \text{2.25}$$

and the equivalent resistance of N resistors in parallel is found from

$$\frac{1}{R_P} = \frac{1}{R_1} + \frac{1}{R_2} + \cdots + \frac{1}{R_N} \qquad \text{2.26}$$

Let us now examine some combinations of these two cases.

LEARNING Example 2.17

We wish to determine the resistance at terminals A-B in the network in Fig. 2.27a.

SOLUTION   Starting at the opposite end of the network from the terminals and combining resistors as shown in the sequence of circuits in Fig. 2.27, we find that the equivalent resistance at the terminals is 5 kΩ.

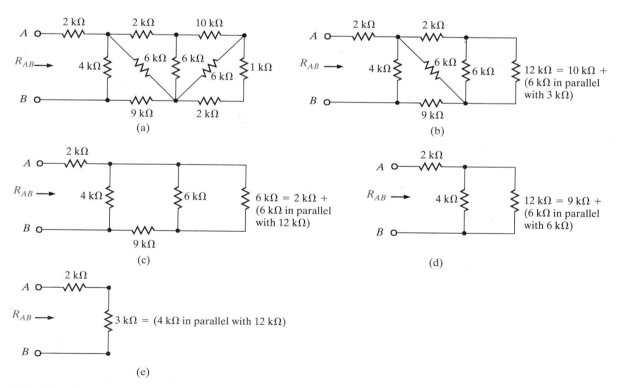

Figure 2.27   *Simplification of a resistance network.*

E2.12 Find the equivalent resistance at the terminals *A-B* in the network in Fig. E2.12.   **ANSWER** $R_{AB} = 22\ k\Omega$.

Figure E2.12

E2.13 Find the equivalent resistance at the terminals *A-B* in the circuit in Fig. E2.13.   **ANSWER** $R_{AB} = 3\ k\Omega$.

Figure E2.13

Problem-Solving Strategy
Simplifying Resistor Combinations

When trying to determine the equivalent resistance at a pair of terminals of a network composed of an interconnection of numerous resistors, it is recommended that the analysis begin at the end of the network opposite the terminals. Two or more resistors are combined to form a single resistor, thus simplifying the network by reducing the number of components as the analysis continues in a steady progression toward the terminals. The simplification involves the following:

1. *Resistors in series.* Resistors R_1 and R_2 are in series if they are connected in tandem and carry exactly the same current. They can then be combined into a single resistor R_S, where $R_S = R_1 + R_2$.

2. *Resistors in parallel.* Resistors R_1 and R_2 are in parallel if they are connected to the same two nodes and have exactly the same voltage across their terminals. They can then be combined into a single resistor R_p, where $R_p = R_1 R_2/(R_1 + R_2)$.

These two combinations are used repeatedly, as needed, to reduce the network to a single resistor at the pair of terminals.

RESISTOR SPECIFICATIONS   Some important parameters that are used to specify resistors are the resistor's value, tolerance, and power rating. The tolerance specifications for resistors are typically 5% and 10%. A listing of standard resistor values with their specified tolerances is shown in Table 2.1.

Table 2.1 Standard resistor values for 5% and 10% tolerances (10% values shown in boldface)

1.0	10	100	1.0k	10k	100k	1.0M	10M
1.1	11	110	1.1k	11k	110k	**1.1M**	11M
1.2	**12**	**120**	**1.2k**	12k	**120k**	1.2M	**12M**
1.3	13	130	1.3k	13k	130k	1.3M	13M
1.5	**15**	**150**	**1.5k**	**15k**	**150k**	**1.5M**	**15M**
1.6	16	160	1.6k	16k	160k	1.6M	16M
1.8	**18**	**180**	**1.8k**	**18k**	**180k**	**1.8M**	**18M**
2.0	20	200	2.0k	20k	200k	2.0M	20M
2.2	**22**	**220**	**2.2k**	**22k**	**220k**	**2.2M**	**22M**
2.4	24	240	2.4k	24k	240k	2.4M	
2.7	**27**	**270**	**2.7k**	**27k**	**270k**	**2.7M**	
3.0	30	300	3.0k	30k	300k	3.0M	
3.3	**33**	**330**	**3.3k**	**33k**	**330k**	**3.3M**	
3.6	36	360	3.6k	36k	360k	3.6M	
3.9	**39**	**390**	**3.9k**	**39k**	**390k**	**3.9M**	
4.3	43	430	4.3k	43k	430k	4.3M	
4.7	**47**	**470**	**4.7k**	**47k**	**470k**	**4.7M**	
5.1	51	510	5.1k	51k	510k	5.1M	
5.6	**56**	**560**	**5.6k**	**56k**	**560k**	**5.6M**	
6.2	62	620	6.2k	62k	620k	6.2M	
6.8	**68**	**680**	**6.8k**	**68k**	**680k**	**6.8M**	
7.5	75	750	7.5k	75k	750k	7.5M	
8.2	**82**	**820**	**8.2k**	**82k**	**820k**	**8.2M**	
9.1	91	910	9.1k	91k	910k	9.1M	

The power rating for a resistor specifies the maximum power that can be dissipated by the resistor. Some typical power ratings for resistors are $\frac{1}{4}$ W, $\frac{1}{2}$ W, 1 W, 2 W, and so forth, up to very high values for high-power applications. Thus in selecting a resistor for some particular application, one important selection criterion is the expected power dissipation.

LEARNING Example 2.18

Given the network in Fig. 2.28, we wish to find the range for both the current and power dissipation in the resistor if R is a 2.7-kΩ resistor with a tolerance of 10%.

Figure 2.28
Circuit used in Example 2.18.

SOLUTION Using the equations $I = V/R = 10/R$ and $P = V^2/R = 100/R$, the minimum and maximum values for the resistor, current, and power are outlined next.

Minimum resistor value = $R(1 - 0.1) = 0.9\,R = 2.43$ kΩ

Maximum resistor value = $R(1 + 0.1) = 1.1\,R = 2.97$ kΩ

Minimum current value = $10/2970 = 3.37$ mA

Maximum current value = $10/2430 = 4.12$ mA

Minimum power value = $100/2970 = 33.7$ mW

Maximum power value = $100/2430 = 41.2$ mW

Thus the range for the current and power are 3.37 mA to 4.12 mA and 33.7 mW to 41.2 mW, respectively.

LEARNING Example 2.19

Given the network shown in Fig. 2.29: (a) find the required value for the resistor R; (b) use Table 2.1 to select a standard 10% tolerance resistor for R; (c) using the resistor selected in (b), determine the voltage across the 3.9-kΩ resistor; (d) calculate the percent error in the voltage V_1, if the standard resistor selected in (b) is used; and (e) determine the power rating for this standard component.

Figure 2.29
Circuit used in Example 2.19.

SOLUTION

(a) Using KVL, the voltage across R is 19 V. Then using Ohm's law, the current in the loop is

$$I = 5/3.9k = 1.282 \text{ mA}$$

The required value of R is then

$$R = 19/0.001282 = 14.82 \text{ k}\Omega$$

(b) As shown in Table 2.1, the nearest standard 10% tolerance resistor is 15 kΩ.

(c) Using the standard 15-kΩ resistor, the actual current in the circuit is

$$I = 24/18.9k = 1.2698 \text{ mA}$$

and the voltage across the 3.9-kΩ resistor is

$$V = IR = (0.0012698)(3.9k) = 4.952 \text{ V}$$

(d) The percent error involved in using the standard resistor is

$$\% \text{ Error} = (4.952 - 5)/5 \times 100 = -0.96\%$$

(e) The power absorbed by the resistor R is then

$$P = IR = (0.0012698)^2(15k) = 24.2 \text{ mW}$$

Therefore, even a quarter-watt resistor is adequate in this application.

LEARNING by Doing

D2.6 Find the possible range of resistance for the following resistors:

(a) A 27-Ω resistor with a tolerance of 5%

(b) A 1.5-kΩ resistor with a tolerance of 10%

ANSWER *(a) 25.65 Ω to 28.35 Ω, (b) 1.35 kΩ to 1.65 kΩ.*

2.6 Circuits with Series-Parallel Combinations of Resistors

At this point we have learned many techniques that are fundamental to circuit analysis. Now we wish to apply them and show how they can be used in concert to analyze circuits. We will illustrate their application through a number of examples that will be treated in some detail.

LEARNING Example 2.20

We wish to find all the currents and voltages labeled in the ladder network shown in Fig. 2.30a.

SOLUTION To begin our analysis of the network, we start at the right end of the circuit and combine the resistors to determine the total resistance seen by the 12-V source. This will allow us to calculate the current I_1. Then employing KVL, KCL, Ohm's law, and/or voltage and current division, we will be able to calculate all currents and voltages in the network.

At the right end of the circuit, the 9-kΩ and 3-kΩ resistors are in series and, thus, can be combined into one equivalent 12-kΩ resistor. This resistor is in parallel with the 4-kΩ resistor, and their combination yields an equivalent 3-kΩ resistor, shown at the right edge of the circuit in Fig. 2.30b. In Fig. 2.30b the two 3-kΩ resistors are in series and their combination is in parallel with the 6-kΩ resistor. Combining all three resistances yields the circuit shown in Fig. 2.30c.

Applying Kirchhoff's voltage law to the circuit in Fig. 2.30c yields

$$I_1(9k + 3k) = 12$$

$$I_1 = 1 \text{ mA}$$

LEARNING by Doing

D 2.7 Find V_o in the following network:

ANSWER $V_o = 1$ V

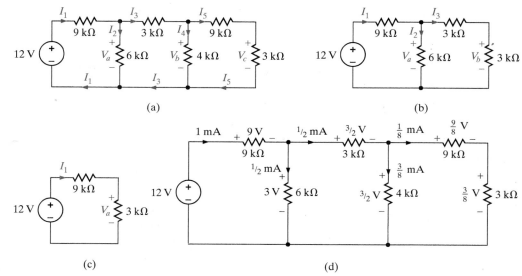

Figure 2.30
Analysis of a ladder network.

V_a can be calculated from Ohm's law as

$$V_a = I_1(3k)$$
$$= 3 \text{ V}$$

or, using Kirchhoff's voltage law,

$$V_a = 12 - 9kI_1$$
$$= 12 - 9$$
$$= 3 \text{ V}$$

Knowing I_1 and V_a, we can now determine all currents and voltages in Fig. 2.30b. Since $V_a = 3$ V, the current I_2 can be found using Ohm's law as

$$I_2 = \frac{3}{6k}$$
$$= \frac{1}{2} \text{ mA}$$

Then, using Kirchhoff's current law, we have

$$I_1 = I_2 + I_3$$
$$1 \times 10^{-3} = \frac{1}{2} \times 10^{-3} + I_3$$
$$I_3 = \frac{1}{2} \text{ mA}$$

Note that the I_3 could also be calculated using Ohm's law:

$$V_a = (3k + 3k)I_3$$
$$I_3 = \frac{3}{6k}$$
$$= \frac{1}{2} \text{ mA}$$

Applying Kirchhoff's voltage law to the right-hand loop in Fig. 2.30b yields

$$V_a - V_b = 3kI_3$$
$$3 - V_b = \frac{3}{2}$$
$$V_b = \frac{3}{2} \text{ V}$$

or, since V_b is equal to the voltage drop across the 3-kΩ resistor, we could use Ohm's law as

$$V_b = 3kI_3$$
$$= \frac{3}{2} \text{ V}$$

We are now in a position to calculate the final unknown currents and voltages in Fig. 2.30a. Knowing V_b, we can calculate I_4 using Ohm's law as

$$V_b = 4kI_4$$
$$I_4 = \frac{\frac{3}{2}}{4k}$$
$$= \frac{3}{8} \text{ mA}$$

LEARNING by Doing

D 2.8   Find I_o in the following circuit:

ANSWER $I_o = 1$ A

(continued)

Then, from Kirchhoff's current law, we have

$$I_3 = I_4 + I_5$$

$$\frac{1}{2} \times 10^{-3} = \frac{3}{8} \times 10^{-3} + I_5$$

$$I_5 = \frac{1}{8}\,\text{mA}$$

We could also have calculated I_5 using the current-division rule. For example,

$$I_5 = \frac{4k}{4k + (9k + 3k)}\,I_3$$

$$= \frac{1}{8}\,\text{mA}$$

Finally, V_c can be computed as

$$V_c = I_5(3k)$$

$$= \frac{3}{8}\,\text{V}$$

V_c can also be found using voltage division (i.e., the voltage V_b will be divided between the 9-kΩ and 3-kΩ resistors). Therefore,

$$V_c = \left[\frac{3k}{3k + 9k}\right]V_b$$

$$= \frac{3}{8}\,\text{V}$$

Note that Kirchhoff's current law is satisfied at every node and Kirchhoff's voltage law is satisfied around every loop, as shown in Fig. 2.30d.

The following example is, in essence, the reverse of the previous example in that we are given the current in some branch in the network and are asked to find the value of the input source.

LEARNING Example 2.21

Given the circuit in Fig. 2.31 and $I_4 = \frac{1}{2}\,\text{mA}$, let us find the source voltage V_o.

Figure 2.31   *Example circuit for analysis.*

SOLUTION   If $I_4 = \frac{1}{2}\,\text{mA}$, then from Ohm's law, $V_b = 3\,\text{V}$. V_b can now be used to calculate $I_3 = 1\,\text{mA}$. Kirchhoff's current law applied at node y yields

$$I_2 = I_3 + I_4$$

$$= 1.5\,\text{mA}$$

Then, from Ohm's law, we have

$$V_a = (1.5 \times 10^{-3})(2k)$$

$$= 3\,\text{V}$$

Since $V_a + V_b$ is now known, I_5 can be obtained:

$$I_5 = \frac{V_a + V_b}{3k + 1k}$$

$$= 1.5\,\text{mA}$$

Applying Kirchhoff's current law at node x yields

$$I_1 = I_2 + I_5$$

$$= 3\,\text{mA}$$

Now KVL applied to any closed path containing V_o will yield the value of this input source. For example, if the path is the outer loop, KVL yields

$$-V_o + 6kI_1 + 3kI_5 + 1kI_5 + 4kI_1 = 0$$

Since $I_1 = 3\,\text{mA}$ and $I_5 = 1.5\,\text{mA}$,

$$V_o = 36\,\text{V}$$

If we had selected the path containing the source and the points x, y, and z, we would obtain

$$-V_o + 6kI_1 + V_a + V_b + 4kI_1 = 0$$

Once again, this equation yields

$$V_o = 36\,\text{V}.$$

E2.14   Find V_o in the network in Fig. E2.14.

ANSWER $V_o = 2$ V.

Figure E2.14

E2.15   Find V_S in the circuit in Fig. E2.15.

ANSWER $V_S = 9$ V.

Figure E2.15

E2.16   Find I_S in the circuit in Fig. E2.16.

ANSWER $I_S = 0.3$ mA.

Figure E2.16

Problem-Solving Strategy
Analyzing Circuits Containing a Single Source and a Series-Parallel Interconnection of Resistors

Step 1.   Systematically reduce the resistive network so that the resistance seen by the source is represented by a single resistor.

Step 2.   Determine the source current for a voltage source or the source voltage if a current source is present.

Step 3.   Expand the network, retracing the simplification steps, and apply Ohm's law, KVL, KCL, voltage division, and current division to determine all currents and voltages in the network.

2.7 Wye ⇌ Delta Transformations

To provide motivation for this topic, consider the circuit in Fig. 2.32. Note that this network has essentially the same number of elements as contained in our recent examples. However, when we attempt to reduce the circuit to an equivalent network containing the source V_1 and an equivalent resistor R, we find that nowhere is a resistor in series or parallel with another. Therefore, we cannot attack the problem directly using the techniques that we have learned thus far. We can, however, replace one portion of the network with an equivalent circuit, and this conversion will permit us, with ease, to reduce the combination of resistors to a single equivalent resistance. This conversion is called the wye-to-delta or delta-to-wye transformation.

Figure 2.32
Network used to illustrate the need for the wye ⇌ delta transformation.

Consider the networks shown in Fig. 2.33. Note that the resistors in Fig. 2.33a form a Δ (delta) and the resistors in Fig. 2.33b form a Y (wye). If both of these configurations are connected at only three terminals a, b, and c, it would be very advantageous if an equivalence could be established between them. It is, in fact, possible to relate the resistances of one network to those of the other such that their terminal characteristics are the same. This relationship between the two network configurations is called the Y-Δ transformation.

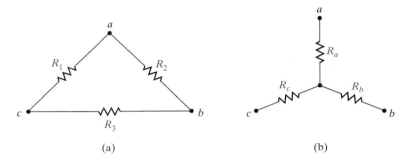

Figure 2.33
Delta and wye resistance networks.

(a)   (b)

The transformation that relates the resistances R_1, R_2, and R_3 to the resistances R_a, R_b, and R_c is derived as follows. For the two networks to be equivalent at each corresponding pair of terminals, it is necessary that the resistance at the corresponding terminals be equal (e.g., the resistance at terminals a and b with c open-circuited must be the same for both networks). Therefore, if we equate the resistances for each corresponding set of terminals, we obtain the following equations:

$$R_{ab} = R_a + R_b = \frac{R_2(R_1 + R_3)}{R_2 + R_1 + R_3}$$

$$R_{bc} = R_b + R_c = \frac{R_3(R_1 + R_2)}{R_3 + R_1 + R_2}$$

$$R_{ca} = R_c + R_a = \frac{R_1(R_2 + R_3)}{R_1 + R_2 + R_3}$$

2.27

Solving this set of equations for R_a, R_b, and R_c yields

$$R_a = \frac{R_1 R_2}{R_1 + R_2 + R_3}$$

$$R_b = \frac{R_2 R_3}{R_1 + R_2 + R_3}$$

$$R_c = \frac{R_1 R_3}{R_1 + R_2 + R_3}$$

2.28

Similarly, if we solve Eq. (2.27) for R_1, R_2, and R_3, we obtain

$$R_1 = \frac{R_a R_b + R_b R_c + R_a R_c}{R_b}$$

$$R_2 = \frac{R_a R_b + R_b R_c + R_a R_c}{R_c}$$

2.29

$$R_3 = \frac{R_a R_b + R_b R_c + R_a R_c}{R_a}$$

Equations (2.28) and (2.29) are general relationships and apply to any set of resistances connected in a Y or Δ. For the balanced case where $R_a = R_b = R_c$ and $R_1 = R_2 = R_3$, the equations above reduce to

$$R_Y = \tfrac{1}{3} R_\Delta$$

2.30

and

$$R_\Delta = 3R_Y$$

2.31

It is important to note that it is not necessary to memorize the formulas in Eqs. (2.28) and (2.29). Close inspection of these equations and Fig. 2.33 illustrates a definite pattern to the relationships between the two configurations. For example, the resistance connected to point a in the wye (i.e., R_a) is equal to the product of the two resistors in the Δ that are connected to point a divided by the sum of all the resistances in the delta. R_b and R_c are determined in a similar manner. Similarly, there are geometrical patterns associated with the equations for calculating the resistors in the delta as a function of those in the wye.

Let us now examine the use of the delta ⇌ wye transformation in the solution of a network problem.

LEARNING Example 2.22

Given the network in Fig. 2.34a, let us find the source current I_S.

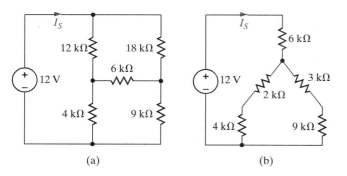

(a)   (b)

Figure 2.34 *Circuits used in Example 2.22.*

SOLUTION   Note that none of the resistors in the circuit are in series or parallel. However, careful examination of the network indicates that the 12k-, 6k-, and 18k-ohm resistors, as well as the 4k-, 6k-, and 9k-ohm resistors each form a delta that can be converted to a wye. Furthermore, the 12k-, 6k-, and 4k-ohm resistors, as well as the 18k-, 6k-, and 9k-ohm resistors each form a wye that can be converted to a delta. Any one of these conversions will lead to a solution. We will perform a delta-to-wye transformation on the 12k-, 6k-, and 18k-ohm resistors, which leads to the circuit in Fig. 2.34b. The 2k- and 4k-ohm resistors, like the 3k- and 9k-ohm resistors, are in series and their parallel combination yields a 4k-ohm resistor. Thus, the source current is

$$I_S = 12/(6k + 4k)$$

$$= 1.2 \text{ mA}.$$

2.8 Circuits with Dependent Sources

In Chapter 1 we outlined the different kinds of dependent sources. These controlled sources are extremely important because they are used to model (see Chapter 16) physical devices such as *npn* and *pnp* bipolar junction transistors (BJTs) and field-effect transistors (FETs) that are either metal-oxide-semiconductor field-effect transistors (MOSFETs) or insulated-gate field-effect transistors (IGFETs). These basic structures are, in turn, used to make analog and digital devices. A typical analog device is an operational amplifier (op-amp). Typical digital devices are random access memories (RAMs), read-only memories (ROMs), and microprocessors. We will now show how to solve simple one-loop and one-node circuits that contain these dependent sources. Although the following examples are fairly simple, they will serve to illustrate the basic concepts.

Problem-Solving Strategy
Circuits with Dependent Sources

Step 1. When writing the KVL and/or KCL equations for the network, treat the dependent source as though it were an independent source.

Step 2. Write the equation that specifies the relationship of the dependent source to the controlling parameter.

Step 3. Solve the equations for the unknowns. Be sure that the number of linearly independent equations matches the number of unknowns.

The following four examples will each illustrate one of the four types of dependent sources: current-controlled voltage source, current-controlled current source, voltage-controlled voltage source, and voltage-controlled current source.

LEARNING Example 2.23

Let us determine the voltage V_o in the circuit in Fig. 2.35.

Figure 2.35   Circuit used in Example 2.23.

SOLUTION   Applying KVL, we obtain

$$-12 + 3kI_1 - V_A + 5kI_1 = 0$$

where

$$V_A = 2000I_1$$

and the units of the multiplier, 2000, are ohms. Solving these equations yields

$$I_1 = 2 \text{ mA}$$

Then

$$V_o = (5 \text{ k})I_1$$
$$= 10 \text{ V}$$

LEARNING Example 2.24

Given the circuit in Fig. 2.36 containing a current-controlled current source, let us find the voltage V_o.

SOLUTION   Applying KCL at the top node, we obtain

$$10 \times 10^{-3} + \frac{V_S}{2k + 4k} + \frac{V_S}{3k} - 4I_o = 0$$

Figure 2.36   Circuit used in Example 2.24.

where

$$I_o = \frac{V_S}{3k}$$

Substituting this expression for the controlled source into the KCL equation yields

$$10^{-2} + \frac{V_S}{6k} + \frac{V_S}{3k} - \frac{4V_S}{3k} = 0$$

Solving this equation for V_S, we obtain

$$V_S = 12 \text{ V}$$

The voltage V_o can now be obtained using a simple voltage divider; that is,

$$V_o = \left[\frac{4k}{2k + 4k} \right] V_S$$

$$= 8 \text{ V}$$

LEARNING Example 2.25

The network in Fig. 2.37 contains a voltage-controlled voltage source. We wish to find V_o in this circuit.

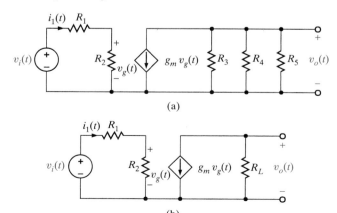

Figure 2.37
Circuit used in
Example 2.25.

SOLUTION   Applying KVL to this network yields

$$-12 + 3kI + 2V_o + 1kI = 0$$

where

$$V_o = 1kI$$

Hence, the KVL equation can be written as

$$-12 + 3kI + 2kI + 1kI = 0$$

or

$$I = 2 \text{ mA}$$

Therefore,

$$V_o = 1kI$$

$$= 2 \text{ V}$$

LEARNING Example 2.26

An equivalent circuit for a FET common-source amplifier or BJT common-emitter amplifier can be modeled by the circuit shown in Fig. 2.38a. We wish to determine an expression for the gain of the amplifier, which is the ratio of the output voltage to the input voltage.

Figure 2.38   *Example circuit containing a voltage-controlled current source.*

SOLUTION   Note that although this circuit, which contains a voltage-controlled current source, appears to be somewhat complicated, we are actually in a position now to solve it with techniques we have studied up to this point. The loop on the left, or input to the amplifier, is essentially detached from the output portion of the amplifier on the right. The voltage across R_2 is $v_g(t)$, which controls the dependent current source.

To simplify the analysis, let us replace the resistors R_3, R_4, and R_5 with R_L such that

$$\frac{1}{R_L} = \frac{1}{R_3} + \frac{1}{R_4} + \frac{1}{R_5}$$

Then the circuit reduces to that shown in Fig. 2.38b. Applying Kirchhoff's voltage law to the input portion of the amplifier yields

$$v_i(t) = i_1(t)(R_1 + R_2)$$

and

$$v_g(t) = i_1(t)R_2$$

Solving these equations for $v_g(t)$ yields

$$v_g(t) = \frac{R_2}{R_1 + R_2} v_i(t)$$

(continued)

From the output circuit, note that the voltage $v_o(t)$ is given by the expression

$$v_o(t) = -g_m v_g(t) R_L$$

Combining this equation with the preceding one yields

$$v_o(t) = \frac{-g_m R_L R_2}{R_1 + R_2} v_i(t)$$

Therefore, the amplifier gain, which is the ratio of the output voltage to the input voltage, is given by

$$\frac{v_o(t)}{v_i(t)} = -\frac{g_m R_L R_2}{R_1 + R_2}$$

Reasonable values for the circuit parameters in Fig. 2.38a are $R_1 = 100\ \Omega$, $R_2 = 1\ \text{k}\Omega$, $g_m = 0.04\ \text{S}$, $R_3 = 50\ \text{k}\Omega$, and $R_4 = R_5 = 10\ \text{k}\Omega$. Hence, the gain of the amplifier under these conditions is

$$\frac{v_o(t)}{v_i(t)} = \frac{-(0.04)(4.545)(10^3)(1)(10^3)}{(1.1)(10^3)}$$

$$= -165.29$$

Thus, the magnitude of the gain is 165.29.

At this point it is perhaps helpful to point out again that when analyzing circuits with dependent sources, we first treat the dependent source as though it were an independent source when we write a Kirchhoff's current or voltage law equation. Once the equation is written, we then write the controlling equation that specifies the relationship of the dependent source to the unknown variable. For instance, the first equation in Example 2.24 treats the dependent source like an independent source. The second equation in the example specifies the relationship of the dependent source to the voltage, which is the unknown in the first equation.

LEARNING EXTENSIONS

E2.17 Find V_o in the circuit in Fig. E2.17.

ANSWER $V_o = 12$ V.

Figure E2.17

E2.18 Find V_o in the network in Fig. E2.18.

ANSWER $V_o = 8$ V.

Figure E2.18

Learning by Application

Throughout this book we endeavor to present a wide variety of examples that demonstrate the usefulness of the material under discussion in a practical environment. To enhance our presentation of the practical aspects of circuit analysis and design, we have dedicated sections, such as this one, in most chapters for the specific purpose of presenting additional application-oriented examples.

LEARNING Example 2.27

An equivalent circuit for a transistor amplifier used in a portable tape player is shown in Fig. 2.39. The ideal independent source, V_S, and R_S represent the magnetic head playback circuitry. The dependent source, R_o and R_{in} model the transistor. Finally, the resistor R_L models the load, which in this case is another transistor circuit. Let us find the voltage gain of the network.

$R_S = 2\ k\Omega$

V_S $R_{in} = 8\ k\Omega$ V V/100 $R_o = 100\ k\Omega$ $R_L = 25\ k\Omega$ V_o

Figure 2.39 *Transistor amplifier circuit model.*

SOLUTION The gain of the transistor amplifier can be derived as follows. The output voltage can be expressed as

$$V_o = -\frac{V}{100}(R_o//R_L)$$

where $(R_o//R_L)$ represents R_o in parallel with R_L.

From the input voltage divider, we can express V as a function of V_S.

$$V = V_S\left(\frac{R_{in}}{R_{in} + R_S}\right)$$

Therefore,

$$V_o = -\frac{V_S}{100}\left(\frac{R_{in}}{R_{in} + R_S}\right)(R_o//R_L)$$

Given the component values in Fig. 2.39, the voltage gain is

$$A_V = \frac{V_o}{V_S} = -160$$

LEARNING Example 2.28

A Wheatstone Bridge circuit is an accurate device for measuring resistance. The circuit, shown in Fig. 2.40, is used to measure the unknown resistor R_x. The center leg of the circuit contains a galvanometer, which is a very sensitive device that can be used to measure current in the microamp range. When the unknown resistor is connected to the bridge, R_3 is adjusted until the current in the galvanometer is zero, at which point the bridge is balanced. In this balanced condition

$$\frac{R_1}{R_3} = \frac{R_2}{R_x}$$

R_1 R_2 I_1 G I_2 I_G I_x R_3 I_3 R_x

Figure 2.40
The Wheatstone bridge circuit.

so that

$$R_x = \left(\frac{R_2}{R_1}\right)R_3$$

Engineers also use this bridge circuit to measure strain in solid material. For example, a system used to determine the weight of a truck is shown in Fig. 2.41a. The platform is supported by cylinders on which strain gauges are mounted. The strain gauges, which measure strain when the cylinder deflects under load, are connected to a Wheatstone bridge as shown in Fig. 2.41b. The strain gauge has a resistance of 120 Ω under no-load conditions and changes value under load. The variable resistor in the bridge is a calibrated precision device.

Weight is determined in the following manner. The ΔR_3 required to balance the bridge represents the Δ strain, which when multiplied by the modulus of elasticity yields the Δ stress. The Δ stress multiplied by the cross-sectional area of the cylinder produces the Δ load, which is used to determine weight.

Let us determine the value of R_3 under no load when the bridge is balanced and its value when the resistance of the strain gauge changes to 120.24 Ω under load.

(continued)

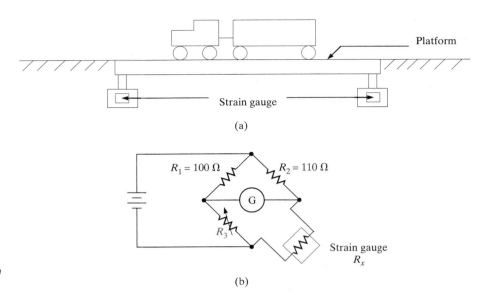

Figure 2.41
*Diagrams used in
Example 2.28.*

SOLUTION   Using the balance equation for the bridge, the value
of R_3 at no load is

$$R_3 = \left(\frac{R_1}{R_2}\right)R_x$$

$$= \left(\frac{100}{110}\right)(120)$$

$$= 109.0909 \ \Omega$$

Under load, the value of R_3 is

$$R_3 = \left(\frac{100}{110}\right)(120.24)$$

$$= 109.3091 \ \Omega$$

Therefore, the ΔR_3 is

$$\Delta R_3 = 109.3091 - 109.0909$$

$$= 0.2182 \ \Omega$$

Learning by Design

Most of this text is concerned with circuit analysis; that is, given a circuit in which all the components are specified, analysis involves finding such things as the voltage across some element or the current through another. Furthermore, the solution of an analysis problem is generally unique. In contrast, design involves determining the circuit configuration that will meet certain specifications. In addition, the solution is generally not unique in that there may be many ways to satisfy the circuit/performance specifications. It is also possible that there is no solution that will meet the design criteria.

In addition to meeting certain technical specifications, designs normally must also meet other criteria, such as economic, environmental, and safety constraints. For example, if a circuit design that meets the technical specifications is either too expensive or unsafe, it is not viable regardless of its technical merit.

At this point, the number of elements that we can employ in circuit design is limited primarily to the linear resistor and the active elements we have presented. However, as we progress through the text we will introduce a number of other elements (for example, the op-amp, capacitor, and inductor), which will significantly enhance our design capability.

We begin our discussion of circuit design by considering a couple of simple examples that demonstrate the selection of specific components to meet certain circuit specifications.

LEARNING Example 2.29

A circuit board within a stereo amplifier requires the use of three different voltages at the circuit nodes. The voltages needed are 4 V, 16 V, and 24 V. Criteria such as weight, cost, and size dictate that these voltages be supplied with a single voltage source. Furthermore, heat constraints dictate that the supply must dissipate less than 5 W. Assuming no loading effect from the balance of the network, design a voltage string (multiple-resistor voltage divider) that will satisfy the requirements.

SOLUTION   A 24-V source can directly provide one of the voltages, and the voltage divider in Fig. 2.42 can be used to generate the other two. The power requirement limits our choices to current less than 208.33 mA. The exact value we choose is somewhat arbitrary and, therefore, we select $I = 200$ mA. The total resistance is then

$$R_1 + R_2 + R_3 = \frac{24}{0.2} = 120 \ \Omega$$

Using voltage division, we find that

$$\frac{R_3}{R_1 + R_2 + R_3} = \frac{R_3}{120} = \frac{4}{24}$$

or

$$R_3 = 20 \ \Omega$$

Similarly,

$$\frac{R_2}{120} = \frac{16 - 4}{24}$$

or

$$R_2 = 60 \ \Omega$$

Finally, since the total resistance is 120 Ω, $R_1 = 40 \ \Omega$.

Figure 2.42
Voltage string for generating intermediate voltages.

LEARNING Example 2.30

The network in Fig. 2.43 is an equivalent circuit for a transistor amplifier used in a stereo preamplifier. The input circuitry, consisting of a 2-mV source in series with a 500-Ω resistor, models the output of a compact disk player. The dependent source, R_{in}, and R_o model the transistor, which amplifies the signal and then sends it to the power amplifier. The 10-kΩ load resistor models the input to the power amplifier that actually drives the speakers. We must design a transistor amplifier as shown in Fig. 2.43 that will provide an overall gain of -200. In practice we do not actually vary the device parameters to achieve the desired gain; rather we select a transistor from the manufacturer's data books that will satisfy the

required specification. The model parameters for three different transistors are listed as follows:

Figure 2.43   *Transistor amplifier circuit model.*

Manufacturer's transistor parameter values

Part Number	R_{in} (kΩ)	R_o (kΩ)	g_m (mA/V)
1	1.0	50	50
2	2.0	75	30
3	8.0	80	20

Design the amplifier by choosing the transistor that produces the most accurate gain. What is the percent error of your choice?

SOLUTION The output voltage can be written

$$V_o = -g_m V (R_o // R_L)$$

Using voltage division at the input to find V,

$$V = V_S \left(\frac{R_{in}}{R_{in} + R_S} \right)$$

Combining these two expressions, we can solve for the gain:

$$A_V = \frac{V_o}{V_S} = -g_m \left(\frac{R_{in}}{R_{in} + R_S} \right) (R_o // R_L)$$

Using the parameter values for the three transistors, we find that Obviously, the best alternative is transistor number 2, which has a gain error of

$$\text{Percent error} = \left(\frac{211.8 - 200}{200} \right) \times 100\% = 5.9\%$$

LEARNING Check

Summary

▶ **Ohm's law** $V = IR$

▶ **The passive sign convention with Ohm's law** The current enters the resistor terminal with the positive voltage reference.

▶ **Kirchhoff's current law (KCL)** The algebraic sum of the currents leaving (entering) a node is zero.

▶ **Kirchhoff's voltage law (KVL)** The algebraic sum of the voltages around any closed path is zero.

▶ **Solving a single-loop circuit** Determine the loop current by applying KVL and Ohm's law.

▶ **Solving a single-node-pair circuit** Determine the voltage between the pair of nodes by applying KCL and Ohm's law.

▶ **The voltage-division rule** The voltage is divided between two series resistors in direct proportion to their resistance.

▶ **The current-division rule** The current is divided between two parallel resistors in reverse proportion to their resistance.

▶ **The equivalent resistance of a network of resistors** Combine resistors in series by adding their resistances. Combine resistors in parallel by adding their conductances. The wye-to-delta and delta-to-wye transformations are also an aid in reducing the complexity of a network.

▶ **Short circuit** Zero resistance, zero voltage; the current in the short is determined by the rest of the circuit.

▶ **Open circuit** Zero conductance, zero current; the voltage across the open terminals is determined by the rest of the circuit.

Problems

SECTION 2.1 ────────────────────────────

2.1 Find the current I and the power supplied by the source in the network in Fig. P2.1.

Figure P2.1

2.2 In the circuit in Fig. P2.2, find the voltage across the current source and the power absorbed by the resistor.

Figure P2.2

2.3 If the 10-kΩ resistor in the network in Fig. P2.3 absorbs 2.5 mW, find V_S.

Figure P2.3

2.4 In the network in Fig. P2.4, the power absorbed by R_x is 5 mW. Find R_x.

Figure P2.4

2.5 In the network in Fig. P2.5, the power absorbed by R_x is 20 mW. Find R_x.

Figure P2.5

2.6 A model for a standard two D-cell flashlight is shown in Fig. P2.6. Find the power dissipated in the lamp.

Figure P2.6

2.7 An automobile uses two halogen headlights connected as shown in Fig. P2.7. Determine the power supplied by the battery if each headlight draws 3 A of current.

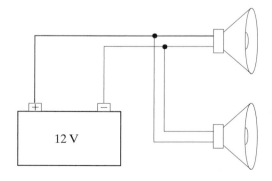

Figure P2.7

2.8 Many years ago a string of Christmas tree lights was manufactured in the form shown in Fig. P2.8a. Today the lights are manufactured as shown in Fig. P2.8b. Is there a good reason for this change?

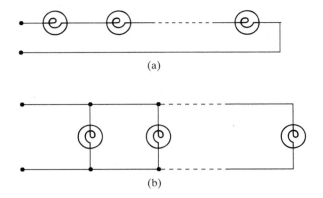

(a)

(b)

Figure P2.8

SECTION 2.2

2.9 Find I_1 in the network in Fig. P2.9.

Figure P2.9

2.10 Find I_1 and I_2 in the circuit in Fig. P2.10.

Figure P2.10

2.11 Find I_x and I_y in the network in Fig. P2.11.

Figure P2.11

2.12 Find I_x, I_y, and I_z in the circuit in Fig. P2.12.

Figure P2.12

2.13 Find I_x in the circuit in Fig. P2.13.

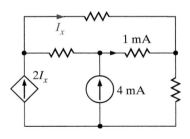

Figure P2.13

2.14 Find I_x, I_y, and I_z in the network in Fig. P2.14.

Figure P2.14

2.15 Find I_x in the circuit in Fig. P2.15.

Figure P2.15

2.16 Find I_x in the network in Fig. P2.16.

Figure P2.16

2.17 Find V_x in the circuit in Fig. P2.17.

Figure P2.17

2.18 Find V_{bd} in the circuit in Fig. P2.18.

Figure P2.18

2.19 Find V_{ad} in the network in Fig. P2.19.

Figure P2.19

2.20 Find V_{af} and V_{ec} in the circuit in Fig. P2.20.

Figure P2.20

2.21 Find V_{ac} in the circuit in Fig. P2.21.

Figure P2.21

2.22 Find V_{ad} and V_{ce} in the circuit in Fig. P2.22.

Figure P2.22

SECTION 2.3

2.23 Find V_{ab} in the network in Fig. P2.23.

Figure P2.23

2.24 Find V_{bd} in the network in Fig. P2.24.

Figure P2.24

2.25 Find V_x in the circuit in Fig. P2.25.

Figure P2.25

2.26 Find V_x in the circuit in Fig. P2.26.

Figure P2.26

2.27 Find V_x in the network in Fig. P2.27.

Figure P2.27

2.28 Find V_1 in the network in Fig. P2.28.

Figure P2.28

Hollond
America
sons

2.29 Find the power absorbed by the 30-kΩ resistor in the circuit in Fig. P2.29.

Figure P2.29

SECTION 2.4

2.30 Find I_o in the network in Fig. P2.30.

Figure P2.30

2.31 Find I_o in the circuit in Fig. P2.31.

Figure P2.31

2.32 Find I_o in the network in Fig. P2.32.

Figure P2.32

2.33 Find V_o in the circuit in Fig. P2.33.

Figure P2.33

2.34 Find I_o in the network in Fig. P2.34.

Figure P2.34

2.35 Determine I_L in the circuit in Fig. P2.35.

Figure P2.35

2.36 Determine I_L in the circuit in Fig. P2.36.

Figure P2.36

SECTION 2.5

2.37 Find R_{AB} in the circuit in Fig. P2.37.

Figure P2.37

2.38 Find R_{AB} in the circuit in Fig. P2.38.

·**Figure P2.38**

2.39 Find R_{AB} in the network in Fig. P2.39.

Figure P2.39

2.40 Find R_{AB} in the circuit in Fig. P2.40.

Figure P2.40

2.41 Find R_{AB} in the network in Fig. P2.41.

Figure P2.41

2.42 Find R_{AB} in the circuit in Fig. P2.42.

Figure P2.42

2.43 Find R_{AB} in the circuit in Fig. P2.43.

Figure P2.43

2.44 Find the range of resistance for the following resistors.
(a) 1 kΩ with a tolerance of 5%
(b) 470 Ω with a tolerance of 2%
(c) 22 kΩ with a tolerance of 10%

2.45 Given the network in Fig. P2.45, find the possible range of values for the current and power dissipated by the following resistors.
(a) 390 Ω with a tolerance of 1%
(b) 560 Ω with a tolerance of 2%

Figure P2.45

2.46 Given the circuit in Fig. P2.46.
(a) find the required value of R.
(b) use Table 2.1 to select a standard 10% tolerance resistor for R.
(c) calculate the actual value of I.
(d) determine the percent error between the actual value of I and that shown in the circuit.
(e) determine the power rating for the resistor R.

Figure P2.46

2.47 The resistors R_1 and R_2 shown in the circuit in Fig. P2.47 are 1 Ω with a tolerance of 5% and 2 Ω with a tolerance of 10%, respectively.
(a) What is the nominal value of the equivalent resistance?
(b) Determine the positive and negative tolerance for the equivalent resistance.

Figure P2.47

2.48 Find I_1 and V_o in the circuit in Fig. P2.48.

Figure P2.48

2.49 Find I_1 and V_o in the circuit in Fig. P2.49.

Figure P2.49

2.50 Find I_1 in the circuit in the Fig. P2.50.

Figure P2.50

2.51 Determine V_o in the network in Fig. P2.51.

Figure P2.51

2.52 Find V_o in the network in Fig. P2.52.

Figure P2.52

2.53 Find I_o in the circuit in Fig. P2.53.

Figure P2.53

2.54 Find V_o in the circuit in Fig. P2.54.

Figure P2.54

2.55 Find I_o in the network in Fig. P2.55.

Figure P2.55

2.56 Find V_o in the network in Fig. P2.56.

Figure P2.56

2.57 Find I_o in the circuit in Fig. P2.57.

Figure P2.57

2.58 If $V_o = 4$ V in the network in Fig. P2.58, find V_S.

Figure P2.58

2.59 If the power absorbed by the 4-kΩ resistor in the network in Fig. P2.59 is 36 mW, find I_o.

Figure P2.59

2.63 In the circuit in Fig. P2.63, $I = 4$ mA. Find V_S.

Figure P2.63

2.60 If the power absorbed by the 4-kΩ resistor in the circuit in Fig. P2.60 is 36 mW, find V_S.

Figure P2.60

2.64 In the circuit in Fig. P2.64, $I_o = 2$ mA. Find I_S.

Figure P2.64

2.61 In the network in Fig. P2.61, the power absorbed by the 4-Ω resistor is 100 W. Find V_S.

Figure P2.61

2.65 In the network in Fig. P2.65, $V_1 = 12$ V. Find V_S.

Figure P2.65

2.62 In the network in Fig. P2.62, $V_o = 6$ V. Find I_S.

Figure P2.62

2.66 In the circuit in Fig. P2.66, $V_o = 2$ V. Find I_S.

Figure P2.66

2.67 In the network in Fig. P2.67, $V_o = 6$ V. Find I_S.

Figure P2.67

2.68 In the circuit in Fig. P2.68, $I_o = 2$ A. Find I_S.

Figure P2.68

2.69 If $I_o = 4$ mA in the circuit in Fig. P2.69, find I_S.

Figure P2.69

2.70 Find I_o in the circuit in Fig. P2.70.

Figure P2.70

2.71 Find V_o in the circuit in Fig. P2.71.

Figure P2.71

2.72 Find the power absorbed by the network in Fig. P2.72.

Figure P2.72

2.73 Find I_o in the circuit in Fig. P2.73.

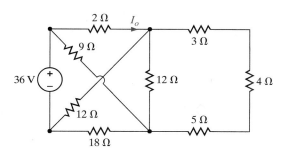

Figure P2.73

2.74 Find I_o in the circuit in Fig. P2.74.

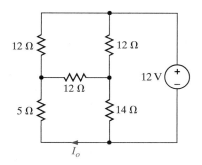

Figure P2.74

2.75 Find I_o in the circuit in Fig. P2.75.

Figure P2.75

2.76 Find I_o in the circuit in Fig. P2.76.

Figure P2.76

2.77 Find I_o in the circuit in Fig. P2.77.

Figure P2.77

2.78 Find V_o in the circuit in Fig. P2.78.

Figure P2.78

2.79 Find V_o in the network in Fig. P2.79.

Figure P2.79

2.80 Find V_o in the network in Fig. P2.80.

Figure P2.80

2.81 Find I_o in the network in Fig. P2.81.

Figure P2.81

2.82 Find V_o in the circuit in Fig. P2.82.

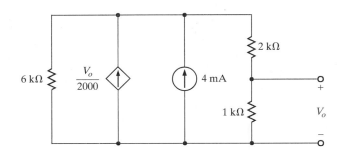

Figure P2.82

2.83 A single-stage transistor amplifier is modeled as shown in Fig. P2.83. Find the current in the load R_L.

Figure P2.83

2.84 A typical transistor amplifier is shown in Fig. P2.84. Find the amplifier gain G (i.e., the ratio of the output voltage to the input voltage).

Figure P2.84

2.85 For the network in Fig. P2.85, choose the values of R_{in} and R_o such that V_o is maximized. What is the resulting ratio, V_o/V_S?

Figure P2.85

2.86 In many amplifier applications we are concerned not only with voltage gain, but also with power gain.

Power gain = A_p = (power delivered to the load)/(power delivered by the input)

Find the power gain for the circuit in Fig. P2.86, where $R_L = 50$ kΩ.

Figure P2.86

2.87 Find the power absorbed by the 10-kΩ resistor in the circuit in Fig. P2.87.

Figure P2.87

2.88 Find the power absorbed by the 12-kΩ resistor in the circuit in Fig. P2.88.

Figure P2.88

Typical Problems Found on the FE Exam

2FE-1 Find the power generated by the source in the network in Fig. 2PFE-1.

Fig. 2PFE-1

2FE-2 Find the equivalent resistance of the circuit in Fig. 2PFE-2 at the terminals A-B.

Fig. 2PFE-2

2FE-3 Find the voltage V_o in the network in Fig. 2PFE-3.

Fig. 2PFE-3

2FE-4 Find the current I_o in the circuit in Fig. 2PFE-4.

Fig. 2PFE-4

Nodal and Loop Analysis Techniques

3

In Chapter 2 we analyzed the simplest possible circuits, those containing only a single-node pair or a single loop. We found that these circuits can be completely analyzed via a single algebraic equation. In the case of the single-node-pair circuit (i.e., one containing two nodes, one of which is a reference node), once the node voltage is known, we can calculate all the currents. In a single-loop circuit, once the loop current is known, we can calculate all the voltages.

In this chapter we extend our capabilities in a systematic manner so that we can calculate all currents and voltages in circuits that contain multiple nodes and loops. Our analyses are based primarily on two laws with which we are already familiar: Kirchhoff's current law (KCL) and Kirchhoff's voltage law (KVL). In a nodal analysis we employ KCL to determine the node voltages, and in a loop analysis we use KVL to determine the loop currents.

We present and discuss a very important commercially available circuit known as the operational amplifier, or op-amp. Op-amps are used in literally thousands of applications, including such things as compact disk (CD) players, random access memories (RAMs), analog-to-digital (A/D) and digital-to-analog (D/A) converters, headphone amplifiers, and electronic instrumentation of all types. Finally, we discuss the terminal characteristics of this circuit and demonstrate its use in practical applications as well as circuit design.

3.1 Nodal Analysis

In a nodal analysis the variables in the circuit are selected to be the node voltages. The node voltages are defined with respect to a common point in the circuit. One node is selected as the reference node, and all other node voltages are defined with respect to that node. Quite often this node is the one to which the largest number of branches are connected. It is commonly called *ground* because it is said to be at ground-zero potential, and it sometimes represents the chassis or ground line in a practical circuit.

We will select our variables as being positive with respect to the reference node. If one or more of the node voltages are actually negative with respect to the reference node, the analysis will indicate it.

In order to understand the value of knowing all the node voltages in a network, we consider once again the network in Fig. 2.30, which is redrawn in Fig. 3.1. The voltages, V_S, V_a, V_b, and V_c, are all measured with respect to the bottom node, which is selected as the reference and labeled with the ground symbol $\perp$. Therefore, the voltage at node 1 is $V_S = 12$ V with respect to the reference node 5; the voltage at node 2 is $V_a = 3$ V with respect to the reference node 5, and so on. Now note carefully that once these node voltages are known, we can immediately calculate any branch current or the power supplied or absorbed by any element, since we know the voltage across every element in the network. For example, the voltage V_1 across the leftmost 9-kΩ resistor is the difference in potential between the two ends of the resistor; that is,

$$V_1 = V_S - V_a$$
$$= 12 - 3$$
$$= 9 \text{ V}$$

Figure 3.1
Circuit with known node voltages.

This equation is really nothing more than an application of KVL around the left-most loop; that is,

$$-V_S + V_1 + V_a = 0$$

In a similar manner, we find that

$$V_3 = V_a - V_b$$

and

$$V_5 = V_b - V_c$$

Then the currents in the resistors are

$$I_1 = \frac{V_1}{9\text{k}} = \frac{V_S - V_a}{9\text{k}}$$

$$I_3 = \frac{V_3}{3\text{k}} = \frac{V_a - V_b}{3\text{k}}$$

$$I_5 = \frac{V_5}{9\text{k}} = \frac{V_b - V_c}{9\text{k}}$$

In addition,

$$I_2 = \frac{V_a - 0}{6k}$$

$$I_4 = \frac{V_b - 0}{4k}$$

since the reference node 5 is at zero potential.

Thus, as a general rule, if we know the node voltages in a circuit, we can calculate the current through any resistive element using Ohm's law; that is,

$$i = \frac{v_m - v_N}{R} \qquad\qquad \textbf{3.1}$$

as illustrated in Fig. 3.2.

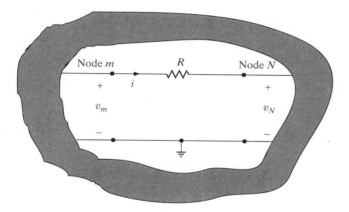

Figure 3.2
Circuit used to illustrate Ohm's law in a multiple-node network.

In Example 2.3 we illustrated that the number of linearly independent KCL equations for an N-node network was $N - 1$. Furthermore, we found that in a two-node circuit, in which one node was the reference node, only one equation was required to solve for the unknown node voltage. What is illustrated in this simple case is true in general; that is, in an N-node circuit one linearly independent KCL equation is written for each of the $N - 1$ nonreference nodes, and this set of $N - 1$ linearly independent simultaneous equations, when solved, will yield the $N - 1$ unknown node voltages.

It is instructive to treat nodal analysis by examining several different types of circuits and illustrating the salient features of each. We begin with the simplest case. However, as a prelude to our discussion of the details of nodal analysis, experience indicates that it is worthwhile to digress for a moment to ensure that the concept of node voltage is clearly understood.

At the outset it is important to specify a reference. For example, to state that the voltage at node A is 12 V means nothing unless we provide the reference point; that is, the voltage at node A is 12 V with respect to what. The circuit in Fig. 3.3 illustrates a portion of a network containing three nodes, one of which is the reference node.

Figure 3.3
An illustration of node voltages.

The voltage $V_1 = 4$ V is the voltage at node 1 with respect to the reference node 3. Similarly, the voltage $V_2 = -2$ V is the voltage at node 2 with respect to node 3. In addition, however, the voltage at node 1 with respect to node 2 is +6 V and the voltage at node 2 with respect to node 1 is −6 V. Furthermore, since the current will flow from the node of higher potential to the node of lower potential, the current in R_1 is from top to bottom, the current in R_2 is from left to right, and the current in R_3 is from bottom to top.

These concepts have important ramifications in our daily lives. If a man were hanging in midair with one hand on one line and one hand on another and the dc line voltage of each line was exactly the same, the voltage across his heart would be zero and he would be safe. If, however, he let go of one line and let his feet touch the ground, the dc line voltage would then exist from his hand to his foot with his heart in the middle. He would probably be dead the instant his foot hit the ground.

In the town where I live, a young man tried to retrieve his parakeet that had escaped its cage and was outside sitting on a power line. He stood on a metal ladder and with a metal pole reached for the parakeet; when the metal pole touched the power line, the man was killed instantly. Electric power is vital to our standard of living, but it is also very dangerous. The material in this book *does not* qualify you to handle it safely. Therefore, always be extremely careful around electric circuits.

Now as we begin our discussion of nodal analysis, our approach will be to begin with simple cases and proceed in a systematic manner to those that are more challenging. Numerous examples will be the vehicle used to demonstrate each facet of this approach. Finally, at the end of this section, we will outline a strategy for attacking any circuit using nodal analysis.

CIRCUITS CONTAINING ONLY INDEPENDENT CURRENT SOURCES

Consider the network shown in Fig. 3.4. There are three nodes, and the bottom node is selected as the reference node. The branch currents are assumed to flow in the directions indicated in the figures. If one or more of the branch currents are actually flowing in a direction opposite to that assumed, the analysis will simply produce a branch current that is negative.

Applying KCL at node 1 yields

$$-i_A + i_1 + i_2 = 0$$

LEARNING Hint

Employing the passive sign convention.

Using Ohm's law $(i = Gv)$ and noting that the reference node is at zero potential, we obtain

$$-i_A + G_1(v_1 - 0) + G_2(v_1 - v_2) = 0$$

or

$$(G_1 + G_2)v_1 - G_2v_2 = i_A$$

KCL at node 2 yields

$$-i_2 + i_B + i_3 = 0$$

Figure 3.4
A three-node circuit.

or

$$-G_2(v_1 - v_2) + i_B + G_3(v_2 - 0) = 0$$

which can be expressed as

$$-G_2 v_1 + (G_2 + G_3)v_2 = -i_B$$

Therefore, the two equations for the two unknown node voltages v_1 and v_2 are

$$(G_1 + G_2)v_1 - G_2 v_2 = i_A$$
$$-G_2 v_1 + (G_2 + G_3)v_2 = -i_B \qquad \textbf{3.2}$$

Note that the analysis has produced two simultaneous equations in the unknowns v_1 and v_2. They can be solved using any convenient technique, and modern calculators and personal computers are very efficient tools for their application.

LEARNING by Doing

D 3.1 For the following network, write the KCL equations for nodes 1 and 2.

ANSWER

$$i_1 + \frac{v_1}{R_2} + \frac{v_1 - v_2}{R_3} + \frac{v_1 - v_2}{R_4} = 0$$

$$i_2 + \frac{v_1 - v_2}{R_3} + \frac{v_1 - v_2}{R_4} = 0$$

$$i_1 + \frac{v_1}{R_2} + \frac{v_1 - v_2}{R_3} + \frac{v_1 - v_2}{R_4}$$

$$i_2 + \frac{v_1 - v_2}{R_3} + \frac{v_1 - v_2}{R_4}$$

In what follows, we will demonstrate three techniques for solving linearly independent simultaneous equations: Gaussian elimination, matrix analysis, and the MATLAB mathematical software package. A brief refresher that illustrates the use of both Gaussian elimination and matrix analysis in the solution of these equations is provided in the Problem-Solving Companion for this text. The use of the MATLAB software is straightforward, and we will demonstrate its use as we encounter the application.

The KCL equations at nodes 1 and 2 produced two linearly independent simultaneous equations:

$$-i_A + i_1 + i_2 = 0$$
$$-i_2 + i_B + i_3 = 0$$

The KCL equation for the third node (reference) is

$$+i_A - i_1 - i_B - i_3 = 0$$

Note that if we add the first two equations, we obtain the third. Furthermore, any two of the equations can be used to derive the remaining equation. Therefore, in this $N = 3$ node circuit, only $N - 1 = 2$ of the equations are linearly independent and required to determine the $N - 1 = 2$ unknown node voltages.

Note that a nodal analysis employs KCL in conjunction with Ohm's law. Once the direction of the branch currents has been *assumed*, then Ohm's law, as illustrated by Fig. 3.2 and expressed by Eq. (3.1), is used to express the branch currents in terms of the unknown node voltages. We can assume the currents to be in any direction. However, once we assume a particular direction, we must be very careful to write the currents correctly in terms of the node voltages using Ohm's law.

LEARNING Example 3.1

Suppose that the network in Fig. 3.4 has the following parameters: $I_A = 1$ mA, $R_1 = 12$ kΩ, $R_2 = 6$ kΩ, $I_B = 4$ mA, and $R_3 = 6$ kΩ. Let us determine all node voltages and branch currents.

SOLUTION For purposes of illustration we will solve this problem using Gaussian elimination, matrix analysis, and MATLAB. Using the parameter values Eq. (3.2) becomes

$$V_1\left[\frac{1}{12k} + \frac{1}{6k}\right] - V_2\left[\frac{1}{6k}\right] = 1 \times 10^{-3}$$

$$-V_1\left[\frac{1}{6k}\right] + V_2\left[\frac{1}{6k} + \frac{1}{6k}\right] = -4 \times 10^{-3}$$

where we employ capital letters because the voltages are constant. The equations can be written as

$$\frac{V_1}{4k} - \frac{V_2}{6k} = 1 \times 10^{-3}$$

$$-\frac{V_1}{6k} + \frac{V_2}{3k} = -4 \times 10^{-3}$$

Using Gaussian elimination, we solve the first equation for V_1 in terms of V_2:

$$V_1 = V_2\left(\frac{2}{3}\right) + 4$$

This value is then substituted into the second equation to yield

$$\frac{-1}{6k}\left(\frac{2}{3}V_2 + 4\right) + \frac{V_2}{3k} = -4 \times 10^{-3}$$

or

$$V_2 = -15 \text{ V}$$

This value for V_2 is now substituted back into the equation for V_1 in terms of V_2, which yields

$$V_1 = \frac{2}{3}V_2 + 4$$

$$= -6 \text{ V}$$

The circuit equations can also be solved using matrix analysis. The general form of the matrix equation is

$$\mathbf{GV} = \mathbf{I}$$

where in this case

$$\mathbf{G} = \begin{bmatrix} \dfrac{1}{4k} & -\dfrac{1}{6k} \\ -\dfrac{1}{6k} & \dfrac{1}{3k} \end{bmatrix}, \mathbf{V} = \begin{bmatrix} V_1 \\ V_2 \end{bmatrix}, \text{ and } \mathbf{I} = \begin{bmatrix} 1 \times 10^{-3} \\ -4 \times 10^{-3} \end{bmatrix}$$

The solution to the matrix equation is

$$\mathbf{V} = \mathbf{G}^{-1}\mathbf{I}$$

and therefore,

$$\begin{bmatrix} V_1 \\ V_2 \end{bmatrix} = \begin{bmatrix} \dfrac{1}{4k} & \dfrac{-1}{6k} \\ \dfrac{-1}{6k} & \dfrac{1}{3k} \end{bmatrix}^{-1} \begin{bmatrix} 1 \times 10^{-3} \\ -4 \times 10^{-3} \end{bmatrix}$$

To calculate the inverse of $\mathbf{G}$, we need the adjoint and the determinant. The adjoint is

$$\text{Adj } \mathbf{G} = \begin{bmatrix} \dfrac{1}{3k} & \dfrac{1}{6k} \\ \dfrac{1}{6k} & \dfrac{1}{4k} \end{bmatrix}$$

and the determinant is

$$|\mathbf{G}| = \left(\frac{1}{3k}\right)\left(\frac{1}{4k}\right) - \left(\frac{-1}{6k}\right)\left(\frac{-1}{6k}\right)$$

$$= \frac{1}{18k^2}$$

Therefore,

$$\begin{bmatrix} V_1 \\ V_2 \end{bmatrix} = 18k^2 \begin{bmatrix} \dfrac{1}{3k} & \dfrac{1}{6k} \\ \dfrac{1}{6k} & \dfrac{1}{4k} \end{bmatrix} \begin{bmatrix} 1 \times 10^{-3} \\ -4 \times 10^{-3} \end{bmatrix}$$

$$= 18k^2 \begin{bmatrix} \dfrac{1}{3k^2} - \dfrac{4}{6k^2} \\ \dfrac{1}{6k^2} - \dfrac{1}{k^2} \end{bmatrix}$$

$$= \begin{bmatrix} -6 \\ -15 \end{bmatrix}$$

The MATLAB solution begins with the set of equations expressed in matrix form as

```
G*V=I
```

where the symbol * denotes the multiplication of the voltage vector $\mathbf{V}$ by the coefficient matrix $\mathbf{G}$. Then once the MATLAB software is loaded into the PC, the coefficient matrix ($\mathbf{G}$) and the vector $\mathbf{V}$ can be expressed in MATLAB notation by typing in the rows of the matrix or vector at the prompt >>. Use semicolons to separate rows and spaces to separate columns. Brackets are used to denote vectors or matrices. When the matrix $\mathbf{G}$ and the vector $\mathbf{I}$ have been defined, then the solution equation

```
V=inv(G)*I
```

which is also typed in at the prompt >>, will yield the unknown vector **V**.

The matrix equation for our circuit expressed in decimal notation is

$$\begin{bmatrix} 0.00025 & -0.00016666 \\ -0.00016666 & 0.0003333 \end{bmatrix} \begin{bmatrix} V_1 \\ V_2 \end{bmatrix} = \begin{bmatrix} 0.001 \\ -0.004 \end{bmatrix}$$

If we now input the coefficient matrix **G**, then the vector **I** and finally the equation **V** = inv(**G**)*I, the computer screen containing these data and the solution vector **V** appears as follows:

```
>> G = [0.00025 -0.000166666;

-0.000166666 0.00033333]

   G =

        1.0e-003 *

        0.2500        -0.1667
       -0.1667         0.3333

>> I = [0.001 ; -0.004]

   I =

        0.0010
       -0.0040

>> V = inv(G)*I
```

V =

    -6.0001
    -15.0002

Knowing the node voltages, we can determine all the currents using Ohm's law:

$$I_1 = \frac{V_1}{R_1} = \frac{-6}{12k} = -\frac{1}{2} \, mA$$

$$I_2 = \frac{V_1 - V_2}{6k} = \frac{-6 - (-15)}{6k} = \frac{3}{2} \, mA$$

and

$$I_3 = \frac{V_2}{6k} = \frac{-15}{6k} = -\frac{5}{2} \, mA$$

Figure 3.5 illustrates the results of all the calculations. Note that KCL is satisfied at every node.

Figure 3.5   *Circuit used in Example 3.1.*

Let us now examine the circuit in Fig. 3.6. The current directions are assumed as shown in the figure.

At node 1, KCL yields

$$i_1 - i_A + i_2 - i_3 = 0$$

or

$$\frac{v_1}{R_1} - i_A + \frac{v_1 - v_2}{R_2} - \frac{v_3 - v_1}{R_3} = 0$$

$$v_1 \left(\frac{1}{R_1} + \frac{1}{R_2} + \frac{1}{R_3} \right) - v_2 \frac{1}{R_2} - v_3 \frac{1}{R_3} = i_A$$

At node 2, KCL yields

$$-i_2 + i_4 - i_5 = 0$$

Figure 3.6
A four-node circuit.

or

$$-\frac{v_1 - v_2}{R_2} + \frac{v_2}{R_4} - \frac{v_3 - v_2}{R_5} = 0$$

$$-v_1\frac{1}{R_2} + v_2\left(\frac{1}{R_2} + \frac{1}{R_4} + \frac{1}{R_5}\right) - v_3\frac{1}{R_5} = 0$$

At node 3, the equation is

$$i_3 + i_5 + i_B = 0$$

or

$$\frac{v_3 - v_1}{R_3} + \frac{v_3 - v_2}{R_5} + i_B = 0$$

$$-v_1\frac{1}{R_3} - v_2\frac{1}{R_5} + v_3\left(\frac{1}{R_3} + \frac{1}{R_5}\right) = -i_B$$

Grouping the node equations together, we obtain

$$v_1\left(\frac{1}{R_1} + \frac{1}{R_2} + \frac{1}{R_3}\right) - v_2\frac{1}{R_2} - v_3\frac{1}{R_3} = i_A$$

$$-v_1\frac{1}{R_2} + v_2\left(\frac{1}{R_2} + \frac{1}{R_4} + \frac{1}{R_5}\right) - v_3\frac{1}{R_5} = 0 \qquad \textbf{3.3}$$

$$-v_1\frac{1}{R_3} - v_2\frac{1}{R_5} + v_3\left(\frac{1}{R_3} + \frac{1}{R_5}\right) = -i_B$$

Note that our analysis has produced three simultaneous equations in the three unknown node voltages v_1, v_2, and v_3. The equations can also be written in matrix form as

$$\begin{bmatrix} \dfrac{1}{R_1} + \dfrac{1}{R_2} + \dfrac{1}{R_3} & -\dfrac{1}{R_2} & -\dfrac{1}{R_3} \\ -\dfrac{1}{R_2} & \dfrac{1}{R_2} + \dfrac{1}{R_4} + \dfrac{1}{R_5} & -\dfrac{1}{R_5} \\ -\dfrac{1}{R_3} & -\dfrac{1}{R_5} & \dfrac{1}{R_3} + \dfrac{1}{R_5} \end{bmatrix} \begin{bmatrix} v_1 \\ v_2 \\ v_3 \end{bmatrix} = \begin{bmatrix} i_A \\ 0 \\ -i_B \end{bmatrix} \qquad \textbf{3.4}$$

At this point it is important that we note the symmetrical form of the equations that describe the two previous networks. Equations (3.2) and (3.3) exhibit the same type of symmetrical form. The **G** matrix for each network is a symmetrical matrix. This symmetry is not accidental. The node equations for networks containing only resistors and independent current sources can always be written in this symmetrical form. We can take advantage of this fact and learn to write the equations by inspection. Note in the first equation of (3.2) that the coefficient of v_1 is the sum of all the conductances connected to node 1 and the coefficient of v_2 is the negative of the conductances connected between node 1 and node 2. The right-hand side of the equation is the sum of the currents entering node 1 through current sources. This equation is KCL at node 1. In the second equation in (3.2), the coefficient of v_2 is the sum of all the conductances connected to node 2, the coefficient of v_1 is the negative of the conductance connected between node 2 and node 1, and the right-hand side of the equation is the sum of the currents entering node 2 through current sources. This equation is KCL at node 2. Similarly, in the first equation in (3.3) the coefficient of v_1 is the sum of the conductances connected to node 1, the coefficient of v_2 is the negative of the conductance connected between node 1 and node 2, the coefficient of v_3 is the negative of the conductance connected between node 1 and node 3, and the right-hand side of the equation is the sum of the currents entering node 1 through current sources. The other two equations in (3.3) are obtained in a similar manner. In general, if KCL is applied to node j with node voltage v_j, the coefficient of v_j is the sum of all the conductances connected to node j and the coefficients of the other node voltages (e.g., v_{j-1}, v_{j+1}) are the negative of the sum of the conductances connected directly between these nodes and node j. The right-hand side of the equation is equal to the sum of the currents entering the node via current sources. Therefore, the left-hand side of the equation represents the sum of the currents leaving node j and the right-hand side of the equation represents the currents entering node j.

LEARNING Example 3.2

Let us apply what we have just learned to write the equations for the network in Fig. 3.7 by inspection. Then given the following parameters, we will determine the node voltages using MATLAB: $R_1 = R_2 = 2\ \text{k}\Omega$, $R_3 = R_4 = 4\ \text{k}\Omega$, $R_5 = 1\ \text{k}\Omega$, $i_A = 4\ \text{mA}$, and $i_B = 2\ \text{mA}$.

Figure 3.7   Circuit used in Example 3.2.

SOLUTION   The equations are

$$v_1\left(\frac{1}{R_1} + \frac{1}{R_2}\right) - v_2(0) - v_3\left(\frac{1}{R_1}\right) = -i_A$$

$$-v_1(0) + v_2\left(\frac{1}{R_3} + \frac{1}{R_4}\right) - v_3\left(\frac{1}{R_4}\right) = i_A - i_B$$

$$-v_1\left(\frac{1}{R_1}\right) - v_2\left(\frac{1}{R_4}\right) + v_3\left(\frac{1}{R_1} + \frac{1}{R_4} + \frac{1}{R_5}\right) = 0$$

which can also be written directly in matrix form as

$$\begin{bmatrix} \frac{1}{R_1} + \frac{1}{R_2} & 0 & -\frac{1}{R_1} \\ 0 & \frac{1}{R_3} + \frac{1}{R_4} & -\frac{1}{R_4} \\ -\frac{1}{R_1} & -\frac{1}{R_4} & \frac{1}{R_1} + \frac{1}{R_4} + \frac{1}{R_5} \end{bmatrix} \begin{bmatrix} v_1 \\ v_2 \\ v_3 \end{bmatrix}$$

$$= \begin{bmatrix} -i_A \\ i_A - i_B \\ 0 \end{bmatrix}$$

Both the equations and the **G** matrix exhibit the symmetry that will always be present in circuits that contain only resistors and current sources.

If the component values are now used, the matrix equation becomes

$$\begin{bmatrix} \frac{1}{2k} + \frac{1}{2k} & 0 & -\frac{1}{2k} \\ 0 & \frac{1}{4k} + \frac{1}{4k} & -\frac{1}{4k} \\ -\frac{1}{2k} & -\frac{1}{4k} & \frac{1}{2k} + \frac{1}{4k} + \frac{1}{1k} \end{bmatrix} \begin{bmatrix} v_1 \\ v_2 \\ v_3 \end{bmatrix} = \begin{bmatrix} -0.004 \\ 0.002 \\ 0 \end{bmatrix}$$

or

$$\begin{bmatrix} 0.001 & 0 & -0.0005 \\ 0 & 0.0005 & -0.00025 \\ -0.0005 & -0.00025 & 0.00175 \end{bmatrix} \begin{bmatrix} v_1 \\ v_2 \\ v_3 \end{bmatrix} = \begin{bmatrix} -0.004 \\ 0.002 \\ 0 \end{bmatrix}$$

If we now employ these data with the MATLAB software, the computer screen containing the data and the results of the MATLAB analysis is as shown next.

```
>> G = [0.001 0 -0.0005 ; 0 0.0005
-0.00025 ; -0.0005 -0.00025 0.00175]

G =

        0.0010            0      -0.0005
             0       0.0005      -0.0003
       -0.0005      -0.0003       0.0018

>> I = [-0.004 ; 0.002 ; 0]

I =

       -0.0040
        0.0020
             0

>> V = inv(G)*I

V =

       -4.3636
        3.6364
       -0.7273
```

LEARNING EXTENSIONS

E3.1 Write the node equations for the circuit in Fig. E3.1.

Figure E3.1

ANSWER

$$\frac{1}{4k}V_1 - \frac{1}{12k}V_2 = 4 \times 10^{-3},$$

$$\frac{-1}{12k}V_1 + \frac{1}{4k}V_2 = -2 \times 10^{-3}.$$

E3.2 Find all the node voltages in the network in Fig. E3.2 using MATLAB.

Figure E3.2

ANSWER $V_1 = 5.4286$ V,
$V_2 = 2.000$ V, $V_3 = 3.1429$ V.

CIRCUITS CONTAINING DEPENDENT CURRENT SOURCES The pres-
ence of a dependent source may destroy the symmetrical form of the nodal equations that define
the circuit. Consider the circuit shown in Fig. 3.8, which contains a current-controlled current
source. The KCL equations for the nonreference nodes are

$$\beta i_o + \frac{v_1}{R_1} + \frac{v_1 - v_2}{R_2} = 0$$

Figure 3.8
Circuit with a dependent source.

and

$$\frac{v_2 - v_1}{R_2} + i_o - i_A = 0$$

where $i_o = v_2/R_3$. Simplifying the equations, we obtain

$$(G_1 + G_2)v_1 - (G_2 - \beta G_3)v_2 = 0$$

$$-G_2 v_1 + (G_2 + G_3)v_2 = i_A$$

or in matrix form

$$\begin{bmatrix} (G_1 + G_2) & -(G_2 - \beta G_3) \\ -G_2 & (G_2 + G_3) \end{bmatrix} \begin{bmatrix} v_1 \\ v_2 \end{bmatrix} = \begin{bmatrix} 0 \\ i_A \end{bmatrix}$$

Note that the presence of the dependent source has destroyed the symmetrical nature of the
node equation.

LEARNING Example 3.3

Let us determine the node voltages for the network in Fig. 3.8 given the following parameters:

$$\beta = 2 \qquad R_2 = 6 \text{ k}\Omega \qquad i_A = 2 \text{ mA}$$
$$R_1 = 12 \text{ k}\Omega \qquad R_3 = 3 \text{ k}\Omega$$

SOLUTION   Using these values with the equations for the network yields

$$\frac{1}{4k} V_1 + \frac{1}{2k} V_2 = 0$$

$$-\frac{1}{6k} V_1 + \frac{1}{2k} V_2 = 2 \times 10^{-3}$$

Solving these equations using any convenient method yields $V_1 = -24/5$ V and $V_2 = 12/5$ V. We can check these answers by determining the branch currents in the network and then using that information to test KCL at the nodes. For example, the current from top to bottom through R_3 is

$$I_o = \frac{V_2}{R_3} = \frac{12/5}{3k} = \frac{4}{5k} \text{ A}$$

Similarly, the current from right to left through R_2 is

$$I_2 = \frac{V_2 - V_1}{R_2} = \frac{12/5 - (-24/5)}{6k} = \frac{6}{5k} \text{ A}$$

All the results are shown in Fig. 3.9. Note that KCL is satisfied at every node.

Figure 3.9   *Circuit used in Example 3.3.*

LEARNING Example 3.4

Let us determine the set of linearly independent equations that when solved will yield the node voltages in the network in Fig. 3.10. Then given the following component values, we will compute the node voltages using MATLAB: $R_1 = 1 \text{ k}\Omega$, $R_2 = R_3 = 2 \text{ k}\Omega$, $R_4 = 4 \text{ k}\Omega$, $i_A = 2 \text{ mA}$, $i_B = 4 \text{ mA}$, and $\alpha = 2$.

Figure 3.10
*Circuit containing a
voltage-controlled
current source.*

SOLUTION   Applying KCL at each of the nonreference nodes yields the equations

$$G_3 v_1 + G_1(v_1 - v_2) - i_A = 0$$
$$i_A + G_1(v_2 - v_1) + \alpha v_x + G_2(v_2 - v_3) = 0$$
$$G_2(v_3 - v_2) + G_4 v_3 - i_B = 0$$

where $v_x = v_2 - v_3$. Simplifying these equations, we obtain

$$(G_1 + G_3)v_1 - G_1 v_2 = i_A$$
$$-G_1 v_1 + (G_1 + \alpha + G_2)v_2 - (\alpha + G_2)v_3 = -i_A$$
$$-G_2 v_2 + (G_2 + G_4)v_3 = i_B$$

Given the component values, the equations become

$$\begin{bmatrix} \dfrac{1}{1k} + \dfrac{1}{2k} & -\dfrac{1}{k} & 0 \\[2mm] -\dfrac{1}{k} & \dfrac{1}{k} + 2 + \dfrac{1}{2k} & -\left(2 + \dfrac{1}{2k}\right) \\[2mm] 0 & -\dfrac{1}{2k} & \dfrac{1}{2k} + \dfrac{1}{4k} \end{bmatrix} \begin{bmatrix} v_1 \\ v_2 \\ v_3 \end{bmatrix}$$

$$= \begin{bmatrix} 0.002 \\ -0.002 \\ 0.004 \end{bmatrix}$$

or

$$\begin{bmatrix} 0.0015 & -0.001 & 0 \\ -0.001 & 2.0015 & -2.0005 \\ 0 & -0.0005 & 0.00075 \end{bmatrix} \begin{bmatrix} v_1 \\ v_2 \\ v_3 \end{bmatrix} = \begin{bmatrix} 0.002 \\ -0.002 \\ 0.004 \end{bmatrix}$$

The MATLAB input and output listings are shown next.

```
>> G = [0.0015 -0.001 0 ; -0.001
2.0015 -2.0005 ; 0 -0.0005 0.00075]

G =

        0.0015       -0.0010             0
       -0.0010        2.0015       -2.0005
             0       -0.0005        0.0008
```

```
>> I = [0.002 ; -0.002 ; 0.004]              >> V = inv(G)*I

   I =                                           V =

        0.0020                                      11.9940
       -0.0020                                      15.9910
        0.0040                                      15.9940
```

LEARNING EXTENSIONS

E3.3   Find the node voltages in the circuit in Fig. E3.3.

Figure E3.3

ANSWER $V_1 = 16$ V,
$V_2 = -8$ V.

E3.4   Find the voltages V_o in the network in Fig. E3.4.

Figure E3.4

ANSWER $V_o = 4$ V.

CIRCUITS CONTAINING INDEPENDENT VOLTAGE SOURCES   As is our practice, in our discussion of this topic we will proceed from the simplest case to those that are more complicated. The simplest case is that in which an independent voltage source is connected to the reference node. The following example illustrates this case.

LEARNING Example 3.5

Consider the circuit shown in Fig. 3.11a. Let us determine all node voltages and branch currents.

SOLUTION   This network has three nonreference nodes with labeled node voltages V_1, V_2, and V_3. Based on our previous discussions, we would assume that in order to find all the node voltages we would need to write a KCL equation at each of the nonreference nodes. The resulting three linearly independent simultaneous equations would produce the unknown node voltages. However, note that V_1 and V_3 are known quantities because an independent voltage source is connected directly between the nonreference node and each of these nodes. Therefore, $V_1 = 12$ V and $V_3 = -6$ V. Furthermore, note that the current through the 9-kΩ resistor is $[12 - (-6)]/9k = 2$ mA from left to right. We do not know V_2 or the current in the remaining resistors. However,

since only one node voltage is unknown, a single-node equation will produce it. Applying KCL to this center node yields

$$\frac{V_2 - V_1}{12k} + \frac{V_2 - 0}{6k} + \frac{V_2 - V_3}{12k} = 0$$

or

$$\frac{V_2 - 12}{12k} + \frac{V_2}{6k} + \frac{V_2 - (-6)}{12k} = 0$$

from which we obtain

$$V_2 = \frac{3}{2} \text{ V}$$

Once all the node voltages are known, Ohm's law can be used to find the branch currents shown in Fig. 3.11b. The diagram illustrates that KCL is satisfied at every node.

Note that the presence of the voltage sources in this example has simplified the analysis, since two of the three linear independent equations are $V_1 = 12$ V and $V_3 = -6$ V. We will find that as a general rule, any time voltage sources are present between nodes, the node voltage equations that describe the network will be simpler.

LEARNING Hint

Any time an independent voltage source is connected between the reference node and a nonreference node, the nonreference node voltage is known.

Figure 3.11
Circuit used in Example 3.5.

(a)   (b)

LEARNING EXTENSION

E3.5   Use nodal analysis to find the current I_o in the network in Fig. E3.5.

ANSWER $I_o = \dfrac{3}{4}$ mA.

Figure E3.5

Next let us consider the case in which an independent voltage source is connected between two nonreference nodes. Once again, we will use an example to illustrate the approach.

LEARNING Example 3.6

We wish to find the currents in the two resistors in the circuit in Fig. 3.12a.

SOLUTION   If we try to attack this problem in a brute force manner, we immediately encounter a problem. Thus far, branch cur-

rents were either known source values or could be expressed as the branch voltage divided by the branch resistance. However, the branch current through the 6-V source is certainly not known and cannot be directly expressed using Ohm's law. We can, of course, give this current a name and write the KCL equations

Figure 3.12
Circuits used in Example 3.6.

(a)   (b)

(continued)

at the two nonreference nodes in terms of this current. However, this approach is no panacea because this technique will result in *two* linearly independent simultaneous equations in terms of *three* unknowns; that is, the two node voltages and the current in the voltage source.

To solve this dilemma, we recall that $N - 1$ linearly independent equations are required to determine the $N - 1$ nonreference node voltages in an N-node circuit. Since our network has three nodes, we need two linearly independent equations. Now note that if somehow one of the node voltages is known, we immediately know the other; that is, if V_1 is known, then $V_2 = V_1 - 6$. If V_2 is known, then $V_1 = V_2 + 6$. Therefore, the difference in potential between the two nodes is *constrained* by the voltage source and, hence,

$$V_1 - V_2 = 6$$

This constraint equation is one of the two linearly independent equations needed to determine the node voltages.

Next consider the network in Fig. 3.12b, in which the 6-V source is completely enclosed within the dashed surface. The constraint equation governs this dashed portion of the network. The remaining equation is obtained by applying KCL to this dashed surface, which is commonly called a *supernode*. Recall that in Chapter 2 we demonstrated that KCL must hold for a surface, and this technique eliminates the problem of dealing with a current through a voltage source. KCL for the supernode is

$$-6 \times 10^{-3} + \frac{V_1}{6k} + \frac{V_2}{12k} + 4 \times 10^{-3} = 0$$

Solving these equations yields $V_1 = 10$ V and $V_2 = 4$ V and, hence, $I_1 = 5/3$ mA and $I_2 = 1/3$ mA. A quick check indicates that KCL is satisfied at every node.

LEARNING Hint

The supernode technique

▸ Use it when a branch between two nonreference nodes contains a voltage source.
▸ First encircle the voltage source and the two connecting nodes to form the supernode.
▸ Write the equation that defines the voltage relationship between the two nonreference nodes as a result of the presence of the voltage source.
▸ Write the KCL equation for the supernode.
▸ If the voltage source is dependent, then the controlling equation for the dependent source is also needed.

LEARNING Example 3.7

Let us determine the current I_o in the network in Fig. 3.13a.

(a)

(b)

Figure 3.13
Example circuit with supernodes.

SOLUTION Examining the network, we note that node voltages V_2 and V_4 are known and the node voltages V_1 and V_3 are constrained by the equation

$$V_1 - V_3 = 12$$

The network is redrawn in Fig. 3.13b.

Since we want to find the current I_o, V_1 (in the supernode containing V_1 and V_3) is written as $V_3 + 12$. The KCL equation at the supernode is then

$$\frac{V_3 + 12 - (-6)}{2k} + \frac{V_3 + 12 - 12}{2k} +$$
$$\frac{V_3 - (-6)}{1k} + \frac{V_3 - 12}{1k} + \frac{V_3}{2k} = 0$$

Solving the equation for V_3 yields

$$V_3 = -\frac{6}{7} \text{ V}$$

I_o can then be computed immediately as

$$I_o = \frac{-\frac{6}{7}}{2k} = -\frac{3}{7} \text{ mA}$$

E3.6 Use nodal analysis to find I_o in the network in Fig. E3.6.

ANSWER $I_o = 3.8$ mA.

Figure E3.6

CIRCUITS CONTAINING DEPENDENT VOLTAGE SOURCES
As the following examples will indicate, networks containing dependent (controlled) sources are treated in the same manner as described earlier.

LEARNING Example 3.8

We wish to find I_o in the network in Fig. 3.14.

Figure 3.14 *Circuit used in Example 3.8.*

SOLUTION Since the dependent voltage source is connected between the node labeled V_1 and the reference node,

$$V_1 = 2kI_x$$

KCL at the node labeled V_2 is

$$\frac{V_2 - V_1}{2k} - \frac{4}{k} + \frac{V_2}{1k} = 0$$

where

$$I_x = \frac{V_2}{1k}$$

Solving these equations yields $V_2 = 8$ V and $V_1 = 16$ V. Therefore

$$I_o = \frac{V_1 - V_2}{2k}$$

$$= 4 \text{ mA}$$

LEARNING Example 3.9

Let us find the current I_o in the network in Fig. 3.15.

Figure 3.15
Circuit used in Example 3.9.

SOLUTION This circuit contains both an independent voltage source and a voltage-controlled voltage source. Note that $V_3 = 6$ V, $V_2 = V_x$, and a supernode exists between the nodes labeled V_1 and V_2.

Applying KCL to the supernode, we obtain

$$\frac{V_1 - V_3}{6k} + \frac{V_1}{12k} + \frac{V_2}{6k} + \frac{V_2 - V_3}{12k} = 0$$

where the constraint equation for the supernode is

$$V_1 - V_2 = 2V_x$$

The final equation is

$$V_3 = 6$$

Solving these equations, we find that

$$V_1 = \frac{9}{2} \text{ V}$$

and, hence,

$$I_o = \frac{V_1}{12k} = \frac{3}{8} \text{ mA}$$

E3.7 Use nodal analysis to find I_o in the circuit in Fig. E3.7.

ANSWER $I_o = \dfrac{4}{3}$ mA.

Figure E3.7

Problem-Solving Strategy
Nodal Analysis

▶ Select one node in the N-node circuit as the reference node. Assume that the node voltage is zero and measure all node voltages with respect to this node.

▶ If only independent current sources are present in the network, write the KCL equations at the $N - 1$ nonreference nodes. If dependent current sources are present, write the KCL equations as is done for networks with only independent current sources; then write the controlling equations for the dependent sources.

▶ If voltage sources are present in the network, they may be connected (1) between the reference node and a nonreference node or (2) between two nonreference nodes. In the former case, if the voltage source is an independent source, then the voltage at one of the nonreference nodes is known. If the source is dependent, it is treated as an independent source when writing the KCL equation, but an additional constraint equation is necessary, as described previously.

In the latter case, if the source is independent, the voltage between the two nodes is constrained by the value of the voltage source, and an equation describing this constraint represents one of the $N - 1$ linearly independent equations required to determine the N-node voltages. The surface of the network described by the constraint equation (i.e., the source and two connecting nodes) is called a supernode. One of the remaining $N - 1$ linearly independent equations is obtained by applying KCL at this supernode. If the voltage source is dependent, it is treated as an independent source when writing the KCL equations, but an additional constraint equation is necessary, as described previously.

3.2 Loop Analysis

In a nodal analysis the unknown parameters are the node voltages, and KCL is employed to determine them. In contrast to this approach, a loop analysis uses KVL to determine currents in the circuit. Once the currents are known, Ohm's law can be used to calculate the voltages. Recall that, in Chapter 2, we found that a single equation was sufficient to determine the current in a circuit containing a single loop. If the circuit contains N independent loops, we will show that N independent simultaneous equations will be required to describe the network. For now we will assume that the circuits are planar, which simply means that we can draw the circuit on a sheet of paper in a way such that no conductor crosses another conductor.

Our approach to loop analysis will mirror that used in nodal analysis (i.e., we will begin with simple cases and systematically proceed to those that are more difficult). Then at the end of this section we will outline a general strategy for employing loop analysis.

CIRCUITS CONTAINING ONLY INDEPENDENT VOLTAGE SOURCES To

begin our analysis, consider the circuit shown in Fig. 3.16. Let us also identify two loops, *A-B-E-F-A* and *B-C-D-E-B*. We now define a new set of current variables called *loop currents*, which can be used to find the physical currents in the circuit. Let us assume that current i_1 flows in the first loop and that current i_2 flows in the second loop. Then the branch current flowing from *B* to *E* through R_3 is $i_1 - i_2$. The directions of the currents have been assumed. As was the case in the nodal analysis, if the actual currents are not in the direction indicated, the values calculated will be negative.

Applying KVL to the first loop yields

$$+v_1 + v_3 + v_2 - v_{S1} = 0$$

KVL applied to loop 2 yields

$$+v_{S2} + v_4 + v_5 - v_3 = 0$$

where $v_1 = i_1 R_1$, $v_2 = i_1 R_2$, $v_3 = (i_1 - i_2)R_3$, $v_4 = i_2 R_4$, and $v_5 = i_2 R_5$.

Substituting these values into the two KVL equations produces the two simultaneous equations required to determine the two loop currents; that is,

$$i_1(R_1 + R_2 + R_3) - i_2(R_3) = v_{S1}$$
$$-i_1(R_3) + i_2(R_3 + R_4 + R_5) = -v_{S2}$$

or in matrix form

$$\begin{bmatrix} R_1 + R_2 + R_3 & -R_3 \\ -R_3 & R_3 + R_4 + R_5 \end{bmatrix} \begin{bmatrix} i_1 \\ i_2 \end{bmatrix} = \begin{bmatrix} v_{S1} \\ -v_{S2} \end{bmatrix}$$

At this point it is important to define what is called a *mesh*. A mesh is a special kind of loop that does not contain any loops within it. Therefore, as we traverse the path of a mesh, we do not encircle any circuit elements. For example, the network in Fig. 3.16 contains two meshes defined by the paths *A-B-E-F-A* and *B-C-D-E-B*. The path *A-B-C-D-E-F-A* is a loop, but it is not a mesh. Since the majority of our analysis in this section will involve writing KVL equations for meshes, we will refer to the currents as mesh currents and the analysis as a *mesh analysis*.

LEARNING Hint

The equations employ the passive sign convention.

LEARNING by Doing

D 3.2 Write the mesh equations for the following circuit.

ANSWER
$$-v_{S1} + i_1 R_1 - v_{S2}$$
$$+ (i_1 - i_2)R_2 = 0$$
$$i_2 R_3 + (i_2 - i_1)R_2$$
$$+ i_2 R_5 + i_2 R_4 + v_{S3} = 0$$

Figure 3.16
A two-loop circuit.

LEARNING Example 3.10

Consider the network in Fig. 3.17a. We wish to find the current I_o.

SOLUTION We will begin the analysis by writing mesh equations. Note that there are no + and − signs on the resistors. However, they are not needed, since we will apply Ohm's law to each resistive element as we write the KVL equations. The equation for the first mesh is

$$-12 + 6kI_1 + 6k(I_1 - I_2) = 0$$

The KVL equation for the second mesh is

$$6k(I_2 - I_1) + 3kI_2 + 3 = 0$$

where $I_o = I_1 - I_2$.

Solving the two simultaneous equations yields $I_1 = 5/4$ mA and $I_2 = 1/2$ mA. Therefore, $I_o = 3/4$ mA. All the voltages and currents in the network are shown in Fig. 3.17b. Recall from nodal analysis that once the node voltages were determined, we could

(continued)

check our analysis using KCL at the nodes. In this case we know the branch currents and can use KVL around any closed path to check our results. For example, applying KVL to the outer loop yields

$$-12 + \frac{15}{2} + \frac{3}{2} + 3 = 0$$

$$0 = 0$$

Since we want to calculate the current I_o, we could use loop analysis, as shown in Fig. 3.17c. Note that the loop current I_1 passes through the center leg of the network and, therefore, $I_1 = I_o$. The two loop equations in this case are

$$-12 + 6k(I_1 + I_2) + 6kI_1 = 0$$

and

$$-12 + 6k(I_1 + I_2) + 3kI_2 + 3 = 0$$

Solving these equations yields $I_1 = 3/4$ mA and $I_2 = 1/2$ mA. Since the current in the 12-V source is $I_1 + I_2 = 5/4$ mA, these results agree with the mesh analysis.

Finally, for purposes of comparison, let us find I_o using nodal analysis. The presence of the two voltage sources would indicate that this is a viable approach. Applying KCL at the top center node, we obtain

$$\frac{V_o - 12}{6k} + \frac{V_o}{6k} + \frac{V_o - 3}{3k} = 0$$

and hence,

$$V_o = \frac{9}{2} \text{ V}$$

and then

$$I_o = \frac{V_o}{6k} = \frac{3}{4} \text{ mA}$$

Note that in this case we had to solve only one equation instead of two.

Figure 3.17 Circuits used in Example 3.10.

Once again we are compelled to note the symmetrical form of the mesh equations that describe the circuit in Fig. 3.16. Note that the coefficient matrix for this circuit is symmetrical.

Since this symmetry is generally exhibited by networks containing resistors and independent voltage sources, we can learn to write the mesh equations by inspection. In the first equation, the coefficient of i_1 is the sum of the resistances through which mesh current 1 flows, and the coefficient of i_2 is the negative of the sum of the resistances common to mesh current 1 and mesh current 2. The right-hand side of the equation is the algebraic sum of the voltage sources in mesh 1. The sign of the voltage source is positive if it aids the assumed direction of the current flow and negative if it opposes the assumed flow. The first equation is KVL for mesh 1. In the second equation, the coefficient of i_2 is the sum of all the resistances in mesh 2, the coefficient of i_1 is the negative of the sum of the resistances common to mesh 1 and mesh 2, and the right-hand side of the equation is the algebraic sum of the voltage sources in mesh 2. In general, if we assume all of the mesh currents to be in the same direction (clockwise or counterclockwise), then if KVL is applied to mesh j with mesh current i_j, the coefficient of i_j is the sum of the resistances in mesh j and the coefficients of the other mesh currents $(e.g., i_{j-1}, i_{j+1})$ are the negatives of the resistances common to these meshes and mesh j. The right-hand side of the equation is equal to the algebraic sum of the voltage sources in mesh j. These voltage sources have a positive sign if they aid the current flow i_j and a negative sign if they oppose it.

LEARNING Example 3.11

Let us write the mesh equations by inspection for the network in Fig. 3.18. Then we will use MATLAB to solve for the mesh currents.

Figure 3.18 Circuit used in Example 3.11.

SOLUTION The three linearly independent simultaneous equations are

$$(4k + 6k)I_1 - (0)I_2 - (6k)I_3 = -6$$
$$-(0)I_1 + (9k + 3k)I_2 - (3k)I_3 = 6$$
$$-(6k)I_1 - (3k)I_2 + (3k + 6k + 12k)I_3 = 0$$

or in matrix form

$$\begin{bmatrix} 10k & 0 & -6k \\ 0 & 12k & -3k \\ -6k & -3k & 21k \end{bmatrix} \begin{bmatrix} I_1 \\ I_2 \\ I_3 \end{bmatrix} = \begin{bmatrix} -6 \\ 6 \\ 0 \end{bmatrix}$$

Note the symmetrical form of the equations. The general form of the matrix equation is

$$\mathbf{RI = V}$$

and the solution of this matrix equation is

$$\mathbf{I = R^{-1}V}$$

The input/output data for a MATLAB solution are as follows:

```
>> R = [10e3 0 -6e3; 0 12e3 -3e3;
   -6e3 -3e3 21e3]
   R =
          10000          0      -6000
              0      12000      -3000
          -6000      -3000      21000
>> V = [ -6 ; 6 ; 0]
   V =
          -6
           6
           0
>> I = inv(R)*V
   I =
      1.0e-003  *
          -0.6757
           0.4685
          -0.1261
```

CIRCUITS CONTAINING INDEPENDENT CURRENT SOURCES Just as the presence of a voltage source in a network simplified the nodal analysis, the presence of a current source simplifies a loop analysis. The following examples illustrate the point.

E3.8 Use mesh equations to find V_o in the circuit in Fig. E3.8.

ANSWER $V_o = \dfrac{33}{5}$ V.

Figure E3.8

LEARNING Example 3.12

We wish to find V_o in the network in Fig. 3.19.

Figure 3.19   *Circuit used in Example 3.12.*

SOLUTION   Since the currents I_1 and I_2 pass directly through a current source, two of the three required equations are

$$I_1 = 4 \times 10^{-3}$$
$$I_2 = -2 \times 10^{-3}$$

The third equation is KVL for the mesh containing the voltage source; that is,

$$4k(I_3 - I_2) + 2k(I_3 - I_1) + 6kI_3 - 3 = 0$$

These equations yield

$$I_3 = \frac{1}{4} \text{ mA}$$

and hence,

$$V_o = 6kI_3 - 3 = \frac{-3}{2} \text{ V}$$

What we have demonstrated in the previous example is the general approach for dealing with independent current sources when writing KVL equations; that is, use one loop through each current source. The number of "window panes" in the network tells us how many equations we need. Additional KVL equations are written to cover the remaining circuit elements in the network. The following example illustrates this approach.

LEARNING Example 3.13

Let us find I_o in the network in Fig. 3.20a.

(a)

(b)

(c)

(d)

(e)

(f)

Figure 3.20   *Circuits used in Example 3.13.*

LEARNING Hint

In this case the 4-mA current source is located on the boundary between two meshes. Thus we will demonstrate two techniques for dealing with this type of situation. One is a special loop technique and the other is known as the supermesh approach.

SOLUTION   First, we select two loop currents I_1 and I_2 such that I_1 passes directly through the 2-mA source, and I_2 passes directly through the 4-mA source, as shown in Fig. 3.20b. Therefore, two of our three linearly independent equations are

$$I_1 = 2 \times 10^{-3}$$
$$I_2 = 4 \times 10^{-3}$$

The remaining loop current I_3 must pass through the circuit elements not covered by the two previous equations and cannot, of course, pass through the current sources. The path for this remaining loop current can be obtained by open-circuiting the current sources, as shown in Fig. 3.20c. When all currents are labeled on the original circuit, the KVL equation for this last loop, as shown in Fig. 3.20d, is

$$-6 + 1kI_3 + 2k(I_2 + I_3) + 2k(I_3 + I_2 - I_1) + 1k(I_3 - I_1) = 0$$

Solving the equations yields

$$I_3 = \frac{-2}{3} \text{ mA}$$

and therefore,

$$I_o = I_1 - I_2 - I_3 = \frac{-4}{3} \text{ mA}$$

Next consider the supermesh technique. In this case the three mesh currents are specified as shown in Fig. 3.20e, and since the voltage across the 4-mA current source is unknown, it is assumed to be V_x. The mesh currents constrained by the current sources are

$$I_1 = 2 \times 10^{-3}$$
$$I_2 - I_3 = 4 \times 10^{-3}$$

The KVL equations for meshes 2 and 3, respectively, are

$$2kI_2 + 2k(I_2 - I_1) - V_x = 0$$
$$-6 + 1kI_3 + V_x + 1k(I_3 - I_1) = 0$$

Adding the last two equations yields

$$-6 + 1kI_3 + 2kI_2 + 2k(I_2 - I_1) + 1k(I_3 - I_1) = 0$$

Note that the unknown voltage V_x has been eliminated. The two constraint equations, together with this latter equation, yield the desired result.

The purpose of the supermesh approach is to avoid introducing the unknown voltage V_x. The supermesh is created by mentally removing the 4-mA current source, as shown in Fig. 3.20f. Then writing the KVL equation around the dotted path, which defines the supermesh, using the original mesh currents as shown in Fig. 3.20e, yields

$$-6 + 1kI_3 + 2kI_2 + 2k(I_2 - I_1) + 1k(I_3 - I_1) = 0$$

Note that this supermesh equation is the same as that obtained earlier by introducing the voltage V_x.

LEARNING EXTENSIONS

E3.9 Find V_o in the network in Fig. E3.9.

ANSWER $V_o = \dfrac{33}{5}$ V.

Figure E3.9

3.10 Find V_o in the network in Fig. E3.10.

ANSWER $V_o = \dfrac{32}{5}$ V.

Figure E3.10

CIRCUITS CONTAINING DEPENDENT SOURCES
We deal with circuits containing dependent sources just as we have in the past. First, we treat the dependent source as though it were an independent source when writing the KVL equations. Then we write the controlling equation for the dependent source. The following examples illustrate the point.

LEARNING Example 3.14

The network in Fig. 3.21 contains both a current-controlled voltage source and a voltage-controlled current source. Let us use MATLAB to determine the loop currents.

SOLUTION The equations for the loop currents shown in the figure are

$$I_1 = \frac{4}{k}$$
$$I_2 = \frac{V_x}{2k}$$
$$-1kI_x + 2k(I_3 - I_1) + 1k(I_3 - I_4) = 0$$
$$1k(I_4 - I_3) + 1k(I_4 - I_2) + 12 = 0$$

(continued)

Figure 3.21
Circuit used in
Example 3.14.

where

$$V_x = 2k(I_3 - I_1)$$

$$I_x = I_4 - I_2$$

Combining these equations yields

$$I_1 = \frac{4}{k}$$

$$I_1 + I_2 - I_3 = 0$$

$$1kI_2 + 3kI_3 - 2kI_4 = 8$$

$$1kI_2 + 1kI_3 - 2kI_4 = 12$$

In matrix form the equations are

$$
\begin{bmatrix}
1 & 0 & 0 & 0 \\
1 & 1 & -1 & 0 \\
0 & 1k & 3k & -2k \\
0 & 1k & 1k & -2k
\end{bmatrix}
\begin{bmatrix}
I_1 \\
I_2 \\
I_3 \\
I_4
\end{bmatrix}
=
\begin{bmatrix}
\dfrac{4}{k} \\
0 \\
8 \\
12
\end{bmatrix}
$$

The input and output data for the MATLAB solution are as follows:

```
>> R = [ 1 0 0 0 ; 1 1 -1 0;
0 1000 3000 -2000;
0 1000 1000 -2000]

R =

        1        0        0        0
        1        1       -1        0
        0     1000     3000    -2000
        0     1000     1000    -2000

>> V = [ 0.004; 0; 8; 12]

V =

    0.0040
         0
    8.0000
   12.0000

>> I = inv(R)*V

I =

    0.0040
   -0.0060
   -0.0020
   -0.0100
```

As a final point, it is very important to examine the circuit carefully before selecting an analysis approach. One method could be much simpler than another, and a little time invested up front may save a lot of time in the long run.

E3.11 Use mesh analysis to find V_o in the circuit in Fig. E3.11.

ANSWER $V_o = 12$ V.

Figure E3.11

Problem-Solving Strategy
Loop Analysis

▶ One loop current is assigned to each independent loop in a circuit that contains N independent loops.

▶ If only independent voltage sources are present in the network, write the N linearly independent KVL equations, one for each loop. If dependent voltage sources are present, write the KVL equation as is done for circuits with only independent voltage sources; then write the controlling equations for the dependent sources.

▶ If current sources are present in the network, either of two techniques can be used. In the first case, one loop current is selected to pass through one of the current sources. This is done for each current source in the network. The remaining loop currents (N − the number of current sources) are determined by open-circuiting the current sources in the network and using this modified network to select them. Once all these currents are defined in the original network, the N loop equations can be written. The second approach is similar to the first with the exception that if two mesh currents pass through a particular current source, a supermesh is formed around this source. The two required equations for the meshes containing this source are the constraint equations for the two mesh currents that pass through the source and the supermesh equation. As indicated earlier, if dependent current sources are present, the controlling equations for these sources are also necessary.

LEARNING EXTENSIONS

E3.12   Use loop analysis to solve the network in Example 3.5 and compare the time and effort involved in the two solution techniques.

E3.13   Use nodal analysis to solve the circuit in Example 3.12 and compare the time and effort involved in the two solution strategies.

3.3 Circuits with Operational Amplifiers

It can be argued that the operational amplifier, or op-amp as it is commonly known, is the single most important integrated circuit for analog circuit design. It is a versatile interconnection of transistors and resistors that vastly expands our capabilities in circuit design, from engine control systems to cellular phones. Early op-amps were built with vacuum tubes, making them bulky and power hungry. The invention of the transistor at Bell Labs in 1947 allowed engineers to create op-amps that were much smaller and more efficient. Still, the op-amp itself consisted of individual transistors and resistors interconnected on a printed circuit board (PCB). When the manufacturing process for integrated circuits (ICs) was developed around 1970, engineers could finally put all of the op-amps transistors and resistors onto a single IC chip. Today, it is common to find as many as four high quality op-amps on a single IC for as little as $0.40. A sample of commercial op-amps is shown in Fig. 3.22.

Let us first examine the origin of the term operational amplifier. Originally, the op-amp was designed to perform mathematical operations such as addition, subtraction, differentiation, and integration. By adding simple networks to the op-amp, we can create these "building blocks" as well as other functions such as voltage scaling, current-to-voltage conversion, and a myriad of more complex applications.

(a)   (b)

Figure 3.22   *A selection of op-amps. On the left in (a) is a discrete op-amp assembled on a printed circuit board (PCB). On the right from top to bottom, a LM324 DIP (dual in-line pack), LMC6294 DIP, and MAX4240 in a SO-5 package (small outline/5 pins). A penny is shown for purposes of comparison. In (b) is the APEX PA03 with its lid removed showing individual transistors and resistors.*

How can we, understanding only sources and resistors, hope to comprehend the performance of the op-amp? The answer is modeling. When all the bells and whistles are removed, an op-amp is simply a very good voltage amplifier. In other words, the output voltage is a scaled replica of the input voltage. Modern op-amps are such good amplifiers that it is easy to create an accurate, first-order model. As mentioned earlier, the op-amp is very popular and is used extensively in circuit design at all levels. We should not be surprised to find that op-amps are available for every application—low voltage, high voltage, micro-power, high speed, high current, and so forth. Fortunately, the topology of our model is independent of these issues.

We begin our discussion with the general purpose LM324 quad (four in a pack) op-amp from National Semiconductor. The pinout for the LM324 is shown in Fig. 3.23 for a DIP (dual in-

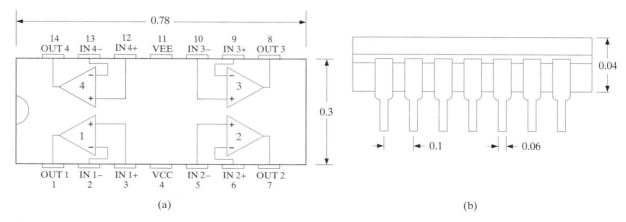

(a)   (b)

Figure 3.23   *(a) The pinout and (b) the dimensional diagram of one side of the LM324 quad op-amp. Note the pin pitch (distance pin-to-pin) is 0.1 inches, a standard for DIP packages.*

line pack) style package with the dimensions specified in inches. Recognizing there are four identical op-amps in the package, we will focus on amplifier 1. Pins 3 and 2 are the input pins, IN_+ and IN_-, and are called the noninverting and inverting inputs, respectively. The output is at pin 1. The relationship that exists between the output and input voltages is

$$V_o = A_o(IN_+ - IN_-)$$ **3.5**

where all voltages are measured with respect to ground and A_o is the gain of the op-amp. From Eq. (3.5), we see that when IN_+, increases, so will V_o. However, if IN_- increases, then V_o will decrease—hence the names noninverting and inverting inputs. We mentioned earlier that op-amps are very good voltage amplifiers. How good? Typical values for A_o are between 10,000 and 1,000,000!

To provide amplification, we need power. This power is obtained from dc voltage sources connected to pins 4 and 11, called V_{CC} and V_{EE}, respectively. Actual values for these power supplies can vary widely depending on the application, from as little as one volt up to several hundred volts. Traditionally, V_{CC} is a positive dc voltage with respect to ground and V_{EE} is either a negative voltage or ground itself.

We can model the input/output relationship of the op-amp, as specified in Eq. (3.5), using a dependent voltage source. The currents into and out of the op-amp terminals (pins 3, 2, and 1) are fairly proportional to the pin voltages; that is, the relationship is essentially that specified by Ohm's law. Thus, we model the I-V performance with two resistors, one at the input terminals (R_i) and another at the output (R_o). The resultant circuit is shown in Fig. 3.24.

Let us now examine the values for A_o, R_i, and R_o. Consider the network in Fig. 3.25, where we have modeled the driving circuit with V_S and a resistance R_{Th1}, and the output load with a resistor R_L.

Since the op-amp is designed to be an excellent voltage amplifier, let us write an equation for the overall gain of the circuit V_{out}/V_{in}. Using voltage division at the input and again at the output, we quickly produce the expression

$$\frac{V_{out}}{V_{in}} = \left[\frac{R_i}{R_i + R_{Th1}} \right] A_o \left[\frac{R_L}{R_o + R_L} \right]$$

Figure 3.24
A simple model of the gain characteristics of an op-amp.

Figure 3.25
A network that depicts an op-amp circuit. V_S and R_{Th1} model the driving circuit, while the load is modeled by R_L.

Table 3.1 A list of commercial op-amps and their model values

Manufacturer	Part No.	A_o (V/V)	R_i (MΩ)	R_o (Ω)	Comments
National	LM324	100,000	1.0	20	General purpose, up to ±16 V supplies, very inexpensive
National	LMC6492	50,000	10^7	150	Low voltage, rail-to-rail inputs and outputs
Maxim	MAX4240	20,000	45	160	Micro-power (1.8 V supply @ 10 μA), rail-to-rail in inputs and outputs
Apex	PA03	125,000	10^5	2	High voltage, ±75 V, and high output current capability, 30 A. That's 2 kW!

To maximize the gain, regardless of the input resistance and load values, we make A_o very large and the voltage division ratios as close to unity as possible. The ideal scenario requires A_o be infinite, R_i to be infinite, and R_o to be zero, yielding a large overall gain of A_o. Table 3.1 shows the actual values of A_o, R_i, and R_o for a sampling of commercial op-amps intended for very different applications. Although A_o, R_i, and R_o are not ideal, they do approximate the ideal conditions.

The power supplies affect performance in two ways. First, each op-amp has minimum and maximum supply ranges, sometimes called rail-to-rail (a trademark of Motorola Corporation) over which the op-amp is guaranteed to function. Second, for proper operation, the input and output voltages are limited to no more than the supply voltages (Op-amps are available that have input and/or output voltage ranges beyond the supply rails: however, these devices constitute a very small percentage of the op-amp market and will not be discussed here). If the inputs and output can reach within a few dozen millivolts of the supplies, then the inputs and output are called rail-to-rail. Otherwise, the inputs/output limits are more severe—usually a volt or so away from the supply values. Combining the model in Fig. 3.25, the values in Table 3.1, and these I/O limitations, we can produce the graphs in Fig. 3.26, which show the output–input relationship for each op-amp outlined in Table 3.1. From the graph we see that the LM6492 and MAX 4240 have rail-to-rail outputs and the LM324 and PA03 do not.

Figure 3.26
Transfer plots for the op-amps listed in Table 3.1. The supply voltages are listed in the plot legends. Note that the LMC6492 and MAX4240 have rail-to-rail output voltages, and the LM324 and PA03 do not.

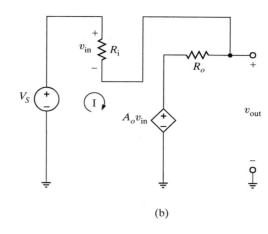

(a)   (b)

Figure 3.27
Circuit (a) and model (b) for the unity gain buffer.

In order to examine the performance of the op-amp in a practical circuit, consider the network shown in Fig. 3.27a called a unity-gain buffer. Note that the op-amp schematic symbol includes the power supplies. Employing the model in Fig. 3.25 yields the circuit in Fig. 3.27b, containing just resistors and dependent sources, which we can easily analyze. The loop equations for the network are

$$V_S = IR_i + IR_o + A_oV_{in}$$

$$V_{out} = IR_o + A_oV_{in}$$

$$V_{in} = IR_i$$

Solving for the gain, V_{out}/V_S, we find

$$\frac{V_{out}}{V_S} = \frac{1}{1 + \dfrac{R_i}{R_o + A_oR_i}}$$

For $R_o \ll R_i$, we have

$$\frac{V_{out}}{V_S} \approx \frac{1}{1 + \dfrac{1}{A_o}}$$

And, if A_o is indeed $\gg 1$,

$$\frac{V_{out}}{V_S} \approx 1$$

Thus, the origin of the name, unity gain buffer, should be apparent. Table 3.2 shows the actual gain values for $V_S = 1$ V using the op-amps listed in Table 3.1. Note how close the gain is to unity and how small the input voltage and current are. These results lead us to simplify

Table 3.2 Unity gain buffer performance for the op-amps listed in Table 3.1

Op-Amp	Buffer Gain	V_{in} (μV)	I (pA)
LM324	0.999990	9.9999	9.9998
LMC6492	0.999980	19.999	1.9999×10^{-6}
MAX4240	0.999950	49.998	1.1111
PA05	0.999992	7.9999	7.9999×10^{-5}

Table 3.3 Consequences of the ideal op-amp model on input terminal I/V values

Model Assumption	Terminal Result
$A_o \to \infty$	Input voltage $\to 0$ V
$R_i \to \infty$	Input current $\to 0$ A

the op-amp in Fig. 3.24 significantly. Hence, we introduce the *ideal op-amp model*, where A_o and R_i are infinite and R_o is zero. This selection of parameter values produces two important results for analyzing op-amp circuits, which are listed in Table 3.3.

From Table 3.3 we find that the ideal model for the op-amp is reduced to that shown in Fig. 3.28. The important characteristics of the model are as follows: (1) Since R_i is extremely large, the input currents to the op-amp are approximately zero (i.e., $i_+ \approx i_- \approx 0$); and (2) if the output voltage is to remain bounded, then as the gain becomes very large and approaches infinity, the voltage across the input terminals must simultaneously become infinitesimally small so that as $A \to \infty$, $v_+ - v_- \to 0$ (i.e., $v_+ - v_- = 0$ or $v_+ = v_-$). The difference between these input voltages is often called the *error signal* for the op-amp (i.e., $v_+ - v_- = v_e$).

Figure 3.28
Ideal model for an operational amplifier. Model parameters:
$i_+ = i_- = 0, v_+ = v_-.$

The ground terminal ⏚ shown on the op-amp is necessary for signal current return, and it guarantees that Kirchhoff's current law is satisfied at both the op-amp and the ground node in the circuit.

In summary, then, our ideal model for the op-amp is simply stated by the following conditions:

$$i_+ = i_- = 0$$
$$v_+ = v_-$$

3.6

These simple conditions are extremely important because they form the basis of our analysis of op-amp circuits.

Let us now use the ideal model to re-examine the unity gain buffer, redrawn again in Fig. 3.29, where the input voltage and currents are shown as zero. Given that $v_{\text{in}} = v_+ - v_-$ is zero, the voltage at both op-amp inputs is V_S. Since the inverting input is physically connected to the output, V_{out} is also V_S—therefore, unity gain!

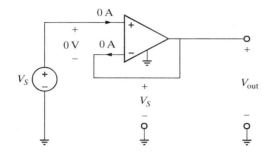

Figure 3.29
*An ideal op-amp
configured as a unity
gain buffer.*

An obvious question at this point is this: If $v_o = v_S$, why not just connect v_S to v_o via two parallel connection wires; why do we need to place an op-amp between them? The answer to this question is fundamental and provides us with some insight that will aid us in circuit analysis and design.

Consider the circuit shown in Fig. 3.30a. In this case v_o is not equal to v_S because of the voltage drop across R_S:

$$v_o = v_S - iR_S$$

However, in Fig. 3.30b, the input current to the op-amp is zero and, therefore, v_S appears at the op-amp input. Since the gain of the op-amp configuration is 1, $v_o = v_S$. In Fig. 3.30a the resistive net-

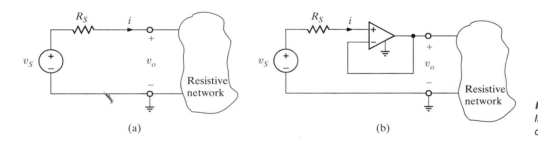

(a) (b)

Figure 3.30
*Illustration of the isolation
capability of a voltage follower.*

work's interaction with the source caused the voltage v_o to be less than v_S. In other words, the re-sistive network loads the source voltage. However, in Fig. 3.30b the op-amp isolates the source from the resistive network, and therefore the voltage follower is referred to as a *buffer amplifier* because it can be used to isolate one circuit from another. The energy supplied to the resistive network in the first case must come from the source v_S, whereas in the second case it comes from the power supplies that supply the amplifier, and little or no energy is drawn from v_S.

As a general rule, when analyzing op-amp circuits we write nodal equations at the op-amp input terminals, using the ideal op-amp model conditions. The following example demonstrates the simplicity of this approach.

LEARNING Example 3.15

Let us determine the gain of the basic inverting op-amp configuration shown in Fig. 3.31.

Figure 3.31
The basic inverting gain stage.

SOLUTION Using the ideal op-amp model conditions, we see that $v_+ = 0$ and, therefore $v_- = 0$. If we now write a node equation at the negative terminal of the op-amp, we obtain

$$\frac{v_S - 0}{R_1} + \frac{v_o - 0}{R_2} = 0$$

or

$$\frac{v_o}{v_S} = -\frac{R_2}{R_1}$$

Note that the gain is a simple resistor ratio. This fact makes the amplifier very versatile in that we can control the gain accurately and alter its value by changing only one resistor. Also, the gain is essentially independent of op-amp parameters. Since the precise values of A, R_i, and R_o are sensitive to such factors as temperature, radiation, and age, their elimination results in a gain that is stable regardless of the immediate environment. Since it is much easier to employ the ideal op-amp model rather than the nonideal model, unless otherwise stated we will use the ideal op-amp assumptions to analyze circuits that contain operational amplifiers.

LEARNING Example 3.16

Consider the op-amp circuit shown in Fig. 3.32. Let us determine an expression for the output voltage.

Figure 3.32 *Differential amplifier operational amplifier circuit.*

SOLUTION The node equation at the inverting terminal is

$$\frac{v_1 - v_-}{R_1} + \frac{v_o - v_-}{R_2} = i_-$$

At the noninverting terminal KCL yields

$$\frac{v_2 - v_+}{R_3} = \frac{v_+}{R_4} + i_+$$

However, $i_+ = i_- = 0$ and $v_+ = v_-$. Substituting these values into the two preceding equations yields

$$\frac{v_1 - v_-}{R_1} + \frac{v_o - v_-}{R_2} = 0$$

and

$$\frac{v_2 - v_-}{R_3} = \frac{v_-}{R_4}$$

Solving these two equations for v_o results in the expression

$$v_o = \frac{R_2}{R_1}\left(1 + \frac{R_1}{R_2}\right)\frac{R_4}{R_3 + R_4}v_2 - \frac{R_2}{R_1}v_1$$

Note that if $R_4 = R_2$ and $R_3 = R_1$, the expression reduces to

$$v_o = \frac{R_2}{R_1}(v_2 - v_1)$$

Therefore, this op-amp can be employed to subtract two input voltages.

LEARNING Example 3.17

The circuit shown in Fig. 3.33a is a precision differential voltage-gain device. It is used to provide a single-ended input for an analog-to-digital converter. We wish to derive an expression for the output of the circuit in terms of the two inputs.

SOLUTION To accomplish this, we draw the equivalent circuit shown in Fig. 3.33b. Recall that the voltage across the input terminals of the op-amp is approximately zero and the currents into the op-amp input terminals are approximately zero. Note that we can write node equations for node voltages v_1 and v_2 in terms of v_o and v_a. Since we are interested in an expression for v_o in terms of the voltages v_1 and v_2, we simply eliminate the v_a terms from the two node equations. The node equations are

$$\frac{v_1 - v_o}{R_2} + \frac{v_1 - v_a}{R_1} + \frac{v_1 - v_2}{R_G} = 0$$

$$\frac{v_2 - v_a}{R_1} + \frac{v_2 - v_1}{R_G} + \frac{v_2}{R_2} = 0$$

Combining the two equations to eliminate v_a, and then writing v_o in terms of v_1 and v_2, yields

$$v_o = (v_1 - v_2)\left(1 + \frac{R_2}{R_1} + \frac{2R_2}{R_G}\right)$$

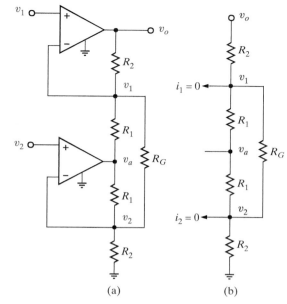

(a)   (b)

Figure 3.33 *Instrumentation amplifier circuit.*

LEARNING EXTENSIONS

E3.14 Find I_o in the network in Fig. E3.14.

ANSWER $I_o = 8.4$ mA.

Figure E3.14

E3.15 Determine the gain of the op-amp circuit in Fig. E3.15.

ANSWER $\dfrac{V_o}{V_S} = 1 + \dfrac{R_2}{R_1}$.

Figure E3.15

E3.16 Determine both the gain and the output voltage of the op-amp configuration shown in Fig. E3.16.

ANSWER $V_o = 0.101$ V, gain = 101.

Figure E3.16

COMPARATORS A comparator, a variant of the op-amp, is designed to compare the noninverting and inverting input voltages. As shown in Fig. 3.34, when the noninverting input voltage is greater, the output goes as high as possible, at or near V_{CC}. On the other hand, if the inverting input voltage is greater, the output goes as low as possible, at or near V_{EE}. Of course, an ideal op-amp can do the same thing, that is, swing the output voltage as far as possible. However, op-amps are not designed to operate with the outputs saturated; whereas comparators are. As a result, comparators are faster and less expensive than op-amps.

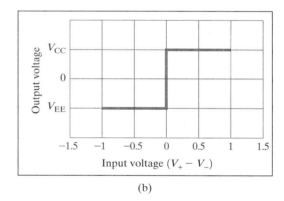

Figure 3.34
(a) An ideal comparator and
(b) its transfer curve.

(a)

(b)

We will present two very different quad comparators in this text, National Semiconductor's LM339 and Maxim's MAX917. Note that the LM339 requires a resistor, called a pull-up resistor, connected between the output pin and V_{CC}. The salient features of these products are listed in Table 3.4. From Table 3.4, it is easy to surmise that the LM339 is a general purpose comparator whereas the MAX917 is intended for low-power applications such as hand-held products.

A common comparator application is the zero-crossing detector, shown in Fig. 3.35a using a LM339 with ±5 V supplies. As seen in Fig. 3.35b, when V_S is positive, V_{out} should be near +5 V and when V_S is negative, V_{out} should be near −5 V. The output changes value on every zero crossing!

Table 3.4 A listing of some of the features of the LM339 and MAX917 comparators

Product	Min. Supply	Max. Supply	Supply Current	Max. Output Current	Typical $R_{PULL-UP}$
LM339	2 V	36 V	3 mA	50 mA	3 kΩ
MAX919	1.8 V	5.5 V	0.8 μA	8 mA	NA

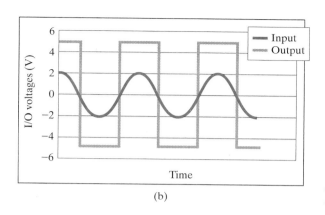

(a)   (b)

Figure 3.35
(a) A zero-crossing detector and (b) the corresponding input/output waveforms.

Learning by Application

At this point, we have a new element, the op-amp, which we can effectively employ in both applications and circuit design. This device is an extremely useful element that vastly expands our capability in these areas. Because of its ubiquitous nature, the addition of the op-amp to our repertoire of circuit elements permits us to deal with a wide spectrum of practical circuits. Thus, we will employ it here, and also use it throughout this text.

LEARNING Example 3.18

The circuit in Fig. 3.36 is an electronic ammeter. It operates as follows: The unknown current, I through R_I produces a voltage, V_I. V_I is amplified by the op-amp to produce a voltage, V_o, which is proportional to I. The output voltage is measured with a simple voltmeter. We want to find the value of R_2 such that 10 V appears at V_o for each milliamp of unknown current.

SOLUTION   Since the current into the op-amp + terminal is zero, the relationship between V_I and I is

$$V_I = IR_I$$

The relationship between the input and output voltages is

$$V_o = V_I \left(1 + \frac{R_2}{R_1} \right)$$

or, solving the equation for V_o/I, we obtain

$$\frac{V_o}{I} = R_I \left(1 + \frac{R_2}{R_1} \right)$$

Using the required ratio V_o/I of 10^4 and resistor values from Fig. 3.36, we can find that

$$R_2 = 9 \text{ k}\Omega$$

Figure 3.36
Electronic ammeter.

Learning by Design

LEARNING Example 3.19

A typical stereo system is shown in block form in Fig. 3.37a. The phonograph output signal is only about 2 μV. The standard input voltage for stereo power amplifiers is about 2 mV. Therefore, the phonograph signal must be amplified by a factor of 1000 before it reaches the poweramp. To accomplish this, a special-purpose amp, called the phono preamp, is used. Stereo manufacturers place them within the preamplifier cabinet, as shown in Fig. 3.37a.

Let us design a phono preamp using an ideal op-amp that has an input resistance of at least 1 MΩ and a voltage gain of 1000. All resistors must be less than 10 MΩ to limit noise.

SOLUTION One possible network is shown in Fig. 3.37b. The input resistance requirement can be easily met with a voltage follower as the first stage of the amplifier. The second stage, or gain stage, can be a noninverting op-amp configuration. We will show in later chapters that the overall voltage gain is the product of the gains of the two stages,

$$A_V = A_{V1}A_{V2} = (1)(1 + (R_2/R_1))$$

To achieve a gain of 1000, we select $R_1 = 1 \text{ k}\Omega$ and $R_2 = 999 \text{ k}\Omega$.

Figure 3.37 *Multistage phonograph amplifier.*

LEARNING Example 3.20

Let us design a temperature sensor that operates from a 3-V supply, and has a visual display consisting of five LEDs—that is, light-emitting diodes. Only one LED should be on at any time, indicating one of the following temperature ranges: less than 65, 65 to 70, 70 to 75, 75 to 80, and greater than 80°F.

SOLUTION In our proposed sensor, shown in Fig. 3.38, resistor R_X and a thermistor, R_T (a temperature sensitive resistor), form a voltage divider to produce the voltage V_T.

$$V_T = 3\left[\frac{R_X}{R_T + R_X}\right] \qquad \textbf{3.6}$$

In the temperature range of interest, a curve fit to a particular commercial thermistor's R-T data yields

$$R_T = 57.45e^{-0.0227T} \qquad \textbf{3.7}$$

with R_T in kΩ and T in degrees Fahrenheit. From Eqs. (3.6) and (3.7), we see that increasing temperature causes R_T to decrease

and V_T to increase. The voltage V_T appears at the unity gain buffer output, where it is divided between R_1, R_2, R_3, and R_4, yielding intermediate voltages V_2, V_3, and V_4. All four comparators and the dc voltage reference, V_{ref}, are contained in the MAX919 package listed in Table 3.4. In this package, $V_{ref} = 1.245$ V.

Based on fundamental comparator operation, when $V_T < V_{ref}$, the output voltage of comparator C_1 will be low, near zero volts. Since $V_4 < V_3 < V_2 < V_T$, all the other comparator output voltages will be low as well. Thus, there is no voltage difference across LED_2, LED_3, LED_4, or LED_5, and these LEDs are off. However, the voltage across LED_1 is not zero. Current will flow from the 3-V supply, through R_{LED} and LED_1, into the output terminal of C_1, through its on-chip circuitry, out of the MAX 919 ground pin and back to the 3-V source, turning on LED_1. This is the desired display for $T < 65°F$.

Then, at exactly 65°F, LED_1 should turn off and LED_2 should turn on. This requires the output of C_1 to go high, near 3 V, while all other comparator outputs remain low. Now, only LED_2 has a nonzero voltage across it. This can be seen in the plots in Fig. 3.39.

Both display scenarios (<65° and 65°-70°) will occur properly if $V_T = V_{ref}$ precisely at $T = 65°F$. Thus, we must find R_X such that $V_T = V_{ref}$ at 65°F. From Eqs. (3.6) and (3.7),

$$1.245 = 3\left[\frac{R_X}{57.45e^{-0.0227(65)} + R_X}\right]$$

which yields $R_X = 9.32 \text{ k}\Omega$. This is the case until T reaches 70°F, where the output of C_2 must go high, turning off LED$_2$ and turning on LED$_3$. Thus, we require $V_2 = V_{ref}$ at exactly $T = 70°F$. Now, $V_4 < V_3 < V_2 = V_{ref}$ and $V_T > V_{ref}$. Repeating this idea at $T = 75$ and 80°F yields the voltage division equations

$$V_2|_{T=70} = 1.245 = \left[\frac{R_2 + R_3 + R_4}{R_\Sigma}\right]V_T|_{T=70}$$

$$V_3|_{T=75} = 1.245 = \left[\frac{R_3 + R_4}{R_\Sigma}\right]V_T|_{T=75}$$

$$V_4|_{T=80} = 1.245 = \left[\frac{R_4}{R_\Sigma}\right]V_T|_{T=80}$$

where $R_\Sigma = R_1 + R_2 + R_3 + R_4$. Arbitrarily selecting R_Σ to be 100 kΩ and using Eqs. (3.6) and (3.7), the required resistor values are $R_4 = 83.11$ kΩ, $R_3 = 5.01$ kΩ, $R_2 = 5.60$ kΩ, and $R_1 = 6.28$ kΩ.

The LEDs employed in this design have a voltage of 2 V when on and a desired current of 1 mA. Consider again the case where LED$_1$ is on; that is, the output of comparator C_1 is zero. By KVL, we have

$$3 = I_{LED}R_{LED} + V_{LED} = (0.001)R_{LED} + 2.0$$

yielding $R_{LED} = 1 \text{ k}\Omega$. Since only one LED is on at any time, there is exactly one resistor, R_{LED}, for every possible display scenario.

Figure 3.38
The temperature sensor schematic diagram.

(a)

(b)

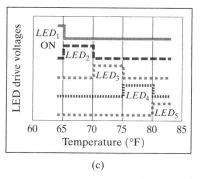

(c)

Figure 3.39
Critical voltages in the temperature sensor of Fig. 3.38. Resistor string voltages (a) are compared to V_{ref} to produce the comparator output voltages in (b). Resistors R_X and R_1–R_4 have been selected such that the comparator outputs change exactly at the desired temperature boundaries. The LEDs drive voltages (c) are such that only one LED is on at any time.

LEARNING Check

Summary

▶ **Nodal analysis for an N-node circuit**

▶ Select one node in the N-node circuit as the reference node. Assume that the node voltage is zero and measure all node voltages with respect to this node.

▶ If only independent current sources are present in the network, write the KCL equations at the $N - 1$ nonreference nodes. If dependent current sources are present, write the KCL equations as is done for networks with only independent current sources; then write the controlling equations for the dependent sources.

▶ If voltage sources are present in the network, they may be connected (1) between the reference node and a nonreference node or (2) between two nonreference nodes. In the former case, if the voltage source is an independent source, then the voltage at one of the nonreference nodes is known. If the source is dependent, it is treated as an independent source when writing the KCL equations, but an additional constraint equation is necessary.

In the latter case, if the source is independent, the voltage between the two nodes is constrained by the value of the voltage source and an equation describing this constraint represents one of the $N - 1$ linearly independent equations required to determine the N-node voltages. The surface of the network described by the constraint equation (i.e., the source and two connecting nodes) is called a supernode. One of the remaining $N - 1$ linearly independent equations is obtained by applying KCL at this supernode. If the voltage source is dependent, it is treated as an independent source when writing the KCL equations, but an additional constraint equation is necessary.

▶ **Loop analysis for an N-loop circuit**

▶ One loop current is assigned to each independent loop in a circuit that contains N independent loops.

▶ If only independent voltage sources are present in the network, write the N linearly independent KVL equations, one for each loop. If dependent voltage sources are present, write the KVL equations as is done for circuits with only independent voltage sources; then write the controlling equations for the dependent sources.

▶ If current sources are present in the network, either of two techniques can be used. In the first case, one loop current is selected to pass through one of the current sources. This is done for each current source in the network. The remaining loop currents (N - the number of current sources) are determined by open-circuiting the current sources in the network and using this modified network to select them. Once all these currents are defined in the original network, the N-loop equations can be written. The second approach is similar to the first with the exception that if two mesh currents pass through a particular current source, a supermesh is formed around this source. The two required equations for the meshes containing this source are the constraint equations for the two mesh currents that pass through the source and the supermesh equation. If dependent current sources are present, the controlling equations for these sources are also necessary.

▶ **Ideal op-amp model** For an ideal op-amp, $i_+ = i_- = 0$ and $v_+ = v_-$. Both nodal and loop analysis are useful in solving circuits containing operational amplifiers.

Problems

SECTION 3.1

3.1 Find I_o in the circuit in Fig. P3.1 using nodal analysis.

Figure P3.1

3.2 Find I_o in the circuit in Fig. P3.2 using nodal analysis.

Figure P3.2

3.3 Find V_2 in the circuit in Fig. P3.3 using nodal analysis.

Figure P3.3

3.4 Use nodal analysis to find V_o in the circuit in Fig. P3.4.

Figure P3.4

3.5 Find I_o in the circuit in Fig. P3.5 using nodal analysis.

Figure P3.5

3.6 Use nodal analysis to find both V_1 and V_o in the circuit in Fig. P3.6.

Figure P3.6

3.7 Find V_1 and V_2 in the circuit in Fig. P3.7 using nodal analysis. Then solve the problem using MATLAB and compare your answers.

Figure P3.7

3.8 Find I_o in the network in Fig. P3.8 using nodal analysis.

Figure P3.8

3.9 Find V_o in the network in Fig. P3.9 using nodal analysis.

Figure P3.9

3.10 Use nodal analysis to find I_o and I_1 in the network in Fig. P3.10.

Figure P3.10

3.11 Find I_o in the circuit in Fig. P3.11 using nodal analysis.

Figure P3.11

3.12 Use nodal analysis to find V_o in the network in Fig. P3.12. Then solve the problem using MATLAB and compare your answers.

Figure P3.12

3.13 Find I_o in the network in Fig. P3.13.

Figure P3.13

3.14 Find V_o in the network in Fig. P3.14.

Figure P3.14

3.15 Use nodal analysis to find V_o in the circuit in Fig. P3.15.

Figure P3.15

3.16 Write the node equations for the circuit in Fig. P3.16 in matrix form, and find all the node voltages using MATLAB.

Figure P3.16

3.17 Find I_1 in the network in Fig. P3.17.

Figure P3.17

3.18 Find I_o in the network in Fig. P3.18.

Figure P3.18

3.19 Find I_o in the circuit in Fig. P3.19.

Figure P3.19

3.20 Find I_o in the network in Fig. P3.20 using nodal analysis.

Figure P3.20

3.21 Use nodal analysis to find V_o in the network in Fig. P3.21. Then solve this problem using MATLAB and compare your answers.

Figure P3.21

3.22 Find V_o in the circuit in Fig. P3.22 using nodal analysis.

Figure P3.22

3.23 Use nodal analysis to find I_o in the network in Fig. P3.23.

Figure P3.23

3.24 Find I_o in the network in Fig. P3.24 using nodal analysis.

Figure P3.24

3.25 Use nodal analysis to find V_o in the circuit in Fig. P3.25.

Figure P3.25

3.26 Find V_o in the circuit in Fig. P3.26 using nodal analysis.

Figure P3.26

3.27 Find V_o in the network in Fig. P3.27 using nodal analysis.

Figure P3.27

3.28 Find V_o in the network in Fig. P3.28 using nodal analysis.

Figure P3.28

3.29 Find V_o in the network in Fig. P3.29 using nodal analysis.

Figure P3.29

3.30 Find I_o in the circuit in Fig. P3.30 using nodal analysis.

Figure P3.30

3.31 Find V_o in the circuit in Fig. P3.31 using nodal analysis.

Figure P3.31

3.32 Use nodal analysis to find V_o in the network in Fig. P3.32.

Figure P3.32

3.33 Find V_o in the circuit in Fig. P3.33 using nodal analysis. Then solve the problem using MATLAB and compare your answers.

Figure P3.33

3.34 Find I_o in the network in Fig. P3.34.

Figure P3.34

3.35 Find V_o in the circuit in Fig. P3.35 using nodal analysis.

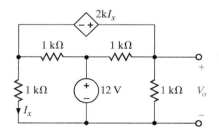

Figure P3.35

3.36 Find I_o in the circuit in Fig. P3.36 using nodal analysis.

Figure P3.36

3.37 Find V_o in the network in Fig. P3.37 using nodal analysis.

Figure P3.37

3.38 Use nodal analysis to find V_o in the network in Fig. P3.38. In addition, determine all branch currents and check KCL at every node.

Figure P3.38

3.39 Use nodal analysis to find V_o in the circuit in Fig. P3.39.

Figure P3.39

3.40 Use nodal analysis to find V_o in the circuit in Fig. P3.40.

Figure. P3.40

3.41 Use MATLAB to find the node voltages in the network in Fig. P3.41.

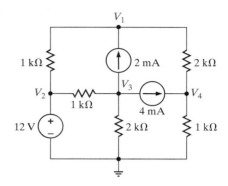

Figure P3.41

SECTION 3.2

3.42 Use mesh equations to find V_o in the circuit in Fig. P3.42.

Figure P3.42

3.44 Use mesh analysis to find V_o in the circuit in Fig. P3.44.

Figure P3.44

3.43 Find V_o in the network in Fig. P3.43 using mesh equations.

Figure P3.43

3.45 Use mesh analysis to find V_o in the network in Fig. P3.45.

Figure P3.45

3.46 Use loop analysis to find V_o in the circuit in Fig. P3.46.

Figure P3.46

3.47 Use loop analysis to find I_o in the circuit in Fig. P3.47.

Figure P3.47

3.48 Use both nodal analysis and mesh analysis to find I_o in the circuit in Fig. P3.48.

Figure P3.48

3.49 Find I_o in the network in Fig. P3.49 using mesh analysis.

Figure P3.49

3.50 Find I_o in the network in Fig. P3.50 using mesh analysis.

Figure P3.50

3.51 Find V_o in the circuit in Fig. P3.51 using mesh analysis.

Figure P3.51

3.52 Use loop analysis to find V_o in the network in Fig. P3.52.

Figure P3.52

3.53 Find I_o in the network in Fig. P3.53 using loop analysis. Then solve the problem using MATLAB and compare your answers.

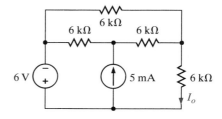

Figure P3.53

3.54 Use loop analysis to find I_o in the circuit in Fig. P3.54.

Figure P3.54

3.55 Find V_o in the network in Fig. P3.55 using both mesh and nodal analyses.

Figure P3.55

3.56 Use loop analysis to find V_o in the circuit in Fig. P3.56.

Figure P3.56

3.57 Use loop analysis to find I_o in the network in Fig. P3.57.

Figure P3.57

3.58 Use loop analysis to find V_o in the network in Fig. P3.58.

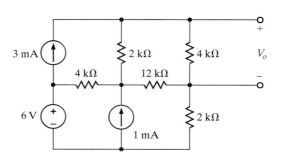

Figure P3.58

3.59 Find V_o in the network in Fig. P3.59.

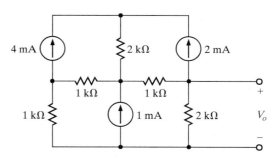

Figure P3.59

3.60 Find V_o in the circuit in Fig. P3.60.

Figure P3.60

3.61 Find I_o in the circuit in Fig. P3.61.

Figure P3.61

3.62 Use loop analysis to find V_o in the network in Fig. P3.62.

Figure P3.62

3.63 Use mesh analysis to find V_o in the circuit in Fig. P3.63.

Figure P3.63

3.64 Use loop analysis to find V_o in the circuit in Fig. P3.64.

Figure P3.64

3.65 Find V_o in the circuit in Fig. P3.65 using mesh analysis.

Figure P3.65

3.66 Use both nodal analysis and mesh analysis to find V_o in the circuit in Fig. P3.66.

Figure P3.66

3.67 Using mesh analysis, find V_o in the circuit in Fig. P3.67.

Figure P3.67

3.68 Find V_o in the network in Fig. P3.68.

Figure P3.68

3.69 Use MATLAB to find the mesh currents in the network in Fig. P3.69.

Figure P3.69

SECTION 3.3 ────────────────────────────────────

Assume that all op-amps in this section are ideal.

3.70 Find V_o in the circuit in Fig. P3.70.

Figure P3.70

3.71 Find V_o in the network in Fig. P3.71 and explain what effect R_1 has on the output.

Figure P3.71

3.72 Find V_o in the network in Fig. P3.72.

Figure P3.72

3.74 Find V_o in the circuit in Fig. P3.74.

Figure P3.74

3.73 The network in Fig. P3.73 is a current-to-voltage converter or transresistance amplifier. Find v_o/i_S for this network.

Figure P3.73

3.75 Find V_o in the circuit in Fig. P3.75.

Figure P3.75

Typical Problems Found on the FE Exam

3FE-1 Find V_o in the circuit in Fig. 3PFE-1.

Figure 3PFE-1

3FE-2 Determine the power dissipated in the 6-ohm resistor in the network in Fig. 3PFE-2.

Figure 3PFE-2

3FE-3 Find the current I_x in the 4-ohm resistor in the circuit in Fig. 3PFE-3.

Figure 3PFE-3

3FE-4 Determine the voltage V_o in the circuit in Fig. 3PFE-4.

Figure 3PFE-4

3FE-5 Given the summing amplifier shown in Fig. 3PFE-5, select the values of R_2 that will produce an output voltage of -3 V.

Figure 3PFE-5

3FE-6 Determine the output voltage V_o of the summing op-amp circuit shown in Fig. 3PFE-6.

Figure 3PFE-6

Additional Analysis Techniques

4

At this point, we have mastered the ability to solve networks containing both independent and dependent sources using either nodal or loop analysis. In this chapter we introduce several new analysis techniques that bolster our arsenal of circuit analysis tools. We will find that in some situations these techniques lead to a quick solution and in other cases they do not. However, these new techniques in many cases do provide an insight into the circuit's operation that cannot be gained from a nodal or loop analysis.

In many practical situations we are interested in the analysis of some portion of a much larger network. If we can model the remainder of the network with a simple equivalent circuit, then our task will be much simpler. For example, consider the problem of analyzing some simple electronic device that is connected to the ac wall plug in our house. In this case the complete circuit includes not only the electronic device but the utility's power grid, which is connected to the device through the circuit breakers in the home. However, if we can accurately model everything outside the device with a simple equivalent circuit, then our analysis will be tractable. Two of the theorems that we present in this chapter will permit us to do just that.

4.1 Introduction

Before introducing additional analysis techniques, let us review some of the topics we have used either explicitly or implicitly in our analyses thus far.

EQUIVALENCE Table 4.1 is a short compendium of some of the equivalent circuits that have been employed in our analyses. This listing serves as a quick review as we begin to look at other techniques that can be used to find a specific voltage or current somewhere in a network and provide additional insight into the network's operation. In addition to the forms listed in the table, it is important to note that a series connection of current sources or a parallel

Table 4.1 Equivalent circuit forms

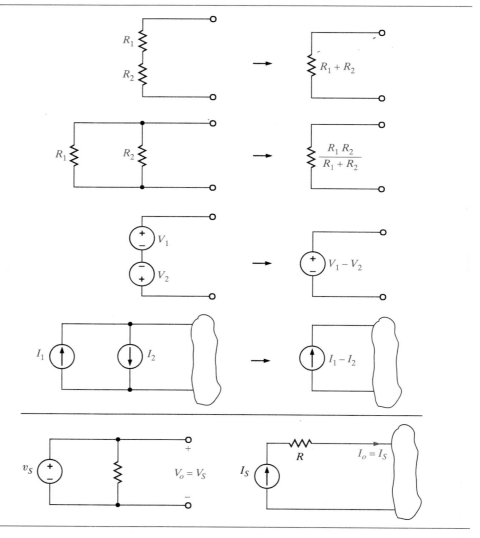

connection of voltage sources is forbidden unless the sources are pointing in the same direction and have exactly the same values.

LINEARITY All the circuits we have analyzed thus far have been linear circuits. Most of the circuits we will analyze in the remainder of the book will also be linear circuits, and any deviation from this type of network will be specifically identified as such.

Linearity requires both additivity and homogeneity (scaling). It can be shown [see the previous edition of this book] that the circuits that we are examining satisfy this important property. The following example illustrates one way in which this property can be used.

LEARNING Example 4.1

For the circuit shown in Fig. 4.1, we wish to determine the output voltage V_{out}. However, rather than approach the problem in a straightforward manner and calculate I_o, then I_1, then I_2, and so on, we will use linearity and simply assume that the output voltage is $V_{out} = 1$ V. This assumption will yield a value for the source voltage. We will then use the actual value of the source voltage and linearity to compute the actual value of V_{out}.

Figure 4.1 Circuit used in Example 4.1.

SOLUTION If we assume that $V_{out} = V_2 = 1$ V, then

$$I_2 = \frac{V_2}{2k} = 0.5 \text{ mA}$$

V_1 can then be calculated as

$$V_1 = 4kI_2 + V_2$$
$$= 3 \text{ V}$$

Hence,

$$I_1 = \frac{V_1}{3k} = 1 \text{ mA}$$

Now, applying KCL,

$$I_o = I_1 + I_2 = 1.5 \text{ mA}$$

Then

$$V_o = 2kI_o + V_1$$
$$= 6 \text{ V}$$

Therefore, the assumption that $V_{out} = 1$ V produced a source voltage of 6 V. However, since the actual source voltage is 12 V, the actual output voltage is 1 V(12/6) = 2 V.

E4.1 Use linearity and the assumption that $I_o = 1$ mA to compute the correct current I_o in the circuit in Fig. E4.1 if $I = 6$ mA.

ANSWER $I_o = 3$ mA.

Figure E4.1

4.2 Superposition

To provide motivation for this subject, let us examine a simple circuit in which two sources contribute to the current in the network.

LEARNING Example 4.2

Consider the circuit in Fig. 4.2a in which the actual values of the voltage sources are left unspecified. Let us use this network to examine the concept of superposition.

SOLUTION The mesh equations for this network are

$$6ki_1(t) - 3ki_2(t) = v_1(t)$$

$$-3ki_1(t) + 9ki_2(t) = -v_2(t)$$

Solving these equations for $i_1(t)$ yields

$$i_1(t) = \frac{v_1(t)}{5k} - \frac{v_2(t)}{15k}$$

In other words, the current $i_1(t)$ has a component due to $v_1(t)$ and a component due to $v_2(t)$. In view of the fact that $i_1(t)$ has two components, one due to each independent source, it would be interesting to examine what each source acting alone would contribute to $i_1(t)$. For $v_1(t)$ to act alone, $v_2(t)$ must be zero. As we pointed out in Chapter 2, $v_2(t) = 0$ means that the source $v_2(t)$ is replaced with a short circuit. Therefore, to determine the value of $i_1(t)$ due to $v_1(t)$ only, we employ the circuit in Fig. 4.2b and refer to this value of $i_1(t)$ as $i_1'(t)$.

$$i_1'(t) = \frac{v_1(t)}{3k + \dfrac{(3k)(6k)}{3k + 6k}} = \frac{v_1(t)}{5k}$$

Let us now determine the value of $i_1(t)$ due to $v_2(t)$ acting alone and refer to this value as $i_1''(t)$. Using the network in Fig. 4.2c,

$$i_2''(t) = -\frac{v_2(t)}{6k + \dfrac{(3k)(3k)}{3k + 3k}} = \frac{-2v_2(t)}{15k}$$

Then, using current division, we obtain

$$i_1''(t) = \frac{-2v_2(t)}{15k}\left(\frac{3k}{3k + 3k}\right) = \frac{-v_2(t)}{15k}$$

Now, if we add the values of $i_1'(t)$ and $i_1''(t)$, we obtain the value computed directly; that is,

$$i_1(t) = i_1'(t) + i_1''(t) = \frac{v_1(t)}{5k} - \frac{v_2(t)}{15k}$$

Note that we have *superposed* the value of $i_1'(t)$ on $i_1''(t)$, or vice versa, to determine the unknown current.

Figure 4.2 Circuits used to illustrate superposition.

(a)   (b)   (c)

What we have demonstrated in Example 4.2 is true in general for linear circuits and is a direct result of the property of linearity. *The principle of superposition*, which provides us with this ability to reduce a complicated problem to several easier problems—each containing only a single independent source—states that

> In any linear circuit containing multiple independent sources, the current or voltage at any point in the network may be calculated as the algebraic sum of the individual contributions of each source acting alone.

When determining the contribution due to an independent source, any remaining voltage sources are made zero by replacing them with short circuits, and any remaining current sources are made zero by replacing them with open circuits.

Although superposition can be used in linear networks containing dependent sources, it is not useful in this case since the dependent source is never made zero.

It is interesting to note that, as the previous example indicates, superposition provides some insight in determining the contribution of each source to the variable under investigation.

We will now demonstrate superposition with two examples and then provide a problem-solving strategy for the use of this technique. For purposes of comparison, we will also solve the networks using both node and loop analyses. Furthermore, we will employ these same networks when demonstrating subsequent techniques, if applicable.

LEARNING Example 4.3

Let us use superposition to find V_o in the circuit in Fig. 4.3a.

(a)

(b)

(c)

(d)

Figure 4.3   *Circuits used in Example 4.3.*

SOLUTION   The contribution of the 2-mA source to the output voltage is found from the network in Fig. 4.3b, using current division

$$I_o = (2 \times 10^{-3})\left(\frac{1k + 2k}{1k + 2k + 6k}\right) = \frac{2}{3}\,\text{mA}$$

and

$$V'_o = I_o(6k) = 4\,\text{V}$$

The contribution of the 3-V source to the output voltage is found from the circuit in Fig. 4.3c. Using voltage division,

$$V''_o = 3\left(\frac{6k}{1k + 2k + 6k}\right)$$

$$= 2\,\text{V}$$

Therefore,

$$V_o = V'_o + V''_o = 6\,\text{V}$$

Although we used two separate circuits to solve the problem, both were very simple.

If we use nodal analysis and Fig. 4.3a to find V_o and recognize that the 3-V source and its connecting nodes form a supernode, V_o can be found from the node equation

$$\frac{V_o - 3}{1k + 2k} - 2 \times 10^{-3} + \frac{V_o}{6k} = 0$$

which yields $V_o = 6$ V. In addition, loop analysis applied as shown in Fig. 4.3d produces the equations

$$I_1 = -2 \times 10^{-3}$$

and

$$3k(I_1 + I_2) - 3 + 6kI_2 = 0$$

which yield $I_2 = 1$ mA and hence $V_o = 6$ V.

LEARNING by Doing

D 4.1   Use superposition to find V_o in the following network.

ANSWER $V_o = V'_o + V''_o = (9 - 3) = 6$ V

D 4.2 Use superposition to find I_o in the following network.

ANSWER

$I_o = I'_o + I''_o = (8 - 2) = 6$ mA

LEARNING Example 4.4

Consider now the network in Fig. 4.4a. Let us use superposition to find V_o.

SOLUTION The contribution of the 6-V source to V_o is found from the network in Fig. 4.4b, which is redrawn in Fig. 4.4c.

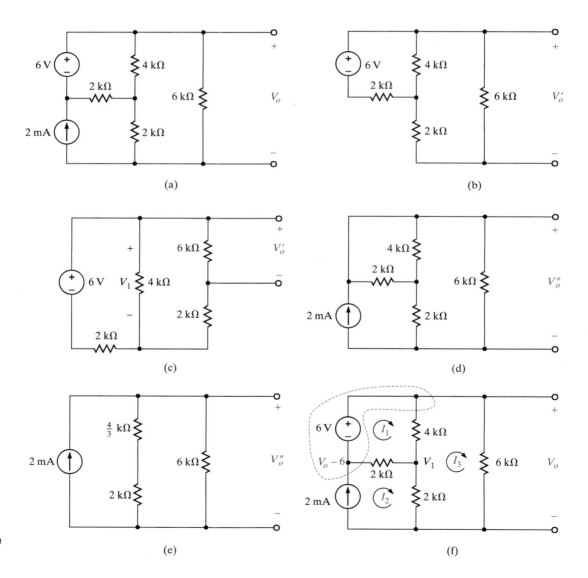

Figure 4.4
Circuits used in
Example 4.4.

The $2 \text{ k}\Omega + 6 \text{ k}\Omega = 8\text{-k}\Omega$ resistor and $4\text{-k}\Omega$ resistor are in parallel, and their combination is an $8/3\text{-k}\Omega$ resistor. Then, using voltage division,

$$V_1 = 6 \left(\frac{\frac{8}{3}\text{k}}{\frac{8}{3}\text{k} + 2\text{k}} \right) = \frac{24}{7} \text{ V}$$

Applying voltage division again,

$$V'_o = V_1 \left(\frac{6\text{k}}{6\text{k} + 2\text{k}} \right) = \frac{18}{7} \text{ V}$$

The contribution of the 2-mA source is found from Fig. 4.4d, which is redrawn in Fig. 4.4e. V''_o is simply equal to the product of the current source and the parallel combination of the resistors; that is,

$$V''_o = \left(2 \times 10^{-3} \right) \left(\frac{10}{3}\text{k}//6\text{k} \right) = \frac{30}{7} \text{ V}$$

Then

$$V_o = V'_o + V''_o = \frac{48}{7} \text{ V}$$

A nodal analysis of the network can be performed using Fig. 4.4f. The equation for the supernode is

$$-2 \times 10^{-3} + \frac{(V_o - 6) - V_1}{2\text{k}} + \frac{V_o - V_1}{4\text{k}} + \frac{V_o}{6\text{k}} = 0$$

The equation for the node labeled V_1 is

$$\frac{V_1 - V_o}{4\text{k}} + \frac{V_1 - (V_o - 6)}{2\text{k}} + \frac{V_1}{2\text{k}} = 0$$

Solving these two equations, which already contain the constraint equation for the supernode, yields $V_o = 48/7$ V.

Once again, referring to the network in Fig. 4.4f, the mesh equations for the network are

$$-6 + 4\text{k}(I_1 - I_3) + 2\text{k}(I_1 - I_2) = 0$$

$$I_2 = 2 \times 10^{-3}$$

$$2\text{k}(I_3 - I_2) + 4\text{k}(I_3 - I_1) + 6\text{k}I_3 = 0$$

Solving these equations, we obtain $I_3 = 8/7$ mA and, hence, $V_o = 48/7$ V.

Problem-Solving Strategy
Applying Superposition

▶ In a network containing multiple independent sources, each source can be applied independently with the remaining sources turned off.

▶ To turn off a voltage source, replace it with a short circuit, and to turn off a current source, replace it with an open circuit.

▶ When the individual sources are applied to the circuit, all the circuit laws and techniques we have learned, or will soon learn, can be applied to obtain a solution.

▶ The results obtained by applying each source independently are then added together algebraically to obtain a solution.

Superposition can be applied to a circuit with any number of dependent and independent sources. In fact, superposition can be applied to such a network in a variety of ways. For example, a circuit with three independent sources can be solved using each source acting alone, as we have just demonstrated, or we could use two at a time and sum the result with that obtained from the third acting alone. In addition, the independent sources do not have to assume their actual value or zero. However, it is mandatory that the sum of the different values chosen add to the total value of the source.

Superposition is a fundamental property of linear equations and, therefore, can be applied to any effect that is linearly related to its cause. In this regard it is important to point out that although superposition applies to the current and voltage in a linear circuit, it cannot be used to determine power because power is a nonlinear function.

E4.2 Compute V_o in the circuit in Fig. E4.2 using superposition.

ANSWER $V_o = \dfrac{4}{3}$ V.

Figure E4.2

4.3 Thévenin's and Norton's Theorems

Thus far we have presented a number of techniques for circuit analysis. At this point we will add two theorems to our collection of tools that will prove to be extremely useful. The theorems are named after their authors, M. L. Thévenin, a French engineer, and E. L. Norton, a scientist formerly with Bell Telephone Laboratories.

Suppose that we are given a circuit and that we wish to find the current, voltage, or power that is delivered to some resistor of the network, which we will call the load. *Thévenin's theorem* tells us that we can replace the entire network, exclusive of the load, by an equivalent circuit that contains only an independent voltage source in series with a resistor in such a way that the current–voltage relationship at the load is unchanged. *Norton's theorem* is identical to the preceding statement except that the equivalent circuit is an independent current source in parallel with a resistor.

Note that this is a very important result. It tells us that if we examine any network from a pair of terminals, we know that with respect to those terminals, the entire network is equivalent to a simple circuit consisting of an independent voltage source in series with a resistor or an independent current source in parallel with a resistor.

In developing the theorems, we will assume that the circuit shown in Fig. 4.5a can be split into two parts, as shown in Fig. 4.5b. In general, circuit B is the load and may be linear or nonlinear. Circuit A is the balance of the original network exclusive of the load and must be linear. As such, circuit A may contain independent sources, dependent sources and resistors, or any other linear element. We require, however, that a dependent source and its control variable appear in the same circuit.

Circuit A delivers a current i to circuit B and produces a voltage v_o across the input terminals of circuit B. From the standpoint of the terminal relations of circuit A, we can replace circuit B by a voltage source of v_o volts (with the proper polarity), as shown in Fig. 4.5c. Since the terminal voltage is unchanged and circuit A is unchanged, the terminal current i is unchanged.

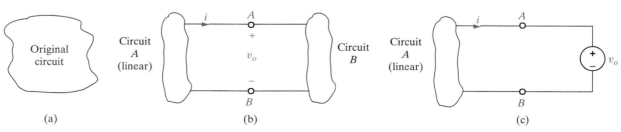

(a) (b) (c)

Figure 4.5
Concepts used to develop Thévenin's theorem.

Now, applying the principle of superposition to the network shown in Fig. 4.5c, the total current i shown in the figure is the sum of the currents caused by all the sources in circuit A and the source v_o that we have just added. Therefore, via superposition the current i can be written

$$i = i_o + i_{sc} \qquad\qquad \textbf{4.1}$$

where i_o is the current due to v_o with all independent sources in circuit A made zero (i.e., voltage sources replaced by short circuits and current sources replaced by open circuits), and i_{sc} is the short-circuit current due to all sources in circuit A with v_o replaced by a short circuit.

The terms i_o and v_o are related by the equation

$$i_o = \frac{-v_o}{R_{\text{Th}}} \qquad\qquad \textbf{4.2}$$

where R_{Th} is the equivalent resistance looking back into circuit A from terminals A-B with all independent sources in circuit A made zero.

Substituting Eq. (4.2) into Eq. (4.1) yields

$$i = -\frac{v_o}{R_{\text{Th}}} + i_{sc} \qquad\qquad \textbf{4.3}$$

This is a general relationship and, therefore, must hold for any specific condition at terminals A-B. As a specific case, suppose that the terminals are open circuited. For this condition, $i = 0$ and v_o is equal to the open-circuit voltage v_{oc}. Thus, Eq. (4.3) becomes

$$i = 0 = \frac{-v_{oc}}{R_{\text{Th}}} + i_{sc} \qquad\qquad \textbf{4.4}$$

Hence,

$$v_{oc} = R_{\text{Th}} i_{sc} \qquad\qquad \textbf{4.5}$$

This equation states that the open-circuit voltage is equal to the short-circuit current times the equivalent resistance looking back into circuit A with all independent sources made zero. We refer to R_{Th} as the Thévenin equivalent resistance.

Substituting Eq. (4.5) into Eq. (4.3) yields

$$i = \frac{-v_o}{R_{\text{Th}}} + \frac{v_{oc}}{R_{\text{Th}}}$$

or

$$v_o = v_{oc} - R_{\text{Th}} i \qquad\qquad \textbf{4.6}$$

Let us now examine the circuits that are described by these equations. The circuit represented by Eq. (4.6) is shown in Fig. 4.6a. The fact that this circuit is equivalent at terminals A-B to circuit A in Fig. 4.5 is a statement of *Thévenin's theorem*. The circuit represented by Eq. (4.3) is shown in Fig. 4.6b. The fact that this circuit is equivalent at terminals A-B to circuit A in Fig. 4.5 is a statement of *Norton's theorem*.

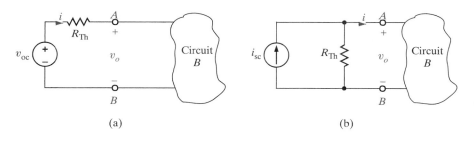

Figure 4.6
Thévenin and Norton equivalent circuits.

The relationships specified in Fig. 4.6 and Eq. (4.5) have added significance because they represent what is called a *source transformation* or *source exchange*. What these relationships tell us is that if we have embedded within a network a current source i in parallel with a resistor R, we can replace this combination with a voltage source of value $v = iR$ in series with the resistor R. The reverse is also true; that is, a voltage source v in series with a resistor R can be replaced with a current source of value $i = v/R$ in parallel with the resistor R. Parameters within the circuit (e.g., an output voltage) are unchanged under these transformations.

We must emphasize that the two equivalent circuits in Fig. 4.6 *are equivalent only at the two external nodes*. For example, if we disconnect circuit B from both networks in Fig. 4.6, the equivalent circuit in Fig. 4.6b dissipates power, but the one in Fig. 4.6a does not.

LEARNING Example 4.5

We will now demonstrate how to find V_o in the circuit in Fig. 4.7a using the repeated application of source transformation.

SOLUTION If we begin at the left end of the network in Fig. 4.7a, the series combination of the 12-V source and 3-kΩ resistor is converted to a 4-mA current source in parallel with the 3-kΩ resistor. If we combine this 3-kΩ resistor with the 6-kΩ resistor, we obtain the circuit in Fig. 4.7b. Note that at this point we have eliminated one circuit element. Continuing the reduction, we convert the 4-mA source and 2-kΩ resistor into an 8-V source in series with this same 2-kΩ resistor. The two 2-kΩ resistors that are in series are now combined to produce the network in Fig. 4.7c. If we now convert the combination of the 8-V source

and 4-kΩ resistor into a 2-mA source in parallel with the 4-kΩ resistor and combine the resulting current source with the other 2-mA source, we arrive at the circuit shown in Fig. 4.7d. At this point, we can simply apply current division to the two parallel resistance paths and obtain

$$I_o = (4 \times 10^{-3})\left(\frac{4k}{4k + 4k + 8k}\right) = 1 \text{ mA}$$

and hence,

$$V_o = (1 \times 10^{-3})(8k) = 8 \text{ V}$$

The reader is encouraged to consider the ramifications of working this problem using any of the other techniques we have presented.

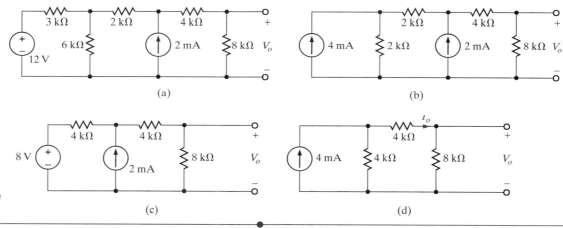

Figure 4.7
Circuits used in
Example 4.5.

Note that this systematic, sometimes tedious, transformation allows us to reduce the network methodically to a simpler equivalent form with respect to some other circuit element. However, we should also realize that this technique is worthless for circuits of the form shown in Fig. 4.4. Furthermore, although applicable to networks containing dependent sources, it is not as useful as other techniques, and care must be taken not to transform the part of the circuit that contains the control variable.

Having demonstrated that there is an inherent relationship between the Thévenin equivalent circuit and the Norton equivalent circuit, we now proceed to apply these two important and useful theorems. The manner in which these theorems are applied depends on the structure of the original network under investigation. For example, if only independent sources are present, we can calculate the open-circuit voltage or short-circuit current and the Thévenin equivalent resistance. However, if dependent sources are also present, the Thévenin equivalent will be determined by calculating v_{oc} and i_{sc}, since this is normally the best approach for determining R_{Th} in a network containing dependent sources. Finally, if circuit A contains no *independent* sources, then both v_{oc} and i_{sc} will necessarily be zero. (Why?) Thus, we cannot determine R_{Th} by v_{oc}/i_{sc}, since the ratio is indeterminate. We must look for another approach. Notice that if $v_{oc} = 0$, then the equivalent circuit is merely the unknown resistance R_{Th}. If we apply an external source to circuit A—a test source v_t—and determine the current, i_t, which flows into circuit A from v_t, then R_{Th} can be determined from $R_{Th} = v_t/i_t$. Although the numerical value of v_t need not be specified, we could let $v_t = 1$ V and then $R_{Th} = 1/i_t$. Alternatively, we could use a current source as a test source and let $i_t = 1$ A; then $v_t = (1)R_{Th}$.

Before we begin our analysis of several examples that will demonstrate the utility of these theorems, remember that these theorems, in addition to an alternate method of attack, often permit us to solve several small problems rather than one large one. They allow us to replace a network, no matter how large, *at a pair of terminals* with a Thévenin or Norton equivalent circuit. In fact, we could represent the entire U.S. power grid at a pair of terminals with one of the equivalent circuits. Once this is done we can quickly analyze the effect of different loads on a network. Thus, these theorems provide us with additional insight into the operation of a specific network.

CIRCUITS CONTAINING ONLY INDEPENDENT SOURCES

LEARNING Example 4.6

Let us use Thévenin's and Norton's theorems to find V_o in the network in Example 4.3.

SOLUTION The circuit is redrawn in Fig. 4.8a. To determine the Thévenin equivalent, we break the network at the 6-kΩ load as shown in Fig. 4.8b. KVL indicates that the open-circuit voltage, V_{oc}, is equal to 3 V plus the voltage V_1, which is the voltage across the current source. The 2 mA from the current source flows through the two resistors (where else could it possibly go!) and, therefore, $V_1 = (2 \times 10^{-3})(1k + 2k) = 6$ V. Therefore, $V_{oc} = 9$ V. By making both sources zero, we can find the Thévenin equivalent resistance, R_{Th}, using the circuit in Fig. 4.8c. Obviously, $R_{Th} = 3$ kΩ.

Now our Thévenin equivalent circuit, consisting of V_{oc} and R_{Th}, is connected back to the original terminals of the load, as shown in Fig. 4.8d. Using a simple voltage divider, we find that $V_o = 6$ V.

To determine the Norton equivalent circuit at the terminals of the load, we must find the short-circuit current as shown in Fig. 4.8e. Note that the short circuit causes the 3-V source to be directly across (i.e., in parallel with) the resistors and the current source. Therefore, $I_1 = 3/(1k + 2k) = 1$ mA. Then, using KCL, $I_{sc} = 3$ mA. We have already determined R_{Th} and, therefore, connecting the Norton equivalent to the load results in the circuit in Fig. 4.8f. Hence, V_o is equal to the source current multiplied by the parallel resistor combination, which is 6 V.

Figure 4.8 Circuits used in Example 4.6.

Consider for a moment some salient features of this example. Note that in applying the theorems there is no point in breaking the network to the left of the 3-V source, since the resistors in parallel with the current source are already a Norton equivalent, which can be immediately changed to a Thévenin equivalent using source transformation! Furthermore, once the network has been simplified using a Thévenin or Norton equivalent, we simply have a new network with which we can apply the theorems again. The following example illustrates this approach.

LEARNING Example 4.7

Let us use Thévenin's theorem to find V_o in the network in Fig. 4.7a, which is redrawn in Fig. 4.9a.

SOLUTION If we break the network to the left of the current source, the open-circuit voltage V_{oc_1} is as shown in Fig. 4.9b. Since there is no current in the 2-kΩ resistor and therefore no voltage across it, V_{oc_1} is equal to the voltage across the 6-kΩ resistor, which can be determined by voltage division as

$$V_{oc_1} = 12\left(\frac{6k}{6k + 3k}\right) = 8 \text{ V}$$

The Thévenin equivalent resistance, R_{Th_1}, is found from Fig. 4.9c as

$$R_{Th_1} = 2k + \frac{(3k)(6k)}{3k + 6k} = 4 \text{ k}\Omega$$

Connecting this Thévenin equivalent back to the original network produces the circuit shown in Fig. 4.9d. We can now apply Thévenin's theorem again, and this time we break the network to the right of the current source as shown in Fig. 4.9e. In this case V_{oc_2} is

$$V_{oc_2} = (2 \times 10^{-3})(4k) + 8 = 16 \text{ V}$$

and R_{Th_2} obtained from Fig. 4.9f is 4 kΩ. Connecting this Thévenin equivalent to the remainder of the network produces the circuit shown in Fig. 4.9g. Simple voltage division applied to this final network yields $V_o = 8$ V. Norton's theorem can be applied in a similar manner to solve this network; however, we save that solution as an exercise.

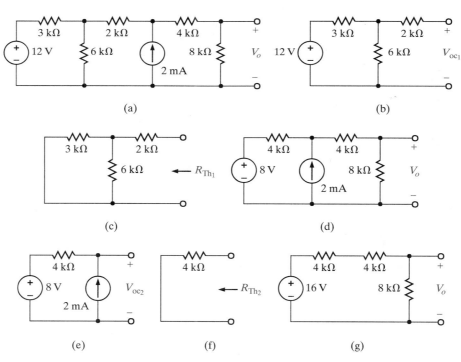

(a)

(b)

(c)

(d)

(e)

(f)

(g)

Figure 4.9 Circuits used in Example 4.7.

LEARNING by Doing

D 4.3 Find V_o in the following network using Thévenin's theorem.

ANSWER

LEARNING Example 4.8

It is instructive to examine the use of Thévenin's and Norton's theorems in the solution of the network in Fig. 4.4a, which is redrawn in Fig. 4.10a.

SOLUTION If we break the network at the 6-kΩ load, the open-circuit voltage is found from Fig. 4.10b. The equations for the mesh currents are

$$-6 + 4kI_1 + 2k(I_1 - I_2) = 0$$

and

$$I_2 = 2 \times 10^{-3}$$

from which we easily obtain $I_1 = 5/3$ mA. Then, using KVL, V_{oc} is

$$V_{oc} = 4kI_1 + 2kI_2$$

$$= 4k\left(\frac{5}{3} \times 10^{-3}\right) + 2k(2 \times 10^{-3})$$

$$= \frac{32}{3} \text{ V}$$

R_{Th} is derived from Fig. 4.10c and is

$$R_{Th} = (2k//4k) + 2k = \frac{10}{3} \text{ k}\Omega$$

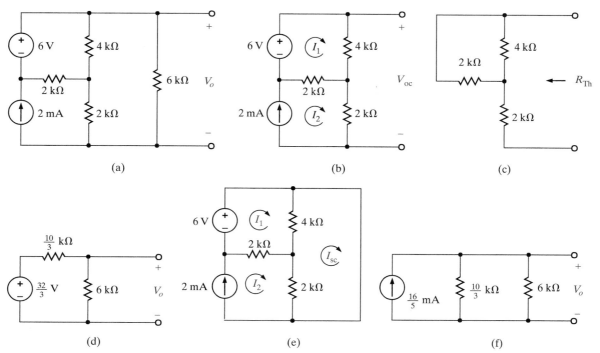

(a)

(b)

(c)

(d)

(e)

(f)

Figure 4.10 Circuits used in Example 4.8.

(continued)

Attaching the Thévenin equivalent to the load produces the network in Fig. 4.10d. Then using voltage division, we obtain

$$V_o = \frac{32}{3}\left(\frac{6k}{6k + \frac{10}{3}k}\right)$$

$$= \frac{48}{7}\ \text{V}$$

In applying Norton's theorem to this problem, we must find the short-circuit current shown in Fig. 4.10e. At this point the quick-thinking reader stops immediately! Three mesh equations applied to the original circuit will immediately lead to the solution, but the three mesh equations in the circuit in Fig. 4.10e will provide only part of the answer, specifically the short-circuit current. Some-

times the use of the theorems is more complicated than a straight-forward attack using node or loop analysis. This would appear to be one of those situations. Interestingly, it is not. We can find I_{sc} from the network in Fig. 4.10e without using the mesh equations. The technique is simple, but a little tricky, and so we ignore it at this time. Having said all these things, let us now finish what we have started. The mesh equations for the network in Fig. 4.10e are

$$-6 + 4\,k(I_1 - I_{sc}) + 2k(I_1 - 2 \times 10^{-3}) = 0$$

$$2k(I_{sc} - 2 \times 10^{-3}) + 4k(I_{sc} - I_1) = 0$$

where we have incorporated the fact that $I_2 = 2 \times 10^{-3}$ A. Solving these equations yields $I_{sc} = 16/5$ mA. R_{Th} has already been determined in the Thévenin analysis. Connecting the Norton equivalent to the load results in the circuit in Fig. 4.10f. Applying Ohm's law to this circuit yields $V_o = 48/7$ V.

LEARNING EXTENSIONS

E4.3 Use Thévenin's theorem to find V_o in the network in Fig. E4.3.

ANSWER $V_o = -3$V.

Figure E4.3

E4.4 Find V_o in the circuit in Fig. E4.2 using (a) source exchange and (b) both Thévenin's and Norton's theorems. When deriving the Norton equivalent circuit, break the network to the left of the 4-kΩ resistor. Why?

ANSWER $V_o = \dfrac{4}{3}$ V.

CIRCUITS CONTAINING ONLY DEPENDENT SOURCES As we have stated earlier, the Thévenin or Norton equivalent of a network containing only dependent sources is R_{Th}. The following examples will serve to illustrate how to determine this Thévenin equivalent resistance.

LEARNING Example 4.9

We wish to determine the Thévenin equivalent of the network in Fig. 4.11a at the terminals A-B.

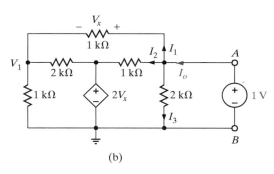

Figure 4.11
Networks employed in
Example 4.9.

(a)   (b)

SOLUTION Our approach to this problem will be to apply a 1-V source at the terminals as shown in Fig. 4.11b and then compute the current I_o and $R_{Th} = 1/I_o$.

The equations for the network in Fig. 4.11b are as follows. KVL around the outer loop specifies that

$$V_1 + V_x = 1$$

The KCL equation at the node labeled V_1 is

$$\frac{V_1}{1k} + \frac{V_1 - 2V_x}{2k} + \frac{V_1 - 1}{1k} = 0$$

Solving the equations for V_x yields $V_x = 3/7$ V. Knowing V_x, we can compute the currents I_1, I_2, and I_3. Their values are

$$I_1 = \frac{V_x}{1k} = \frac{3}{7} \text{ mA}$$

$$I_2 = \frac{1 - 2V_x}{1k} = \frac{1}{7} \text{ mA}$$

$$I_3 = \frac{1}{2k} = \frac{1}{2} \text{ mA}$$

Therefore,

$$I_o = I_1 + I_2 + I_3$$
$$= \frac{15}{14} \text{ mA}$$

and

$$R_{Th} = \frac{1}{I_o}$$
$$= \frac{14}{15} \text{ k}\Omega$$

LEARNING Example 4.10

Let us determine R_{Th} at the terminals A-B for the network in Fig. 4.12a.

(a)

(b)

Figure 4.12 Networks used in Example 4.10.

SOLUTION Our approach to this problem will be to apply a 1-mA current source at the terminals A-B and compute the terminal voltage V_2 as shown in Fig. 4.12b. Then $R_{Th} = V_2/0.001$.

The node equations for the network are

$$\frac{V_1 - 2000I_x}{2k} + \frac{V_1}{1k} + \frac{V_1 - V_2}{3k} = 0$$

$$\frac{V_2 - V_1}{3k} + \frac{V_2}{2k} = 1 \times 10^{-3}$$

and

$$I_x = \frac{V_1}{1k}$$

Solving these equations yields

$$V_2 = \frac{10}{7} \text{ V}$$

and hence,

$$R_{Th} = \frac{V_2}{1 \times 10^{-3}}$$
$$= \frac{10}{7} \text{ k}\Omega$$

CIRCUITS CONTAINING BOTH INDEPENDENT AND DEPENDENT SOURCES In these types of circuits we must calculate both the open-circuit voltage and short-circuit current to calculate the Thévenin equivalent resistance. Furthermore, we must remember that we cannot split the dependent source and its controlling variable when we break the network to find the Thévenin or Norton equivalent.

We now illustrate this technique with a circuit containing a current-controlled voltage source.

LEARNING Example 4.11

Let us use Thévenin's theorem to find V_o in the network in Fig. 4.13a.

SOLUTION To begin, we break the network at points A-B. Could we break it just to the right of the 12-V source? No! Why? The open-circuit voltage is calculated from the network in Fig. 4.13b. Note that we now use the source $2000I'_x$ because this circuit is different from that in Fig. 4.13a. KCL for the super-node around the 12-V source is

$$\frac{(V_{oc} + 12) - (-2000I'_x)}{1k} + \frac{V_{oc} + 12}{2k} + \frac{V_{oc}}{2k} = 0$$

where

$$I'_x = \frac{V_{oc}}{2k}$$

yielding $V_{oc} = -6\ V$.

I_{sc} can be calculated from the circuit in Fig. 4.13c. Note that the presence of the short circuit forces I''_x to zero and, therefore, the network is reduced to that shown in Fig. 4.13d.

Therefore,

$$I_{sc} = \frac{-12}{\frac{2}{3}k} = -18\ mA$$

Then

$$R_{Th} = \frac{V_{oc}}{I_{sc}} = \frac{1}{3}k\Omega$$

Connecting the Thévenin equivalent circuit to the remainder of the network at terminals A-B produces the circuit in Fig. 4.13e. At this point, simple voltage division yields

$$V_o = (-6)\left(\frac{1k}{1k + 1k + \frac{1}{3}k}\right) = \frac{-18}{7}\ V$$

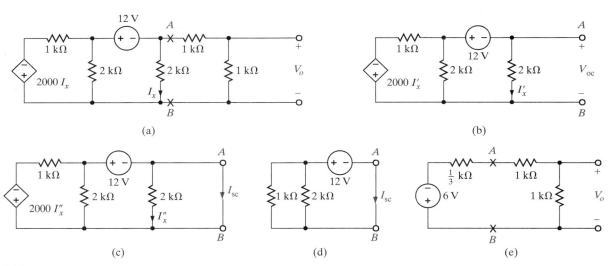

(a)

(b)

(c)

(d)

(e)

Figure 4.13 Circuits used in Example 4.11.

Problem-Solving Strategy
Applying Thévenin's Theorem

▶ Remove the load and find the voltage across the open-circuit terminals, V_{oc}. All the circuit analysis techniques presented here can be used to compute this voltage.

▶ Determine the Thévenin equivalent resistance of the network at the open terminals with the load removed. Three different types of circuits may be encountered in determining the resistance, R_{Th}.

▶ If the circuit contains only independent sources, they are made zero by replacing the voltage sources with short circuits and the current sources with open circuits. R_{Th} is then found by computing the resistance of the purely resistive network at the open terminals.

▶ If the circuit contains only dependent sources, an independent voltage or current source is applied at the open terminals and the corresponding current or voltage at these terminals is measured. The voltage/current ratio at the terminals is the Thévenin equivalent resistance. Since there is no energy source, the open-circuit voltage is zero in this case.

▶ If the circuit contains both independent and dependent sources, the open-circuit terminals are shorted and the short-circuit current between these terminals is determined. The ratio of the open-circuit voltage to the short-circuit current is the resistance R_{Th}.

▶ If the load is now connected to the Thévenin equivalent circuit, consisting of V_{oc} in series with R_{Th}, the desired solution can be obtained.

The problem-solving strategy for Norton's theorem is essentially the same as that for Thévenin's theorem with the exception that we are dealing with the short-circuit current instead of the open-circuit voltage.

LEARNING EXTENSION

E4.5 Find V_o in the circuit in Fig. E4.5 using Thévenin's theorem.

ANSWER $V_o = \dfrac{36}{13}$ V.

Figure E4.5

At this point it is worthwhile to pause for a moment and reflect about what we have learned; that is, let us compare the use of node or loop analysis with that of the theorems discussed in this chapter. When we examine a network for analysis, one of the first things we should do is count the number of nodes and loops. Next we consider the number of sources. For example, are there a number of voltage sources or current sources present in the network? All these data, together with the information that we expect to glean from the network, give a basis for selecting the simplest approach. With the current level of computational power available to us, we can solve the node or loop equations that define the network in a flash.

With regard to the theorems, we have found that in some cases the theorems do not necessarily simplify the problem and a straightforward attack using node or loop analysis is as good an approach as any. This is a valid point provided that we are simply looking for some particular voltage or current. However, the real value of the theorems is the insight and understanding that they provide about the physical nature of the network. For example, superposition tells us what each source contributes to the quantity under investigation. However, a computer solution of the node or loop equations does not tell us the effect of changing certain parameter values in the circuit. It does not help us understand the concept of loading a network or the ramifications of interconnecting networks or the idea of matching a network for maximum power transfer. The theorems help us to understand the effect of using a transducer at the input of an amplifier with a given input resistance. They help us explain the effect of a load, such as a speaker, at the output of an amplifier. We derive none of this information from a node or loop analysis. In fact, as a simple example, suppose that a network at a specific pair of terminals has a Thévenin equivalent circuit consisting of a voltage source in series with a 2-kΩ resistor. If we connect a 2-Ω resistor to the network at these terminals, the voltage across the 2-Ω resistor will be essentially nothing. This result is fairly obvious using the Thévenin theorem approach; however, a node or loop analysis gives us no clue as to why we have obtained this result.

We have studied networks containing only dependent sources. This is a very important topic because all electronic devices, such as transistors, are modeled in this fashion as illustrated in Chapter 16. Motors in power systems are also modeled in this way. We use these amplification devices for many different purposes, such as speed control for automobiles.

In addition, it is interesting to note that when we employ source transformation as we did in Example 4.5, we are simply converting back and forth between a Thévenin equivalent circuit and a Norton equivalent circuit.

Finally, we have a powerful tool at our disposal that can be used to provide additional insight and understanding for both circuit analysis and design. That tool is Microsoft EXCEL, and it permits us to study the effects, on a network, of varying specific parameters. The following example will illustrate the simplicity of this approach.

LEARNING Example 4.12

We wish to use Microsoft EXCEL to plot the Thévenin equivalent parameters V_{oc} and R_{Th} for the circuit in Fig. 4.14 over the R_x range 0 to 10 kΩ.

Figure 4.14
Circuit used in Example 4.12.

SOLUTION The Thévenin resistance is easily found by replacing the voltage sources with short circuits. The result is

$$R_{Th} = 4//R_x = \frac{4R_x}{4 + R_x} \qquad \textbf{4.7}$$

where R_x and R_{Th} are in kΩ. Superposition can be used effectively to find V_{oc}. If the 12-V source is replaced by a short circuit

$$V_{oc_1} = -6\left[\frac{R_x}{R_x + 4}\right]$$

Applying this same procedure for the 6-V source yields

$$V_{oc_2} = 12$$

and the total open-circuit voltage is

$$V_{oc} = 12 - 6\left[\frac{R_x}{R_x + 4}\right] \qquad \textbf{4.8}$$

In EXCEL we wish to (1) vary R_x between 0 and 10 kΩ, (2) calculate R_{Th} and V_{oc} at each R_x value, and (3) plot V_{oc} and R_{Th} versus R_x. We begin by opening EXCEL and entering column headings as shown in Fig. 4.15a. Next, we enter a zero in the first cell of the R_x column at column-row location A4. To automatically fill the column with values, go to the Edit menu and select Fill/Series to open the window shown in Fig. 4.15b, which has already been edited appropriately for 101 data points. The result is a series of R_x values from 0 to 10 kΩ in 100 Ω steps. To enter Eq. (4.8), go to location B4 (right under the V_{oc} heading). Enter the following text and do not forget the equal sign:

```
=12-6*A4/(A4+4)
```

This is Eq. (4.8) with R_x replaced by the first value for R_x, which is at column-row location A4. Similarly for R_{Th}, enter the following expression at C4.

```
=4*A4/(A4+4)
```

To replicate the expression in cell B4 for all R_x values, select cell B4, grab the lower right corner of the cell, hold and drag down to cell B104, and release. Repeat for R_{Th} by replicating cell C4.

To plot the data, first drag the cursor across all cells between A4 and C104. Next, from the Insert menu, select Chart. We recommend strongly that you choose the XY (Scatter) chart type. EXCEL will take you step by step through the basic formatting of your chart, which, after some manipulations, might look similar to the chart in Fig. 4.15c.

Figure 4.15
(a) The EXCEL spreadsheet for Example 4.12 showing the desired column headings. (b) The Fill/Series window edited for varying R_x and (c) the final plot of V_{oc} and R_{Th}.

4.4 Maximum Power Transfer

In circuit analysis we are sometimes interested in determining the maximum power that can be delivered to a load. By employing Thévenin's theorem, we can determine the maximum power that a circuit can supply and the manner in which to adjust the load to effect maximum power transfer.

Suppose that we are given the circuit shown in Fig. 4.16. The power that is delivered to the load is given by the expression

$$P_{\text{load}} = i^2 R_L = \left(\frac{v}{R + R_L} \right)^2 R_L$$

We want to determine the value of R_L that maximizes this quantity. Hence, we differentiate this expression with respect to R_L and equate the derivative to zero.

$$\frac{dP_{\text{load}}}{dR_L} = \frac{(R + R_L)^2 v^2 - 2v^2 R_L (R + R_L)}{(R + R_L)^4} = 0$$

which yields

$$R_L = R$$

In other words, maximum power transfer takes place when the load resistance $R_L = R$. Although this is a very important result, we have derived it using the simple network in Fig. 4.16. However, we should recall that v and R in Fig. 4.16 could represent the Thévenin equivalent circuit for any linear network.

Figure 4.16
Equivalent circuit for examining maximum power transfer.

LEARNING Example 4.13

Let us find the value of R_L for maximum power transfer in the network in Fig. 4.17a and the maximum power that can be transferred to this load.

SOLUTION To begin, we derive the Thévenin equivalent circuit for the network exclusive of the load. V_{oc} can be calculated from the circuit in Fig. 4.17b. The mesh equations for the network are

$$I_1 = 2 \times 10^{-3}$$

$$3k(I_2 - I_1) + 6kI_2 + 3 = 0$$

Solving these equations yields $I_2 = 1/3$ mA and, hence,

$$V_{oc} = 4kI_1 + 6kI_2$$

$$= 10 \text{ V}$$

R_{Th}, shown in Fig. 4.17c, is 6 kΩ; therefore, $R_L = R_{Th} = 6$ kΩ for maximum power transfer. The maximum power transferred to the load is

$$P_L = \left(\frac{10}{12k}\right)^2 (6k) = \frac{25}{6} \text{ mW}$$

LEARNING by Doing

D 4.4 In the network in Fig. 4.16, $v = 12$ V and $R = 2$ kΩ. Determine the R_L for maximum power transfer and the maximum power transferred.

ANSWER $R_L = 2$ kΩ
$P_{max} = 18$ mW

(a)   (b)   (c)   (d)

Figure 4.17 *Circuits used in Example 4.13.*

LEARNING Example 4.14

Let us find R_L for maximum power transfer and the maximum power transferred to this load in the circuit in Fig. 4.18a.

SOLUTION We wish to reduce the network to the form shown in Fig. 4.16. We could form the Thévenin equivalent circuit by breaking the network at the load. However, a close examination of the network indicates that our analysis will be simpler if we break the network to the left of the 4-kΩ resistor. When we do this, however, we must realize that for maximum power transfer $R_L = R_{Th} + 4$ kΩ. V_{oc} can be calculated from the network in Fig. 4.18b. Forming a supernode around the dependent source and its connecting nodes, the KCL equation for this supernode is

$$\frac{V_{oc} - 2000I'_x}{1k + 3k} + (-4 \times 10^{-3}) + \frac{V_{oc}}{2k} = 0$$

where

$$I'_x = \frac{V_{oc}}{2k}$$

These equations yield $V_{oc} = 8$ V. The short-circuit current can be found from the network in Fig. 4.18c. It is here that we find the advantage of breaking the network to the left of the 4-kΩ resistor. The short circuit shorts the 2-kΩ resistor and, therefore, $I''_x = 0$. Hence, the circuit is reduced to that in Fig. 4.18d, where clearly $I_{sc} = 4$ mA. Then

$$R_{Th} = \frac{V_{oc}}{I_{sc}} = 2 \text{ k}\Omega$$

Connecting the Thévenin equivalent to the remainder of the original circuit produces the network in Fig. 4.18e. For maximum power transfer $R_L = R_{Th} + 4$ k$\Omega = 6$ kΩ, and the maximum power transferred is

$$P_L = \left(\frac{8}{12k}\right)^2 (6k) = \frac{8}{3} \text{ mW}$$

(a)   (b)

(c)   (d)   (e)

Figure 4.18   Circuit used in Example 4.14.

E4.6   Given the circuit in Fig. E4.6, find R_L for maximum power transfer and the maximum power transferred.

ANSWER $R_L = 6\ \text{k}\Omega$, $P_L = 2/3\ \text{mW}$.

Figure E4.6

4.5 dc Spice Analysis Using Schematic Capture

INTRODUCTION   The original version of SPICE (Simulation Program with Integrated Circuit Emphasis) was developed at the University of California at Berkeley. It quickly became an industry standard for simulating integrated circuits. With the advent and development of the PC industry, several companies began selling PC- and Macintosh-compatible versions of SPICE. One company, OrCAD Corporation, a division of Cadence Design Systems, Inc., produces a PC-compatible version called PSPICE, which we will discuss in some detail in this text.

In SPICE, circuit information such as the names and values of resistors and sources as well as how they are interconnected is input using data statements with a specific format. The particulars for every element must be typed in a precise order. This makes debugging difficult since you must know the proper format to recognize formatting errors.

PSPICE, as well as other SPICE-based simulators, now employs a feature known as schematic capture. Using an editor, we bypass the cryptic formatting, and simply draw the circuit diagram, assigning element values via dialog boxes. The editor then converts the circuit diagram to an original SPICE format for actual simulation. In the student version of Release 9.1, the most current software available from ORCAD, there are two editors, CAPTURE and *Schematics*. CAPTURE is intended for printed circuit board (PCB) design, while *Schematics* is more generic. When installing PSPICE, the window in Fig. 4.19 will appear, giving you the option of choosing either or both editors. In this text, we will use *Schematics* exclusively.

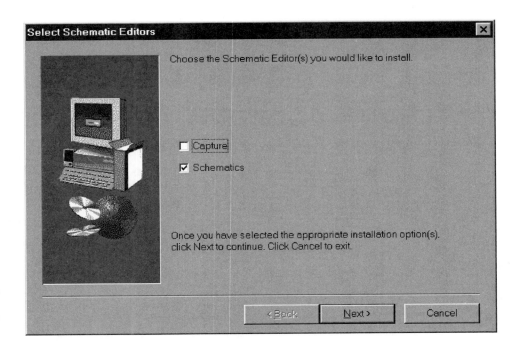

Figure 4.19
In the student version of Release 9.1, you can choose to install either or both schematic capture editors.

When elements such as resistors and voltage sources are given values, it is convenient to use scale factors. PSPICE supports the following list.

T	tera	10^{12}	K	kilo	10^3	N	nano	10^{-9}
G	giga	10^9	M	milli	10^{-3}	P	pico	10^{-12}
MEG	mega	10^6	U	micro	10^{-6}	F	femto	10^{-15}

Component values must be immediately followed by a character—spaces are not allowed. Any text can follow the scale factor. Finally, PSPICE is case insensitive, so there is no difference between 1 Mohm and l mohm in PSPICE. Figure 4.20 shows the five-step procedure for simulating circuits using *Schematics* and PSPICE. We will demonstrate this procedure in the following example.

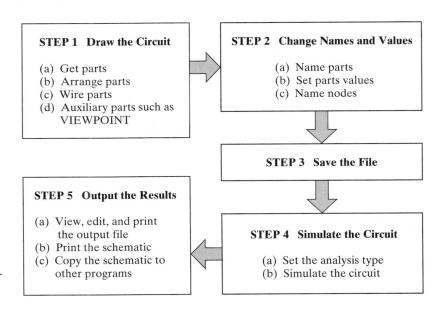

STEP 1 Draw the Circuit

(a) Get parts
(b) Arrange parts
(c) Wire parts
(d) Auxiliary parts such as VIEWPOINT

STEP 2 Change Names and Values

(a) Name parts
(b) Set parts values
(c) Name nodes

STEP 3 Save the File

STEP 5 Output the Results

(a) View, edit, and print the output file
(b) Print the schematic
(c) Copy the schematic to other programs

STEP 4 Simulate the Circuit

(a) Set the analysis type
(b) Simulate the circuit

Figure 4.20
The five-step procedure for PSPICE simulations.

A dc SIMULATION EXAMPLE   The following font conventions will be used through-out this tutorial. Uppercase text refers to PSPICE programs, menus, dialog boxes, and utilities. All boldface text denotes keyboard or mouse inputs. In each instance, the case in boldface text matches that in PSPICE. Let us simulate the circuit in Fig. 4.21 using PSPICE. Following the flowchart procedure in Fig. 4.20, the first step is to open *Schematics* using the **Start/Programs/Pspice Student/Schematics** sequence of pop-up menus. When *Schematics* opens, our screen will change to the *Schematics* editor window shown in Fig. 4.22.

Figure 4.21
A dc circuit used for simulation.

Figure 4.22   *The* Schematics *editor window.*

Step 1: Drawing the Schematic

Next, we obtain and place the required parts: three resistors, two dc voltage sources, and a dc cur-rent source. To get the voltage sources, left-click on the **Draw** menu and select **Get New Part** as shown in Fig. 4.23. The **Part Browser** in Fig. 4.24 appears, listing all the parts available in PSPICE. Since we do not know the *Schematics* name for a dc voltage source, select **Libraries**, and the **Library Browser** in Fig. 4.25 appears. This browser lists all the parts libraries available to us. The dc voltage source is called VDC and is located in the SOURCE.slb library, as seen in Fig. 4.25. Thus, we select the part VDC and click OK. The box in Fig. 4.24 reappears.

If we now select **Place & Close** we revert back to the *Schematics* editor in Fig. 4.22 with one difference: the mouse pointer has become a dc voltage source symbol. The source can be posi-tioned within the drawing area by moving the mouse and then placed by left-clicking *once*. To place the second source, *move the mouse some distance*, and left-click again to place V2. To stop adding sources, we right-click once. Moving the mouse between part placements keeps the parts from stacking atop one another in the diagram. In Fig. 4.21, V2 is oriented horizontally. To rotate

Figure 4.23   *Getting a new part in* Schematics.

Figure 4.24   *The Part Browser window.*

Figure 4.25   *The Library Browser window.*

Figure 4.26
The Schematics *editor after placing the dc voltage sources.*

V2 in your schematic, select it by clicking on it once. The part should turn red. In the **Edit** menu, choose **Rotate**. This causes V2 to spin 90° counterclockwise. Next, we click on V2 and drag it to the desired location. A diagram similar to that shown in Fig. 4.26 should result.

Next, we place the resistors. Repeat the process of **Draw/Get New Part**, except this time, when the **Part Browser Basic** dialog box appears, we type in **R** and select **Place & Close**. The mouse pointer then becomes a resistor. Place each resistor and right-click when done. Note that the resistors are automatically assigned default values of 1 kΩ. Current sources are in the SOURCE.slb library and are called IDC. Get one, place it, and rotate it twice. The resulting schematic is shown in Fig. 4.27.

The parts can now be interconnected. Go to the **Draw** menu and select **Wire**. The mouse pointer will turn into a symbolic pencil. To connect the top of V1 to R1, point the pencil at the end of the wire stub protruding from V1, click once and release. Next, we move the mouse up and over to the left end of R1. A line is drawn up and over a 90° angle, appearing dashed as

Figure 4.27
The schematic after part placement and ready for wiring.

Figure 4.28
*Caught in the act of connecting
V1 to R1.*

shown in Fig. 4.28. *Dashed lines are not yet wires!* We must left-click again to complete and "cut" the connection. The dashed lines become solid and the wiring connection is made. Excess wire fragments (extended dashed lines) can be removed by selecting **Re**_d_**raw** from the **View** menu. The remaining wires can be drawn using a *Schematics* shortcut. To reactivate the wiring pencil, double right-click. This shortcut reactivates the most recent mouse use. We simply repeat the steps listed above to complete the wiring.

PSPICE requires that all schematics have a ground or reference terminal. The ground node voltage will be zero and all other node voltages are referenced to it. We can use either the analog ground (AGND) or the earth ground (EGND) part from the PORT.sIb library shown in Fig. 4.25. Get this part and place it at the bottom of the schematic as shown in Fig. 4.29. Make sure that the part touches the bottom wire in the diagram so that the node dot appears in your schematic. The wiring is now completed.

Figure 4.29
*Schematic with all parts and
wiring completed.*

Step 2: Changing Component Names and Values

To change the name of R1 to Ra, double-click on the text "R1." The **Edit Reference Designator** dialog box appears as shown in Fig. 4.30. Simply type in the new name for the resistor, **Ra**, and select **OK**. Next we will change the resistor's value by double-clicking on the value, "1k." Now the **Attributes** dialog box in Fig. 4.31 appears. Type in **2k** and select **OK**. In a similar manner, edit the names and values of the other parts. The circuit shown in Fig. 4.32 is now ready to be saved.

Figure 4.30
Changing the name of R1 to Ra.

Figure 4.31
Changing the value of Ra from 1k to 2k.

Figure 4.32
The finished schematic, ready for simulation.

Step 3: Saving the Schematic

To save the schematic, simply go to the **File** menu and select **Save**. All *Schematics* files are automatically given the extension .sch.

The Netlist   The netlist is the old-fashioned SPICE code listing for circuit diagrams drawn in *Schematics*. A netlist can be created directly or as part of the simulation process. To create the netlist directly, go to the **Analysis** menu in Fig. 4.22 and select **Create Netlist**. You should receive either the message *Netlist Created* or a dialog box informing you of netlist errors. To view the netlist, return to the **Analysis** menu and select **Examine Netlist**, which opens the file shown in Fig. 4.33. All six elements appear along with their proper values. The text $N_001 $N_002 and so on are the node numbers that *Schematics* created when it converted the diagram to a netlist.

Figure 4.33
The netlist for our circuit.

By tracing through the node numbers we can be assured that our circuit is properly connected. The source V1 is 12 volts positive at node 2 with respect to node 0. EGND is always node number zero. Ra connects node 2 to node 1 and so forth.

How did PSPICE generate these node numbers in this particular order? In PSPICE, the terminals of a part symbol are called pins and are numbered. For example, the pin numbers for resistor, dc voltage, and current source symbols are shown in Fig. 4.34. None of the parts in Fig. 4.34 have been rotated. When the **Get New Part** sequence generates a part with a horizontal orientation, like the resistor, pin 1 is on the left. All vertically oriented parts (the sources) have pin 1 at the top. In the netlist, the order of the node numbers is always pin 1 then pin 2 for each component. Furthermore, when a part is rotated, the pin numbers also rotate. The most critical consequence of the pin numbers is current direction. PSPICE simulations always report the current flowing into pin 1 and out of pin 2.

To change the node number order of a part in the netlist, we must rotate that part 180°. Note that the order of node numbers for Rb in Fig. 4.33 is node 0 (EGND) then node 1. Accordingly, simulation results for the current in Rb will be the negative of Io as defined in Fig. 4.21. To force the netlist to agree with Fig. 4.21, we must rotate Rb twice. To do this,

Figure 4.34
A selection of parts showing the pin-numbering format used in Schematics.

remove the wiring, click on Rb and select **Rotate** in the **Edit** menu twice, rewire and resave the file.

Step 4: Simulating the Circuit

The node voltages in the circuit are typically of interest to us, and *Schematics* permits us to identify each node with a unique name. For example, if we wish to call the node at Rb, Vo, we simply double-click on the wiring at the output node and the dialog box in Fig. 4.35 will appear. Then type **Vo** in the space shown and select **OK**.

Figure 4.35
Creating a custom node name.

Simulation results for dc node voltages and branch currents can be displayed directly on the schematic. In the **Analysis/Display Results on Schematic** menu, choose **Enable Voltage Display** and **Enable Current Display**. This will display all voltages and currents. Unwanted voltage and current displays can be deleted by selecting the data and pressing the DELETE key.

Individual node voltages can also be displayed using the VIEWPOINT (dc voltmeter) from the SPECIAL library. VIEWPOINT parts are placed at nodes and display dc node voltages with respect to ground. We will use a VIEWPOINT part to display Vo and **Analysis/Display/Enable Current Display** for the current, Io. The completed diagram will appear as shown in Fig. 4.36.

Simulation begins by choosing the type of analysis we wish to perform. This is done by selecting **Setup** from the **Analysis** menu. The SETUP dialog box is shown in Fig. 4.37. A *Schematics* dc analysis is requested by selecting **Bias Point Detail** then **Close**. The simulation results will include all node voltages, the currents through all voltage sources, and the total power dissipation. These data will be found in the output text file, accessible at **Analysis/Examine Output**. All VIEWPOINT voltages and currents will also be displayed on the schematic page.

Figure 4.36
The finalized circuit, ready for simulation.

Figure 4.37
The ANALYSIS SETUP window showing the kinds of simulation we can request.

Clearly, a number of different analyses could be requested—for example, a **DC Sweep**. In this case, *Schematics* will ask for the dc source's value you wish to sweep, the start and stop values, and the increment. The simulation results will contain all dc node voltages and branch currents as a function of the varying source value. Two additional analyses, **Transient** and **AC Sweep**, will be discussed in subsequent chapters.

After exiting the SETUP dialog box, select **Simulate** from the **Analysis** menu. When the simulation is finished, your schematic should look much like that shown in Fig. 4.38, where Vo and Io are displayed automatically.

Figure 4.38
The simulation results:
Vo = 6.75 V and
Io = 2.25 mA.

Step 5: Viewing and Printing the Results

All node voltages, voltage source currents, and total power dissipation are contained in the output file. To view these data, select **Examine Output** from the **Analysis** menu. Within the output file is a section containing the node voltages, voltage source currents, and total power dissipation as shown in Fig. 4.39. Indeed, Vo is 6.75 V and the total power dissipation is 5.25 mW. The current bears closer inspection. PSPICE says the currents through V1 and V2 are 1.75 mA and −4.375 mA, respectively. Recall from the netlist discussion that in PSPICE, current flows from a part's pin number 1 to pin 2. As seen in Fig. 4.34, pin 1 is at the positive end of the voltage source. PSPICE is telling us that the current flowing top-down through V1 in Fig. 4.21 is 1.75 mA, while 4.375 mA flows through V2 left-to-right. Based on the passive sign convention, V1 is consuming power while V2 generates power! Since the output file is simply a text editor, the file can be edited, saved, printed, or copied and pasted to other programs.

Figure 4.39
The output file simulation results.

To print the circuit diagram, point the mouse pointer above and left of the upper left hand corner of the diagram, click left, hold, and drag the mouse beyond the lower right edge of the drawing. A box will grow as you drag, eventually surrounding the circuit. Go to the **File** menu and select **Print**. The dialog box in Fig. 4.40 will appear. Select the options **Only Print Selected Area** and **User Definable Zoom Factor**. For most of the small schematics you will create, a scale of 125% to 200% will do fine. Other options are self-explanatory. Finally, select **OK** to print.

If you prefer that the grid dots not appear in the printout, go to the **Options/Display Options** menu and de-select the GRID ON option. To change the grid color, go to the **Options/Display Preferences** menu, select **Grid** and select a color from the drop-down edit box.

To incorporate your schematic into other applications such as text processors, draw a box around the diagram as described above, then under the **Edit** menu, select **Copy to Clipboard**. You can now paste the circuit into other programs.

Figure 4.40
The Schematics *Print window.*

LEARNING Example 4.15

Let us use PSPICE to find the voltage V_o and the current I_x in the circuit in Fig. 4.41.

SOLUTION The PSPICE *Schematics* diagram is shown in Fig. 4.42. From the diagram, we find that $V_o = 150$ V and $I_x = -1.25$ A.

Figure 4.41 Circuit used in Example 4.15.

Figure 4.42 The Schematics diagram for the network in Figure 4.41.

Learning by Application

LEARNING Example 4.16

On Monday afternoon, Connie suddenly remembers that she has a term paper due Tuesday morning. When she sits at her computer to start typing, she discovers that the computer mouse doesn't work. After disassembly and some inspection, she finds that the mouse contains a printed circuit board that is powered by a 5-V supply contained inside the computer case. Furthermore, the board is found to contain several resistors, some op-amps, and one unidentifiable device, which is connected directly to the computer's 5-V supply as shown in Fig. 4.43a. Using a voltmeter to measure the node voltages, Connie confirms that all resistors and op-amps are functioning properly and the power supply voltage reaches the mouse board. However, without knowing the mystery device's function within the circuit, she cannot determine its condition. A phone call to the manufacturer reveals that the device is indeed linear but is also proprietary. With some persuasion, the manufacturer's representative agrees that if Connie can find the Thévenin equivalent circuit for the element at nodes A-B with the computer on, he will tell her if it is functioning properly. Armed with a single 1-kΩ resistor and a voltmeter, Connie attacks the problem.

SOLUTION To find the Thévenin equivalent for the unknown device, together with the 5-V source, Connie first isolates nodes A and B from the rest of the devices on the board to measure the open-circuit voltage. The resulting voltmeter reading is $V_{AB} = 2.4$ V. Thus, the Thévenin equivalent voltage is 2.4 V. Then she connects the 1-kΩ resistor at nodes A-B as shown in Fig. 4.43b. The voltmeter reading is now $V_{AB} = 0.8$ V. Using voltage division to express V_{AB} in terms of V_{Th}, R_{Th}, and R_{test} in Fig. 4.43b yields the expression

$$0.8 = V_{Th}\left(\frac{1k}{1k + R_{Th}}\right)$$

Solving the equations for R_{Th}, we obtain

$$R_{Th} = 2.0 \text{ k}\Omega$$

Therefore, the unknown device and the 5-V source can be represented at the terminals A-B by the Thévenin equivalent circuit shown in Fig. 4.43c. When Connie phones the manufacturer with the data, the representative informs her that the device has indeed failed.

Figure 4.43
Networks used in
Example 4.16.

(a)

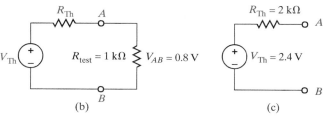

(b)   (c)

Learning by Design

LEARNING Example 4.17

If the battery current in a laptop computer becomes too large, damage may occur to the system components. We wish to design an overcurrent sensor to protect a laptop from this condition.

SOLUTION   We propose the circuit, shown in Fig. 4.44, to monitor the battery current and output a control signal, OVER, that can be used to disconnect the battery. The major components employed are a differential amplifier, a comparator, and a resistor used to sense the battery current that supplies the laptop components. Under normal operating conditions, the control signal voltage is low, typically zero, but if the battery current exceeds a predetermined trip point, this voltage becomes high, near V_{batt}. The comparator circuit will compare the sensed voltage to a dc reference, V_{trip}, to generate the overcurrent signal. Since the power lost in R_{sense} is not available to the computer components, V_{sense} must be small. However, to provide a reasonably sized voltage to the comparator, we will amplify V_{sense} before comparing it to V_{trip}. A differential amplifier is employed because V_{sense} is not referenced to ground. When this amplifier output, V_A, exceeds V_{trip}, the comparator output will be approximately equal to the dc supply voltage, V_{batt}.

Using a typical laptop battery voltage of 12 V and a maximum power consumption of about 54 W yields a corresponding

battery current of 4.5 A. If we select the current value for the trip point to be 9 A, and further select the reference voltage to be halfway between the battery voltage and ground, that is, $V_{trip} = 6$ V, then the resistors R_A and R_B form a voltage divider and each is chosen to be 10 kΩ. Next, if we choose R_{sense} such that the power it consumes during normal operation is 0.5% of the total power, then

$$R_{sense} = \frac{(0.005)P}{I_{batt}^2} = \frac{(0.005)(54)}{4.5^2} = 13.3 \text{ m}\Omega$$

Ohm's law relates V_{sense} at the trip point to the trip point current, and thus

$$V_{sense} = (I_{trip})(R_{sense}) = (9)(0.0133) = 0.12 \text{ V}$$

Since, at the trip point, $V_A = V_{trip} = 6$ V, V_{sense} must be amplified by $6/0.12 = 50$. As indicated earlier, the gain of a differential amplifier is R_2/R_1. If we select $R_1 = 1$ kΩ, then $R_2 = 50$ kΩ will produce the proper gain. A PSPICE plot of simulated overvoltage versus battery current is shown in Fig. 4.45. Note that OVER is low until $I_{batt} = 9$ A, at which point OVER goes high. Clearly, other circuitry is needed to interpret the OVER output signal, but the design of the sensing circuitry has been completed.

Figure 4.44   *A simple overcurrent sensor circuit.*

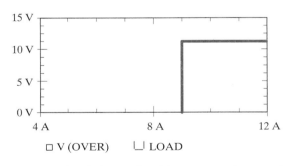

Figure 4.45   *Simulation results of the overcurrent sensor circuit in Figure 4.44.*

LEARNING Example 4.18

A chemical sensor outputs a voltage proportional to carbon monoxide concentration. The defining equation is

$$V_S = C/250$$

where C is carbon monoxide in ppm. In a particular application,

the concentration range of interest is 125 to 250 ppm, which corresponds to a sensor voltage of 0.5 to 1 V. We want to expand this range to produce a new voltage, V_o, between 0 and 5 V. Let us design a circuit that performs this conversion.

Figure 4.46

(a) The general level shifter. Superposition subcircuits for contributions of (b) V_S and (c) V_{ref}.

SOLUTION The linear relationship between V_o and V_S is of the form

$$V_o = mV_S + b$$

the parameters are set such that a concentration of 125 ppm produces $V_S = 0.5$ V and $V_o = 0$ V, while 250 ppm yields $V_S = 1$ V and $V_o = 5$ V, then

$$0 = m(0.5) + b \quad \text{and} \quad 5 = m(1.0) + b$$

Solving for m and b, we find $m = 10$ and $b = -5$. Therefore,

$$V_o = 10V_S - 5 \qquad \textbf{4.9}$$

Now consider the op-amp circuit in Fig. 4.46a, called a level shifter. Using superposition, the output voltage can be easily determined. First, we find the contribution of V_S to V_o by setting V_{ref} to zero as shown in Fig. 4.46b. This is the classic noninverting gain stage, where

$$V_{o1} = \left[1 + \frac{R_2}{R_1}\right]V_S$$

Next, we find the effect of V_{ref} by setting V_S to zero. This yields the classic inverting gain stage, shown in Fig. 4.46c, where

$$V_{o2} = -\left[\frac{R_2}{R_1}\right]V_{ref}$$

The output voltage is the sum of these contributions

$$V_o = \left[1 + \frac{R_2}{R_1}\right]V_S - \left[\frac{R_2}{R_1}\right]V_{ref} \qquad \textbf{4.10}$$

Comparing Eqs. (4.9) and (4.10), we see that

$$1 + \frac{R_2}{R_1} = 10 \qquad \frac{R_2}{R_1} = 9 \qquad V_{ref} = \frac{5}{9} \text{ V}$$

Arbitrarily choosing $R_1 = 10 \text{ k}\Omega$ yields $R_2 = 90 \text{ k}\Omega$. Voltage division can produce V_{ref} from a 5-V supply voltage. The final circuit is shown in Fig. 4.47, where the unity gain buffer isolates the voltage division resistor string from the level shifter.

Figure 4.47 The final level shifter circuit.

Summary

▶ Linearity: This property requires both additivity and homogeneity. Using this property, the voltage or current somewhere in a network can be determined by assuming a specific value for the variable and then determining what source value is required to produce it. The ratio of the specified source value to that computed from the assumed value of the variable, together with the assumed value of the variable, can be used to obtain a solution.

▶ In a linear network containing multiple independent sources, the principle of superposition allows us to compute any current or voltage in the network as the algebraic sum of the individual contributions of each source acting alone.

▶ Superposition is a linear property and does not apply to nonlinear functions such as power.

▶ Using Thévenin's theorem, we can replace some portion of a network at a pair of terminals with a voltage source V_{oc} in series with a resistor R_{Th}. V_{oc} is the open-circuit voltage at the terminals, and R_{Th} is the Thévenin equivalent resistance obtained by looking into the terminals with all independent sources made zero.

▶ Using Norton's theorem, we can replace some portion of a network at a pair of terminals with a current source I_{sc} in parallel with a resistor R_{Th}. I_{sc} is the short-circuit current at the terminals and R_{Th} is the Thévenin equivalent resistance.

▶ Source transformation permits us to replace a voltage source V in series with a resistance R by a current source $I = V/R$ in parallel with the resistance R. The reverse is also true.

This is an interchange relationship between Thévenin and Norton equivalent circuits.

▶ Maximum power transfer can be achieved by selecting the load R_L to be equal to R_{Th} found by looking into the network from the load terminals.

▶ dc PSPICE with Schematic Capture is an effective tool in analyzing dc circuits.

Problems

SECTION 4.1

4.1 Find I_o in the circuit in Fig. P4.1 using linearity and the assumption that $I_o = 1$ mA.

Figure P4.1

4.2 Find I_o in the network in Fig. P4.2 using linearity and the assumption that $I_o = 1$ mA.

Figure P4.2

4.3 Find V_o in the network in Fig. P4.3 using linearity and the assumption that $V_o = 1$ mV.

Figure P4.3

4.4 Find V_o in the circuit in Fig. P4.4 using linearity and the assumption that $V_o = 1$ V.

Figure P4.4

SECTION 4.2

4.5 In the network in Fig. P4.5, find I_o using superposition.

Figure P4.5

4.6 Find I_o in the circuit in Fig. P4.6 using superposition.

Figure P4.6

4.7 In the network in Fig. P4.7, find V_o using superposition.

Figure P4.7

4.8 Find V_o in the network in Fig. P4.8 using superposition.

Figure P4.8

4.9 Find I_o in the network in Fig. P4.9 using superposition.

Figure P4.9

4.10 Find I_o in the network in Fig. P4.10 using superposition.

Figure P4.10

4.11 Find I_o in the network in Fig. P4.11 using superposition.

Figure P4.11

4.12 Find I_o in the network in Fig. P4.12 using superposition.

Figure P4.12

4.13 Find I_o in the circuit in Fig. P4.13 using superposition.

Figure P4.13

4.14 Use superposition to find I_o in the circuit in Fig. P4.14.

Figure P4.14

4.15 Find I_o in the network in Fig. P4.15 using superposition.

Figure P4.15

4.16 Find I_o in the network in Fig. P4.16 using superposition.

Figure P4.16

4.17 Find I_o in the network in Fig. P4.17 using superposition.

Figure P4.17

4.18 Find V_o in the network in Fig. P4.18 using superposition.

Figure P4.18

SECTION 4.3

4.19 Use source transformation to find I_o in the circuit in Fig. P4.19.

Figure P4.19

4.20 Find V_o in the network in Fig. P4.20 using source transformation.

Figure P4.20

4.21 Use source transformation to find V_o in the network in Fig. P4.21.

Figure P4.21

4.22 Find V_o in the network in Fig. P4.22 using source transformation.

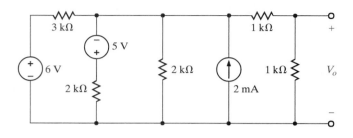

Figure P4.22

4.23 Find I_o in the circuit in Fig. P4.23 using source transformation.

Figure P4.23

4.24 Find I_o in the network in Fig. P4.24 using source transformation.

Figure P4.24

4.25 Use source transformation to find I_o in the circuit in Fig. P4.25.

Figure P4.25

4.26 Find V_o in the network in Fig. P4.26 using source transformation.

Figure P4.26

4.27 Find I_o in the circuit in Fig. P4.27 using source transformation.

Figure P4.27

4.28 Find I_o in the network in Fig. P4.28 using source transformation.

Figure P4.28

4.29 Find I_o in the network in Fig. P4.29 using source transformation.

Figure P4.29

4.30 Find I_o in the network in Fig. P4.30 using source transformation.

Figure P4.30

4.31 Find I_o in the network in Fig. P4.31 using source transformation.

Figure P4.31

4.34 Use Thévenin's theorem to find I_o in the network in Fig. P4.34.

Figure P4.34

4.32 Use Thévenin's theorem to find V_o in the network in Fig. P4.32.

Figure P4.32

4.35 Find I_o in the network in Fig. P4.35 using Thévenin's theorem.

Figure P4.35

4.33 Use Thévenin's theorem to find V_o in the network in Fig. P4.33.

Figure P4.33

4.36 Find V_o in the circuit in Fig. P4.36 using Thévenin's theorem.

Figure P4.36

4.37 Find I_o in the circuit in Fig. P4.37 using Thévenin's theorem.

Figure P4.37

4.38 Find I_o in the network in Fig. P4.38 using Thévenin's theorem.

Figure P4.38

4.39 Find V_o in the circuit in Fig. P4.39 using Thévenin's theorem.

Figure P4.39

4.40 Find V_o in the circuit in Fig. P4.40 using Thévenin's theorem.

Figure P4.40

4.41 Find I_o in the network in Fig. P4.41 using Thévenin's theorem.

Figure P4.41

4.42 Find I_o in the network in Fig. P4.42 using Thévenin's theorem.

Figure P4.42

4.43 Find I_o in the network in Fig. P4.43 using Thévenin's theorem.

Figure P4.43

4.44 Find I_o in the network in Fig. P4.44 using Thévenin's theorem.

Figure P4.44

4.45 Find V_o in the network in Fig. P4.45 using Thévenin's theorem.

Figure P4.45

4.46 Find V_o in the network in Fig. P4.46 using Thévenin's theorem.

Figure P4.46

4.47 Use a combination of Thévenin's theorem and superposition to find V_o in the circuit in Fig. P4.47.

Figure P4.47

4.48 In the network in Fig. P4.48, find V_o using Thévenin's theorem.

Figure P4.48

4.49 Find the Thévenin equivalent of the network in Fig. P4.49 at the terminals *A-B*.

Figure P4.49

4.50 Find the Thévenin equivalent of the network in Fig. P4.50 at the terminals *A-B*.

Figure P4.50

4.51 Find V_o in the network in Fig. P4.51 using Thévenin's theorem.

Figure P4.51

4.52 Find V_o in the network in Fig. P4.52 using Thévenin's theorem.

Figure P4.52

4.53 Find V_o in the circuit in Fig. P4.53 using Thévenin's theorem.

Figure P4.53

4.54 Use Thévenin's theorem to find V_o in the circuit in Fig. P4.54.

Figure P4.54

4.55 Use Thévenin's theorem to find I_o in the circuit in Fig. P4.55.

Figure P4.55

4.56 Find V_o in the network in Fig. P4.56 using Thévenin's theorem.

Figure P4.56

4.57 Use Thévenin's theorem to find V_o in the network in Fig. P4.57.

Figure P4.57

4.58 Find V_o in the network in Fig. P4.58 using Thévenin's theorem.

Figure P4.58

4.59 Use Thévenin's theorem to find V_o in the circuit in Fig. P4.59.

Figure P4.59

4.60 Find V_o in the network in Fig. P4.60 using Thévenin's theorem.

Figure P4.60

4.61 Use Norton's theorem to find V_o in the network in Fig. P4.61.

Figure P4.61

4.62 Find I_o in the network in Fig. P4.62 using Norton's theorem.

Figure P4.62

4.63 Use Norton's theorem to find I_o in the circuit in Fig. P4.63.

Figure P4.63

4.64 Find I_o in the network in Fig. P4.64 using Norton's theorem.

Figure P4.64

4.65 Find I_o in the network in Fig. P4.65 using Norton's theorem.

Figure P4.65

4.66 Use Norton's theorem to find V_o in the network in Fig. P4.66.

Figure P4.66

4.67 Find V_o in the network in Fig. P4.67 using Norton's theorem.

Figure P4.67

4.68 Given the linear circuit in Fig. P4.68, it is known that when a 2-kΩ load is connected to the terminals A-B, the load current is 10 mA. If a 10-kΩ load is connected to the terminals, the load current is 6 mA. Find the current in a 20-kΩ load.

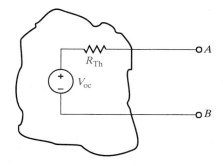

Figure P4.68

4.69 If an 8-kΩ load is connected to the terminals of the network in Fig. P4.69, $V_{AB} = 16$ V. If a 2-kΩ load is connected to the terminals, $V_{AB} = 8$ V. Find V_{AB} if a 20-kΩ load is connected to the terminals.

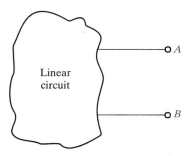

Figure P4.69

SECTION 4.4

4.70 Find R_L for maximum power transfer and the maximum power that can be transferred in the network in Fig. P4.70.

Figure P4.70

4.71 Find R_L for maximum power transfer and the maximum power that can be transferred in the network in Fig. P4.71.

Figure P4.71

4.72 In the network in Fig. P4.72, find R_L for maximum power transfer and the maximum power that can be transferred to this load.

Figure P4.72

4.73 Find R_L for maximum power transfer and the maximum power that can be transferred in the network in Fig. P4.73.

Figure P4.73

SECTION 4.5

4.74 Solve problem 4.5 using PSPICE.

4.75 Solve problem 4.21 using PSPICE.

4.76 Solve problem 4.32 using PSPICE.

4.77 Solve problem 4.36 using PSPICE.

Typical Problems Found on the FE Exam

4FE-1 Determine the maximum power that can be delivered to the load R_L in the network in Fig. 4PFE-1.

Fig. 4PFE-1

4FE-2 Find the value of the load R_L in the network in Fig. 4PFE-2 that will achieve maximum power transfer, and determine the value of the maximum power.

Fig. 4PFE-2

4FE-3 Find the value of R_L in the network in Fig. 4PFE-3 for maximum power transfer to this load.

Fig. 4PFE-3

Capacitance and Inductance

5

Have you ever wondered how a tiny camera battery is able to produce a blinding flash or how a hand-held "stun gun" can deliver 50,000 V? The answer is energy storage, and in this chapter we introduce two elements that possess this property: the capacitor and the inductor. Both capacitors and inductors are linear elements; however, unlike the resistor, their terminal characteristics are described by linear differential equations. Another distinctive feature of these elements is their ability to absorb energy from the circuit, store it temporarily, and later return it. Elements that possess this energy storage capability are referred to simply as *storage elements*.

Capacitors are capable of storing energy when a voltage is present across the element. The energy is actually stored in an electric field not unlike that produced by sliding across a car seat on a dry winter day. Conversely, inductors are capable of storing energy when a current is passing through them, causing a magnetic field to form. This phenomenon can be demonstrated by placing a needle compass in the vicinity of a current. The current causes a magnetic field whose energy deflects the compass needle.

A very important circuit, which employs a capacitor in a vital role, is also introduced. This circuit, known as an op-amp integrator, produces an output voltage that is proportional to the integral of the input voltage. The significance of this circuit is that any system (for example, electrical, mechanical, hydraulic, biological, social, economic, and so on) that can be described by a set of linear differential equations with constant coefficients can be modeled by a network consisting of op-amp integrators. Thus, very complex and costly systems can be tested safely and inexpensively prior to construction and implementation.

Finally, we examine some practical circuits where capacitors and inductors are normally found or can be effectively used in circuit design.

5.1 Capacitors

A *capacitor* is a circuit element that consists of two conducting surfaces separated by a non-conducting, or *dielectric*, material. A simplified capacitor and its electrical symbol are shown in Fig. 5.1.

Figure 5.1
Capacitor and its electrical symbol.

(a) (b)

There are many different kinds of capacitors, and they are categorized by the type of dielectric material that is used between the conducting plates. Although any good insulator can serve as a dielectric, each type has characteristics that make it more suitable for particular applications.

For general applications in electronic circuits (e.g., coupling between stages of amplification) the dielectric material may be paper impregnated with oil or wax, mylar, polystyrene, mica, glass, or ceramic.

Ceramic dielectric capacitors constructed of barium titanates have a large capacitance-to-volume ratio because of their high dielectric constant. Mica, glass, and ceramic dielectric capacitors will operate satisfactorily at high frequencies.

Aluminum electrolytic capacitors, which consist of a pair of aluminum plates separated by a moistened borax paste electrolyte, can provide high values of capacitance in small volumes. They are typically used for filtering, bypassing, and coupling, and in power supplies and motor-starting applications. Tantalum electrolytic capacitors have lower losses and more stable characteristics than those of aluminum electrolytic capacitors. Figure 5.2 shows a variety of typical discrete capacitors. A capacitor constructed with integrated circuit processing is shown in Chapter 16.

Figure 5.2
*Some typical capacitors
(courtesy of Cornell Dubilier).*

In addition to these capacitors, which we deliberately insert in a network for specific applications, stray capacitance is present any time there is a difference in potential between two conducting materials separated by a dielectric. Because this stray capacitance can cause unwanted coupling between circuits, extreme care must be exercised in the layout of electronic systems on printed circuit boards.

Capacitance is measured in coulombs per volt or farads. The unit *farad* (F) is named after Michael Faraday, a famous English physicist. Capacitors may be fixed or variable and typically range from thousands of microfarads (μF) to a few picofarads (pF).

However, capacitor technology, initially driven by the modern interest in electric vehicles, is rapidly changing. For example, the capacitor in the photograph in Fig. 5.3 was designed and built at the Space Power Institute at Auburn University to achieve high energy and high power density capacitors for application in electric vehicles and space. This capacitor, which is specially constructed, is rated at 55 F, 500 joules (J). It is interesting to calculate the dimensions of a simple equivalent capacitor consisting of two parallel plates each of area A, separated by a distance d as shown in Fig. 5.1. We learned in basic physics that the capacitance of two parallel plates of area A, separated by distance d, is

Figure 5.3
A 55-F, 500-J capacitor and a 10-in ruler for comparison.

$$C = \frac{\varepsilon_o A}{d}$$

where ε_o, the permitivity of free space, is 8.85×10^{-12} F/m. If we assume the plates are separated by a distance in air of the thickness of one sheet of oil-impregnated paper, which is about 1.016×10^{-4} m, then

$$55\,\text{F} = \frac{(8.85 \times 10^{-12})A}{1.016 \times 10^{-4}}$$

$$A = 6.3141 \times 10^{8}\,\text{m}^2$$

and since 1 square mile is equal to 2.59×10^{6} square meters, the area is

$$A \approx 244 \text{ square miles}$$

which is the area of a medium-sized city! It would now seem that the capacitor in the photograph is much more impressive than it originally appeared. This capacitor is actually constructed using multiple layers of a combination of pressed carbon and metal fibers, which are heat-treated to adhere to metal foil. The metal foils are connected to form a stack of 18 cells, within this geometry. There are literally millions of pieces of carbon and metal fibers employed to obtain the required surface area.

Suppose now that a source is connected to the capacitor shown in Fig. 5.1; then positive charges will be transferred to one plate and negative charges to the other. The charge on the capacitor is proportional to the voltage across it such that

$$q = Cv \hspace{4cm} \textbf{5.1}$$

where C is the proportionality factor known as the capacitance of the element in farads.

The charge differential between the plates creates an electric field that stores energy. Because of the presence of the dielectric, the conduction current that flows in the wires that connect the capacitor to the remainder of the circuit cannot flow internally between the plates. However, via electromagnetic field theory it can be shown that this conduction current is equal to the displacement current that flows between the plates of the capacitor and is present any time that an electric field or voltage varies with time.

Our primary interest is in the current–voltage terminal characteristics of the capacitor. Since the current is

$$i = \frac{dq}{dt}$$

then for a capacitor

$$i = \frac{d}{dt}Cv$$

which for constant capacitance is

$$i = C\frac{dv}{dt} \qquad \textbf{5.2}$$

Equation (5.2) can be rewritten as

$$dv = \frac{1}{C}i\,dt$$

Now integrating this expression from $t = -\infty$ to some time t and assuming $v(-\infty) = 0$ yields

$$v(t) = \frac{1}{C}\int_{-\infty}^{t} i(x)\,dx \qquad \textbf{5.3}$$

where $v(t)$ indicates the time dependence of the voltage. Equation (5.3) can be expressed as two integrals, so that

$$v(t) = \frac{1}{C}\int_{-\infty}^{t_0} i(x)\,dx + \frac{1}{C}\int_{t_0}^{t} i(x)\,dx \qquad \textbf{5.4}$$

$$= v(t_0) + \frac{1}{C}\int_{t_0}^{t} i(x)\,dx$$

where $v(t_0)$ is the voltage due to the charge that accumulates on the capacitor from time $t = -\infty$ to time $t = t_0$.

The energy stored in the capacitor can be derived from the power that is delivered to the element. This power is given by the expression

$$p(t) = v(t)i(t) = Cv(t)\frac{dv(t)}{dt} \qquad \textbf{5.5}$$

and hence the energy stored in the electric field is

$$w_C(t) = \int_{-\infty}^{t} Cv(x)\frac{dv(x)}{dx}\,dx = C\int_{-\infty}^{t} v(x)\frac{dv(x)}{dx}\,dx$$

$$= C\int_{v(-\infty)}^{v(t)} v(x)\,dv(x) = \tfrac{1}{2}Cv^2(x)\Big|_{v(-\infty)}^{v(t)}$$

$$= \tfrac{1}{2}Cv^2(t)\ \text{J} \qquad \textbf{5.6}$$

since $v(t = -\infty) = 0$. The expression for the energy can also be written using Eq. (5.1) as

$$w_C(t) = \frac{1}{2}\frac{q^2(t)}{C} \qquad \textbf{5.7}$$

Equations (5.6) and (5.7) represent the energy stored by the capacitor, which, in turn, is equal to the work done by the source to charge the capacitor.

The polarity of the voltage across a capacitor being charged is shown in Fig. 5.1b. In the ideal case the capacitor will hold the charge for an indefinite period of time if the source is removed. If at some later time an energy-absorbing device (e.g., a flashbulb) is connected across the capacitor, a discharge current will flow from the capacitor and, therefore, the capacitor will supply its stored energy to the device.

LEARNING Example 5.1

If the charge accumulated on two parallel conductors charged to 12 V is 600 pC, what is the capacitance of the parallel conductors?

SOLUTION Using Eq. (5.1), we find that

$$C = \frac{Q}{V} = \frac{(600)(10^{-12})}{12} = 50 \text{ pF}$$

LEARNING Example 5.2

The voltage across a 5-μF capacitor has the waveform shown in Fig. 5.4a. Determine the current waveform.

SOLUTION Note that

$$v(t) = \frac{24}{6 \times 10^{-3}} t \qquad 0 \le t \le 6 \text{ ms}$$

$$= \frac{-24}{2 \times 10^{-3}} t + 96 \qquad 6 \le t < 8 \text{ ms}$$

$$= 0 \qquad 8 \text{ ms} \le t$$

Using Eq. (5.2), we find that

$$i(t) = C \frac{dv(t)}{dt}$$

$$= 5 \times 10^{-6}(4 \times 10^{3}) \qquad 0 \le t \le 6 \text{ ms}$$

$$= 20 \text{ mA} \qquad 0 \le t \le 6 \text{ ms}$$

$$i(t) = 5 \times 10^{-6}(-12 \times 10^{3}) \qquad 6 \le t \le 8 \text{ ms}$$

$$= -60 \text{ mA} \qquad 6 \le t < 8 \text{ ms}$$

and

$$i(t) = 0 \qquad 8 \text{ ms} \le t$$

Therefore, the current waveform is as shown in Fig. 5.4b and $i(t) = 0$ for $t > 8$ ms.

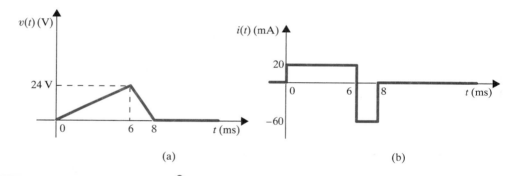

Figure 5.4
Voltage and current waveforms for a 5-μF capacitor.

(a)

(b)

LEARNING Example 5.3

Determine the energy stored in the electric field of the capacitor in Example 5.2 at $t = 6$ ms.

SOLUTION Using Eq. (5.6), we have

$$w(t) = \tfrac{1}{2}Cv^2(t)$$

At $t = 6$ ms,

$$w(6 \text{ ms}) = \tfrac{1}{2}(5 \times 10^{-6})(24)^2$$

$$= 1440 \text{ μJ}$$

LEARNING EXTENSION

E5.1 A 10-μF capacitor has an accumulated charge of 500 nC. Determine the voltage across the capacitor. **ANSWER** 0.05 V.

LEARNING Example 5.4

The current in an initially uncharged 4-μF capacitor is shown in Fig. 5.5a. Let us derive the waveforms for the voltage, power, and energy and compute the energy stored in the electric field of the capacitor at $t = 2$ ms.

(a)

(b)

(c)

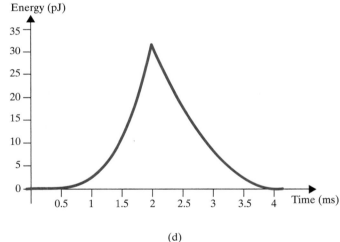

(d)

Figure 5.5 *Waveforms used in Example 5.4.*

SOLUTION The equations for the current waveform in the specific time intervals are

$$i(t) = \frac{16 \times 10^{-6}t}{2 \times 10^{-3}} \qquad 0 \leq t \leq 2 \text{ ms}$$

$$= -8 \times 10^{-6} \qquad 2 \text{ ms} \leq t \leq 4 \text{ ms}$$

$$= 0 \qquad 4 \text{ ms} < t$$

Since $v(0) = 0$, the equation for $v(t)$ in the time interval $0 \leq t \leq 2$ ms is

$$v(t) = \frac{1}{(4)(10^{-6})} \int_0^t 8(10^{-3})x \, dx = 10^3 t^2$$

and hence,

$$v(2 \text{ ms}) = 10^3(2 \times 10^{-3})^2 = 4 \text{ mV}$$

In the time interval 2 ms $\leq t \leq 4$ ms,

$$v(t) = \frac{1}{(4)(10^{-6})} \int_{2(10^{-3})}^t -(8)(10^{-6})dx + (4)(10^{-3})$$

$$= -2t + 8 \times 10^{-3}$$

The waveform for the voltage is shown in Fig. 5.5b.

Since the power is $p(t) = v(t)i(t)$, the expression for the power in the time interval $0 \le t \le 2$ ms is $p(t) = 8t^3$. In the time interval 2 ms $\le t \le 4$ ms, the equation for the power is

$$p(t) = -(8)(10^{-6})(-2t + 8 \times 10^{-3})$$
$$= 16(10^{-6})t - 64(10^{-9})$$

The power waveform is shown in Fig. 5.5c. Note that during the time interval $0 \le t \le 2$ ms, the capacitor is absorbing energy and during the interval 2 ms $\le t \le 4$ ms, it is delivering energy.

The energy is given by the expression

$$w(t) = \int_{t_0}^{t} p(x)\, dx + w(t_0)$$

In the time interval $0 \le t \le 2$ ms,

$$w(t) = \int_{0}^{t} 8x^3 \, dx = 2t^4$$

Hence,

$$w(2 \text{ ms}) = 32 \text{ pJ}$$

In the time interval $2 \le t \le 4$ ms,

$$w(t) = \int_{2 \times 10^{-3}}^{t} \left[(16 \times 10^{-6})x - (64 \times 10^{-9}) \right] dx + 32 \times 10^{-12}$$
$$= \left[(8 \times 10^{-6})x^2 - (64 \times 10^{-9})x \right]_{2 \times 10^{-3}}^{t} + 32 \times 10^{-12}$$
$$= (8 \times 10^{-6})t^2 - (64 \times 10^{-9})t + 128 \times 10^{-12}$$

From this expression we find that $w(2 \text{ ms}) = 32$ pJ and $w(4 \text{ ms}) = 0$. The energy waveform is shown in Fig. 5.5d.

LEARNING EXTENSIONS

E5.2   The voltage across a 2-μF capacitor is shown in Fig. E5.2. Determine the waveform for the capacitor current.

Figure E5.2

ANSWER

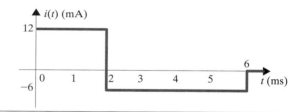

E5.3   Compute the energy stored in the electric field of the capacitor in Extension Exercise E5.2 at $t = 2$ ms.

ANSWER $w = 144 \ \mu$J.

5.2 Inductors

An *inductor* is a circuit element that consists of a conducting wire usually in the form of a coil. Two typical inductors and their electrical symbols are shown in Fig. 5.6. Inductors are typically categorized by the type of core on which they are wound. For example, the core material may be air or any nonmagnetic material, iron, or ferrite. Inductors made with air or nonmagnetic materials are widely used in radio, television, and filter circuits. Iron-core inductors are used in electrical power supplies and filters. Ferrite-core inductors are widely used in high-frequency applications. Note that in contrast to the magnetic core that confines the flux, as shown in Fig. 5.6b, the flux lines for nonmagnetic inductors extend beyond the inductor itself, as illustrated in Fig. 5.6a. Like stray capacitance, stray inductance can result from any element carrying current surrounded by flux linkages. Figure 5.7 shows a variety of typical inductors. An inductor constructed with integrated circuit processing is shown in Chapter 16.

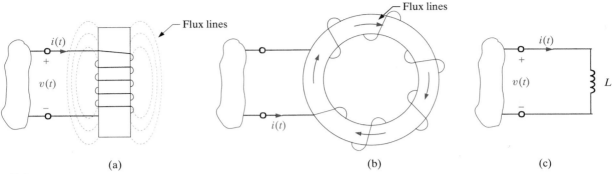

(a)

(b)

(c)

Figure 5.6
Two inductors and their electrical symbol.

LEARNING **Hint**

Note the use of the passive sign convention, as illustrated in Fig. 5.6.

Figure 5.7
Some typical inductors (courtesy of Coilcraft).

LEARNING **by Doing**

D 5.2 Write the *i-v* relationship for the following inductors.

(a)

(b)

ANSWER

(a) $v(t) = -L\dfrac{di(t)}{dt}$

(b) $v(t) = L\dfrac{di(t)}{dt}$

From a historical standpoint, developments that led to the mathematical model we employ to represent the inductor are as follows. It was first shown that a current-carrying conductor would produce a magnetic field. It was later found that the magnetic field and the current that produced it were linearly related. Finally, it was shown that a changing magnetic field produced a voltage that was proportional to the time rate of change of the current that produced the magnetic field; that is,

$$v(t) = L\frac{di(t)}{dt} \qquad \textbf{5.8}$$

The constant of proportionality L is called the inductance and is measured in the unit henry, named after the American inventor Joseph Henry, who discovered the relationship. As seen in Eq. (5.8), 1 henry (H) is dimensionally equal to 1 volt-second per ampere.

Following the development of the mathematical equations for the capacitor, we find that the expression for the current in an inductor is

$$i(t) = \frac{1}{L}\int_{-\infty}^{t} v(x)\,dx \qquad \textbf{5.9}$$

which can also be written as

$$i(t) = i(t_0) + \frac{1}{L}\int_{t_0}^{t} v(x)\,dx \qquad \textbf{5.10}$$

The power delivered to the inductor can be used to derive the energy stored in the element. This power is equal to

$$p(t) = v(t)i(t)$$
$$= \left[L\frac{di(t)}{dt}\right]i(t) \qquad \textbf{5.11}$$

Therefore, the energy stored in the magnetic field is

$$w_L(t) = \int_{-\infty}^{t} \left[L \frac{di(x)}{dx} \right] i(x) \, dx$$

Following the development of Eq. (5.6), we obtain

$$w_L(t) = \tfrac{1}{2} L i^2(t) \text{ J} \qquad\qquad \textbf{5.12}$$

The inductor, like the resistor and capacitor, is a passive element. The polarity of the voltage across the inductor is shown in Fig. 5.6.

Practical inductors typically range from a few microhenrys to tens of henrys. From a circuit design standpoint it is important to note that inductors cannot be easily fabricated on an integrated circuit chip, and therefore chip designs typically employ only active electronic devices, resistors, and capacitors that can be easily fabricated in microcircuit form.

LEARNING Example 5.5

The current in a 10-mH inductor has the waveform shown in Fig. 5.8a. Determine the voltage waveform.

SOLUTION Using Eq. (5.8) and noting that

$$i(t) = \frac{20 \times 10^{-3} t}{2 \times 10^{-3}} \qquad 0 \le t \le 2 \text{ ms}$$

$$i(t) = \frac{-20 \times 10^{-3} t}{2 \times 10^{-3}} + 40 \times 10^{-3} \qquad 2 \le t \le 4 \text{ ms}$$

and

$$i(t) = 0 \qquad 4 \text{ ms} < t$$

we find that

$$v(t) = \left(10 \times 10^{-3}\right) \frac{20 \times 10^{-3}}{2 \times 10^{-3}} \qquad 0 \le t \le 2 \text{ ms}$$

$$= 100 \text{ mV}$$

and

$$v(t) = \left(10 \times 10^{-3}\right) \frac{-20 \times 10^{-3}}{2 \times 10^{-3}} \qquad 2 \le t \le 4 \text{ ms}$$

$$= -100 \text{ mV}$$

and $v(t) = 0$ for $t > 4$ ms. Therefore, the voltage waveform is shown in Fig. 5.8b.

(a)

(b)

Figure 5.8 *Current and voltage waveforms for a 10-mH inductor.*

LEARNING Example 5.6

The current in a 2-mH inductor is

$$i(t) = 2 \sin 377t \text{ A}$$

Determine the voltage across the inductor and the energy stored in the inductor.

SOLUTION From Eq. (5.8), we have

$$v(t) = L \frac{di(t)}{dt}$$

$$= \left(2 \times 10^{-3}\right) \frac{d}{dt} \left(2 \sin 377t\right)$$

$$= 1.508 \cos 377t \text{ V}$$

and from Eq. (5.12),

$$w_L(t) = \tfrac{1}{2} L i^2(t)$$

$$= \tfrac{1}{2}\left(2 \times 10^{-3}\right)(2 \sin 377t)^2$$

$$= 0.004 \sin^2 377t \text{ J}$$

LEARNING Example 5.7

The voltage across a 200-mH inductor is given by the expression

$$v(t) = (1 - 3t)e^{-3t}\text{mV} \quad t \geq 0$$
$$= 0 \qquad\qquad t < 0$$

Let us derive the waveforms for the current, energy, and power.

SOLUTION The waveform for the voltage is shown in Fig. 5.9a. The current is derived from Eq. (5.10) as

$$i(t) = \frac{10^3}{200} \int_0^t (1 - 3x)e^{-3x}\,dx$$

$$= 5\left\{ \int_0^t e^{-3x}\,dx - 3\int_0^t xe^{-3x}\,dx \right\}$$

$$= 5\left\{ \frac{e^{-3x}}{-3}\bigg|_0^t - 3\left[-\frac{e^{-3x}}{9}(3x + 1) \right]_0^t \right\}$$

$$= 5te^{-3t}\text{ mA} \quad t \geq 0$$
$$= 0 \qquad\qquad t < 0$$

A plot of the current waveform is shown in Fig. 5.9b. The power is given by the expression

$$p(t) = v(t)i(t)$$
$$= 5t(1 - 3t)e^{-6t}\ \mu\text{W} \qquad t \geq 0$$
$$= 0 \qquad\qquad\qquad t < 0$$

The equation for the power is plotted in Fig. 5.9c. The expression for the energy is

$$w(t) = \tfrac{1}{2}Li^2(t)$$
$$= 2.5t^2e^{-6t}\ \mu\text{J} \qquad t \geq 0$$
$$= 0 \qquad\qquad\qquad t < 0$$

This equation is plotted in Fig. 5.9d.

(a)

(b)

(c)

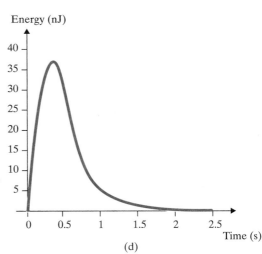

(d)

Figure 5.9
Waveforms used in Example 5.7.

E5.4 The current in a 5-mH inductor has the waveform shown in Fig. E5.4. Compute the waveform for the inductor voltage.

ANSWER

Figure E5.4

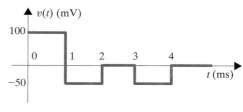

E5.5 Compute the energy stored in the magnetic field of the inductor in Extension Exercise E5.4 at $t = 1.5$ ms.

ANSWER $W = 562.5$ nJ.

CAPACITOR AND INDUCTOR SPECIFICATIONS
There are a couple of important parameters that are used to specify capacitors and inductors. In the case of capacitors, the capacitance value, working voltage, and tolerance are issues that must be considered in their application. Standard capacitor values range from a few pF to about 50 mF. Capacitors larger than 1 F are available, but will not be discussed here. Table 5.1 is a list of standard capacitor values, which are typically given in picofarads or microfarads. Although both smaller and larger ratings are available, the standard working voltage, or dc voltage rating, is typically between 6.3 V and 500 V. Manufacturers specify this working voltage since it is critical to keep the applied voltage below the breakdown point of the dielectric. Tolerance is an adjunct to the capacitance value and is usually listed as a percentage of the nominal value. Standard tolerance values are ±5%, ±10%, and ±20%. Occasionally, tolerances for single-digit pF capacitors are listed in pF. For example, 5 pF ±0.25 pF.

Table 5.1 Standard capacitor values

pF	pF	pF	pF	μF	μF	μF	μF	μF	μF	μF
1	10	100	1000	0.010	0.10	1.0	10	100	1000	10,000
	12	120	1200	0.012	0.12	1.2	12	120	1200	12,000
1.5	15	150	1500	0.015	0.15	1.5	15	150	1500	15,000
	18	180	1800	0.018	0.18	1.8	18	180	1800	18,000
2	20	200	2000	0.020	0.20	2.0	20	200	2000	20,000
	22	220	2200	0.022	0.20	2.2	22	220	2200	22,000
	27	270	2700	0.027	0.27	2.7	27	270	2700	27,000
3	33	330	3300	0.033	0.33	3.3	33	330	3300	33,000
4	39	390	3900	0.039	0.39	3.9	39	390	3900	39,000
5	47	470	4700	0.047	0.47	4.7	47	470	4700	47,000
6	51	510	5100	0.051	0.51	5.1	51	510	5100	51,000
7	56	560	5600	0.056	0.56	5.6	56	560	5600	56,000
8	68	680	6800	0.068	0.68	6.8	68	680	6800	68,000
9	82	820	8200	0.082	0.82	8.2	82	820	8200	82,000

Table 5.2 Standard inductor values

nH	nH	nH	μH	μH	μH	mH	mH	mH
1	10	100	1.0	10	100	1.0	10	100
1.2	12	120	1.2	12	120	1.2	12	
1.5	15	150	1.5	15	150	1.5	15	
1.8	18	180	1.8	18	180	1.8	18	
2	20	200	2.0	20	200	2.0	20	
2.2	22	220	2.2	22	200	2.2	22	
2.7	27	270	2.7	27	270	2.7	27	
3	33	330	3.3	33	330	3.3	33	
4	39	390	3.9	39	390	3.9	39	
5	47	470	4.7	47	470	4.7	47	
6	51	510	5.1	51	510	5.1	51	
7	56	560	5.6	56	560	5.6	56	
8	68	680	6.8	68	680	6.8	68	
9	82	820	8.2	82	820	8.2	82	

There are two principle inductor specifications, inductance and resistance. Standard commercial inductances range from about 1 nH to around 100 mH. Larger inductances can, of course, be custom built for a price. Table 5.2 lists the standard inductor values. The current rating for inductors typically extends from a few dozen mA's to about 1 A. Tolerances are typically 5% or 10% of the specified value.

As indicated in Chapter 2, wire-wound resistors are simply coils of wire, and therefore it is only logical that inductors will have some resistance. The major difference between wire-wound resistors and inductors is the wire material. High-resistance materials such as Nichrome are used in resistors and low-resistance copper is used in inductors. The resistance of the copper wire is dependent on the length and diameter of the wire. Table 5.3 lists the American Wire Gauge (AWG) standard wire diameters and the resulting resistance per foot for copper wire.

Table 5.3 Resistance per foot of solid copper wire

AWG No.	Diameter (in.)	mΩ/ft
12	0.0808	1.59
14	0.0641	2.54
16	0.0508	4.06
18	0.0400	6.50
20	0.0320	10.4
22	0.0253	16.5
24	0.0201	26.2
26	0.0159	41.6
28	0.0126	66.2
30	0.0100	105
32	0.0080	167
34	0.0063	267
36	0.0049	428
38	0.0039	684
40	0.0031	1094

LEARNING Example 5.8

We wish to find the possible range of capacitance values for a 51-mF capacitor that has a tolerance of 20%.

SOLUTION The minimum capacitor value is $0.8C = 40.8$ mF, and the maximum capacitor value is $1.2C = 61.2$ mF.

LEARNING Example 5.9

The capacitor in Fig. 5.10a is a 100-nF capacitor with a tolerance of 20%. If the voltage waveform is as shown in Fig. 5.10b, let us graph the current waveform for the minimum and maximum capacitor values.

SOLUTION The maximum capacitor value is $1.2C = 120$ nF,

and the minimum capacitor value is $0.8C = 80$ nF. The maximum and minimum capacitor currents, obtained from the equation

$$i(t) = C\frac{dv(t)}{dt}$$

are shown in Fig. 5.10c.

(a)

(b)

LEARNING by Doing

D 5.3 Find the possible range of capacitance values for the following capacitors:

(a) 100 pF with a tolerance of 10%

(b) 8.2 nF with a tolerance of 5%

ANSWER *(a) 90 pF to 110 pF, (b) 7.79 nF to 8.61 nF*

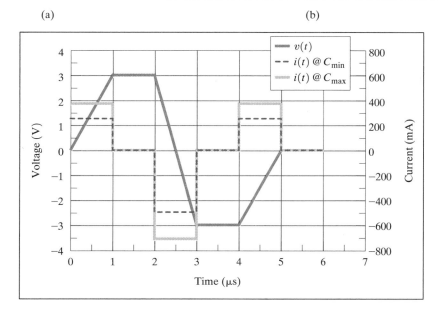

(c)

Figure 5.10
Circuit and graphs used in Example 5.9.

LEARNING Example 5.10

The inductor in Fig. 5.11a is a 100-μH inductor with a tolerance of 10%. If the current waveform is as shown in Fig. 5.11b, let us graph the voltage waveform for the minimum and maximum inductor values.

SOLUTION The maximum inductor value is $1.1L = 110 \mu H$, and the minimum inductor value is $0.9L = 90 \mu H$. The max-

imum and minimum inductor voltages, obtained from the equation

$$v(t) = L \frac{di(t)}{dt}$$

are shown in Fig. 5.11c.

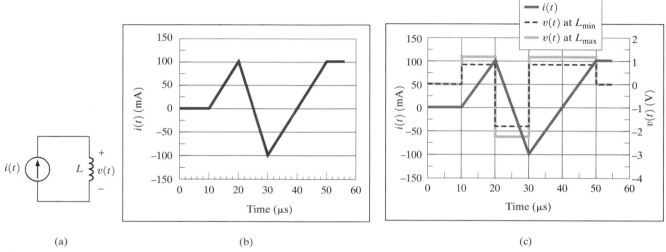

(a) (b) (c)

Figure 5.11
Circuit and graphs used in Example 5.10.

5.3 Capacitor and Inductor Combinations

SERIES CAPACITORS If a number of capacitors are connected in series, their equivalent capacitance can be calculated using KVL. Consider the circuit shown in Fig. 5.12a. For this circuit

$$v(t) = v_1(t) + v_2(t) + v_3(t) + \cdots + v_N(t) \qquad \textbf{5.13}$$

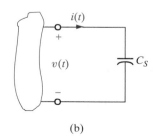

Figure 5.12
Equivalent circuit for N series-connected capacitors.

(a) (b)

but

$$v_i(t) = \frac{1}{C_i} \int_{t_0}^{t} i(t)\, dt + v_i(t_0)$$ **5.14**

Therefore, Eq. (5.13) can be written as follows using Eq. (5.14):

$$v(t) = \left(\sum_{i=1}^{N} \frac{1}{C_i} \right) \int_{t_0}^{t} i(t)\, dt + \sum_{i=1}^{N} v_i(t_0)$$ **5.15**

$$= \frac{1}{C_S} \int_{t_0}^{t} i(t)\, dt + v(t_0)$$ **5.16**

where

$$v(t_0) = \sum_{i=1}^{N} v_i(t_0)$$

and

$$\frac{1}{C_S} = \sum_{i=1}^{N} \frac{1}{C_i} = \frac{1}{C_1} + \frac{1}{C_2} + \cdots + \frac{1}{C_N}$$ **5.17**

LEARNING Hint

Capacitors in series combine like resistors in parallel.

Thus, the circuit in Fig. 5.12b is equivalent to that in Fig. 5.12a under the conditions stated previously.

It is also important to note that since the same current flows in each of the series capacitors, each capacitor gains the same charge in the same time period. The voltage across each capacitor will depend on this charge and the capacitance of the element.

LEARNING Example 5.11

Determine the equivalent capacitance and the initial voltage for the circuit shown in Fig. 5.13.

Figure 5.13
Circuit containing multiple capacitors with initial voltages.

SOLUTION Note that these capacitors must have been charged before they were connected in series or else the charge of each would be equal and the voltages would be in the same direction.

The equivalent capacitance is

$$\frac{1}{C_S} = \frac{1}{2} + \frac{1}{3} + \frac{1}{6}$$

where all capacitance values are in microfarads.

Therefore, $C_S = 1$ μF and, as seen from the figure, $v(t_0) = -3$ V. Note that the total energy stored in the circuit is

$$w(t_0) = \frac{1}{2}\left[2 \times 10^{-6}(2)^2 + 3 \times 10^{-6}(-4)^2 + 6 \times 10^{-6}(-1)^2 \right]$$

$$= 31\ \mu J$$

However, the energy recoverable at the terminals is

$$w_C(t_0) = \frac{1}{2} C_S v^2(t)$$

$$= \frac{1}{2}\left[1 \times 10^{-6}(-3)^2 \right]$$

$$= 4.5\ \mu J$$

LEARNING Example 5.12

Two previously uncharged capacitors are connected in series and then charged with a 12-V source. One capacitor is 30 μF and the other is unknown. If the voltage across the 30-μF capacitor is 8 V, find the capacitance of the unknown capacitor.

SOLUTION The charge on the 30-μF capacitor is

$$Q = CV = (30\ \mu F)(8\ V) = 240\ \mu C$$

Since the same current flows in each of the series capacitors, each capacitor gains the same charge in the same time period.

$$C = \frac{Q}{V} = \frac{240\ \mu C}{4V} = 60\ \mu F$$

PARALLEL CAPACITORS To determine the equivalent capacitance of N capacitors connected in parallel, we employ KCL. As can be seen from Fig. 5.14a,

$$i(t) = i_1(t) + i_2(t) + i_3(t) + \cdots + i_N(t) \qquad \textbf{5.18}$$

$$= C_1 \frac{dv(t)}{dt} + C_2 \frac{dv(t)}{dt} + C_3 \frac{dv(t)}{dt} + \cdots + C_N \frac{dv(t)}{dt}$$

$$= \left(\sum_{i=1}^{N} C_i \right) \frac{dv(t)}{dt}$$

$$= C_p \frac{dv(t)}{dt} \qquad \textbf{5.19}$$

LEARNING Hint

Capacitors in parallel combine like resistors in series.

where

$$C_p = C_1 + C_2 + C_3 + \cdots + C_N \qquad \textbf{5.20}$$

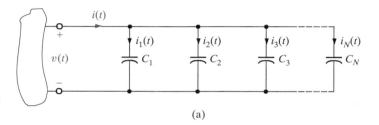

Figure 5.14
Equivalent circuit for N capacitors connected in parallel.

(a)

(b)

LEARNING Example 5.13

Determine the equivalent capacitance at terminals A-B of the circuit shown in Fig. 5.15.

SOLUTION $C_p = 15 \, \mu\text{F}$

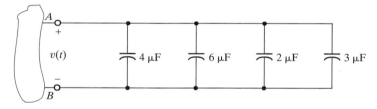

Figure 5.15
Circuit containing multiple capacitors in parallel.

LEARNING EXTENSION

E5.6 Two initially uncharged capacitors are connected as shown in Fig. E5.6. After a period of time, the voltage reaches the value shown. Determine the value of C_1.

ANSWER $C_1 = 4 \, \mu\text{F}$.

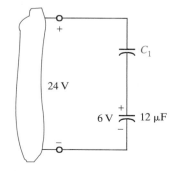

Figure E5.6

E5.7   Compute the equivalent capacitance of the network in Fig. E5.7.

ANSWER $C_{eq} = 1.5\ \mu F.$

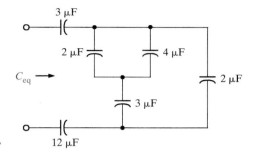

Figure E5.7

SERIES INDUCTORS

SERIES INDUCTORS   If N inductors are connected in series, the equivalent inductance of the combination can be determined as follows. Referring to Fig. 5.16a and using KVL, we see that

$$v(t) = v_1(t) + v_2(t) + v_3(t) + \cdots + v_N(t) \qquad \textbf{5.21}$$

and therefore,

$$v(t) = L_1 \frac{di(t)}{dt} + L_2 \frac{di(t)}{dt} + L_3 \frac{di(t)}{dt} + \cdots + L_N \frac{di(t)}{dt} \qquad \textbf{5.22}$$

$$= \left(\sum_{i=1}^{N} L_i \right) \frac{di(t)}{dt}$$

$$= L_S \frac{di(t)}{dt} \qquad \textbf{5.23}$$

where

$$L_S = \sum_{i=1}^{N} L_i = L_1 + L_2 + \cdots + L_N \qquad \textbf{5.24}$$

Therefore, under this condition the network in Fig. 5.16b is equivalent to that in Fig. 5.16a.

LEARNING Hint

Inductors in series combine like resistors in series.

(a)

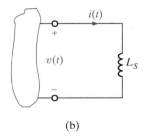

(b)

Figure 5.16
Equivalent circuit for N series-connected inductors.

LEARNING Example 5.14

Find the equivalent inductance of the circuit shown in Fig. 5.17.

SOLUTION   The equivalent inductance of the circuit shown in Fig. 5.17 is

$$L_S = 1H + 2H + 4H$$

$$= 7H$$

Figure 5.17
Circuit containing multiple inductors.

PARALLEL INDUCTORS Consider the circuit shown in Fig. 5.18a, which contains N parallel inductors. Using KCL, we can write

$$i(t) = i_1(t) + i_2(t) + i_3(t) + \cdots + i_N(t) \qquad \textbf{5.25}$$

However,

$$i_j(t) = \frac{1}{L_j} \int_{t_0}^t v(x)\,dx + i_j(t_0) \qquad \textbf{5.26}$$

Substituting this expression into Eq. (5.25) yields

$$i(t) = \left(\sum_{j=1}^N \frac{1}{L_j} \right) \int_{t_0}^t v(x)\,dx + \sum_{j=1}^N i_j(t_0) \qquad \textbf{5.27}$$

$$= \frac{1}{L_p} \int_{t_0}^t v(x)\,dx + i(t_0) \qquad \textbf{5.28}$$

where

$$\frac{1}{L_p} = \frac{1}{L_1} + \frac{1}{L_2} + \frac{1}{L_3} + \cdots + \frac{1}{L_N} \qquad \textbf{5.29}$$

LEARNING Hint

Inductors in parallel combine like resistors in parallel.

and $i(t_0)$ is equal to the current in L_p at $t = t_0$. Thus, the circuit in Fig. 5.18b is equivalent to that in Fig. 5.18a under the conditions stated previously.

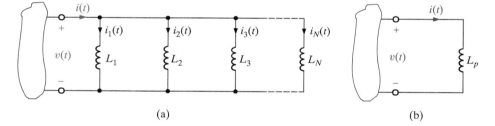

Figure 5.18
Equivalent circuits for N inductors connected in parallel.

(a)   (b)

LEARNING Example 5.15

Determine the equivalent inductance and the initial current for the circuit shown in Fig. 5.19.

Figure 5.19
Circuit containing multiple inductors with initial currents.

SOLUTION The equivalent inductance is

$$\frac{1}{L_p} = \frac{1}{12} + \frac{1}{6} + \frac{1}{4}$$

where all inductance values are in millihenrys

$$L_P = 2 \text{ mH}$$

and the initial current is $i(t_0) = -1$ A.

The previous material indicates that capacitors combine like conductances, whereas inductances combine like resistances.

E5.8   Determine the equivalent inductance of the network in Fig. E5.8 if all inductors are 6 mH.   **ANSWER**   9.429 mH.

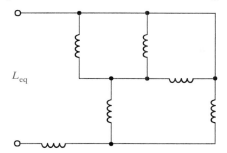

Figure E5.8

5.4 *RC* Operational Amplifier Circuits

Two very important *RC* op-amp circuits are the differentiator and the integrator. These circuits are derived from the circuit for an inverting op-amp by replacing the resistors R_1 and R_2, respectively, by a capacitor. Consider, for example, the circuit shown in Fig. 5.20a. The circuit equations are

$$C_1 \frac{d}{dt}(v_1 - v_-) + \frac{v_o - v_-}{R_2} = i_-$$

However, $v_- = 0$ and $i_- = 0$. Therefore,

$$v_o(t) = -R_2 C_1 \frac{dv_1(t)}{dt} \qquad \textbf{5.30}$$

Thus, the output of the op-amp circuit is proportional to the derivative of the input.

The circuit equations for the op-amp configuration in Fig. 5.20b are

$$\frac{v_1 - v_-}{R_1} + C_2 \frac{d}{dt}(v_o - v_-) = i_-$$

but since $v_- = 0$ and $i_- = 0$, the equation reduces to

$$\frac{v_1}{R_1} = -C_2 \frac{dv_o}{dt}$$

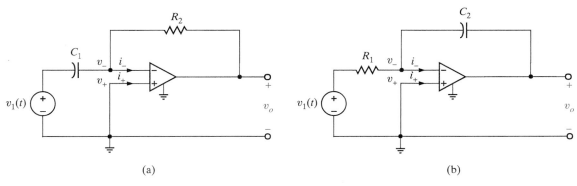

(a)   (b)

Figure 5.20   *Differentiator and integrator operational amplifier circuits.*

or

$$v_o(t) = \frac{-1}{R_1 C_2} \int_{-\infty}^{t} v_1(x)\,dx$$

$$= \frac{-1}{R_1 C_2} \int_{0}^{t} v_1(x)\,dx + v_o(0) \qquad \textbf{5.31}$$

If the capacitor is initially discharged, then $v_o(0) = 0$; hence,

$$v_o(t) = \frac{-1}{R_1 C_2} \int_{0}^{t} v_1(x)\,dx \qquad \textbf{5.32}$$

Thus, the output voltage of the op-amp circuit is proportional to the integral of the input voltage.

LEARNING Example 5.16

The waveform in Fig. 5.21a is applied at the input of the differentiator circuit shown in Fig. 5.20a. If $R_2 = 1\ \text{k}\Omega$ and $C_1 = 2\ \mu\text{F}$, determine the waveform at the output of the op-amp.

SOLUTION Using Eq. (5.30), we find that the op-amp output is

$$v_o(t) = -R_2 C_1 \frac{dv_1(t)}{dt}$$

$$= -(2)10^{-3} \frac{dv_1(t)}{dt}$$

$dv_1(t)/dt = (2)10^3$ for $0 \le t < 5$ ms, and therefore,

$$v_o(t) = -4\ \text{V} \qquad 0 \le t < 5\ \text{ms}$$

$dv_1(t)/dt = -(2)10^3$ for $5 \le t < 10$ ms, and therefore,

$$v_o(t) = 4\ \text{V} \qquad 5 \le t < 10\ \text{ms}$$

Hence, the output waveform of the differentiator is shown in Fig. 5.21b.

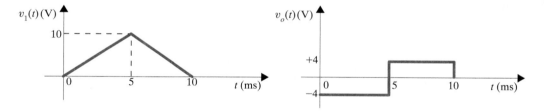

Figure 5.21
Input and output waveforms for a differentiator circuit.

LEARNING Example 5.17

If the integrator shown in Fig. 5.20b has the parameters $R_1 = 5\ \text{k}\Omega$ and $C_2 = 0.2\ \mu\text{F}$, determine the waveform at the op-amp output if the input waveform is given as in Fig. 5.22a and the capacitor is initially discharged.

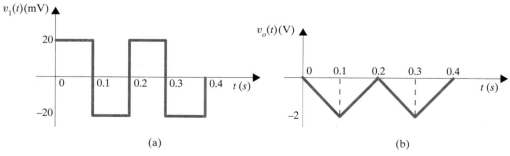

(a) (b)

Figure 5.22 Input and output waveforms for an integrator circuit.

SOLUTION   The integrator output is given by the expression

$$v_o(t) = \frac{-1}{R_1 C_2} \int_0^t v_1(x)\,dx$$

which with the given circuit parameters is

$$v_o(t) = -10^3 \int_0^t v_1(x)\,dx$$

In the interval $0 \le t < 0.1$ s, $v_1(t) = 20$ mV. Hence,

$$v_o(t) = -10^3(20)10^{-3}t \qquad 0 \le t < 0.1 \text{ s}$$
$$= -20t$$

At $t = 0.1$ s, $v_o(t) = -2$ V. In the interval from 0.1 to 0.2 s, the integrator produces a positive slope output of $20t$ from $v_o(0.1) = -2$ V to $v_o(0.2) = 0$ V. This waveform from $t = 0$ to $t = 0.2$ s is repeated in the interval $t = 0.2$ to $t = 0.4$ s, and therefore, the output waveform is shown in Fig. 5.22b.

E5.9   The waveform in Fig. E5.9 is applied to the input terminals of the op-amp differentiator circuit. Determine the differentiator output waveform if the op-amp circuit parameters are $C_1 = 2$ F and $R_2 = 2\ \Omega$.

ANSWER

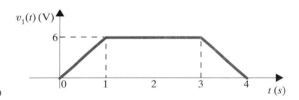

Figure E5.9

Learning by Application

LEARNING Example 5.18

An excellent example of capacitor operation is the memory inside a personal computer. This memory, called dynamic random access memory (DRAM) contains as many as half a billion data storage sites called cells (circa 2000). Expect this number to roughly double every two years for the next decade or two. Let us examine in some detail the operation of a single DRAM cell.

SOLUTION   Figure 5.23a shows a simple model for a DRAM cell. Data are stored on the cell capacitor in true/false (or 1/0) format, where a large capacitor voltage represents a true condition and a low voltage represents a false condition. The switch closes to allow access from the processor to the DRAM cell. Current source I_{leak} is an unintentional, or parasitic, current

(a)   (b)   (c)

Figure 5.23   *A simple circuit model showing (a) the DRAM memory cell, (b) the effect of charge leakage from the cell capacitor, and (c) cell conditions at the beginning of a read operation.*

(continued)

that models charge leakage from the capacitor. Another parasitic model element is the capacitance, C_{out}, the capacitance of the wiring connected to the output side of the cell. Both I_{leak} and C_{out} have enormous impacts on DRAM performance and design.

Consider storing a true condition in the cell. A high voltage of 3.0 V is applied at node I/O and the switch is closed, causing the voltage on C_{cell} to quickly rise to 3.0 V. We open the switch and the data are stored. During the store operation the charge, energy and number of electrons, n, used are

$$Q = CV = (50 \times 10^{-15})(3) = 150 \text{ fC}$$
$$W = \tfrac{1}{2}CV^2 = (0.5)(50 \times 10^{-15})(3^2) = 225 \text{ fJ}$$
$$n = Q/q = 150 \times 10^{-15}/(1.6 \times 10^{-19}) = 937{,}500 \text{ electrons}$$

Once data are written, the switch opens and the capacitor begins to discharge through I_{leak}. A measure of DRAM quality is the time required for the data voltage to drop by half, from 3.0 V to 1.5 V. Let us call this time t_H. For the capacitor, we know

$$v_{cell}(t) = \frac{1}{C_{cell}} \int i_{cell} \, dt \text{ V}$$

where, from Fig. 5.23b, $i_{cell}(t) = -I_{leak}$. Performing the integral yields

$$v_{cell}(t) = \frac{1}{C_{cell}} \int (-I_{leak}) \, dt = -\frac{I_{leak}}{C_{cell}} t + K$$

We know that at $t = 0$, $v_{cell} = 3$ V. Thus, $K = 3$ and the cell voltage is

$$v_{cell}(t) = 3 - \frac{I_{leak}}{C_{cell}} t \text{ V} \qquad \textbf{5.33}$$

Substituting $t = t_H$ and $v_{cell}(t_H) = 1.5$ V into Eq. 5.33 and solving for t_H yields $t_H = 15$ ms. Thus, the cell data are gone in only a few milliseconds! The solution is rewriting the data before it can disappear. This technique, called *refresh*, is a must for all DRAM using this one-transistor cell.

To see the affect of C_{out}, consider reading a fully charged $(v_{cell} = 3.0 \text{ V})$ true condition. The I/O line is usually precharged to half the data voltage. In this example, that would be 1.5 V as seen in Fig. 5.23c. (To isolate the effect of C_{out}, we have removed I_{leak}.) Next, the switch is closed. What happens next is best viewed as a conservation of charge. Just before the switch closes, the total stored charge in the circuit is

$$Q_T = Q_{out} + Q_{cell} = V_{I/O}C_{out} + V_{cell}C_{cell} =$$
$$(1.5)(450 \times 10^{-15}) + (3)(50 \times 10^{-15}) = 825 \text{ fC}$$

When the switch closes, the capacitor voltages are the same (let us call it V_o) and the total charge is unchanged.

$$Q_T = 825 \text{ fC} = V_o C_{out} + V_o C_{cell} =$$
$$V_o(450 \times 10^{-15} + 50 \times 10^{-15})$$

and

$$V_o = 1.65 \text{ V}$$

Thus, the change in voltage at $V_{I/O}$ during the read operation is only 0.15 V. A very sensitive amplifier is required to quickly detect such a small change. In DRAMs, these amplifiers are called *sense amps*. How can v_{cell} change instantaneously when the switch closes? It cannot. In an actual DRAM cell, a transistor, which has a small equivalent resistance, acts as the switch. The resulting RC time constant is very small, indicating a very fast circuit. Recall that we are not analyzing the cell's speed—only the final voltage value, V_o. As long as the power lost in the switch is small compared to the capacitor energy, we can be comfortable in neglecting the switch resistance. By the way, if a false condition (zero volts) were read from the cell, then V_o would drop from its precharged value of 1.5 V to 1.35 V—a negative change of 0.15 V. This symmetric voltage change is the reason for precharging the I/O node to half the data voltage. Review the effects of I_{leak} and C_{out}. You will find that eliminating them would greatly simplify the refresh requirement and improve the voltage swing at node I/O when reading data. DRAM designers earn a very good living trying to do just that.

LEARNING Example 5.19

Traditionally, integrated circuits, or ICs, are built on silicon die and packaged by gluing the die to a mechanical base and connecting small wires called wirebonds (1 μm diameter) from the die to the package's external leads as seen in Fig. 5.24a. The wirebond, like any piece of wire, has inductance (about 2 nH), which can affect circuit performance. Since

$$v_L(t) = L \frac{di_L(t)}{dt} \qquad \textbf{5.34}$$

any IC that has large current transients has undesirable voltage transients across its wirebonds. This is particularly trou-

blesome in the ground wirebond of digital switching ICs since the current there can be large. The voltage transient on the ground connection is called *ground bounce*. From (Eq. 5.34), there are three ways to decrease ground bounce: lower the inductance, decrease the current level, or operate at slower speeds. The last solution is no solution at all since we all want faster and faster electronics. However, the first two ideas have merit. In fact, manufacturers of very-high-speed ICs are presently (circa 2000) changing from wirebond to flip-chip connections where small solder balls provide electrical contact and the die is mounted upside down as seen in

Figs. 5.24b and c. This decreases the parasitic inductance to around 0.1 nH. Let us investigate the effects of flip-chip ball inductance on ground bounce.

SOLUTION Consider the circuit in Fig. 5.25a where $i_G(t)$ and L_{ball} model the current required by a single digital gate (smallest subcircuit) and the flip-chip ball inductance, respectively. The resulting ground bounce, $v_{GB}(t)$, shown in Fig. 5.25b, is ± 0.1 mV, only 0.004% of a typical 2.5-V supply voltage. Now consider a more realistic scenario where the IC contains 100,000 gates with 10% of them switching at the same time, 10,000 switching gates.

The current through L_{ball} increases 10,000 times and, from Eq. (5.34), ground bounce increases by the same factor to ± 1 V, an unacceptable level. Thus, we resort to the second solution: less current. Instead of having all the logic gates connected to the same ground connection, we use many ground connections (balls) and divide the gates among them. For example, if we divide our 100,000 gates equally between 100 ground balls, then there are only 1000 gates/ball. If only 10% of those switch at the same time (100 gates), then the ground bounce, shown in Fig. 5.25b, is only 10 mV, 0.4% of the supply voltage, which can be tolerated.

(a)   (b)   (c)

Figure 5.24
In wirebonded packages, (a) small wirebonds connect the die to the package. In flip-chip attachment, small solder balls are placed on the die's contact points (b), then the die is flipped onto the package (c). Under heat, the solder flows, forming the electrical contacts.

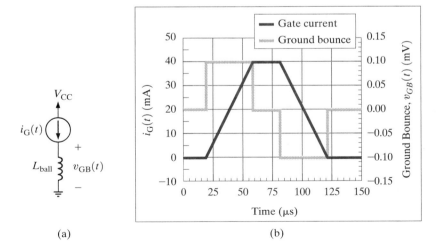

(a)   (b)

Figure 5.25
(a) An equivalent circuit for a single logic gate and (b) the resulting ground bounce.

LEARNING Check

Summary

▶ The important (dual) relationships for capacitors and inductors are as follows:

$$q = Cv$$

$$i(t) = C\frac{dv(t)}{dt} \qquad v(t) = L\frac{di(t)}{dt}$$

$$v(t) = \frac{1}{C}\int_{-\infty}^{t} i(x)\,dx \qquad i(t) = \frac{1}{L}\int_{-\infty}^{t} v(x)\,dx$$

$$p(t) = Cv(t)\frac{dv(t)}{dt} \qquad p(t) = Li(t)\frac{di(t)}{dt}$$

$$w_C(t) = \tfrac{1}{2}Cv^2(t) \qquad w_L(t) = \tfrac{1}{2}Li^2(t)$$

▶ The passive sign convention is used with capacitors and inductors.

▶ In dc steady state a capacitor looks like an open circuit and an inductor looks like a short circuit.

▶ Leakage resistance is present in practical capacitors and inductors.

▶ When capacitors are interconnected, their equivalent capacitance is determined as follows: Capacitors in series combine like resistors in parallel and capacitors in parallel combine like resistors in series.

▶ When inductors are interconnected, their equivalent inductance is determined as follows: Inductors in series combine like resistors in series, and inductors in parallel combine like resistors in parallel.

▶ *RC* operational amplifier circuits can be used to differentiate or integrate an electrical signal.

Problems

SECTION 5.1

5.1 An uncharged 100-μF capacitor is charged by a constant current of 1 mA. Find the voltage across the capacitor after 4 s.

5.2 A 12-μF capacitor has an accumulated charge of 480 μC. Determine the voltage across the capacitor.

5.3 A capacitor has an accumulated charge of 600 μC with 5 V across it. What is the value of capacitance?

5.4 A 25-μF capacitor initially charged to −10 V is charged by a constant current of 2.5 μA. Find the voltage across the capacitor after $2\frac{1}{2}$ min.

5.5 The energy that is stored in a 25-μF capacitor is $w(t) = 12\sin^2 377t$ J. Find the current in the capacitor.

5.6 An uncharged 10-mF capacitor is charged by the current $i(t) = 10\cos 377t$ mA. Find (a) the expression for the voltage across the capacitor and (b) the expression for the power.

5.7 The voltage across a 100-μF capacitor is given by the expression $v(t) = 120\sin 377t$ V. Find (a) the current in the capacitor and (b) the expression for the energy stored in the element.

5.8 A capacitor is charged by a constant current of 2 mA and results in a voltage increase of 12 V in a 10-s interval. What is the value of the capacitance?

5.9 The current in a 100-μF capacitor is shown in Fig. P5.9. Determine the waveform for the voltage across the capacitor if it is initially uncharged.

Figure P5.9

5.10 The voltage across a 100-μF capacitor is shown in Fig. P5.10. Compute the waveform for the current in the capacitor.

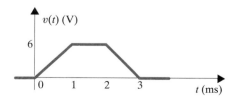

Figure P5.10

5.11 The voltage across a 6-μF capacitor is shown in Fig. P5.11. Compute the waveform for the current in the capacitor.

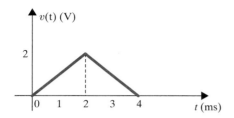

Figure P5.11

5.12 The voltage across a 50-μF capacitor is shown in Fig. P5.12. Determine the current waveform.

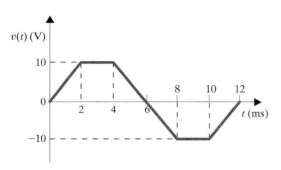

Figure P5.12

5.13 The voltage across a 2-μF capacitor is given by the waveform in Fig. P5.13. Compute the current waveform.

Figure P5.13

5.14 The voltage across a 0.1-F capacitor is given by the waveform in Fig. P5.14. Find the waveform for the current in the capacitor.

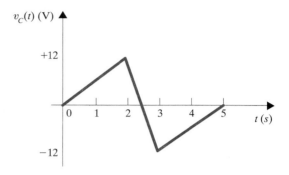

Figure P5.14

5.15 The waveform for the current in a 200-μF capacitor is shown in Fig. P5.15. Determine the waveform for the capacitor voltage.

Figure P5.15

5.16 Draw the waveform for the current in a 12-μF capacitor when the capacitor voltage is as described in Fig. P5.16.

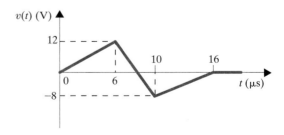

Figure P5.16

5.17 Draw the waveform for the current in a 3-µF capacitor when the voltage across the capacitor is given in Fig. P5.17.

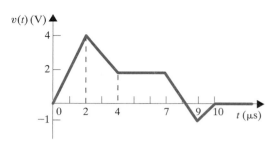

Figure P5.17

5.18 The waveform for the current in a 100-µF initially uncharged capacitor is shown in Fig. P5.18. Determine the waveform for the capacitor's voltage.

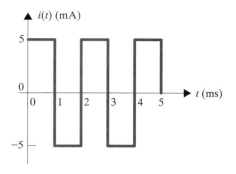

Figure P5.18

5.19 The voltage across a 6-µF capacitor is given by the waveform in Fig. P5.19. Plot the waveform for the capacitor current.

Figure P5.19

SECTION 5.2 ————————————————————

5.20 The current in an inductor changes from 0 to 200 mA in 4 ms and induces a voltage of 100 mV. What is the value of the inductor?

5.21 The current in a 100-mH inductor is $i(t) = 2 \sin 377t$ A. Find (a) the voltage across the inductor and (b) the expression for the energy stored in the element.

5.22 A 10-mH inductor has a sudden current change from 200 mA to 100 mA in 1 ms. Find the induced voltage.

5.23 The induced voltage across a 10-mH inductor is $v(t) = 120 \cos 377t$ V. Find (a) the expression for the inductor current and (b) the expression for the power.

5.24 The current in a 25-mH inductor is given by the expressions

$$i(t) = 0 \qquad\qquad t < 0$$
$$i(t) = 10(1 - e^{-t})\text{mA} \qquad t > 0$$

Find (a) the voltage across the inductor and (b) the expression for the energy stored in it.

5.25 Given the data in the previous problem, find the voltage across the inductor and the energy stored in it after 1 s.

5.26 The current in a 50-µH inductor is specified as follows:

$$i(t) = 0 \qquad\qquad t < 0$$
$$i(t) = 2te^{-4t}\text{A} \qquad t > 0$$

Find (a) the voltage across the inductor, (b) the time at which the current is a maximum, and (c) the time at which the voltage is a minimum.

5.27 The current

$$i(t) = 0 \qquad\qquad t < 0$$
$$i(t) = 100e^{-t/10}\text{A} \qquad t > 0$$

is present in a 150-mH inductor. Find both the voltage across the inductor and the energy stored in it after 5 sec.

5.28 The current in a 10-mH inductor is shown in Fig. P5.28. Find the voltage across the inductor.

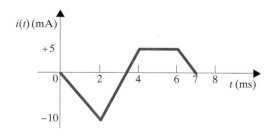

Figure P5.28

5.29 The current in a 50-mH inductor is given in Fig. P5.29. Sketch the inductor voltage.

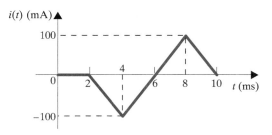

Figure P5.29

5.30 The current in a 16-mH inductor is given by the waveform in Fig. P5.30. Find the waveform for the voltage across the inductor.

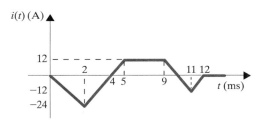

Figure P5.30

5.31 Draw the waveform for the voltage across a 10-mH inductor when the inductor current is given by the waveform shown in Fig. P5.31.

Figure P5.31

5.32 The voltage across a 10-mH inductor is shown in Fig. P5.32. Determine the waveform for the inductor current. $v(t) = 0$, $t < 0$.

Figure P5.32

5.33 The waveform for the voltage across a 20-mH inductor is shown in Fig. P5.33. Compute the waveform for the inductor current. $v(t) = 0, t < 0$.

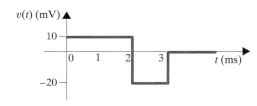

Figure P5.33

5.34 The voltage across a 2-H inductor is given by the waveform shown in Fig. P5.34. Find the waveform for the current in the inductor. $v(t) = 0, t < 0$.

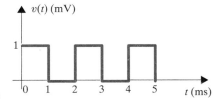

Figure P5.34

5.35 Find the possible capacitance range of the following capacitors.

    (a) 0.068 μF with a tolerance of 10%

    (b) 120 pF with a tolerance of 20%

    (c) 39 μF with a tolerance of 20%

5.36 The capacitor in Fig. P5.36a is 51 nF with a tolerance of 10%. Given the voltage waveform in Fig. P5.36b, graph the current $i(t)$ for the minimum and maximum capacitor values.

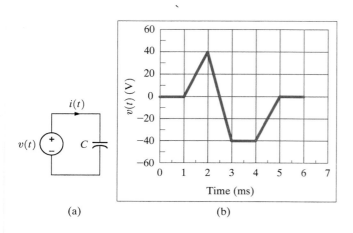

(a)　　　　　　(b)

Figure P5.36

5.37 Find the possible inductance range of the following capacitors.

   (a)　10 mH with a tolerance of 10%

   (b)　2.0 nH with a tolerance of 5%

   (c)　68 μH with a tolerance of 10%

5.38 The inductor in Fig. P5.38a is 330 μH with a tolerance of 5%. Given the current waveform in Fig. 5.38b, graph the voltage $v(t)$ for the minimum and maximum inductor values.

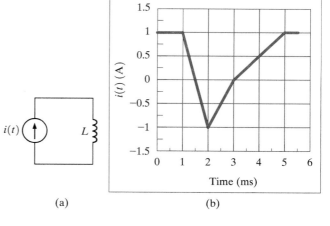

(a)　　　　　　(b)

Figure P5.38

5.39 The inductor in Fig. P5.39a is 4.7 μH with a tolerance of 20%. Given the current waveform in Fig. 5.39b, graph the voltage $v(t)$ for the minimum and maximum inductor values.

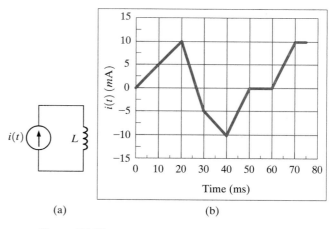

(a)　　　　　　(b)

Figure P5.39

SECTION 5.3

5.40 What values of capacitance can be obtained by interconnecting a 4-μF capacitor, a 6-μF capacitor, and a 12-μF capacitor?

5.41 Given a 1-, 3-, and 4-μF capacitor, can they be interconnected to obtain an equivalent 2-μF capacitor?

5.42 Given four 2-μF capacitors, find the maximum value and minimum value that can be obtained by interconnecting the capacitors in series/parallel combinations.

5.43 The two capacitors in Fig. P5.43 were charged and then connected as shown. Determine the equivalent capacitance, the initial voltage at the terminals, and the total energy stored in the network.

Figure P5.43

5.44 The two capacitors shown in Fig. P5.44 have been connected for some time and have reached their present values. Find V_o.

Figure P5.44

5.45 The three capacitors shown in Fig. P5.45 have been connected for some time and have reached their present values. Find V_1 and V_2.

Figure P5.45

5.46 Select the value of C to produce the desired total capacitance of $C_T = 2$ μF in the circuit in Fig. P5.46.

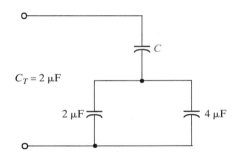

Figure P5.46

5.47 Select the value of C to produce the desired total capacitance of $C_T = 1$ μF in the circuit in Fig. P5.47.

Figure P5.47

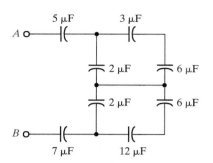

5.48 Find the equivalent capacitance at terminals *A-B* in Fig. P5.48.

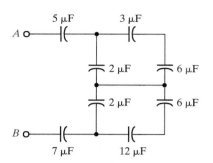

Figure P5.48

5.49 Determine the total capacitance of the network in Fig. P5.49.

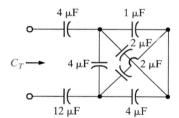

Figure P5.49

5.50 Find C_T in the network in Fig. P5.50 if (a) the switch is open and (b) the switch is closed.

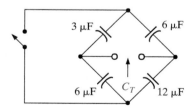

Figure P5.50

5.51 Find the total capacitance C_T of the network in Fig. P5.51.

Figure P5.51

5.52 Compute the equivalent capacitance of the network in Fig. P5.52 if all the capacitors are 4 μF.

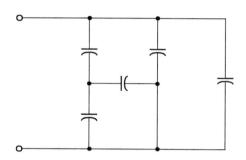

Figure P5.52

5.53 If all the capacitors in Fig. P5.53 are 6 μF, find C_{eq}.

Figure P5.53

5.54 Given the capacitors in Fig. P5.54 are $C_1 = 2.0$ μF with a tolerance of 2% and $C_2 = 2.0$ μF with a tolerance of 20%, find the following.

(a) The nominal value of C_{eq}.

(b) The minimum and maximum possible values of C_{eq}.

(c) The percent errors of the minimum and maximum values.

Figure P5.54

5.55 The capacitor values for the network in Fig. P5.55 are $C_1 = 0.1$ μF with a tolerance of 10%, $C_2 = 0.33$ μF with a tolerance of 20%, and $C_3 = 1$ μF with a tolerance of 10%. Find the following.

(a) The nominal value of C_{eq}.

(b) The minimum and maximum possible values of C_{eq}.

(c) The percent errors of the minimum and maximum values.

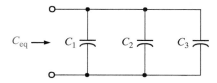

Figure P5.55

5.56 Select the value of L that produces a total inductance of $L_T = 10$ mH in the circuit in Fig. P5.56.

Figure P5.56

5.57 Find the value of L in the network in Fig. P5.57 so that the total inductance L_T will be 2 mH.

Figure P5.57

5.58 Find the value of L in the network in Fig. P5.58 so that the value of L_T will be 2 mH.

Figure P5.58

5.59 Determine the inductance at terminals A-B in the network in Fig. P5.59.

Figure P5.59

5.60 Compute the equivalent inductance of the network in Fig. P5.60 if all inductors are 4 mH.

Figure P5.60

5.61 Determine the inductance at terminals A-B in the network in Fig P5.61.

Figure P5.61

5.62 Find the total inductance at the terminals of the network in Fig. P5.62.

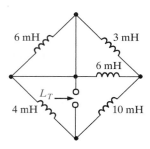

Figure P5.62

5.63 Given the network shown in Fig. P5.63, find (a) the equivalent inductance at terminals A-B with terminals C-D short circuited, and (b) the equivalent inductance at terminals C-D with terminals A-B open circuited.

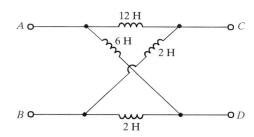

Figure P5.63

SECTION 5.4

5.64 For the network in Fig. P5.64, choose C such that

$$v_o = -10 \int v_S \, dt$$

Figure P5.64

5.65 For the network in Fig. P5.65, $v_S(t) = 120 \cos 377t$ V. Find $v_o(t)$.

5.66 For the network in Fig. P5.66, $v_S(t) = 115 \sin 377t$ V. Find $v_o(t)$.

Figure P5.65

Figure P5.66

Typical Problems Found on the FE Exam

5FE-1 Given three capacitors with values 2 μF, 4 μF, and 6 μF, can the capacitors be interconnected so that the combination is an equivalent 3 μF?

5FE-2 The current pulse shown in Fig. 5PFE-2 is applied to a 1-μF capacitor. Determine the charge on the capacitor and the energy stored.

Figure 5PFE-2

5FE-3 The two capacitors shown in Fig. 5PFE-3 have been connected for some time and have reached their present values. Determine the energy stored in the unknown capacitor C_x.

Figure 5PFE-3

6 — First- and Second-Order Transient Circuits

LEARNING Goals

We now consider circuits that are in transition from one state to another; that is, a circuit may have no forcing function and we suddenly apply one, or we instantly remove from a circuit the source of energy. The study of the circuit behavior in this transition phase we call a transient analysis. This transition is affected by the presence of a capacitor or an inductor, or both, since these two elements are capable of storing energy and releasing it over some interval of time. Our analysis includes both first-order circuits, those containing a single capacitor or inductor, and second-order circuits, those in which both a capacitor and inductor are present.

6.1 Introduction

In this chapter we perform what is normally referred to as a transient analysis. We begin our analysis with first-order circuits—that is, those that contain only a single storage element. When only a single storage element is present in the network, the network can be described by a first-order differential equation.

Our analysis involves an examination and description of the behavior of a circuit as a function of time after a sudden change in the network occurs due to switches opening or closing. Because of the presence of one or more storage elements, the circuit response to a sudden change will go through a transition period prior to settling down to a steady-state value. It is this transition period that we will examine carefully in our transient analysis.

One of the important parameters that we will examine in our transient analysis is the circuit's time constant. This is a very important network parameter because it tells us how fast the circuit will respond to changes. We can contrast two very different systems to obtain a feel for the parameter. For example, consider the model for a room air-conditioning system and the model for a single-transistor stage of amplification in a computer chip. If we change the setting for the air conditioner from 70 degrees to 60 degrees, the unit will come on and the room will begin to cool. However, the temperature measured by a thermometer in the room will fall very slowly and, thus, the time required to reach the desired temperature is long. However, if we send a trigger signal to a transistor to change state, the action may take only a few nanoseconds. These two systems will have vastly different time constants.

Our analysis of first-order circuits begins with the presentation of two techniques for performing a transient analysis: the differential equation approach, in which a differential equation is written and solved for each network, and a step-by-step approach, which takes advantage of the known form of the solution in every case. In the second-order case, both an inductor and capacitor are present simultaneously and the network is described by a second-order differential equation. Although the *RLC* circuits are more complicated than the first-order single storage circuits, we will follow a development similar to that used in the first-order case.

Our presentation will deal only with very simple circuits, since the analysis can quickly become complicated for networks that contain more than one loop or one nonreference node. Furthermore, we will demonstrate a much simpler method for handling these circuits when we cover the Laplace transform later in this book. We will analyze several networks in which the parameters have been chosen to illustrate the different types of circuit response. In addition, we will extend our PSPICE analysis techniques to the analysis of transient circuits. Finally, a number of application-oriented examples are presented and discussed.

We begin our discussion by recalling that in Chapter 5 we found that capacitors and inductors were capable of storing electric energy. In the case of a charged capacitor, the energy is stored in the electric field that exists between the positively and negatively charged plates. This stored energy can be released if a circuit is somehow connected across the capacitor that provides a path through which the negative charges move to the positive charges. As we know, this movement of charge constitutes a current. The rate at which the energy is discharged is a direct function of the parameters in the circuit that is connected across the capacitor's plates.

As an example, consider the flash circuit in a camera. Recall that the operation of the flash circuit, from a user standpoint, involves depressing the push button on the camera that triggers both the shutter and the flash and then waiting a few seconds before repeating the process to take the next picture. This operation can be modeled using the circuit in Fig. 6.1a. The voltage source and resistor R_S model the batteries that power the camera and flash. The capacitor models the energy storage, the switch models the push button, and finally the resistor R models the xenon flash lamp. Thus, if the capacitor is charged, when the switch is closed, the capacitor voltage drops and energy is released through the xenon lamp, producing the flash. In practice this

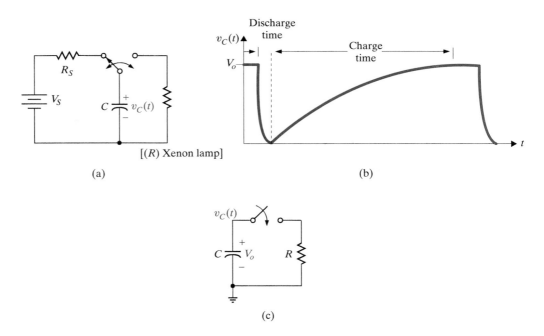

Figure 6.1
Diagrams used to describe a camera's flash circuit.

energy release takes about a millisecond and the discharge time is a function of the elements in the circuit. When the push button is released and the switch is then opened, the battery begins to recharge the capacitor. Once again, the time required to charge the capacitor is a function of the circuit elements. The discharge and charge cycles are graphically illustrated in Fig. 6.1b. Although the discharge time is very fast, it is not instantaneous. To provide further insight into this phenomenon, consider what we might call a *free-body diagram* of the right half of the network in Fig. 6.1a as shown in Fig. 6.1c (that is, a charged capacitor that is discharged through a resistor). When the switch is closed, KCL for the circuit is

$$C \frac{dv_C(t)}{dt} + \frac{v_C(t)}{R} = 0$$

or

$$\frac{dv_C(t)}{dt} + \frac{1}{RC} v_C(t) = 0$$

We will demonstrate in the next section that the solution of this equation is

$$v_C(t) = V_o e^{-t/RC}$$

Note that this function is a decaying exponential and the rate at which it decays is a function of the values of R and C. The product RC is a very important parameter, and we will give it a special name in the following discussions.

6.2 First-Order Circuits

GENERAL FORM OF THE RESPONSE EQUATIONS In our study of first-order transient circuits we will show that the solution of these circuits (i.e., finding a voltage or current) requires us to solve a first-order differential equation of the form

$$\frac{dx(t)}{dt} + ax(t) = f(t) \qquad \textbf{6.1}$$

Although there are a number of techniques for solving an equation of this type, we will obtain a general solution that we will then employ in two different approaches to transient analysis.

A fundamental theorem of differential equations states that if $x(t) = x_p(t)$ is any solution to Eq. (6.1), and $x(t) = x_c(t)$ is any solution to the homogeneous equation

$$\frac{dx(t)}{dt} + ax(t) = 0 \qquad \textbf{6.2}$$

then

$$x(t) = x_p(t) + x_c(t) \qquad \textbf{6.3}$$

is a solution to the original Eq. (6.1). The term $x_p(t)$ is called *the particular integral solution*, or forced response, and $x_c(t)$ is called the *complementary solution*, or natural response.

At the present time we confine ourselves to the situation in which $f(t) = A$ (i.e., some constant). The general solution of the differential equation then consists of two parts that are obtained by solving the two equations

$$\frac{dx_p(t)}{dt} + ax_p(t) = A \qquad \textbf{6.4}$$

$$\frac{dx_c(t)}{dt} + ax_c(t) = 0 \qquad \textbf{6.5}$$

Since the right-hand side of Eq. (6.4) is a constant, it is reasonable to assume that the solution $x_p(t)$ must also be a constant. Therefore, we assume that

$$x_p(t) = K_1 \qquad \textbf{6.6}$$

Substituting this constant into Eq. (6.4) yields

$$K_1 = \frac{A}{a} \qquad \textbf{6.7}$$

Examining Eq. (6.5), we note that

$$\frac{dx_c(t)/dt}{x_c(t)} = -a \qquad \textbf{6.8}$$

This equation is equivalent to

$$\frac{d}{dt}\left[\ln x_c(t)\right] = -a$$

Hence,

$$\ln x_c(t) = -at + c$$

and therefore,

$$x_c(t) = K_2 e^{-at} \qquad\qquad \textbf{6.9}$$

Thus, a solution of Eq. (6.1) is

$$x(t) = x_p(t) + x_c(t)$$

$$= \frac{A}{a} + K_2 e^{-at} \qquad\qquad \textbf{6.10}$$

The constant K_2 can be found if the value of the independent variable $x(t)$ is known at one instant of time.

Equation (6.10) can be expressed in general in the form

$$x(t) = K_1 + K_2 e^{-t/\tau} \qquad\qquad \textbf{6.11}$$

Once the solution in Eq. (6.11) is obtained, certain elements of the equation are given names that are commonly employed in electrical engineering. For example, the term K_1 is referred to as the *steady-state solution*: the value of the variable $x(t)$ as $t \rightarrow \infty$ when the second term becomes negligible. The constant τ is called the *time constant* of the circuit. Note that the second term in Eq. (6.11) is a decaying exponential that has a value, if $\tau > 0$, of K_2 for $t = 0$ and a value of 0 for $t = \infty$. The rate at which this exponential decays is determined by the time constant τ. A graphical picture of this effect is shown in Fig. 6.2a. As can be seen from the figure, the value of $x_c(t)$ has fallen from K_2 to a value of $0.368K_2$ in one time constant, a drop of 63.2%. In two time constants the value of $x_c(t)$ has fallen to $0.135K_2$, a drop of 63.2% from the value at time $t = \tau$. This means that the gap between a point on the curve and the final value of the curve is closed by 63.2% each time constant. Finally, after five time constants, $x_c(t) = 0.0067K_2$, which is less than 1%.

Figure 6.2 *Time-constant illustrations.*

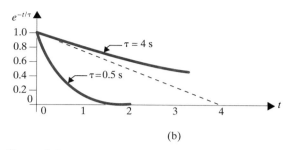

(b)

Figure 6.2   *Continued*

An interesting property of the exponential function shown in Fig. 6.2a is that the initial slope of the curve intersects the time axis at a value of $t = \tau$. In fact, we can take any point on the curve, not just the initial value, and find the time constant by finding the time required to close the gap by 63.2%. Finally, the difference between a small time constant (i.e., fast response) and a large time constant (i.e., slow response) is shown in Fig. 6.2b. These curves indicate that if the circuit has a small time constant, it settles down quickly to a steady-state value. Conversely, if the time constant is large, more time is required for the circuit to settle down or reach steady state. In any case, note that the circuit response essentially reaches steady state within five time constants (i.e., 5τ).

Note that the previous discussion has been very general in that no particular form of the circuit has been assumed—except that it results in a first-order differential equation.

ANALYSIS TECHNIQUES

The Differential Equation Approach   Equation (6.11) defines the general form of the solution of first-order transient circuits; that is, it represents the solution of the differential equation that describes an unknown current or voltage *anywhere in the network*. One of the ways that we can arrive at this solution is to solve the equations that describe the network behavior using what is often called the *state-variable approach*. In this technique we write the equation for the voltage across the capacitor and/or the equation for the current through the inductor. Recall from Chapter 5 that these quantities cannot change instantaneously. Let us first illustrate this technique in the general sense and then examine two specific examples.

Consider the circuit shown in Fig. 6.3a. At time $t = 0$, the switch closes. The KCL equation that describes the capacitor voltage for time $t > 0$ is

$$C\frac{dv(t)}{dt} + \frac{v(t) - V_S}{R} = 0$$

(a)

(b)

Figure 6.3
RC and RL circuits.

LEARNING by Doing

D 6.1 If the capacitor in the network in Fig. 6.3a is initially charged to $V_S/2$, find the complete solutions for $v(t)$.

ANSWER

$$v(t) = V_S - \frac{V_S}{2} e^{-t/RC}$$

D 6.2 Find $i(t)$ for $t > 0$ in the following network:

ANSWER

$$i(t) = I_S\left(1 - e^{-\frac{R}{L}t}\right)$$

CircuitWorks

26

or

$$\frac{dv(t)}{dt} + \frac{v(t)}{RC} = \frac{V_S}{RC}$$

From our previous development, we assume that the solution of this first-order differential equation is of the form

$$v(t) = K_1 + K_2 e^{-t/\tau}$$

Substituting this solution into the differential equation yields

$$-\frac{K_2}{\tau} e^{-t/\tau} + \frac{K_1}{RC} + \frac{K_2}{RC} e^{-t/\tau} = \frac{V_S}{RC}$$

Equating the constant and exponential terms, we obtain

$$K_1 = V_S$$

$$\tau = RC$$

Therefore,

$$v(t) = V_S + K_2 e^{-t/RC}$$

where V_S is the steady-state value and RC is the network's time constant. K_2 is determined by the initial condition of the capacitor. For example, if the capacitor is initially uncharged (that is, the voltage across the capacitor is zero at $t = 0$), then

$$0 = V_S + K_2$$

or

$$K_2 = -V_S$$

Hence, the complete solution for the voltage $v(t)$ is

$$v(t) = V_S - V_S e^{-t/RC}$$

The circuit in Fig. 6.3b can be examined in a similar manner. The KVL equation that describes the inductor current for $t > 0$ is

$$L\frac{di(t)}{dt} + Ri(t) = V_S$$

A development identical to that just used yields

$$i(t) = \frac{V_S}{R} + K_2 e^{-\left(\frac{R}{L}\right)t}$$

where V_S/R is the steady-state value and L/R is the circuit's time constant. If there is no initial current in the inductor, then at $t = 0$

$$0 = \frac{V_S}{R} + K_2$$

and

$$K_2 = \frac{-V_S}{R}$$

Hence,

$$i(t) = \frac{V_S}{R} - \frac{V_S}{R} e^{-\frac{R}{L}t}$$

is the complete solution. Note that if we wish to calculate the voltage across the resistor, then

$$v_R(t) = Ri(t)$$

$$= V_S\left(1 - e^{-\frac{R}{L}t}\right)$$

LEARNING Example 6.1

Consider the circuit shown in Fig. 6.4a. Assuming that the switch has been in position 1 for a long time, at time $t = 0$ the switch is moved to position 2. We wish to calculate the current $i(t)$ for $t > 0$.

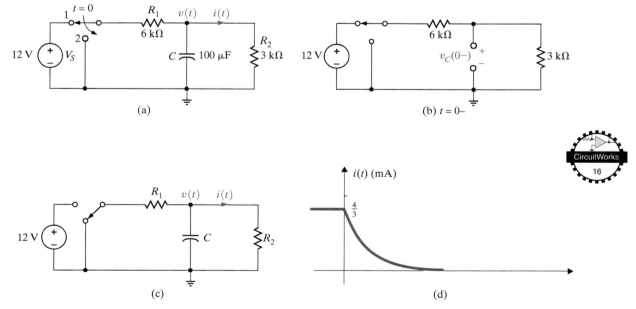

Figure 6.4 *Analysis of RC circuits.*

(continued)

SOLUTION At $t = 0-$ the capacitor is fully charged and conducts no current since the capacitor acts like an open circuit to dc. The initial voltage across the capacitor can be found using voltage division. As shown in Fig. 6.4b,

$$v_C(0-) = 12\left(\frac{3k}{6k + 3k}\right) = 4 \text{ V}$$

The network for $t > 0$ is shown in Fig. 6.4c. The KCL equation for the voltage across the capacitor is

$$\frac{v(t)}{R_1} + C\frac{dv(t)}{dt} + \frac{v(t)}{R_2} = 0$$

Using the component values, the equation becomes

$$\frac{dv(t)}{dt} + 5v(t) = 0$$

The form of the solution to this homogeneous equation is

$$v(t) = K_2 e^{-t/\tau}$$

If we substitute this solution into the differential equation, we find that $\tau = 0.2$ s. Thus,

$$v(t) = K_2 e^{-t/0.2} \text{ V}$$

Using the initial condition $v_C(0-) = v_C(0+) = 4$ V, we find that the complete solution is

$$v(t) = 4e^{-t/0.2} \text{ V}$$

Then $i(t)$ is simply

$$i(t) = \frac{v(t)}{R_2}$$

or

$$i(t) = \frac{4}{3}e^{-t/0.2} \text{ mA}$$

LEARNING Example 6.2

The switch in the network in Fig. 6.5a opens at $t = 0$. Let us find the output voltage $v_o(t)$ for $t > 0$.

SOLUTION At $t = 0-$ the circuit is in steady state and the inductor acts like a short circuit. The initial current through the inductor can be found in many ways; however, we will form a Thévenin equivalent for the part of the network to the left of the inductor, as shown in Fig. 6.5b. From this network we find that $I_1 = 4$ A and $V_{oc} = 4$ V. In addition, $R_{Th} = 1 \ \Omega$. Hence, $i_L(0-)$ obtained from Fig. 6.5c is $i_L(0-) = 4/3$ A.

The network for $t > 0$ is shown in Fig. 6.5d. Note that the 4-V independent source and the 2-ohm resistor in series with it no longer have any impact on the resulting circuit. The KVL equation for the circuit is

$$-V_{S_1} + R_1 i(t) + L\frac{di(t)}{dt} + R_3 i(t) = 0$$

which with the component values reduces to

$$\frac{di(t)}{dt} + 2i(t) = 6$$

The solution to this equation is of the form

$$i(t) = K_1 + K_2 e^{-t/\tau}$$

which when substituted into the differential equation yields

$$K_1 = 3$$

$$\tau = 1/2$$

Therefore,

$$i(t) = \left(3 + K_2 e^{-2t}\right) \text{ A}$$

Evaluating this function at the initial condition, which is

$$i_L(0-) = i_L(0+) = i(0) = 4/3 \text{ A}$$

we find that

$$K_2 = \frac{-5}{3}$$

Hence,

$$i(t) = \left(3 - \frac{5}{3}e^{-2t}\right) \text{ A}$$

and then

$$v_o(t) = 6 - \frac{10}{3}e^{-2t} \text{ V}$$

A plot of the voltage $v_o(t)$ is shown in Fig. 6.5e.

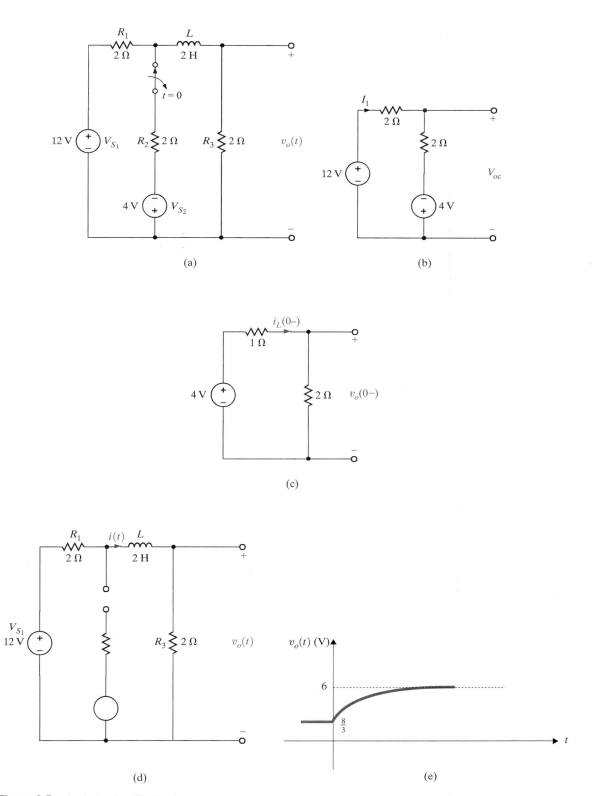

Figure 6.5 *Analysis of an RL circuit.*

LEARNING EXTENSIONS

E6.1   Find $v_C(t)$ for $t > 0$ in the circuit shown in Fig. E6.1.

ANSWER
$v_C(t) = 8e^{-t/0.6}$ V.

Figure E6.1

E6.2   In the circuit shown in Fig. E6.2, the switch opens at $t = 0$. Find $i_1(t)$ for $t > 0$.

ANSWER $i_1(t) = 1e^{-9t}$ A.

Figure E6.2

The Step-by-Step Approach   In the previous analysis technique we derived the differential equation for the capacitor voltage or inductor current, solved the differential equation, and used the solution to find the unknown variable in the network. In the very methodical technique that we will now describe we will use the fact that Eq. (6.11) is the form of the solution and employ circuit analysis to determine the constants K_1, K_2, and τ.

From Eq. (6.11) we note that as $t \to \infty$, $e^{-at} \to 0$ and $x(t) = K_1$. Therefore, if the circuit is solved for the variable $x(t)$ in steady state (i.e., $t \to \infty$) with the capacitor replaced by an open circuit [v is constant and therefore $i = C(dv/dt) = 0$] or the inductor replaced by a short circuit [i is constant and therefore $v = L(di/dt) = 0$], then the variable $x(t) = K_1$. Note that since the capacitor or inductor has been removed, the circuit is a dc circuit with constant sources and resistors, and therefore only dc analysis is required in the steady-state solution.

The constant K_2 in Eq. (6.11) can also be obtained via the solution of a dc circuit in which a capacitor is replaced by a voltage source or an inductor is replaced by a current source. The value of the voltage source for the capacitor or the current source for the inductor is a known value at one instant of time. In general, we will use the initial condition value since it is generally the one known, but the value at any instant could be used. This value can be obtained in numerous ways and is often specified as input data in a statement of the problem. However, a more likely situation is one in which a switch is thrown in the circuit and the initial value of the capacitor voltage or inductor current is determined from the previous circuit (i.e., the circuit before the switch is thrown). It is normally assumed that the previous circuit has reached steady state, and therefore the voltage across the capacitor or the current through the inductor can be found in exactly the same manner as was used to find K_1.

Finally, the value of the time constant can be found by determining the Thévenin equivalent resistance at the terminals of the storage element. Then $\tau = R_{Th}C$ for an RC circuit, and $\tau = L/R_{Th}$ for an RL circuit.

Let us now reiterate this procedure in a step-by-step fashion.

Problem-Solving Strategy
Using the Step-by-Step Approach

Step 1.   We assume a solution for the variable $x(t)$ of the form $x(t) = K_1 + K_2 e^{-t/\tau}$.

Step 2.   Assuming that the original circuit has reached steady state before a switch was thrown (thereby producing a new circuit), draw this previous circuit with the capacitor replaced by an open circuit or the inductor replaced by a short circuit. Solve for the voltage across the capacitor, $v_C(0-)$, or the current through the inductor, $i_L(0-)$, prior to switch action.

Step 3.   Assuming that the energy in the storage element cannot change in zero time, draw the circuit, valid only at $t = 0+$. The switches are in their new positions and the capacitor is replaced by a voltage source with a value of $v_C(0+) = v_C(0-)$ or the inductor is replaced by a current source with value $i_L(0+) = i_L(0-)$. Solve for the initial value of the variable $x(0+)$.

Step 4.   Assuming that steady state has been reached after the switches are thrown, draw the equivalent circuit, valid for $t > 5\tau$, by replacing the capacitor by an open circuit or the inductor by a short circuit. Solve for the steady-state value of the variable

$$x(t)|_{t>5\tau} \doteq x(\infty)$$

Step 5.   Since the time constant for all voltages and currents in the circuit will be the same, it can be obtained by reducing the entire circuit to a simple series circuit containing a voltage source, resistor, and a storage element (i.e., capacitor or inductor) by forming a simple Thévenin equivalent circuit at the terminals of the storage element. This Thévenin equivalent circuit is obtained by looking into the circuit from the terminals of the storage element. The time constant for a circuit containing a capacitor is $\tau = R_{Th}C$, and for a circuit containing an inductor it is $\tau = L/R_{Th}$.

Step 6.   Using the results of steps 3, 4, and 5, we can evaluate the constants in step 1 as

$$x(0+) = K_1 + K_2$$

$$x(\infty) = K_1$$

and therefore, $K_1 = x(\infty)$, $K_2 = x(0+) - x(\infty)$, and hence the solution is

$$x(t) = x(\infty) + [x(0+) - x(\infty)]e^{-t/\tau}$$

Keep in mind that this solution form applies only to a first-order circuit having constant, dc sources. If the sources are not dc, the forced response will be different. Generally, the forced response is of the same form as the forcing functions (sources) and their derivatives.

LEARNING Example 6.3

Consider the circuit shown in Fig. 6.6a. The circuit is in steady state prior to time $t = 0$, when the switch is closed. Let us cal- culate the current $i(t)$ for $t > 0$.

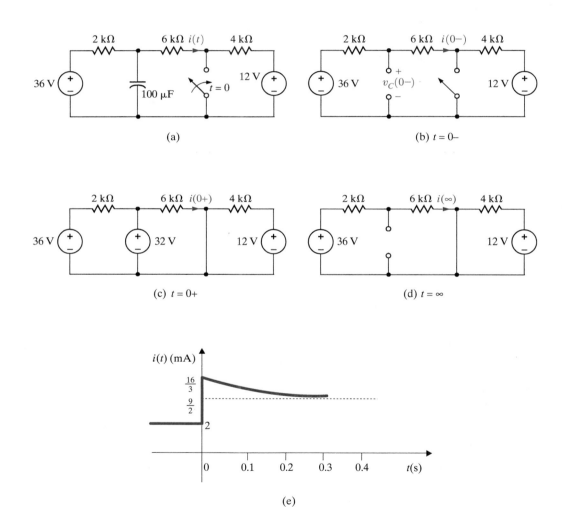

Figure 6.6 Analysis of an RC transient circuit with a constant forcing function.

SOLUTION

Step 1. $i(t)$ is of the form $K_1 + K_2 e^{-t/\tau}$.

Step 2. The initial voltage across the capacitor is calculated from Fig. 6.6b as

$$v_C(0-) = 36 - (2)(2)$$
$$= 32 \text{ V}$$

Step 3. The new circuit, valid only for $t = 0+$, is shown in Fig. 6.6c. The value of the voltage source that replaces the capacitor is $v_C(0-) = v_C(0+) = 32$ V. Hence,

$$i(0+) = \frac{32}{6k}$$
$$= \frac{16}{3} \text{ mA}$$

Step 4. The equivalent circuit, valid for $t > 5\tau$, is shown in Fig. 6.6d. The current $i(\infty)$ caused by the 36-V source is

$$i(\infty) = \frac{36}{2k + 6k}$$
$$= \frac{9}{2} \text{ mA}$$

Step 5. The Thévenin equivalent resistance, obtained by looking into the open-circuit terminals of the capacitor in Fig. 6.6d, is

$$R_{\text{Th}} = \frac{(2k)(6k)}{2k + 6k} = \frac{3}{2} \text{ k}\Omega$$

Therefore, the circuit time constant is

$$\tau = R_{\text{Th}}C$$
$$= \left(\frac{3}{2}\right)(10^3)(100)(10^{-6})$$
$$= 0.15 \text{ s}$$

Step 6.

$$K_1 = i(\infty) = \frac{9}{2} \text{ mA}$$

$$K_2 = i(0+) - i(\infty) = i(0+) - K_1$$

$$= \frac{16}{3} - \frac{9}{2}$$
$$= \frac{5}{6} \text{ mA}$$

Therefore,

$$i(t) = \frac{36}{8} + \frac{5}{6} e^{-t/0.15} \text{ mA}$$

Let us now employ MATLAB to plot the function. First, an interval for the variable t must be specified. The beginning of the interval will be chosen to be $t = 0$. The end of the interval will be chosen to be 10 times the time constant. This is realized in MATLAB as follows:

```
>>tau = 0.15
>>tend = 10*tau
```

Once the time interval has been specified, we can use MATLAB's linspace function to generate an array of evenly spaced points in the interval. The linspace command has the following syntax: `linspace(x1,x2,N)` where `x1` and `x2` denote the beginning and ending points in the interval and `N` represents the number of points. Thus, to generate an array containing 150 points in the interval [0, tend], we execute the following command:

```
>>t = linspace(0, tend, 150)
```

The MATLAB program for generating a plot of the function is

```
>>tau = 0.15;
>>tend = 10*tau;
>>t = linspace(0, tend, 150);
>>i = 9/2 + (5/6)*exp(-t/tau);
>>plot(t,i)
>>xlabel('Time (s)')
>>ylabel('Current (mA)')
```

The MATLAB plot is shown in Fig. 6.7 and can be compared to the sketch in Fig. 6.6(e). Examination of Fig. 6.6(e) indicates once again that although the voltage across the capacitor is continuous at $t = 0$, the current $i(t)$ in the 6-kΩ resistor jumps at $t = 0$ from 2 mA to $5\frac{1}{3}$ mA, and finally decays to $4\frac{1}{2}$ mA.

(continued)

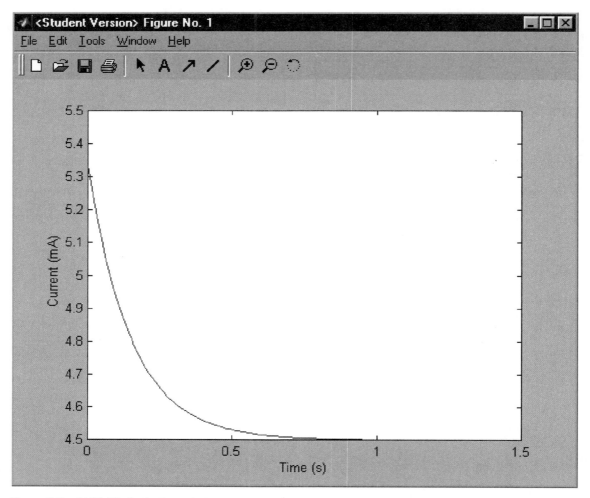

Figure 6.7 *MATLAB plot for Example 6.3.*

LEARNING Example 6.4

The circuit shown in Fig. 6.8a is assumed to have been in a steady-state condition prior to switch closure at $t = 0$. We wish to calculate the voltage $v(t)$ for $t > 0$.

SOLUTION

Step 1. $v(t)$ is of the form $K_1 + K_2 e^{-t/\tau}$.

Step 2. In Fig. 6.8b we see that

$$i_L(0-) = \frac{24}{4 + \dfrac{(6)(3)}{6 + 3}} \left(\frac{6}{6 + 3} \right)$$

$$= \frac{8}{3} \text{ A}$$

Step 3. The new circuit, valid only for $t = 0+$, is shown in Fig. 6.8c, which is equivalent to the circuit shown in Fig. 6.8d. The value of the current source that replaces the inductor is $i_L(0-) = i_L(0+) = \frac{8}{3}$ A. The node voltage $v_1(0+)$ can be determined from the circuit in Fig. 6.8d using a single-node equation, and $v(0+)$ is equal to the difference between the source voltage and $v_1(0+)$. The equation for $v_1(0+)$ is

$$\frac{v_1(0+) - 24}{4} + \frac{v_1(0+)}{6} + \frac{8}{3} + \frac{v_1(0+)}{12} = 0$$

or

$$v_1(0+) = \frac{20}{3} \text{ V}$$

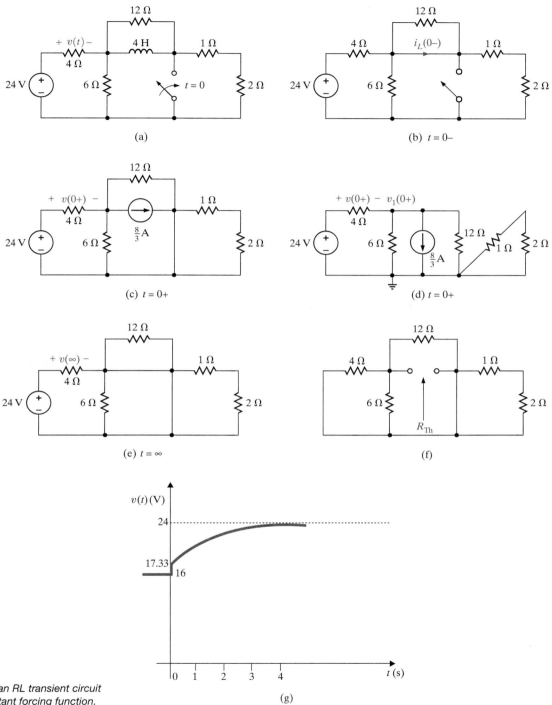

Figure 6.8
Analysis of an RL transient circuit
with a constant forcing function.

Then

$$v(0+) = 24 - v_1(0+)$$

$$= \frac{52}{3} \text{ V}$$

Step 4. The equivalent circuit for the steady-state condition after switch closure is given in Fig. 6.8e. Note that the 6-, 12-, 1-, and 2-Ω resistors are shorted, and therefore $v(\infty) = 24$ V.

(continued)

Step 5. The Thévenin equivalent resistance is found by looking into the circuit from the inductor terminals. This circuit is shown in Fig. 6.8f. Note carefully that R_{Th} is equal to the 4-, 6-, and 12-Ω resistors in parallel. Therefore, $R_{Th} = 2\ \Omega$, and the circuit time constant is

$$\tau = \frac{L}{R_{Th}} = \frac{4}{2} = 2\ \text{s}$$

Step 6. From the previous analysis we find that

$$K_1 = v(\infty) = 24$$

$$K_2 = v(0+) - v(\infty) = -\frac{20}{3}$$

and hence that

$$v(t) = 24 - \frac{20}{3} e^{-t/2}\ \text{V}$$

From Fig. 6.8b we see that the value of $v(t)$ before switch closure is 16 V. This value jumps to 17.33 V at $t = 0$. The MATLAB program for generating the plot (shown in Fig. 6.9) of this function for $t > 0$, is listed next.

```
>>tau = 2;
>>tend = 10*tau;
>>t = linspace(0, tend, 150);
>>v = 24 - (20/3)*exp(-t/tau);
>>plot (t,v)
>>xlabel('Time (s)')
>>ylabel('Voltage (V)')
```

This plot can be compared to the sketch shown in Fig. 6.8g.

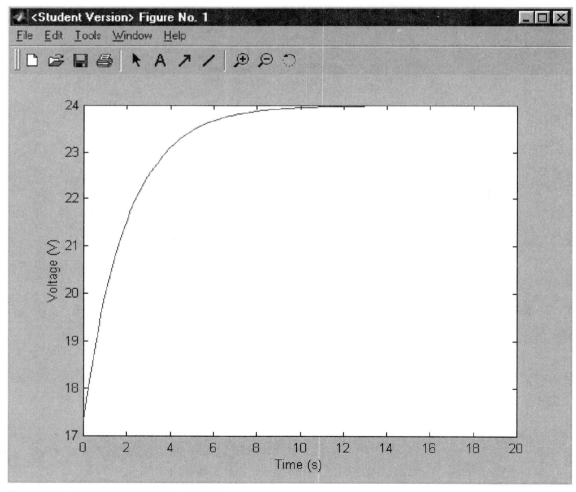

Figure 6.9 *MATLAB plot for Example 6.4.*

LEARNING EXTENSIONS

E6.3 Consider the network in Fig. E6.3. The switch opens at $t = 0$. Find $v_o(t)$ for $t > 0$.

ANSWER

$$v_o(t) = \frac{24}{5} + \frac{1}{5} e^{-(5/8)t} \text{ V.}$$

Figure E6.3

E6.4 Consider the network in Fig. E6.4. If the switch opens at $t = 0$, find the output voltage $v_o(t)$ for $t > 0$.

ANSWER

$$v_o(t) = 6 - \frac{10}{3} e^{-2t} \text{ V.}$$

Figure E6.4

LEARNING Example 6.5

The circuit shown in Fig. 6.10a has reached steady state with the switch in position 1. At time $t = 0$ the switch moves from position 1 to position 2. We want to calculate $v_o(t)$ for $t > 0$.

SOLUTION

    Step 1. $v_o(t)$ is of the form $K_1 + K_2 e^{-t/\tau}$.

    Step 2. Using the circuit in Fig. 6.10b, we can calculate $i_L(0-)$

$$i_A = \frac{12}{4} = 3 \text{ A}$$

Then

$$i_L(0-) = \frac{12 + 2i_A}{6} = \frac{18}{6} = 3 \text{ A}$$

Step 3. The new circuit, valid only for $t = 0+$, is shown in Fig. 6.10c. The value of the current source that replaces the inductor is $i_L(0-) = i_L(0+) = 3$ A. Because of the current source

$$v_o(0+) = (3)(6) = 18 \text{ V}$$

Step 4. The equivalent circuit, for the steady-state condition after switch closure, is given in Fig. 6.10d. Using the voltages

(continued)

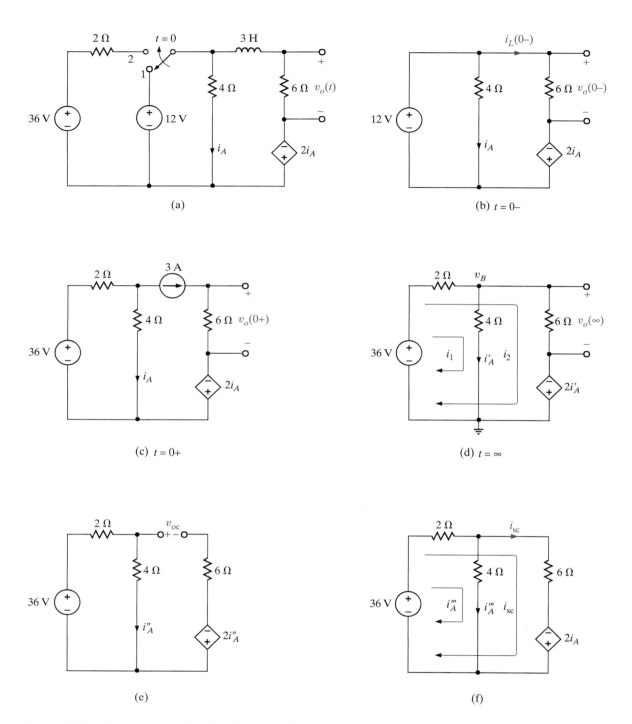

Figure 6.10 Analysis of an RL transient circuit containing a dependent source.

and currents defined in the figure, we can compute $v_o(\infty)$ in a variety of ways. For example, using node equations we can find $v_o(\infty)$ from

$$\frac{v_B - 36}{2} + \frac{v_B}{4} + \frac{v_B + 2i_A'}{6} = 0$$

$$i_A' = \frac{v_B}{4}$$

$$v_o(\infty) = v_B + 2i_A'$$

or, using loop equations,

$$36 = 2(i_1 + i_2) + 4i_1$$

$$36 = 2(i_1 + i_2) + 6i_2 - 2i_1$$

$$v_o(\infty) = 6i_2$$

Using either approach, we find that $v_o(\infty) = 27$ V.

Step 5.   The Thévenin equivalent resistance can be obtained via v_{oc} and i_{sc} because of the presence of the dependent source. From Fig. 6.10e we note that

$$i_A'' = \frac{36}{2 + 4} = 6 \text{ A}$$

Therefore,

$$v_{oc} = (4)(6) + 2(6)$$

$$= 36 \text{ V}$$

From Fig. 6.10f we can write the following loop equations:

$$36 = 2(i_A''' + i_{sc}) + 4i_A'''$$

$$36 = 2(i_A''' + i_{sc}) + 6i_{sc} - 2i_A'''$$

Solving these equations for i_{sc} yields

$$i_{sc} = \frac{9}{2} \text{ A}$$

Therefore,

$$R_{Th} = \frac{v_{oc}}{i_{sc}} = \frac{36}{9/2} = 8 \text{ } \Omega$$

Hence, the circuit time constant is

$$\tau = \frac{L}{R_{Th}} = \frac{3}{8} \text{ s}$$

Step 6.   Using the information just computed, we can derive the final equation for $v_o(t)$:

$$K_1 = v_o(\infty) = 27$$

$$K_2 = v_o(0+) - v_o(\infty) = 18 - 27 = -9$$

Therefore,

$$v_o(t) = 27 - 9e^{-t/(3/8)} \text{ V}$$

LEARNING EXTENSION

E6.5   If the switch in the network in Fig. E6.5 closes at $t = 0$, find $v_o(t)$ for $t > 0$.

ANSWER
$$v_o(t) = 24 + 36e^{-(t/12)} \text{ V}.$$

Figure E6.5

At this point it is appropriate to state that not all switch action will always occur at time $t = 0$. It may occur at any time t_0. In this case the results of the step-by-step analysis yield the following equations:

$$x(t_0) = K_1 + K_2$$

$$x(\infty) = K_1$$

and

$$x(t) = x(\infty) + [x(t_0) - x(\infty)]e^{-(t-t_0)/\tau} \qquad t > t_0$$

The function is essentially time-shifted by t_0 seconds.

Finally, note that if more than one independent source is present in the network, we can simply employ superposition to obtain the total response.

PULSE RESPONSE Thus far we have examined networks in which a voltage or current source is suddenly applied. As a result of this sudden application of a source, voltages or currents in the circuit are forced to change abruptly. A forcing function whose value changes in a discontinuous manner or has a discontinuous derivative is called a *singular function*. Two such singular functions that are very important in circuit analysis are the unit impulse function and the unit step function. We will defer a discussion of the former until a later chapter and concentrate on the latter.

The *unit step function* is defined by the following mathematical relationship:

$$u(t) = \begin{cases} 0 & t < 0 \\ 1 & t > 0 \end{cases}$$

In other words, this function, which is dimensionless, is equal to zero for negative values of the argument and equal to 1 for positive values of the argument. It is undefined for a zero argument where the function is discontinuous. A graph of the unit step is shown in Fig. 6.11a. The unit step is dimensionless, and therefore a voltage step of V_o volts or a current step of I_o amperes is written as $V_o u(t)$ and $I_o u(t)$, respectively. Equivalent circuits for a voltage step are shown in Figs. 6.11b and c. Equivalent circuits for a current step are shown in Figs. 6.11d and e. If we use the definition of the unit step, it is easy to generalize this function by replacing the argument t by $t - t_0$. In this case

$$u(t - t_0) = \begin{cases} 0 & t < t_0 \\ 1 & t > t_0 \end{cases}$$

A graph of this function is shown in Fig. 6.11f. Note that $u(t - t_0)$ is equivalent to delaying $u(t)$ by t_0 seconds, so that the abrupt change occurs at time $t = t_0$.

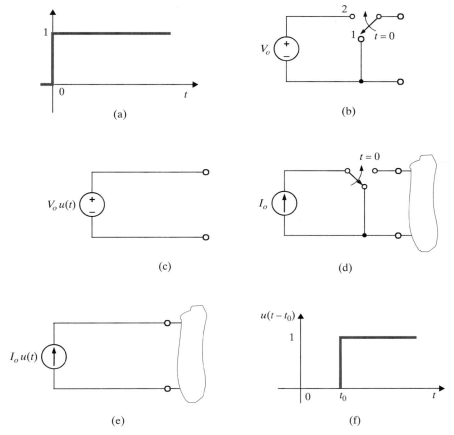

Figure 6.11 *Graphs and models of the unit step function.*

Step functions can be used to construct one or more pulses. For example, the voltage pulse shown in Fig. 6.12a can be formulated by initiating a unit step at $t = 0$ and subtracting one that starts at $t = T$, as shown in Fig. 6.12b. The equation for the pulse is

$$v(t) = A[u(t) - u(t - T)]$$

If the pulse is to start at $t = t_0$ and have width T, the equation would be

$$v(t) = A\{u(t - t_0) - u[t - (t_0 + T)]\}$$

Using this approach, we can write the equation for a pulse starting at any time and ending at any time. Similarly, using this approach, we could write the equation for a series of pulses, called a *pulse train*, by simply forming a summation of pulses constructed in the manner illustrated previously.

The following example will serve to illustrate many of the concepts we have just presented.

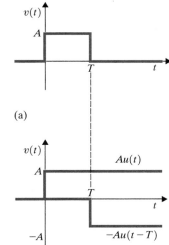

Figure 6.12
Construction of a pulse via two step functions.

LEARNING Example 6.6

Consider the circuit shown in Fig. 6.13a. The input function is the voltage pulse shown in Fig. 6.13b. Since the source is zero for all negative time, the initial conditions for the network are zero $\left[\text{i.e., } v_C(0-) = 0\right]$. The response $v_o(t)$ for $0 < t < 0.3$ s is due to the application of the constant source at $t = 0$ and is not influenced by any source changes that will occur later. At $t = 0.3$ s the forcing function becomes zero and therefore $v_o(t)$ for $t > 0.3$ s is the source-free or natural response of the network.

Let us determine the expression for the voltage $v_o(t)$.

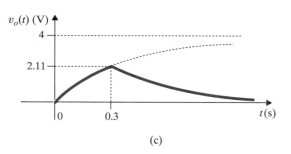

Figure 6.13 *Pulse response of a network.*

SOLUTION Since the output voltage $v_o(t)$ is a voltage division of the capacitor voltage, and the initial voltage across the capacitor is zero, we know that $v_o(0+) = 0$.

If no changes were made in the source after $t = 0$, the steady-state value of $v_o(t)$ $\left[\text{i.e., } v_o(\infty)\right]$ due to the application of the unit step at $t = 0$ would be

$$v_o(\infty) = \frac{9}{6k + 4k + 8k}(8k)$$

$$= 4 \text{ V}$$

The Thévenin equivalent resistance is

$$R_{\text{Th}} = \frac{(6k)(12k)}{6k + 12k}$$

$$= 4 \text{ k}\Omega$$

Therefore, the circuit time constant τ is

$$\tau = R_{\text{Th}}C$$

$$= (4)(10^3)(100)(10^{-6})$$

$$= 0.4 \text{ s}$$

Therefore, the response $v_o(t)$ for the period $0 < t < 0.3$ s is

$$v_o(t) = 4 - 4e^{-t/0.4} \text{ V} \quad 0 < t < 0.3 \text{ s}$$

The capacitor voltage can be calculated by realizing that using voltage division, $v_o(t) = \frac{2}{3}v_C(t)$. Therefore,

$$v_C(t) = \frac{3}{2}\left(4 - 4e^{-t/0.4}\right) \text{ V}$$

Since the capacitor voltage is continuous,

$$v_C(0.3-) = v_C(0.3+)$$

and therefore,

$$v_o(0.3+) = \frac{2}{3}v_C(0.3-)$$

$$= 4\left(1 - e^{-0.3/0.4}\right)$$

$$= 2.11 \text{ V}$$

Since the source is zero for $t > 0.3$ s, the final value for $v_o(t)$ as $t \to \infty$ is zero. Therefore, the expression for $v_o(t)$ for $t > 0.3$ s is

$$v_o(t) = 2.11e^{-(t-0.3)/0.4} \text{ V} \quad t > 0.3 \text{ s}$$

The term $e^{-(t-0.3)/0.4}$ indicates that the exponential decay starts at $t = 0.3$ s. The complete solution can be written by means of superposition as

$$v_o(t) = 4\left(1 - e^{-t/0.4}\right)u(t) - 4\left(1 - e^{-(t-0.3)/0.4}\right)u(t - 0.3) \text{ V}$$

LEARNING by Doing

D 6.3 Plot the following functions:

(a) $i(t) = 10[u(t) - u(t - 0.01)]$ mA

(b) $v(t) = -10[u(t - 1) - u(t - 2)]$ V

ANSWER

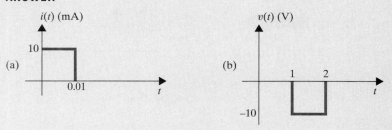

---

or, equivalently, the complete solution is

$$v_o(t) = \begin{cases} 0 & t < 0 \\ 4(1 - e^{-t/0.4}) \text{ V} & 0 < t < 0.3 \text{ s} \\ 2.11e^{-(t-0.3)/0.4} \text{ V} & 0.3 \text{ s} < t \end{cases}$$

which in mathematical form is

$$v_o(t) = 4(1 - e^{-t/0.4})[u(t) - u(t - 0.3)] + 2.11e^{-(t-0.3)/0.4}u(t - 0.3) \text{ V}$$

Note that the term $[u(t) - u(t - 0.3)]$ acts like a gating function that captures only the part of the step response that exists in the time interval $0 < t < 0.3$ s. The output as a function of time is shown in Fig. 6.13c.

E6.6 The voltage source in the network in Fig. E6.6a is shown in Fig. E6.6b. The initial current in the inductor must be zero. (Why?) Determine the output voltage $v_o(t)$ for $t > 0$.

ANSWER $v_o(t) = 0$ for $t < 0$, $4(1 - e^{-(3/2)t})$ V for $0 \leq t \leq 1$, and $3.11e^{-(3/2)(t-1)}$ V for $1 < t$.

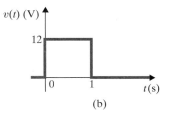

(a)

(b)

Figure E6.6

6.3 Second-Order Circuits

THE BASIC CIRCUIT EQUATION   To begin our development, let us consider the two basic *RLC* circuits shown in Fig. 6.14. We assume that energy may be initially stored in both the inductor and capacitor. The node equation for the parallel *RLC* circuit is

$$\frac{v}{R} + \frac{1}{L}\int_{t_0}^{t} v(x)\, dx + i_L(t_0) + C\frac{dv}{dt} = i_S(t)$$

(a)   (b)

Figure 6.14 *Parallel and series RLC circuits.*

Similarly, the loop equation for the series *RLC* circuit is

$$Ri + \frac{1}{C}\int_{t_0}^{t} i(x)\, dx + v_C(t_0) + L\frac{di}{dt} = v_S(t)$$

Note that the equation for the node voltage in the parallel circuit is of the same form as that for the loop current in the series circuit. Therefore, the solution of these two circuits is dependent on solving one equation. If the two preceding equations are differentiated with respect to time, we obtain

$$C\frac{d^2v}{dt^2} + \frac{1}{R}\frac{dv}{dt} + \frac{v}{L} = \frac{di_S}{dt}$$

and

$$L\frac{d^2i}{dt^2} + R\frac{di}{dt} + \frac{i}{C} = \frac{dv_S}{dt}$$

Since both circuits lead to a second-order differential equation with constant coefficients, we will concentrate our analysis on this type of equation.

LEARNING by Doing

D 6.4 Write the differential equation that describes the node voltage in (a) and the loop current in (b).

(a)

(b)

ANSWER   (a) $\dfrac{d^2v}{dt^2} + \dfrac{1}{RC}\dfrac{dv}{dt} + \dfrac{v}{LC} = 0$ (b) $\dfrac{d^2i}{dt^2} + \dfrac{R}{L}\dfrac{di}{dt} + \dfrac{i}{LC} = 0$

THE RESPONSE EQUATIONS In concert with our development of the solution of a first-order differential equation that results from the analysis of either an *RL* or an *RC* circuit as outlined earlier, we will now employ the same approach here to obtain the solution of a second-order differential equation that results from the analysis of *RLC* circuits. As a general rule, for this case we are confronted with an equation of the form

$$\frac{d^2x(t)}{dt^2} + a_1\frac{dx(t)}{dt} + a_2x(t) = f(t) \qquad \textbf{6.12}$$

Once again we use the fact that if $x(t) = x_p(t)$ is a solution to Eq. (6.12), and if $x(t) = x_c(t)$ is a solution to the homogeneous equation

$$\frac{d^2x(t)}{dt^2} + a_1\frac{dx(t)}{dt} + a_2x(t) = 0$$

then

$$x(t) = x_p(t) + x_c(t)$$

is a solution to the original Eq. (6.12). If we again confine ourselves to a constant forcing function [i.e., $f(t) = A$], the development at the beginning of this chapter shows that the solution of Eq. (6.12) will be of the form

$$x(t) = \frac{A}{a_2} + x_c(t) \qquad \textbf{6.13}$$

Let us now turn our attention to the solution of the homogeneous equation

$$\frac{d^2x(t)}{dt^2} + a_1\frac{dx(t)}{dt} + a_2x(t) = 0$$

where a_1 and a_2 are constants. For simplicity we will rewrite the equation in the form

$$\frac{d^2x(t)}{dt^2} + 2\zeta\omega_0\frac{dx(t)}{dt} + \omega_0^2x(t) = 0 \qquad \textbf{6.14}$$

where we have made the following simple substitutions for the constants $a_1 = 2\zeta\omega_0$ and $a_2 = \omega_0^2$.

Following the development of a solution for the first-order homogeneous differential equation earlier in this chapter, the solution of Eq. (6.14) must be a function whose first- and second-order derivatives have the same form, so that the left-hand side of Eq. (6.14) will become identically zero for all t. Again we assume that

$$x(t) = Ke^{st}$$

Substituting this expression into Eq. (6.14) yields

$$s^2Ke^{st} + 2\zeta\omega_0sKe^{st} + \omega_0^2Ke^{st} = 0$$

LEARNING by Doing

D 6.5 Given the homogeneous differential equation

$$4\frac{d^2x(t)}{dt^2} + 8\frac{dx(t)}{dt} + 16x(t) = 0$$

determine the characteristic equation, the damping ratio, and the undamped natural frequency.

ANSWER *Characteristic equation:* $s^2 + 2s + 4 = 0$

$$\zeta = \tfrac{1}{2}$$
$$\omega_0 = 2$$

Dividing both sides of the equation by Ke^{st} yields

$$s^2 + 2\zeta\omega_0 s + \omega_0^2 = 0 \qquad\qquad \textbf{6.15}$$

This equation is commonly called the *characteristic equation*; ζ is called the exponential *damping ratio*, and ω_0 is referred to as the *undamped natural frequency*. The importance of this terminology will become clear as we proceed with the development. If this equation is satisfied, our assumed solution $x(t) = Ke^{st}$ is correct. Employing the quadratic formula, we find that Eq. (6.15) is satisfied if

$$s = \frac{-2\zeta\omega_0 \pm \sqrt{4\zeta^2\omega_0^2 - 4\omega_0^2}}{2}$$

$$= -\zeta\omega_0 \pm \omega_0\sqrt{\zeta^2 - 1} \qquad\qquad \textbf{6.16}$$

Therefore, there are two values of s, s_1 and s_2 that satisfy Eq. (6.15):

$$s_1 = -\zeta\omega_0 + \omega_0\sqrt{\zeta^2 - 1}$$
$$s_2 = -\zeta\omega_0 - \omega_0\sqrt{\zeta^2 - 1} \qquad\qquad \textbf{6.17}$$

Therefore, in general, the complementary solution of Eq. (6.14) is of the form

$$x_c(t) = K_1 e^{s_1 t} + K_2 e^{s_2 t} \qquad\qquad \textbf{6.18}$$

K_1 and K_2 are constants that can be evaluated via the initial conditions $x(0)$ and $dx(0)/dt$. For example, since

$$x(t) = K_1 e^{s_1 t} + K_2 e^{s_2 t}$$

then

$$x(0) = K_1 + K_2$$

and

$$\left.\frac{dx(t)}{dt}\right|_{t=0} = \frac{dx(0)}{dt} = s_1 K_1 + s_2 K_2$$

Hence, $x(0)$ and $dx(0)/dt$ produce two simultaneous equations, which when solved yield the constants K_1 and K_2.

Close examination of Eqs. (6.17) and (6.18) indicates that the form of the solution of the homogeneous equation is dependent on the value ζ. For example, if $\zeta > 1$, the roots of the characteristic equation, s_1 and s_2, also called the *natural frequencies* because they determine the natural (unforced) response of the network, are real and unequal; if $\zeta < 1$, the roots are complex numbers; and finally, if $\zeta = 1$, the roots are real and equal.

Let us now consider the three distinct forms of the unforced response—that is, the response due to an initial capacitor voltage or initial inductor current.

Case 1, $\zeta > 1$ This case is commonly called *overdamped*. The natural frequencies s_1 and s_2 are real and unequal, and therefore the natural response of the network described by the second-order differential equation is of the form

$$x_c(t) = K_1 e^{-(\zeta\omega_0 - \omega_0\sqrt{\zeta^2-1})t} + K_2 e^{-(\zeta\omega_0 + \omega_0\sqrt{\zeta^2-1})t} \qquad \textbf{6.19}$$

where K_1 and K_2 are found from the initial conditions. This indicates that the natural response is the sum of two decaying exponentials.

Case 2, $\zeta < 1$ This case is called *underdamped*. Since $\zeta < 1$, the roots of the characteristic equation given in Eq. (6.17) can be written as

$$s_1 = -\zeta\omega_0 + j\omega_0\sqrt{1 - \zeta^2} = -\sigma + j\omega_d$$
$$s_2 = -\zeta\omega_0 - j\omega_0\sqrt{1 - \zeta^2} = -\sigma - j\omega_d$$

where $j = \sqrt{-1}$, $\sigma = \zeta\omega_0$ and $\omega_d = \omega_0\sqrt{1 - \zeta^2}$. Thus, the natural frequencies are complex numbers (briefly discussed in the Appendix). The natural response is then of the form

$$x_c(t) = e^{-\zeta\omega_0 t}\left(A_1 \cos \omega_0\sqrt{1 - \zeta^2}\,t + A_2 \sin \omega_0\sqrt{1 - \zeta^2}\,t\right) \qquad \textbf{6.20}$$

where A_1 and A_2, like K_1 and K_2, are constants, which are evaluated using the initial conditions $x(0)$ and $dx(0)/dt$. This illustrates that the natural response is an exponentially damped oscillatory response.

Case 3, $\zeta = 1$ This case, called *critically damped*, results in

$$s_1 = s_2 = -\zeta\omega_0$$

In the case where the characteristic equation has repeated roots, the general solution is of the form

$$x_c(t) = B_1 e^{-\zeta\omega_0 t} + B_2 t e^{-\zeta\omega_0 t} \qquad \textbf{6.21}$$

where B_1 and B_2 are constants derived from the initial conditions.

LEARNING by Doing

D 6.6 Determine the general form of the solution of the equation in Learning by Doing 6.5.

ANSWER $x_c(t) = e^{-t}(A_1\cos\sqrt{3}t + A_2\sin\sqrt{3}t)$.

LEARNING by Doing

D 6.7 Determine the general form of the solution of the equation

$$\frac{d^2x(t)}{dt^2} + \frac{4\,dx(t)}{dt} + 4x(t) = 0$$

ANSWER
$x_c(t) = B_1 e^{-2t} + B_2 t e^{-2t}$

It is informative to sketch the natural response for the three cases we have discussed: overdamped, Eq. (6.19); underdamped, Eq. (6.20); and critically damped, Eq. (6.21). Figure 6.15 graphically illustrates the three cases for the situations in which $x_c(0) = 0$. Note that the critically damped response peaks and decays faster than the overdamped response. The underdamped response is an exponentially damped sinusoid whose rate of decay is dependent on the factor ζ. Actually, the terms $\pm\, e^{-\zeta\omega_0 t}$ define what is called the *envelope* of the response, and the damped oscillations (i.e., the oscillations of decreasing amplitude) exhibited by the waveform in Fig. 6.15b are called *ringing*.

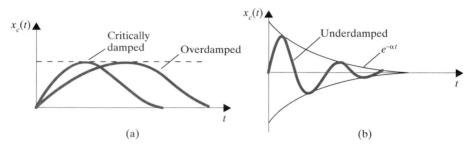

Figure 6.15   *Comparison of overdamped, critically damped, and underdamped responses.*

LEARNING EXTENSIONS

E6.7   A parallel *RLC* circuit has the following circuit parameters: $R = 1\ \Omega$, $L = 2$ H, and $C = 2$ F. Compute the damping ratio and the undamped natural frequency of this network.

ANSWER
$\zeta = 0.5$, $\omega_0 = 0.5$ rad/s.

E6.8   A series *RLC* circuit consists of $R = 2\ \Omega$, $L = 1$ H, and a capacitor. Determine the type of response exhibited by the network if (a) $C = \frac{1}{2}$ F, (b) $C = 1$ F, and (c) $C = 2$ F.

ANSWER   (a) underdamped, (b) critically damped, (c) overdamped.

THE NETWORK RESPONSE   We will now analyze a number of simple *RLC* networks that contain both nonzero initial conditions and constant forcing functions. Circuits that exhibit overdamped, underdamped, and critically damped responses will be considered.

Problem-Solving Strategy
Second-Order Transient Circuits

▶   Write the differential equation that describes the circuit.

▶   Derive the characteristic equation, which can be written in the form $s^2 + 2\zeta\omega_0 s + \omega_0^2 = 0$, where ζ is the damping ratio and ω_0 is the undamped natural frequency.

▶ The two roots of the characteristic equation will determine the type of response. If the roots are real and unequal (i.e., $\zeta > 1$), the network response is overdamped. If the roots are real and equal (i.e., $\zeta = 1$), the network response is critically damped. If the roots are complex (i.e., $\zeta < 1$), the network response is underdamped.

▶ The damping condition and corresponding response for the aforementioned three cases outlined are as follows:

Overdamped: $x(t) = K_1 e^{-(\zeta\omega_0 - \omega_0\sqrt{\zeta^2-1})t} + K_2 e^{-(\zeta\omega_0 + \omega_0\sqrt{\zeta^2-1})t}$

Critically damped: $x(t) = B_1 e^{-\zeta\omega_0 t} + B_2 t e^{-\zeta\omega_0 t}$

Underdamped: $x(t) = e^{-\sigma t}(A_1 \cos\omega_d t + A_2 \sin\omega_d t)$, where $\sigma = \zeta\omega_0$, and $\omega_d = \omega_0\sqrt{1 - \zeta^2}$

▶ Two initial conditions, either given or derived, are required to obtain the two unknown coefficients in the response equation.

The following examples will serve to demonstrate the analysis techniques.

LEARNING Example 6.7

Consider the parallel *RLC* circuit shown in Fig. 6.16. The second-order differential equation that describes the voltage $v(t)$ is

$$\frac{d^2v}{dt^2} + \frac{1}{RC}\frac{dv}{dt} + \frac{v}{LC} = 0$$

$v(t)$

Figure 6.16   *Parallel RLC circuit.*

A comparison of this equation with Eqs. (6.14) and (6.15) indicates that for the parallel *RLC* circuit the damping term is $1/2\,RC$ and the undamped natural frequency is $1/\sqrt{LC}$. If the circuit parameters are $R = 2\,\Omega$, $C = \frac{1}{5}$ F, and $L = 5$ H, the equation becomes

$$\frac{d^2v}{dt^2} + 2.5\frac{dv}{dt} + v = 0$$

Let us assume that the initial conditions on the storage elements are $i_L(0) = -1$ A and $v_C(0) = 4$ V. Let us find the node voltage $v(t)$ and the inductor current.

SOLUTION   The characteristic equation for the network is

$$s^2 + 2.5s + 1 = 0$$

and the roots are

$$s_1 = -2$$
$$s_2 = -0.5$$

Since the roots are real and unequal, the circuit is overdamped, and $v(t)$ is of the form

$$v(t) = K_1 e^{-2t} + K_2 e^{-0.5t}$$

The initial conditions are now employed to determine the constants K_1 and K_2. Since $v(t) = v_C(t)$,

$$v_C(0) = v(0) = 4 = K_1 + K_2$$

The second equation needed to determine K_1 and K_2 is normally obtained from the expression

$$\frac{dv(t)}{dt} = -2K_1 e^{-2t} - 0.5K_2 e^{-0.5t}$$

However, the second initial condition is not $dv(0)/dt$. If this were the case, we would simply evaluate the equation at $t = 0$. This would produce a second equation in the unknowns K_1 and K_2. We can, however, circumvent this problem by noting that the node equation for the circuit can be written as

$$C\frac{dv(t)}{dt} + \frac{v(t)}{R} + i_L(t) = 0$$

(continued)

or

$$\frac{dv(t)}{dt} = \frac{-1}{RC} v(t) - \frac{i_L(t)}{C}$$

At $t = 0$,

$$\frac{dv(0)}{dt} = \frac{-1}{RC} v(0) - \frac{1}{C} i_L(0)$$

$$= -2.5(4) - 5(-1)$$

$$= -5$$

However, since

$$\frac{dv(t)}{dt} = -2K_1 e^{-2t} - 0.5K_2 e^{-0.5t}$$

then when $t = 0$

$$-5 = -2K_1 - 0.5K_2$$

This equation, together with the equation

$$4 = K_1 + K_2$$

produces the constants $K_1 = 2$ and $K_2 = 2$. Therefore, the final equation for the voltage is

$$v(t) = 2e^{-2t} + 2e^{-0.5t} \text{ V}$$

Note that the voltage equation satisfies the initial condition $v(0) = 4$ V. The response curve for this voltage $v(t)$ is shown in Fig. 6.17.

The inductor current is related to $v(t)$ by the equation

$$i_L(t) = \frac{1}{L} \int v(t) \, dt$$

Substituting our expression for $v(t)$ yields

$$i_L(t) = \frac{1}{5} \int \left[2e^{-2t} + 2e^{-0.5t} \right] dt$$

or

$$i_L(t) = -\frac{1}{5} e^{-2t} - \frac{4}{5} e^{-0.5t} \text{ A}$$

Note that in comparison with the RL and RC circuits, the response of this RLC circuit is controlled by two time constants. The first term has a time constant of $\frac{1}{2}$, and the second term has a time constant of 2.

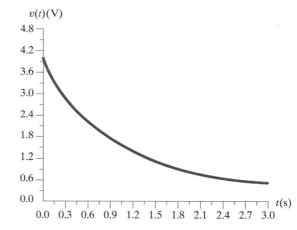

Figure 6.17   *Overdamped response.*

LEARNING Example 6.8

The series RLC circuit shown in Fig. 6.18 has the following parameters: $C = 0.04$ F, $L = 1$ H, $R = 6 \ \Omega$, $i_L(0) = 4$ A, and $v_C(0) = -4$ V. The equation for the current in the circuit is given by the expression

$$\frac{d^2 i}{dt^2} + \frac{R}{L} \frac{di}{dt} + \frac{i}{LC} = 0$$

A comparison of this equation with Eqs. (6.14) and (6.15) illustrates that for a series RLC circuit the damping term is $R/2L$ and the undamped natural frequency is $1/\sqrt{LC}$. Substituting the circuit element values into the preceding equation yields

$$\frac{d^2 i}{dt^2} + 6 \frac{di}{dt} + 25i = 0$$

Let us determine the expression for both the current and the capacitor voltage.

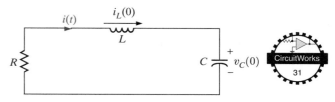

Figure 6.18   *Series RLC circuit.*

SOLUTION   The characteristic equation is then

$$s_2 + 6s + 25 = 0$$

and the roots are

$$s_1 = -3 + j4$$
$$s_2 = -3 - j4$$

Since the roots are complex, the circuit is underdamped, and the expression for $i(t)$ is

$$i(t) = K_1 e^{-3t} \cos 4t + K_2 e^{-3t} \sin 4t$$

Using the initial conditions, we find that

$$i(0) = 4 = K_1$$

and

$$\frac{di}{dt} = -4K_1 e^{-3t} \sin 4t - 3K_1 e^{-3t} \cos 4t$$
$$+ 4K_2 e^{-3t} \cos 4t - 3K_2 e^{-3t} \sin 4t$$

and thus

$$\frac{di(0)}{dt} = -3K_1 + 4K_2$$

Although we do not know $di(0)/dt$, we can find it via KVL. From the circuit we note that

$$Ri(0) + L \frac{di(0)}{dt} + v_C(0) = 0$$

or

$$\frac{di(0)}{dt} = -\frac{R}{L} i(0) - \frac{v_C(0)}{L}$$
$$= -\frac{6}{1} (4) + \frac{4}{1}$$
$$= -20$$

Therefore,

$$-3K_1 + 4K_2 = -20$$

and since $K_1 = 4$, $K_2 = -2$, the expression then for $i(t)$ is

$$i(t) = 4e^{-3t} \cos 4t - 2e^{-3t} \sin 4t \text{ A}$$

Note that this expression satisfies the initial condition $i(0) = 4$. The voltage across the capacitor could be determined via KVL using this current:

$$Ri(t) + L \frac{di(t)}{dt} + v_C(t) = 0$$

or

$$v_C(t) = -Ri(t) - L \frac{di(t)}{dt}$$

Substituting the preceding expression for $i(t)$ into this equation yields

$$v_C(t) = -4e^{-3t} \cos 4t + 22e^{-3t} \sin 4t \text{ V}$$

Note that this expression satisfies the initial condition $v_C(0) = -4$ V.

The MATLAB program for plotting this function in the time interval $t > 0$ is listed as follows:

```
>>tau = 1/3;
>>tend = 10*tau;
>>t = linspace(0, tend, 150);
>>v1 = -4*exp(-3*t).*cos(4*t);
>>v2 = 22*exp(-3*t).*sin(4*t);
>>v = v1+v2;
>>plot (t,v)
>>xlabel('Time (s)')
>>ylabel('Voltage (V)')
```

Note that the functions `exp(-3*t)` and `cos(4*t)` produce arrays the same size as `t`. Hence, when we multiply these functions together, we need to specify that we want the arrays to be multiplied element by element. Element-by-element multiplication of an array is denoted in MATLAB by the `.*` notation. Multiplying an array by a scalar `(-4*exp(-3*t)`, for example, can be performed by using the `*` by itself.

The plot generated by this program is shown in Fig. 6.19.

Figure 6.19   *Underdamped response.*

LEARNING Example 6.9

Let us examine the circuit in Fig. 6.20, which is slightly more complicated than the two we have considered earlier. The two equations that describe the network are

$$L\frac{di(t)}{dt} + R_1 i(t) + v(t) = 0$$

$$i(t) = C\frac{dv(t)}{dt} + \frac{v(t)}{R_2}$$

Figure 6.20 Series-parallel RLC circuit.

Substituting the second equation into the first yields

$$\frac{d^2v}{dt^2} + \left(\frac{1}{R_2 C} + \frac{R_1}{L}\right)\frac{dv}{dt} + \frac{R_1 + R_2}{R_2 LC}v = 0$$

If the circuit parameters and initial conditions are

$$R_1 = 10\ \Omega \qquad C = \tfrac{1}{8}\text{F} \qquad v_C(0) = 1\text{ V}$$
$$R_2 = 8\ \Omega \qquad L = 2\text{ H} \qquad i_L(0) = \tfrac{1}{2}\text{A}$$

the differential equation becomes

$$\frac{d^2v}{dt^2} + 6\frac{dv}{dt} + 9v = 0$$

We wish to find expressions for the current $i(t)$ and the voltage $v(t)$.

SOLUTION The characteristic equation is then

$$s^2 + 6s + 9 = 0$$

and hence the roots are

$$s_1 = -3$$
$$s_2 = -3$$

Since the roots are real and equal, the circuit is critically damped. The term $v(t)$ is then given by the expression

$$v(t) = K_1 e^{-3t} + K_2 t e^{-3t}$$

Since $v(t) = v_C(t)$,

$$v(0) = v_C(0) = 1 = K_1$$

In addition,

$$\frac{dv(t)}{dt} = -3K_1 e^{-3t} + K_2 e^{-3t} - 3K_2 t e^{-3t}$$

However,

$$\frac{dv(t)}{dt} = \frac{i(t)}{C} - \frac{v(t)}{R_2 C}$$

Setting these two expressions equal to one another and evaluating the resultant equation at $t = 0$ yields

$$\frac{1/2}{1/8} - \frac{1}{1} = -3K_1 + K_2$$

$$3 = -3K_1 + K_2$$

Since $K_1 = 1$, $K_2 = 6$ and the expression for $v(t)$ is

$$v(t) = e^{-3t} + 6t e^{-3t}\text{ V}$$

Note that the expression satisfies the initial condition $v(0) = 1$. The current $i(t)$ can be determined from the nodal analysis equation at $v(t)$.

$$i(t) = C\frac{dv(t)}{dt} + \frac{v(t)}{R_2}$$

Substituting $v(t)$ from the preceding equation, we find

$$i(t) = \frac{1}{8}\left[-3e^{-3t} + 6e^{-3t} - 18t e^{-3t}\right] + \frac{1}{8}\left[e^{-3t} + 6t e^{-3t}\right]$$

or

$$i(t) = \frac{1}{2}e^{-3t} - \frac{3}{2}te^{-3t} \text{ A}$$

If this expression for the current is employed in the circuit equation,

$$v(t) = -L\frac{di(t)}{dt} - R_1i(t)$$

we obtain

$$v(t) = e^{-3t} + 6te^{-3t} \text{ V}$$

which is identical to the expression derived earlier.

The MATLAB program for generating a plot of this function, shown in Fig. 6.21, is listed here.

```
>>tau = 1/3;
>>tend = 10*tau;
>>t = linspace(0, tend, 150);
>>v = exp(-3*t) + 6*t.*exp(-3*t);
>>plot(t,v)
>>xlabel('Time (s)')
>>ylabel('Voltage (V)')
```

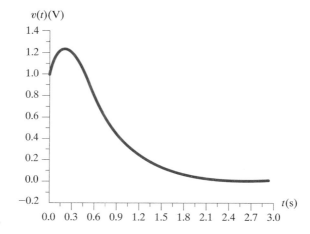

Figure 6.21
Critically damped response.

LEARNING EXTENSION

E6.9 The switch in the network in Fig. E6.9 opens at $t = 0$. Find $i(t)$ for $t > 0$.

ANSWER
$i(t) = -2e^{-t/2} + 4e^{-t} \text{ A}$.

Figure E6.9

E6.10 The switch in the network in Fig. E6.10 moves from position 1 to position 2 at $t = 0$. Find $v_o(t)$ for $t > 0$.

ANSWER
$$v_o(t) = 2(e^{-t} - 3e^{-3t}) \text{ V}.$$

Figure E6.10

LEARNING Example 6.10

Consider the circuit shown in Fig. 6.22. This circuit is the same as that analyzed in Example 6.8, except that a constant forcing function is present. The circuit parameters are the same as those used in Example 6.8:

$$C = 0.04 \text{ F} \qquad i_L(0) = 4 \text{ A}$$
$$L = 1 \text{ H} \qquad v_C(0) = -4 \text{ V}$$
$$R = 6 \text{ }\Omega$$

We want to find an expression for $v_C(t)$ for $t > 0$.

SOLUTION From our earlier mathematical development we know that the general solution of this problem will consist of a particular solution plus a complementary solution. From

Example 6.8 we know that the complementary solution is of the form $K_3 e^{-3t} \cos 4t + K_4 e^{-3t} \sin 4t$. The particular solution is a constant, since the input is a constant and therefore the general solution is

$$v_C(t) = K_3 e^{-3t} \cos 4t + K_4 e^{-3t} \sin 4t + K_5$$

An examination of the circuit shows that in the steady state the final value of $v_C(t)$ is 12 V, since in the steady-state condition, the inductor is a short circuit and the capacitor is an open circuit. Thus, $K_5 = 12$. The steady-state value could also be immediately calculated from the differential equation. The form of the general solution is then

$$v_C(t) = K_3 e^{-3t} \cos 4t + K_4 e^{-3t} \sin 4t + 12$$

Figure 6.22
Series RLC circuit with a step function input.

The initial conditions can now be used to evaluate the constants K_3 and K_4.

$$v_C(0) = -4 = K_3 + 12$$

$$-16 = K_3$$

Since the derivative of a constant is zero, the results of Example 6.8 show that

$$\frac{dv_C(0)}{dt} = \frac{i(0)}{C} = 100 = -3K_3 + 4K_4$$

and since $K_3 = -16$, $K_4 = 13$. Therefore, the general solution for $v_C(t)$ is

$$v_C(t) = 12 - 16e^{-3t} \cos 4t + 13e^{-3t} \sin 4t \text{ V}$$

Note that this equation satisfies the initial condition $v_C(0) = -4$ and the final condition $v_C(\infty) = 12$ V.

LEARNING EXTENSION

E6.11   The switch in the network in Fig. E6.11 moves from position 1 to position 2 at $t = 0$. Compute $i_o(t)$ for $t > 0$ and use this current to determine $v_o(t)$ for $t > 0$.

ANSWER

$i_o(t) = -\frac{11}{6}e^{-3t} + \frac{14}{6}e^{-6t}$ A,

$v_o(t) = 12 + 18i_o(t)$ V.

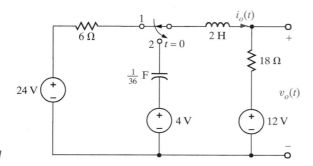

Figure E6.11

6.4 *Transient PSPICE Analysis Using Schematic Capture*

INTRODUCTION   In transient analyses, we determine voltages and currents as functions of time. Typically, the time dependence is demonstrated by plotting the waveforms using time as the independent variable. PSPICE can perform this kind of analysis, called a *Transient* simulation, in which all voltages and currents are determined over a specified time duration. To facilitate plotting, PSPICE uses what is known as the PROBE utility, which will be described later. As an introduction to transient analysis, let us simulate the circuit in Fig. 6.23, plot the voltage $v_C(t)$ and the current $i(t)$, and extract the time constant. Although we will introduce some new PSPICE topics in this section, *Schematics* fundamentals such as getting parts, wiring, and editing part names and values have already been covered in Chapter 4. Also, uppercase text refers to PSPICE utilities and dialog boxes whereas boldface denotes keyboard or mouse inputs.

Figure 6.23 *A circuit used for Transient simulation.*

THE SWITCH PARTS The inductor and capacitor parts are called L and C, respectively, and are in the ANALOG library. The switch, called SW_TCLOSE, is in the EVAL library. There is also a SW_TOPEN part that models an opening switch. After placing and wiring the switch along with the other parts, the *Schematics* circuit appears as that shown in Fig. 6.24.

To edit the switch's attributes, double-click on the switch symbol and the ATTRIBUTES box in Fig. 6.25 will appear. Deselecting the **Include Non-changeable Attributes** and **Include System-defined Attributes** fields limits the attribute list to those we can edit and is highly recommended.

Figure 6.24 *The Schematics circuit.*

Figure 6.25
The switch's ATTRIBUTES box.

The attribute **tClose** is the time at which the switch begins to close, and **ttran** is the time required to complete the closure. Switch attributes **Rclosed** and **Ropen** are the switch's resistance in the closed and open positions, respectively. During simulations, the resistance of the switch changes linearly from **Ropen** at $t = $ **tClose** to **Rclosed** at $t = $ **tClose + ttran.**

When using the SW_TCLOSE and SW_TOPEN parts to simulate ideal switches, care should be taken to ensure that the values for **ttran**, **Rclosed**, and **Ropen** are appropriate for valid simulation results. In our present example, we see that the switch and R_1 are in series, thus, their resistances add. Using the default values listed in Fig. 6.25, we find that when the switch is closed, the switch resistance, **Rclosed**, is 0.01 Ω, 100,000 times smaller than that of the resistor. The resulting series-equivalent resistance is essentially that of the resistor. Alternatively, when the switch is open, the switch resistance is 1 MΩ, 1,000 times larger than that of the resistor. Now, the equivalent resistance is much larger than that of the resistor. Both are desirable scenarios.

To determine a reasonable value for **ttran**, we first estimate the duration of the transient response. The component values yield a time constant of 1 ms, and thus all voltages and currents will reach steady state in about 5 ms. For accurate simulations, **ttran** should be much less than 5 ms. Therefore, the default value of 1 μs, is viable in this case.

THE IMPORTANCE OF PIN NUMBERS   As mentioned in Chapter 4, each component within the various Parts libraries has two or more terminals. Within PSPICE, these terminals are called pins and are numbered sequentially starting with pin 1, as shown in Fig. 6.26 for several two-terminal parts. The significance of the pin numbers is their effect on currents plotted using the PROBE utility. PROBE always plots the current entering pin 1 and exiting pin 2. Thus, if the current through an element is to be plotted, the part should be oriented in the *Schematics* diagram such that the defined current direction enters the part at pin 1. This can be done by using the ROTATE command in the EDIT menu. ROTATE causes the part to spin 90° counterclockwise. In our example, we will plot the current $i(t)$ by plotting the current through the capacitor, I(C1). Therefore, when the *Schematics* circuit in Fig. 6.24 was created, the capacitor was rotated 270°. As a result, pin 1 is at the top of the diagram and the assigned current direction in Fig. 6.23 matches the direction presumed by PROBE. If a component's current direction in PROBE is opposite the desired direction, simply go to the *Schematics* circuit, rotate the part in question 180°, and re-simulate.

Figure 6.26
Pin numbers for common PSPICE parts.

SETTING INITIAL CONDITIONS To set the initial condition of the capacitor voltage, double-click on the capacitor symbol in Fig. 6.24 to open its ATTRIBUTE box, as shown in Fig. 6.27. Click on the **IC** field and set the value to the desired voltage, 0 V in this example. Setting the initial condition on an inductor current is done in a similar fashion. Be forewarned that the initial condition for a capacitor voltage is positive at pin 1 versus pin 2. Similarly, the initial condition for an inductor's current will flow into pin 1 and out of pin 2.

Figure 6.27
Setting the capacitor initial condition.

SETTING UP A TRANSIENT ANALYSIS The simulation duration is selected using **SETUP** from the **ANALYSIS** menu. When the SETUP window shown in Fig. 6.28 appears, double-click on the text **TRANSIENT** and the TRANSIENT window in Fig. 6.29 will appear. The simulation period described by **Final time** is selected as 6 milliseconds. All simulations start at $t = 0$. The **No-Print Delay** field sets the time the simulation runs before data collection begins. **Print Step** is the interval used for printing data to the output file. **Print Step** has no effect on the data used to create PROBE plots. The **Detailed Bias Pt.** option is useful when simulating circuits containing transistors and diodes, and thus will not be used here. When **Skip initial transient solution** is enabled, all capacitors and inductors that do not have specific initial condition values in their ATTRIBUTES boxes will use zero initial conditions.

Sometimes, plots created in PROBE are not smooth. This is caused by an insufficient number of data points. More data points can be requested by inputting a **Step Ceiling** value. A reasonable first guess would be a hundredth of the **Final Time**. If the resulting PROBE plots are still unsatisfactory, reduce the **Step Ceiling** further. As soon as the TRANSIENT window is complete, simulate the circuit by selecting **Simulate** from the **Analysis** menu.

Figure 6.28
The ANALYSIS SETUP window.

Figure 6.29
The TRANSIENT window.

PLOTTING IN PROBE When the PSPICE simulation is finished, the PROBE window shown in Fig. 6.30 will open. If not, select **Run Probe** from the **Analysis** menu. In Fig. 6.30, we see three subwindows: the main display window, the output window, and the simulation status window. The waveforms we choose to plot appear in the main display window. The output window shows messages from PSPICE about the success or failure of the simulation. Run-time information about the simulation appears in the simulation status window. Here we will focus on the main display window.

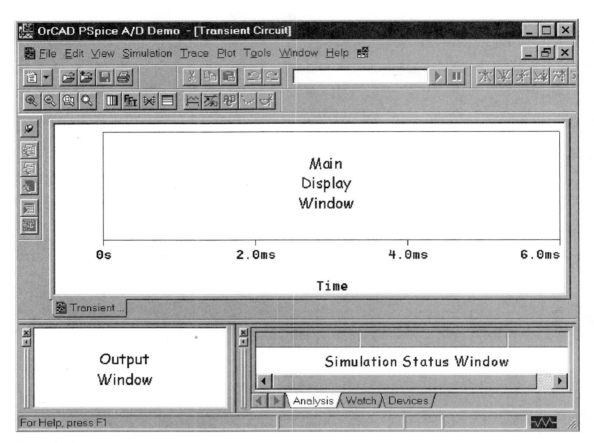

Figure 6.30
The PROBE window.

To plot the voltage, $v_C(t)$, select **Add Trace** from the **Trace** menu. The ADD TRACES window is shown in Fig. 6.31. Note that the options **Alias Names** and **Subcircuit Nodes** have been deselected, which greatly simplifies the ADD TRACES window. The capacitor voltage is obtained by clicking on **V(Vc)** in the left column. The PROBE window should look like that shown in Fig. 6.32.

Before adding the current $i(t)$ to the plot, we note that the dc source is 10 V and the resistance is 1 kΩ, which results in a loop current of a few milliamps. Since the capacitor voltage span is much greater, we will plot the current on a second y axis. From the **Plot** menu, select

Figure 6.31
The Add Traces window.

Figure 6.32
The capacitor voltage.

Figure 6.33
The capacitor voltage and the clockwise loop current.

Add Y Axis. To add the current to the plot, select **Add Trace** from the **Trace** menu, then select **I(C1)**. Figure 6.33 shows the PROBE plot for $v_C(t)$ and $i(t)$.

FINDING THE TIME CONSTANT   Given that the final value of $v_C(t)$ is 10 V, we can write

$$v_C(t) = 10\left[1 - e^{-t/\tau}\right] \text{ V}$$

When $t = \tau$,

$$v_C(\tau) = 10\left(1 - e^{-1}\right) \text{ V} = 6.32 \text{ V}$$

To determine the time at which the capacitor voltage is 6.32 V, we activate the cursors by selecting **Cursor/Display** in the **Trace** menu. There are two cursors that can be used to extract x-y data from the plots. Use the ← and → arrow keys to move the first cursor. By holding the SHIFT key down, the arrow keys move the second cursor. Moving the first cursor along the voltage plot, as shown in Fig. 6.34, we find that 6.32 V occurs at a time of 1 ms. Therefore, the time constant is 1 ms—exactly the *RC* product.

Figure 6.34
PROBE plot for $v_C(t)$ and time constant extraction.

SAVING AND PRINTING PROBE PLOTS   Saving plots within PROBE requires the use of the **Display Control** command in the **Windows** menu. Using the SAVE/RESTORE DISPLAY window shown in Fig. 6.35, we simply name the plot and click on **Save**. Figure 6.35 shows that one plot has already been saved, TranExample. This procedure saves the plot attributes such as axes settings, additional text, and cursor settings in a file with a .prb extension. Additionally, the .prb file contains a reference to the appropriate data file, which has a .dat extension, and contains the actual simulation results. Therefore, in using **Display Control** to save a plot, we do not save the plot itself, only the plot settings and the .dat file's name. To access an old PROBE plot, enter PROBE, and from the **File** menu, open the appropriate .dat file. Next, access the DISPLAY CONTROL window, select the file of interest and click on **RESTORE**. Use the **Save As** option in Fig. 6.35 to save the .prb file to any directory on any disk, hard or floppy.

To copy the PROBE plot to other documents such as word processors, select the **Copy to Clipboard** command in the **Window** menu. The window in Fig. 6.36 will appear showing several options. If your PROBE display screen background is black, it is recommended that you choose the **make window and plot backgrounds transparent** and **change white to black** options. When the plot is pasted, it will have a white background with black text—a better scenario for printing. To print a PROBE plot, select **Print** from the **File** menu and the PRINT window in Fig. 6.37 will open. The options in this window are self-explanatory. Note that printed plots have white backgrounds with black text.

Figure 6.35
The SAVE/RESTORE DISPLAY window used in PROBE to save plots.

Figure 6.36
The SAVE/RESTORE DISPLAY window used in PROBE to save plots.

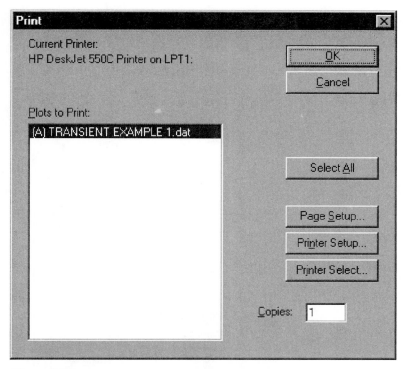

Figure 6.37
The PRINT window used in PROBE.

OTHER PROBE FEATURES   There are several features within PROBE for plot manipulation and data extraction. Within the **Plot** menu resides commands for editing the plot itself. These include altering the axes, adding axes, and adding additional plots to the page. Also in the **Tools** menu is the **Label** command, which allows one to add marks (data point values), explanatory text, lines, and shapes to the plot.

LEARNING Example 6.11

Using the PSPICE *Schematics* editor, draw the circuit in Fig. 6.38 and use the PROBE utility to find the time at which the capacitor and inductor current are equal.

SOLUTION   Figure 6.39a shows the *Schematics* circuit, and the simulation results are shown in Fig. 6.39b. Based on the PROBE plot, the currents are equal at 561.8 ns.

Figure 6.38
Circuit used in Example 6.11.

(continued)

(a)

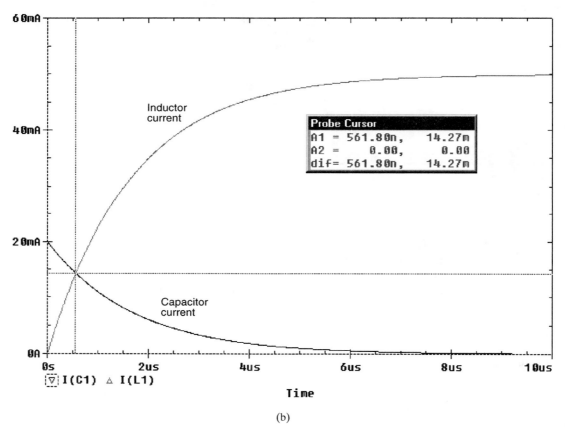

(b)

Figure 6.39
(a) PSPICE network and (b) simulation results for $i_C(t)$ and $i_L(t)$ in Example 6.11.

Learning by Application

There are a wide variety of applications for transient circuits. The following examples will serve to demonstrate some of them.

LEARNING Example 6.12

A heart pacemaker circuit is shown in Fig. 6.40. The SCR (silicon-controlled rectifier) is a solid-state device that has two distinct modes of operation. When the voltage across the SCR is increasing but less than 5 V, the SCR behaves like an open circuit, as shown in Fig. 6.41a. Once the voltage across the SCR reaches 5 V, the device functions like a current source, as shown in Fig. 6.41b. This behavior will continue as long as the SCR voltage remains above 0.2 V. At this voltage, the SCR shuts off and again becomes an open circuit.

Assume that at $t = 0$, $v_C(t)$ is 0 V and the 1-μF capacitor begins to charge toward the 6-V source voltage. Find the resistor value such that $v_C(t)$ will equal 5 V (the SCR firing voltage) at 1 s. At $t = 1$ s, the SCR fires and begins discharging the capac-

itor. Find the time required for $v_C(t)$ to drop from 5 V to 0.2 V. Finally, plot $v_C(t)$ for the three cycles.

SOLUTION For $t < 1$ s, the equivalent circuit for the pacemaker is shown in Fig. 6.42. As indicated earlier, the capacitor voltage has the form

$$v_C(t) = 6 - 6e^{-t/RC}$$

A voltage of 0.2 V occurs at

$$t_1 = 0.034RC$$

Figure 6.40
Heart pacemaker equivalent circuit.

(a)   (b)

Figure 6.41
Equivalent circuits for silicon-controlled rectifer.

(continued)

Figure 6.42
Pacemaker equivalent network during capacitor charge cycle.

whereas a voltage of 5 V occurs at

$$t_2 = 1.792RC$$

We desire that $t_2 - t_1 = 1$ s. Therefore,

$$t_2 - t_1 = 1.758RC = 1 \text{ s}$$

and

$$RC = 0.569 \text{ s} \quad \text{and} \quad R = 569 \text{ k}\Omega$$

At $t = 1$ s the SCR fires and the pacemaker is modeled by the circuit in Fig. 6.43. The form of the discharge waveform is

$$v(t) = K_1 + K_2 e^{-(t-1)/RC}$$

Figure 6.43
Pacemaker equivalent network during capacitor discharge cycle.

The term $(t - 1)$ appears in the exponential to shift the function 1 s, since during that time the capacitor was charging. Just after the SCR fires at $t = 1^+$ s, $v_C(t)$ is still 5 V, whereas at $t = \infty$, $v_C(t) = 6 - IR$. Therefore,

$$K_1 + K_2 = 5 \quad \text{and} \quad K_1 = 6 - IR$$

Our solution, then, is of the form

$$v_C(t) = 6 - IR + (IR - 1)e^{-(t-1)/RC}$$

Let T be the time beyond 1 s necessary for $v(t)$ to drop to 0.2 V. We write

$$v_C(T + 1) = 6 - IR + (IR - 1)e^{-T/RC} = 0.2$$

Substituting for I, R, and C, we find

$$T = 0.11 \text{ s}$$

The output waveform is shown in Fig. 6.44.

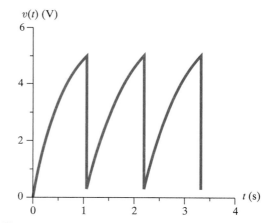

Figure 6.44   *Heart pacemaker output voltage waveform.*

LEARNING Example 6.13

One of the most common and necessary subcircuits that appears in a wide variety of electronic systems—for example, stereos, TVs, radios, and computers—is a quality dc voltage source or power supply. The standard wall socket supplies an alternating current (ac) voltage waveform shown in Fig. 6.45a, and the conversion of this voltage to a desired dc level is done as illustrated in Fig. 6.45b. The ac waveform is converted to a quasi-dc

voltage by an inexpensive ac–dc converter whose output contains remnants of the ac input and is unregulated. A higher quality dc output is created by a switching dc–dc converter. Of the several versions of dc–dc converters, we will focus on a topology called the boost converter, shown in Fig. 6.46. Let us develop an equation relating the output voltage to the switching characteristics.

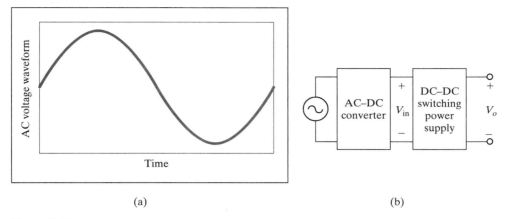

(a)   (b)

Figure 6.45
(a) The ac voltage waveform at a standard wall outlet and (b) a block diagram of a modern dc power supply.

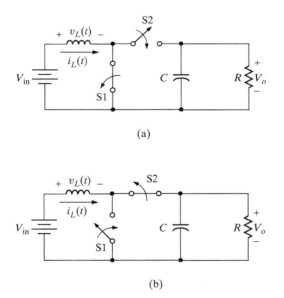

(a)

(b)

Figure 6.46
The boost converter with switch settings for time intervals
(a) t_{on} and (b) t_{off}.

SOLUTION   Consider the boost converter in Fig. 6.46a, where switch 1 (S1) is closed and S2 is open for a time interval t_{on}. This isolates the inductor from the capacitor, creating two sub-circuits that can be analyzed independently. Note that during t_{on} the inductor current and stored energy are increasing while at the output node, the capacitor voltage discharges exponentially into the load. If the capacitor's time constant $(\tau = RC)$ is large, then the output voltage will decrease slowly. Thus, during t_{on} energy is stored in the inductor and the capacitor provides energy to the load.

(continued)

Next, we change both switch positions so that S1 is open and S2 is closed for a time interval t_{off}, as seen in Fig. 6.46b. Since the inductor current cannot change instantaneously, current flows into the capacitor and the load, recharging the capacitor. During t_{off} the energy that was added to the inductor during t_{on} is used to recharge the capacitor and drive the load. When t_{off} has elapsed, the cycle is repeated.

Note that the energy added to the inductor during t_{on} must go to the capacitor and load during t_{off}, otherwise, the inductor energy would increase to the point that the inductor would fail. This requires that the energy stored in the inductor must be the same at the end of each switching cycle. Recalling that the inductor energy is related to the current by

$$w(t) = \frac{1}{2} L i^2(t)$$

we can state that the inductor current must also be the same at the end of each switching cycle, as shown in Fig. 6.47. The inductor current during t_{on} and t_{off} can be written as

$$i_L(t) = \frac{1}{L} \int_0^{t_{on}} v_L(t)\, dt = \frac{1}{L} \int_0^{t_{on}} V_{in}\, dt = \left[\frac{V_{in}}{L}\right] t_{on} + I_0$$
$$0 < t < t_{on}$$

$$i_L(t) = \frac{1}{L} \int_{t_{on}}^{t_{on}+t_{off}} v_L(t)\, dt = \frac{1}{L} \int_{t_{on}}^{t_{on}+t_{off}} (V_{in} - V_o)\, dt =$$

$$\left[\frac{V_{in} - V_o}{L}\right] t_{off} + I_0 \qquad t_{on} < t < t_{off} \quad \textbf{6.22}$$

where I_0 is the initial current at the beginning of each switching cycle. If the inductor current is the same at the beginning and end of each switching cycle, then the integrals in Eq. (6.22) must sum to zero. Or,

$$V_{in} t_{on} = (V_o - V_{in}) t_{off} = (V_o - V_{in})(T - t_{on})$$

where T is the period $(T = t_{on} + t_{off})$. Solving for V_o yields

$$V_o = V_{in}\left[\frac{T}{T - t_{on}}\right] = V_{in}\left[\frac{1}{(T - t_{on})/T}\right] =$$

$$V_{in}\left[\frac{1}{(1 - t_{on}/T)}\right] = V_{in}\left[\frac{1}{1 - D}\right]$$

where D is the duty cycle $(D = t_{on}/T)$. Thus, by controlling the duty cycle, we control the output voltage. Since D is always a positive fraction, V_o is always bigger than V_{in}—thus the name, boost converter. A plot of V_o/V_{in} versus duty cycle is shown in Fig. 6.48.

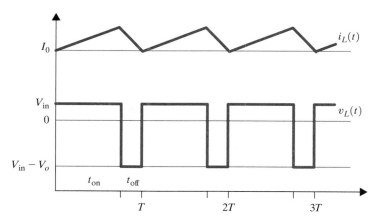

Figure 6.47
Waveform sketches for the inductor voltage and current.

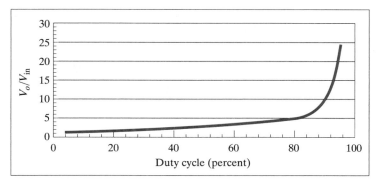

Figure 6.48
Effect of duty cycle on boost converter gain.

Learning by Design

The following example provides an introduction to some typical strategies for transient circuits. In this example the problem involves the selection of the proper circuit parameters to achieve a specified transient response.

LEARNING Example 6.14

The network in Fig. 6.49 models an automobile ignition system. The voltage source represents the standard 12-V battery. The inductor is the ignition coil, which is magnetically coupled to the starter (not shown). The inductor's internal resistance is modeled by the resistor, and the switch is the keyed ignition switch. Initially the switch connects the ignition circuitry to the battery, and thus the capacitor is charged to 12 V. To start the motor, we close the switch, thereby discharging the capacitor through the inductor. Assuming that optimum starter operation requires an overdamped response for $i_L(t)$ that reaches at least 1 A within 100 ms after switching and remains above 1 A for between 1 and 1.5 s, let us find a value for the capacitor that will produce such a current waveform. In addition, let us plot the response including the time interval just prior to moving the switch and verify our design.

Figure 6.49
Circuit model for ignition system.

(continued)

SOLUTION Before the switch is moved at $t = 0$, the capacitor looks like an open circuit and the inductor acts like a short circuit. Thus,

$$i_L(0^-) = i_L(0^+) = 0 \text{ A} \qquad \text{and} \qquad v_C(0^-) = v_C(0^+) = 12 \text{ V}$$

After switching, the circuit is a series RLC unforced network described by the characteristic equation

$$s^2 + \frac{R}{L}s + \frac{1}{LC} = 0$$

with roots at $s = -s_1$ and $-s_2$. The characteristic equation is of the form

$$(s + s_1)(s + s_2) = s^2 + (s_1 + s_2)s + s_1 s_2 = 0$$

Comparing the two expressions, we see that

$$\frac{R}{L} = s_1 + s_2 = 20$$

and

$$\frac{1}{LC} = s_1 s_2$$

Since the network must be overdamped, the inductor current is of the form

$$i_L(t) = K_1 e^{-s_1 t} + K_2 e^{-s_2 t}$$

Just after switching,

$$i_L(0^+) = K_1 + K_2 = 0$$

or

$$K_2 = -K_1$$

Also, at $t = 0^+$, the inductor voltage equals the capacitor voltage because $i_L = 0$ and therefore $i_L R = 0$. Thus, we can write

$$v_L(0^+) = L\frac{di_L(0^+)}{dt} \Rightarrow -s_1 K_1 + s_2 K_1 = \frac{12}{L}$$

or

$$K_1 = \frac{60}{s_2 - s_1}$$

At this point, let us arbitrarily choose $s_1 = 3$ and $s_2 = 17$, which satisfies the condition $s_1 + s_2 = 20$, and furthermore,

$$K_1 = \frac{60}{s_2 - s_1} = \frac{60}{14} = 4.29$$

$$C = \frac{1}{Ls_1 s_2} = \frac{1}{(0.2)(3)(17)} = 98 \text{ mF}$$

Hence, $i_L(t)$ is

$$i_L(t) = 4.29\left[e^{-3t} - e^{-17t}\right] \text{ A}$$

Figure 6.50a shows a plot of $i_L(t)$. At 100 ms the current has increased to 2.39 A, which meets the initial magnitude specifications. However, one second later at $t = 1.1$ s, $i_L(t)$ has fallen to only 0.16 A—well below the magnitude-over-time requirement. Simply put, the current falls too quickly. To make an informed estimate for s_1 and s_2, let us investigate the effect the roots exhibit on the current waveform when $s_2 > s_1$.

Since $s_2 > s_1$, the exponential associated with s_2 will decay to zero faster than that associated with s_1. This causes $i_L(t)$ to rise—the larger the value of s_2, the faster the rise. After $5(1/s_2)$ seconds have elapsed, the exponential associated with s_2 is approximately zero and $i_L(t)$ decreases exponentially with a time constant of $\tau = 1/s_1$. Thus, to slow the fall of $i_L(t)$ we should reduce s_1. Hence, let us choose $s_1 = 1$. Since $s_1 + s_2$ must equal 20, $s_2 = 19$. Under these conditions

$$C = \frac{1}{Ls_1 s_2} = \frac{1}{(0.2)(1)(19)} = 263 \text{ mF}$$

and

$$K_1 = \frac{60}{s_2 - s_1} = \frac{60}{18} = 3.33$$

Thus, the current is

$$i_L(t) = 3.33\left[e^{-t} - e^{-19t}\right] \text{ A}$$

which is shown in Fig. 6.50b. At 100 ms the current is 2.52 A. Also, at $t = 1.1$ s, the current is 1.11 A—above the 1-A requirement. Therefore, the choice of $C = 263$ mF meets all starter specifications.

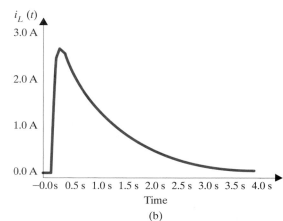

Figure 6.50 *Ignition current as a function of time.*

Summary

First-Order Circuits

▶ An *RC* or *RL* transient circuit is said to be first order if it contains only a single capacitor or single inductor. The voltage or current anywhere in the network can be obtained by solving a first-order differential equation.

▶ The form of a first-order differential equation with a constant forcing function is

$$\frac{dx(t)}{dt} + \frac{x(t)}{\tau} = A$$

and the solution is

$$x(t) = A\tau + K_2 e^{-t/\tau}$$

where $A\tau$ is referred to as the steady-state solution and τ is called the time constant.

▶ The function $e^{-t/\tau}$ decays to a value that is less than 1% of its initial value after a period of 5τ. Therefore, the time constant, τ, determines the time required for the circuit to reach steady state.

▶ The time constant for an *RC* circuit is $R_{\text{Th}}C$ and for an *RL* circuit is L/R_{Th}, where R_{Th} is the Thévenin equivalent resistance looking into the circuit at the terminals of the storage element (i.e., capacitor or inductor).

▶ The two approaches proposed for solving first-order transient circuits are the differential equation approach and the step-by-step method. In the former case, the differential equation that describes the dynamic behavior of the circuit is solved to determine the desired solution. In the latter case, the initial conditions and the steady-state value of the voltage across the capacitor or current in the inductor are used in conjunction with the circuit's time constant and the known form of the desired variable to obtain a solution.

▶ The response of a first-order transient circuit to an input pulse can be obtained by treating the pulse as a combination of two step-function inputs.

Second-Order Circuits

▶ The voltage or current in an *RLC* transient circuit can be described by a constant coefficient differential equation of the form

$$\frac{d^2x(t)}{dt^2} + 2\zeta\omega_0 \frac{dx(t)}{dt} + \omega_0^2 x(t) = f(t)$$

where $f(t)$ is the network forcing function.

▶ The characteristic equation for a second-order circuit is $s^2 + 2\zeta\omega_0 s + \omega_0^2 = 0$, where ζ is the damping ratio and ω_0 is the undamped natural frequency.

▶ If the two roots of the characteristic equation are

  ▶ real and unequal, then $\zeta > 1$ and the network response is overdamped

  ▶ real and equal, then $\zeta = 1$ and the network response is critically damped

  ▶ complex conjugates, then $\zeta < 1$ and the network response is underdamped

▶ The three types of damping together with the corresponding network response are as follows:

  1. Overdamped:
  $$x(t) = K_1 e^{-(\zeta\omega_0 - \omega_0\sqrt{\zeta^2-1})t} + K_2 e^{-(\zeta\omega_0 + \omega_0\sqrt{\zeta^2-1})t}$$

  2. Critically damped: $x(t) = B_1 e^{-\zeta\omega_0 t} + B_2 t e^{-\zeta\omega_0 t}$

  3. Underdamped: $x(t) = e^{-\sigma t}(A_1 \cos \omega_d t + A_2 \sin \omega_d t)$,
  where $\sigma = \zeta\omega_0$ and $\omega_d = \omega_0 \sqrt{1 - \zeta^2}$

▶ Two initial conditions are required to derive the two unknown coefficients in the network response equations.

Problems

SECTION 6.2

6.1 Use the differential equation approach to find $v_C(t)$ for $t > 0$ in the circuit in Fig. P6.1.

Figure P6.1

6.2 Use the differential equation approach to find $i(t)$ for $t > 0$ in the network in Fig. P6.2.

Figure P6.2

6.3 Use the differential equation approach to find $v_C(t)$ for $t > 0$ in the circuit in Fig. P6.3.

Figure P6.3

6.4 Use the differential equation approach to find $i_o(t)$ for $t > 0$ in the network in Fig. P6.4.

Figure P6.4

6.5 In the network in Fig. P6.5, find $i_o(t)$ for $t > 0$ using the differential equation approach.

Figure P6.5

6.6 In the circuit in Fig. P6.6, find $i_o(t)$ for $t > 0$ using the differential equation approach.

Figure P6.6

6.7 Use the differential equation approach to find $v_o(t)$ for $t > 0$ in the circuit in Fig. P6.7 and plot the response including the time interval just prior to switch action.

Figure P6.7

6.8 Use the differential equation approach to find $i(t)$ for $t > 0$ in the circuit in Fig. P6.8 and plot the response

including the time interval just prior to opening the switch.

Figure P6.8

6.9 Use the differential equation approach to find $v_C(t)$ for $t > 0$ in the circuit in Fig. P6.9 and plot the response including the time interval just prior to opening the switch.

Figure P6.9

6.10 Use the differential equation approach to find $i_L(t)$ for $t > 0$ in the circuit in Fig. P6.10 and plot the response including the time interval just prior to opening the switch.

Figure P6.10

6.11 Use the differential equation approach to find $i(t)$ for $t > 0$ in the circuit in Fig. P6.11 and plot the response including the time interval just prior to switch movement.

Figure P6.11

6.12 Use the differential equation approach to find $v_C(t)$ for $t > 0$ in the circuit in Fig. P6.12 and plot the response including the time interval just prior to closing the switch.

Figure P6.12

6.13 Use the differential equation approach to find $v_o(t)$ for $t > 0$ in the circuit in Fig. P6.13 and plot the response including the time interval just prior to opening the switch.

Figure P6.13

6.14 Use the differential equation approach to find $i_o(t)$ for $t > 0$ in the circuit in Fig. P6.14 and plot the response including the time interval just prior to closing the switch.

Figure P6.14

6.15 Use the differential equation approach to find $i(t)$ for $t > 0$ in the circuit in Fig. P6.15 and plot the response including the time interval just prior to switch movement.

Figure P6.15

6.16 Use the differential equation approach to find $v_o(t)$ for $t > 0$ in the circuit in Fig. P6.16 and plot the response including the time interval just prior to switch action.

Figure P6.16

6.17 Use the differential equation approach to find $v_o(t)$ for $t > 0$ in the circuit in Fig. P6.17 and plot the response including the time interval just prior to closing the switch.

Figure P6.17

6.18 Use the differential equation approach to find $v_o(t)$ for $t > 0$ in the circuit in Fig. P6.18 and plot the response including the time interval just prior to opening the switch.

Figure P6.18

6.19 Use the differential equation approach to find $v_o(t)$ for $t > 0$ in the circuit in Fig. P6.19 and plot the response including the time interval just prior to opening the switch.

Figure P6.19

6.20 Use the differential equation approach to find $i_o(t)$ for $t > 0$ in the circuit in Fig. P6.20 and plot the response including the time interval just prior to opening the switch.

Figure P6.20

6.21 Use the differential equation approach to find $i_o(t)$ for $t > 0$ in the circuit in Fig. P6.21 and plot the response including the time interval just prior to opening the switch.

Figure P6.21

6.22 Use the differential equation approach to find $i_o(t)$ for $t > 0$ in the circuit in Fig. P6.22 and plot the response including the time interval just prior to opening the switch.

Figure P6.22

6.23 Using the differential equation approach, find $i_o(t)$ for $t > 0$ in the circuit in Fig. P6.23 and plot the response including the time interval just prior to opening the switch.

Figure P6.23

6.24 Use the differential equation approach to find $v_o(t)$ for $t > 0$ in the circuit in Fig. P6.24 and plot the response including the time interval just prior to opening the switch.

Figure P6.24

6.25 Find $v_C(t)$ for $t > 0$ in the network in Fig. P6.25 using the step-by-step method.

Figure P6.25

6.26 Use the step-by-step method to find $i_o(t)$ for $t > 0$ in the circuit in Fig. P6.26.

Figure P6.26

6.27 Find $i_o(t)$ for $t > 0$ in the network in Fig. P6.27 using the step-by-step method.

Figure P6.27

6.28 Use the step-by-step method to find $i_o(t)$ for $t > 0$ in the circuit in Fig. P6.28.

Figure P6.28

6.29 Use the step-by-step technique to find $i_o(t)$ for $t > 0$ in the network in Fig. P6.29.

Figure P6.29

6.30 Use the step-by-step method to find $v_o(t)$ for $t > 0$ in the network in Fig. P6.30.

Figure P6.30

6.31 Use the step-by-step method to find $i_o(t)$ for $t > 0$ in the circuit in Fig. P6.31.

Figure P6.31

6.32 Find $v_o(t)$ for $t > 0$ in the network in Fig. P6.31 using the step-by-step technique.

6.33 Find $i_o(t)$ for $t > 0$ in the network in Fig. P6.33 using the step-by-step method.

Figure P6.33

6.34 Find $v_o(t)$ for $t > 0$ in the network in Fig. P6.34 using the step-by-step method.

Figure P6.34

6.35 Use the step-by-step method to find $v_o(t)$ for $t > 0$ in the network in Fig. P6.35.

Figure P6.35

6.36 Use the step-by-step technique to find $i_o(t)$ for $t > 0$ in the network in Fig. P6.36.

Figure P6.36

6.37 Find $i_o(t)$ for $t > 0$ in the network in Fig. P6.37 using the step-by-step method.

Figure P6.37

6.38 Use the step-by-step method to find $i_o(t)$ for $t > 0$ in the network in Fig. P6.38.

Figure P6.38

6.39 Find $v_o(t)$ for $t > 0$ in the circuit in Fig. P6.38 using the step-by-step method.

6.40 Find $i_o(t)$ for $t > 0$ in the network in Fig. P6.40 using the step-by-step method.

Figure P6.40

6.41 Find $v_o(t)$ for $t > 0$ in the network in Fig. P6.41 using the step-by-step method.

Figure P6.41

6.42 Find $v_o(t)$ for $t > 0$ in the network in Fig. P6.42 using the step-by-step method.

Figure P6.42

6.43 Find $i_o(t)$ for $t > 0$ in the circuit in Fig. P6.43 using the step-by-step technique.

Figure P6.43

6.44 Find $v_o(t)$ for $t > 0$ in the network in Fig. P6.44 using the step-by-step method.

Figure P6.44

6.45 Use the step-by-step method to find $i_o(t)$ for $t > 0$ in the network in Fig. P6.45.

Figure P6.45

6.46 Find $v_o(t)$ for $t > 0$ in the circuit in Fig. P6.46 using the step-by-step method.

Figure P6.46

6.47 Use the step-by-step technique to find $v_o(t)$ for $t > 0$ in the circuit in Fig. P6.47.

Figure P6.47

6.48 Find $v_o(t)$ for $t > 0$ in the circuit in Fig. P6.48 using the step-by-step method.

Figure P6.48

6.49 Use the step-by-step method to find $v_o(t)$ for $t > 0$ in the network in Fig. P6.49.

Figure P6.49

6.50 Find $v_o(t)$ for $t > 0$ in the network in Fig. P6.50 using the step-by-step method.

Figure P6.50

6.51 Use the step-by-step technique to find $i_o(t)$ for $t > 0$ in the network in Fig. P6.51.

Figure P6.51

6.52 Use the step-by-step method to find $i_o(t)$ for $t > 0$ in the network in Fig. P6.52.

Figure P6.52

6.53 Find $i_o(t)$ for $t > 0$ in the circuit in Fig. P6.53 using the step-by-step method.

Figure P6.53

6.54 Find $i_o(t)$ for $t > 0$ in the circuit in Fig. P6.54 using the step-by-step technique.

Figure P6.54

SECTION 6.3

6.55 The differential equation that describes the current $i_o(t)$ in a network is

$$\frac{d^2 i_o(t)}{dt^2} + 6\left[\frac{di_o(t)}{dt}\right] + 8i_o(t) = 0$$

Find

(a) the characteristic equation of the network.

(b) the network's natural frequencies.

(c) the expression for $i_o(t)$.

6.56 The terminal current in a network is described by the equation

$$\frac{d^2 i_o(t)}{dt^2} + 10\left[\frac{di_o(t)}{dt}\right] + 25i_o(t) = 0$$

Find

(a) the characteristic equation of the network.

(b) the network's natural frequencies.

(c) the equation for $i_o(t)$.

6.57 The voltage $v_1(t)$ in a network is defined by the equation

$$\frac{d^2 v_1(t)}{dt^2} + 2\left[\frac{dv_1(t)}{dt}\right] + 5v_1(t) = 0$$

Find

(a) the characteristic equation of the network.

(b) the circuit's natural frequencies.

(c) the expression for $v_1(t)$.

6.58 The output voltage of a circuit is described by the differential equation

$$\frac{d^2 v_o(t)}{dt^2} + 6\left[\frac{dv_o(t)}{dt}\right] + 10v_o(t) = 0$$

Find

(a) the characteristic equation of the circuit.

(b) the network's natural frequencies.

(c) the equation for $v_o(t)$.

6.59 The parameters for a parallel RLC circuit are $R = 1 \, \Omega$, $L = \frac{1}{5}$ H, and $C = \frac{1}{4}$ F. Determine the type of damping exhibited by the circuit.

6.60 A series RLC circuit contains a resistor $R = 2 \, \Omega$ and a capacitor $C = \frac{1}{8}$ F. Select the value of the inductor so that the circuit is critically damped.

6.61 For the underdamped circuit shown in Fig. 6.61, determine the voltage $v(t)$ if the initial conditions on the storage elements are $i_L(0) = 1$ A and $v_C(0) = 10$ V.

Figure P6.61

6.62 Given the circuit and the initial conditions of Problem 6.61, determine the current through the inductor.

6.63 Find $v_C(t)$ for $t > 0$ in the circuit in Fig. P6.63 if $v_C(0) = 0$.

Figure P6.63

6.64 Find $v_o(t)$ for $t > 0$ in the circuit in Fig. P6.64 and plot the response including the time interval just prior to moving the switch.

Figure P6.64

6.65 Find $v_C(t)$ for $t > 0$ in the circuit in Fig. P6.65.

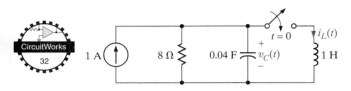

Figure P6.65

6.66 Find $i_L(t)$ for $t > 0$ in the circuit in Fig. P6.65.

6.67 Given the circuit in Fig. P6.67, find the equation for $i(t)$, $t > 0$.

Figure P6.67

6.68 In the circuit shown in Fig. P6.68, find $v(t)$, $t > 0$.

Figure P6.68

6.69 Find $i_o(t)$ for $t > 0$ in the circuit in Fig. P6.69 and plot the response including the time interval just prior to opening the switch.

Figure P6.69

6.70 Find $v_o(t)$ for $t > 0$ in the circuit in Fig. P6.70 and plot the response including the time interval just prior to closing the switch.

Figure P6.70

6.71 Find $v_o(t)$ for $t > 0$ in the circuit in Fig. P6.71 and plot the response including the time interval just prior to moving the switch.

Figure P6.71

SECTION 6.4

6.72 Using the PSPICE *Schematics* editor, draw the circuit in Fig. P6.72, and use the PROBE utility to plot $v_C(t)$ and determine the time constants for $0 < t < 1$ ms and 1 ms $< t < \infty$. Also, find the maximum voltage on the capacitor.

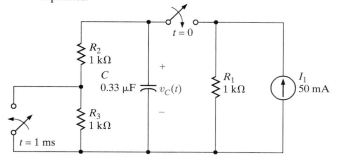

Figure P6.72

6.73 Using the PSPICE *Schematics* editor, draw the circuit in Fig. P6.73, and use the PROBE utility to find the maximum values of $v_L(t)$, $i_C(t)$, and $i(t)$.

Figure P6.73

SECTION 6.6

6.74 Design a series *RCL* circuit with $R \geq 1$ kΩ that has the characteristic equation

$$s^2 + 4 \times 10^7 s + 4 \times 10^{14} = 0$$

6.75 Design a parallel *RLC* circuit with $R \geq 1$ kΩ that has the characteristic equation

$$s^2 + 4 \times 10^7 s + 3 \times 10^{14} = 0$$

Typical Problems Found on the FE Exam

6FE-1 In the circuit in Fig. 6PFE-1, the switch, which has been closed for a long time, opens at $t = 0$. Find the value of the capacitor voltage $v_C(t)$ at $t = 2$ s.

Figure 6PFE-1

6FE-2 In the network in Fig. 6PFE-2, the switch closes at $t = 0$. Find $v_o(t)$ at $t = 1$ s.

Figure 6PFE-2

6FE-3 Assume that the switch in the network in Fig. 6PFE-3 has been closed for some time. At $t = 0$ the switch opens. Determine the time required for the capacitor voltage to decay to one-half of its initially charged value.

Figure 6PFE-3

7 | *AC Steady-State Analysis*

LEARNING Goals

In the preceding chapters we have considered in some detail both the natural and forced response of a network. We found that the natural response was a characteristic of the network and was independent of the forcing function. The forced response, however, depends directly on the type of forcing function, which until now has generally been a constant. At this point we will diverge from this tack to consider an extremely important excitation: the *sinusoidal forcing function*. Nature is replete with examples of sinusoidal phenomena, and although this is important to us as we examine many different types of physical systems, one reason that we can appreciate at this point for studying this forcing function is that it is the dominant waveform in the electric power industry. The signal present at the ac outlets in our home, office, laboratory, and so on is sinusoidal. In addition, we will demonstrate in Chapter 14 that via Fourier analysis we can represent any periodic electrical signal by a sum of sinusoids.

In this chapter we concentrate on the steady-state forced response of networks with sinusoidal driving functions. We will ignore the initial conditions and the transient or natural response, which will eventually vanish for the type of circuits with which we will be dealing. We refer to this as an *ac steady-state analysis*.

Our approach will be to begin by first studying the characteristics of a sinusoidal function as a prelude to using it as a forcing function for a circuit. We will mathematically relate this sinusoidal forcing function to a complex forcing function, which will lead us to define a phasor. By employing phasors we effectively transform a set of differential equations with sinusoidal forcing functions in the time domain into a set of algebraic equations containing complex numbers in the frequency domain. We will show that in this frequency domain Kirchhoff's laws are valid and, thus, all the analysis techniques that we have learned for dc analysis are applicable in ac steady-state analysis. Finally, we demonstrate the power of PSPICE in the solution of ac steady-state circuits.

7.1 Sinusoids

Let us begin our discussion of sinusoidal functions by considering the sine wave

$$x(\omega t) = X_M \sin \omega t \qquad \textbf{7.1}$$

where $x(t)$ could represent either $v(t)$ or $i(t)$. X_M is the *amplitude* or *maximum value*, ω is the *radian* or *angular frequency*, and ωt is the *argument* of the sine function. A plot of the function in Eq. (7.1) as a function of its argument is shown in Fig. 7.1a. Obviously, the function repeats itself every 2π radians. This condition is described mathematically as $x(\omega t + 2\pi) = x(\omega t)$ or in general for period T as

$$x\big[\omega(t + T)\big] = x(\omega t) \qquad \textbf{7.2}$$

meaning that the function has the same value at time $t + T$ as it does at time t.

(a)

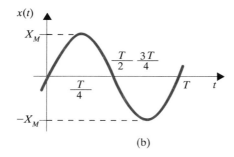

(b)

Figure 7.1
Plots of a sine wave as a function of both ωt and t.

The waveform can also be plotted as a function of time, as shown in Fig. 7.1b. Note that this function goes through one period every T seconds, or in other words, in 1 second it goes through $1/T$ periods or cycles. The number of cycles per second, called Hertz, is the frequency f, where

LEARNING Hint

The relationship between frequency and period

$$f = \frac{1}{T} \qquad \textbf{7.3}$$

Now since $\omega T = 2\pi$, as shown in Fig. 7.1a, we find that

$$\omega = \frac{2\pi}{T} = 2\pi f \qquad\qquad \textbf{7.4}$$

which is, of course, the general relationship among period in seconds, frequency in Hertz, and radian frequency.

Now that we have discussed some of the basic properties of a sine wave, let us consider the following general expression for a sinusoidal function:

$$x(t) = X_M \sin(\omega t + \theta) \qquad\qquad \textbf{7.5}$$

In this case $(\omega t + \theta)$ is the argument of the sine function, and θ is called the *phase angle*. A plot of this function is shown in Fig. 7.2, together with the original function in Eq. (7.1) for comparison. Because of the presence of the phase angle, any point on the waveform $X_M \sin(\omega t + \theta)$ occurs θ radians earlier in time than the corresponding point on the waveform $X_M \sin \omega t$. Therefore, we say that $X_M \sin \omega t$ *lags* $X_M \sin(\omega t + \theta)$ by θ radians. In the more general situation, if

$$x_1(t) = X_{M_1} \sin(\omega t + \theta)$$

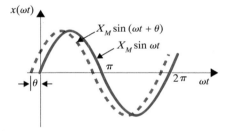

Figure 7.2
Graphical illustration of $X_M \sin(\omega t + \theta)$ leading $X_M \sin \omega t$ by θ radians.

and

$$x_2(t) = X_{M_2} \sin(\omega t + \phi)$$

then $x_1(t)$ leads $x_2(t)$ by $\theta - \phi$ radians and $x_2(t)$ lags $x_1(t)$ by $\theta - \phi$ radians. If $\theta = \phi$, the waveforms are identical and the functions are said to be *in phase*. If $\theta \neq \phi$, the functions are *out of phase*.

The phase angle is normally expressed in degrees rather than radians, and therefore we will simply state at this point that we will use the two forms interchangeably; that is,

$$x(t) = X_M \sin\left(\omega t + \frac{\pi}{2}\right) = X_M \sin(\omega t + 90°) \qquad\qquad \textbf{7.6}$$

Rigorously speaking, since ωt is in radians, the phase angle should be also. However, it is common practice and convenient to use degrees for phase, and therefore, that will be our practice in this text.

In addition, it should be noted that adding to the argument integer multiples of either 2π radians or 360° does not change the original function. This can easily be shown mathematically but is visibly evident when examining the waveform, as shown in Fig. 7.2.

Although our discussion has centered on the sine function, we could just as easily have used the cosine function, since the two waveforms differ only by a phase angle; that is,

LEARNING Hint

A very important point

$$\cos \omega t = \sin\left(\omega t + \frac{\pi}{2}\right) \qquad \textbf{7.7}$$

$$\sin \omega t = \cos\left(\omega t - \frac{\pi}{2}\right) \qquad \textbf{7.8}$$

LEARNING Hint

Some trigonometric identities that are useful in phase angle calculations

It should be noted that when comparing one sinusoidal function with another *of the same frequency* to determine the phase difference, it is necessary to express both functions as either sines or cosines with positive amplitudes. Once in this format, the phase angle between the functions can be computed as outlined previously. Two other trigonometric identities that normally prove useful in phase angle determination are

$$-\cos(\omega t) = \cos(\omega t \pm 180°) \qquad \textbf{7.9}$$

$$-\sin(\omega t) = \sin(\omega t \pm 180°) \qquad \textbf{7.10}$$

Finally, the angle-sum and angle-difference relationships for sines and cosines may be useful in the manipulation of sinusoidal functions. These relations are

$$\sin(\alpha + \beta) = \sin\alpha\cos\beta + \cos\alpha\sin\beta$$
$$\cos(\alpha + \beta) = \cos\alpha\cos\beta - \sin\alpha\sin\beta$$
$$\sin(\alpha - \beta) = \sin\alpha\cos\beta - \cos\alpha\sin\beta$$
$$\cos(\alpha - \beta) = \cos\alpha\cos\beta + \sin\alpha\sin\beta$$

$$\textbf{7.11}$$

LEARNING Example 7.1

We wish to plot the waveforms for the following functions:

(a) $v(t) = 1\cos(\omega t + 45°)$,

(b) $v(t) = 1\cos(\omega t + 225°)$, and

(c) $v(t) = 1\cos(\omega t - 315°)$.

SOLUTION Figure 7.3a shows a plot of the function $v(t) = 1\cos\omega t$. Figure 7.3b is a plot of the function $v(t) = 1\cos(\omega t + 45°)$. Figure 7.3c is a plot of the function $v(t) = 1\cos(\omega t + 225°)$. Note that since

$$v(t) = 1\cos(\omega t + 225°) = 1\cos(\omega t + 45° + 180°)$$

this waveform is 180° out of phase with the waveform in Fig. 7.3b; that is, $\cos(\omega t + 225°) = -\cos(\omega t + 45°)$, and Fig. 7.3c is the negative of Fig. 7.3b. Finally, since the function

$$v(t) = 1\cos(\omega t - 315°) = 1\cos(\omega t - 315° + 360°) = 1\cos(\omega t + 45°)$$

this function is identical to that shown in Fig. 7.3b.

(continued)

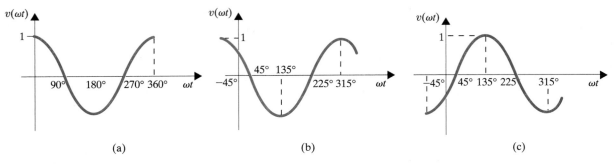

Figure 7.3   Cosine waveforms with various phase angles.

LEARNING Example 7.2

Determine the frequency and the phase angle between the two voltages $v_1(t) = 12 \sin(1000t + 60°)$ V and $v_2(t) = -6 \cos(1000t + 30°)$ V.

SOLUTION   The frequency in Hertz (Hz) is given by the expression

$$f = \frac{\omega}{2\pi} = \frac{1000}{2\pi} = 159.2 \text{ Hz}$$

Using Eq. (7.9), $v_2(t)$ can be written as

$$v_2(t) = -6 \cos(\omega t + 30°) = 6 \cos(\omega t + 210°) \text{ V}$$

Then employing Eq. (7.7), we obtain

$$6 \sin(\omega t + 300°) \text{ V} = 6 \sin(\omega t - 60°) \text{ V}$$

Now that both voltages of the same frequency are expressed as sine waves with positive amplitudes, the phase angle between $v_1(t)$ and $v_2(t)$ is $60° - (-60°) = 120°$; that is, $v_1(t)$ leads $v_2(t)$ by 120° or $v_2(t)$ lags $v_1(t)$ by 120°.

LEARNING by Doing

D 7.1   The local power company supplies voltage to the home at a radian frequency of $\omega = 377$ r/s. What is the frequency in Hertz, or equivalently, cycles per second?

ANSWER   *60 Hz*

LEARNING EXTENSIONS

E7.1   Given the voltage $v(t) = 120 \cos(314t + \pi/4)$ V, determine the frequency of the voltage in Hertz and the phase angle in degrees.

ANSWER $f = 50$ Hz, $\theta = 45°$.

E7.2   Three branch currents in a network are known to be

$$i_1(t) = 2 \sin(377t + 45°) \text{ A}$$
$$i_2(t) = 0.5 \cos(377t + 10°) \text{ A}$$
$$i_3(t) = -0.25 \sin(377t + 60°) \text{ A}$$

Determine the phase angles by which $i_1(t)$ leads $i_2(t)$ and $i_1(t)$ leads $i_3(t)$.

ANSWER i_1 leads i_2 by $-55°$, i_1 leads i_3 by 165°.

7.2 Sinusoidal and Complex Forcing Functions

In the preceding chapters we applied a constant forcing function to a network and found that the steady-state response was also constant.

In a similar manner, if we apply a sinusoidal forcing function to a linear network, the steady-state voltages and currents in the network will also be sinusoidal. This should also be clear from the KVL and KCL equations. For example, if one branch voltage is a sinusoid of some frequency, the other branch voltages must be sinusoids of the same frequency if KVL is to apply around any closed path. This means, of course, that the forced solutions of the differential equations that describe a network with a sinusoidal forcing function are sinusoidal functions of time. For example, if we assume that our input function is a voltage $v(t)$ and our output response is a current $i(t)$ as shown in Fig. 7.4, then if $v(t) = A\sin(\omega t + \theta)$, $i(t)$ will be of the form $i(t) = B\sin(\omega t + \phi)$. The critical point here is that we know the form of the output response, and therefore the solution involves simply determining the values of the two parameters B and ϕ.

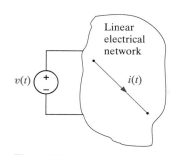

Figure 7.4
Current response to an applied voltage in an electrical network.

LEARNING Example 7.3

Consider the circuit in Fig. 7.5. Let us derive the expression for the current.

Figure 7.5
A simple RL circuit.

SOLUTION   The KVL equation for this circuit is

$$L\frac{di(t)}{dt} + Ri(t) = V_M \cos\omega t$$

Since the input forcing function is $V_M \cos\omega t$, we assume that the forced response component of the current $i(t)$ is of the form

$$i(t) = A\cos(\omega t + \phi)$$

which can be written using Eq. (7.11) as

$$i(t) = A\cos\phi\cos\omega t - A\sin\phi\sin\omega t$$
$$= A_1 \cos\omega t + A_2 \sin\omega t$$

Note that this is, as we observed in Chapter 6, of the form of the forcing function $\cos\omega t$ and its derivative $\sin\omega t$. Substituting this form for $i(t)$ into the preceding differential equation yields

$$L\frac{d}{dt}(A_1\cos\omega t + A_2\sin\omega t)$$
$$+ R(A_1\cos\omega t + A_2\sin\omega t) = V_M\cos\omega t$$

Evaluating the indicated derivative produces

$$-A_1\omega L\sin\omega t + A_2\omega L\cos\omega t + RA_1\cos\omega t$$
$$+ RA_2\sin\omega t = V_M\cos\omega t$$

By equating coefficients of the sine and cosine functions, we obtain

$$-A_1\omega L + A_2 R = 0$$
$$A_1 R + A_2\omega L = V_M$$

that is, two simultaneous equations in the unknowns A_1 and A_2. Solving these two equations for A_1 and A_2 yields

$$A_1 = \frac{RV_M}{R^2 + \omega^2 L^2}$$

$$A_2 = \frac{\omega L V_M}{R^2 + \omega^2 L^2}$$

Therefore,

$$i(t) = \frac{RV_M}{R^2 + \omega^2 L^2}\cos\omega t + \frac{\omega L V_M}{R^2 + \omega^2 L^2}\sin\omega t$$

which, using the last identity in Eq. (7.11), can be written as

$$i(t) = A\cos(\omega t + \phi)$$

(continued)

where A and ϕ are determined as follows:

$$A \cos\phi = \frac{RV_M}{R^2 + \omega^2 L^2}$$

$$A \sin\phi = \frac{-\omega L V_M}{R^2 + \omega^2 L^2}$$

Hence,

$$\tan\phi = \frac{A \sin\phi}{A \cos\phi} = -\frac{\omega L}{R}$$

and therefore,

$$\phi = -\tan^{-1}\frac{\omega L}{R}$$

and since

$$(A \cos\phi)^2 + (A \sin\phi)^2 = A^2(\cos^2\phi + \sin^2\phi) = A^2$$

$$A^2 = \frac{R^2 V_M^2}{\left(R^2 + \omega^2 L^2\right)^2} + \frac{(\omega L)^2 V_M^2}{\left(R^2 + \omega^2 L^2\right)^2}$$

$$= \frac{V_M^2}{R^2 + \omega^2 L^2}$$

$$A = \frac{V_M}{\sqrt{R^2 + \omega^2 L^2}}$$

Hence, the final expression for $i(t)$ is

$$i(t) = \frac{V_M}{\sqrt{R^2 + \omega^2 L^2}} \cos\left(\omega t - \tan^{-1}\frac{\omega L}{R}\right)$$

The preceding analysis indicates that ϕ is zero if $L = 0$ and hence $i(t)$ is in phase with $v(t)$. If $R = 0$, $\phi = -90°$, and the current lags the voltage by 90°. If L and R are both present, the current lags the voltage by some angle between 0° and 90°.

This example illustrates an important point—solving even a simple one-loop circuit containing one resistor and one inductor is very complicated when compared to the solution of a single-loop circuit containing only two resistors. Imagine for a moment how laborious it would be to solve a more complicated circuit using the procedure we have employed in Example 7.3. To circumvent this approach, we will establish a correspondence between sinusoidal time functions and complex numbers. We will then show that this relationship leads to a set of algebraic equations for currents and voltages in a network (e.g., loop currents or node voltages) in which the coefficients of the variables are complex numbers. Hence, once again we will find that determining the currents or voltages in a circuit can be accomplished by solving a set of algebraic equations; however, in this case, their solution is complicated by the fact that variables in the equations have complex, rather than real, coefficients.

The vehicle we will employ to establish a relationship between time-varying sinusoidal functions and complex numbers is Euler's equation, which for our purposes is written as

$$e^{j\omega t} = \cos\omega t + j\sin\omega t \qquad \textbf{7.12}$$

This complex function has a real part and an imaginary part:

$$\mathrm{Re}\!\left(e^{j\omega t}\right) = \cos\omega t$$

$$\mathrm{Im}\!\left(e^{j\omega t}\right) = \sin\omega t \qquad \textbf{7.13}$$

where $\mathrm{Re}(\cdot)$ and $\mathrm{Im}(\cdot)$ represent the real part and the imaginary part, respectively, of the function in the parentheses.

Now suppose that we select as our forcing function in Fig. 7.4 the nonrealizable voltage

$$v(t) = V_M e^{j\omega t} \qquad\qquad \textbf{7.14}$$

which because of Euler's identity can be written as

$$v(t) = V_M \cos \omega t + j V_M \sin \omega t \qquad\qquad \textbf{7.15}$$

The real and imaginary parts of this function are each realizable. We think of this complex forcing function as two forcing functions, a real one and an imaginary one, and as a consequence of linearity, the principle of superposition applies and thus the current response can be written as

$$i(t) = I_M \cos(\omega t + \phi) + j I_M \sin(\omega t + \phi) \qquad\qquad \textbf{7.16}$$

where $I_M \cos(\omega t + \phi)$ is the response due to $V_M \cos \omega t$ and $j I_M \sin(\omega t + \phi)$ is the response due to $j V_M \sin \omega t$. This expression for the current containing both a real and an imaginary term can be written via Euler's equation as

$$i(t) = I_M e^{j(\omega t + \phi)} \qquad\qquad \textbf{7.17}$$

Because of the preceding relationships we find that rather than applying the forcing function $V_M \cos \omega t$ and calculating the response $I_M \cos(\omega t + \phi)$, we can apply the complex forcing function $V_M e^{j\omega t}$ and calculate the response $I_M e^{j(\omega t + \phi)}$, the real part of which is the desired response $I_M \cos(\omega t + \phi)$. Although this procedure may initially appear to be more complicated, it is not. It is via this technique that we will convert the differential equation to an algebraic equation that is much easier to solve.

LEARNING Example 7.4

Once again, let us determine the current in the *RL* circuit examined in Example 7.3. However, rather than applying $V_M \cos \omega t$ we will apply $V_M e^{j\omega t}$.

SOLUTION The forced response will be of the form

$$i(t) = I_M e^{j(\omega t + \phi)}$$

where only I_M and ϕ are unknown. Substituting $v(t)$ and $i(t)$ into the differential equation for the circuit, we obtain

$$RI_M e^{j(\omega t + \phi)} + L \frac{d}{dt}\left(I_M e^{j(\omega t + \phi)}\right) = V_M e^{j\omega t}$$

Taking the indicated derivative, we obtain

$$RI_M e^{j(\omega t + \phi)} + j\omega L I_M e^{j(\omega t + \phi)} = V_M e^{j\omega t}$$

Dividing each term of the equation by the common factor $e^{j\omega t}$ yields

$$RI_M e^{j\phi} + j\omega L I_M e^{j\phi} = V_M$$

which is an algebraic equation with complex coefficients. This equation can be written as

$$I_M e^{j\phi} = \frac{V_M}{R + j\omega L}$$

Converting the right-hand side of the equation to exponential or polar form produces the equation

$$I_M e^{j\phi} = \frac{V_M}{\sqrt{R^2 + \omega^2 L^2}} e^{j[-\tan^{-1}(\omega L/R)]}$$

(A quick refresher on complex numbers is given in the Appendix for readers who need to sharpen their skills in this area.) The

(continued)

preceding form clearly indicates that the magnitude and phase of the resulting current are

$$I_M = \frac{V_M}{\sqrt{R^2 + \omega^2 L^2}}$$

and

$$\phi = -\tan^{-1} \frac{\omega L}{R}$$

However, since our actual forcing function was $V_M \cos \omega t$ rather than $V_M e^{j\omega t}$, our actual response is the real part of the complex response:

$$i(t) = I_M \cos(\omega t + \phi)$$

$$= \frac{V_M}{\sqrt{R^2 + \omega^2 L^2}} \cos\left(\omega t - \tan^{-1} \frac{\omega L}{R}\right)$$

Note that this is identical to the response obtained in the previous example by solving the differential equation for the current $i(t)$.

LEARNING Hint

Summary of complex number relationships:

$$x + jy = re^{j\theta}$$

$$r = \sqrt{x^2 + y^2}$$

$$\theta = \tan^{-1} \frac{y}{x}$$

$$x = r \cos \theta$$

$$y = r \sin \theta$$

$$\frac{1}{e^{j\theta}} = e^{-j\theta}$$

7.3 Phasors

Once again let us assume that the forcing function for a linear network is of the form

$$v(t) = V_M e^{j\omega t}$$

Then every steady-state voltage or current in the network will have the same form and the same frequency ω; for example, a current $i(t)$ will be of the form $i(t) = I_M e^{j(\omega t + \phi)}$.

As we proceed in our subsequent circuit analyses, we will simply note the frequency and then drop the factor $e^{j\omega t}$ since it is common to every term in the describing equations. Dropping the term $e^{j\omega t}$ indicates that every voltage or current can be fully described by a magnitude and phase. For example, a voltage $v(t)$ can be written in exponential form as

$$v(t) = V_M \cos(\omega t + \theta) = \text{Re}\left[V_M e^{j(\omega t + \theta)}\right] \qquad \textbf{7.18}$$

or as a complex number

$$v(t) = \text{Re}\left(V_M \underline{/\theta}\, e^{j\omega t}\right) \qquad \textbf{7.19}$$

LEARNING Hint

If $v(t) = V_M \cos(\omega t + \theta)$ and $i(t) = I_M \cos(\omega t + \phi)$, then in phasor notation

$$\mathbf{V} = V_M \underline{/\theta}$$

and

$$\mathbf{I} = I_M \underline{/\phi}$$

Since we are working with a complex forcing function, the real part of which is the desired answer, and each term in the equation will contain $e^{j\omega t}$, we can drop $\text{Re}(\cdot)$ and $e^{j\omega t}$ and work only with the complex number $V_M \underline{/\theta}$. This complex representation is commonly called a *phasor*. As a distinguishing feature, phasors will be written in boldface type. In a completely identical manner a voltage $v(t) = V_M \cos(\omega t + \theta) = \text{Re}\left[V_M e^{j(\omega t + \theta)}\right]$ and a current $i(t) = I_M \cos(\omega t + \phi) = \text{Re}\left[I_M e^{j(\omega t + \phi)}\right]$ are written in phasor notation as $\mathbf{V} = V_M \underline{/\theta}$ and $\mathbf{I} = I_M \underline{/\phi}$, respectively.

LEARNING Example 7.5

Again, we consider the *RL* circuit in Example 7.3. Let us use phasors to determine the expression for the current.

SOLUTION The differential equation is

$$L \frac{di(t)}{dt} + Ri(t) = V_M \cos \omega t$$

The forcing function can be replaced by a complex forcing function that is written as $\mathbf{V}e^{j\omega t}$ with phasor $\mathbf{V} = V_M \underline{/0°}$. Similarly, the forced response component of the current $i(t)$ can be replaced by a complex function that is written as $\mathbf{I}e^{j\omega t}$ with phasor $\mathbf{I} = I_M \underline{/\phi}$. From our previous discussions we recall that the solution of the differential equation is the real part of this current.

Using the complex forcing function, we find that the differential equation becomes

$$L \frac{d}{dt} \left(\mathbf{I}e^{j\omega t} \right) + R\mathbf{I}e^{j\omega t} = \mathbf{V}e^{j\omega t}$$

$$j\omega L \mathbf{I}e^{j\omega t} + R\mathbf{I}e^{j\omega t} = \mathbf{V}e^{j\omega t}$$

Note that $e^{j\omega t}$ is a common factor and, as we have already indicated, can be eliminated, leaving the phasors; that is,

$$j\omega L \mathbf{I} + R\mathbf{I} = \mathbf{V}$$

Therefore,

$$\mathbf{I} = \frac{\mathbf{V}}{R + j\omega L} = I_M \underline{/\phi} = \frac{V_M}{\sqrt{R^2 + \omega^2 L^2}} \underline{\left/ -\tan^{-1} \frac{\omega L}{R} \right.}$$

Thus,

$$i(t) = \frac{V_M}{\sqrt{R^2 + \omega^2 L^2}} \cos \left(\omega t - \tan^{-1} \frac{\omega L}{R} \right)$$

which once again is the function we obtained earlier.

LEARNING Hint

The differential equation is reduced to a phasor equation.

We define relations between phasors after the $e^{j\omega t}$ term has been eliminated as "phasor, or frequency domain, analysis." Thus we have transformed a set of differential equations with sinusoidal forcing functions in the time domain into a set of algebraic equations containing complex numbers in the frequency domain. In effect, we are now faced with solving a set of algebraic equations for the unknown phasors. The phasors are then simply transformed back to the time domain to yield the solution of the original set of differential equations. In addition, we note that the solution of sinusoidal steady-state circuits would be relatively simple if we could write the phasor equation directly from the circuit description. In Section 7.4 we will lay the groundwork for doing just that.

Note that in our discussions we have tacitly assumed that sinusoidal functions would be represented as phasors with a phase angle based on a cosine function. Therefore, if sine functions are used, we will simply employ the relationship in Eq. (7.7) to obtain the proper phase angle.

In summary, while $v(t)$ represents a voltage in the time domain, the phasor $\mathbf{V}$ represents the voltage in the frequency domain. The phasor contains only magnitude and phase information, and the frequency is implicit in this representation. The transformation from the time domain to the frequency domain, as well as the reverse transformation, is shown in Table 7.1.

LEARNING Hint

Phasor analysis

1. Using phasors, transform a set of differential equations in the time domain into a set of algebraic equations in the frequency domain.

2. Solve the algebraic equations for the unknown phasors.

3. Transform the now-known phasors back to the time domain.

Table 7.1 Phasor representation

Time Domain	Frequency Domain
$A \cos (\omega t \pm \theta)$	$A \underline{/\pm\theta}$
$A \sin (\omega t \pm \theta)$	$A \underline{/\pm\theta - 90°}$

Recall that the phase angle is based on a cosine function and, therefore, if a sine function is involved, a 90° shift factor must be employed, as shown in the table.

The following examples illustrate the use of the phasor transformation.

It is important to note, however, that if a network contains only sine sources, there is no need to perform the 90° shift. We simply perform the normal phasor analysis and then the *imaginary* part of the time-varying complex solution is the desired response. Simply put, cosine sources generate a cosine response and sine sources generate a sine response.

LEARNING EXTENSIONS

E7.3   Convert the following voltage functions to phasors.

$$v_1(t) = 12 \cos(377t - 425°) \text{ V}$$

$$v_2(t) = 18 \sin(2513t + 4.2°) \text{ V}$$

ANSWER $\mathbf{V}_1 = 12 \underline{/-425°} \text{ V}$,
$\mathbf{V}_2 = 18 \underline{/-85.8°} \text{ V}$.

E7.4   Convert the following phasors to the time domain if the frequency is 400 Hz.

$$\mathbf{V}_1 = 10 \underline{/20°} \text{ V}$$

$$\mathbf{V}_2 = 12 \underline{/-60°} \text{ V}$$

ANSWER
$v_1(t) = 10 \cos(800\pi t + 20°) \text{ V}$,
$v_2(t) = 12 \cos(800\pi t - 60°) \text{ V}$.

7.4 Phasor Relationships for Circuit Elements

As we proceed in our development of the techniques required to analyze circuits in the sinusoidal steady-state, we are now in a position to establish the phasor relationships between voltage and current for the three passive elements R, L, and C.

In the case of a resistor as shown in Fig. 7.6a, the voltage–current relationship is known to be

LEARNING Hint

Current and voltage are in phase.

$$v(t) = Ri(t) \tag{7.20}$$

Applying the complex voltage $V_M e^{j(\omega t + \theta_v)}$ results in the complex current $I_M e^{j(\omega t + \theta_i)}$, and therefore Eq. (7.20) becomes

$$V_M e^{j(\omega t + \theta_v)} = R I_M e^{j(\omega t + \theta_i)}$$

which reduces to

$$V_M e^{j\theta_v} = R I_M e^{j\theta_i} \tag{7.21}$$

Equation (7.21) can be written in phasor form as

$$\mathbf{V} = R\mathbf{I} \tag{7.22}$$

where

$$\mathbf{V} = V_M e^{j\theta_v} = V_M \underline{/\theta_v} \quad \text{and} \quad \mathbf{I} = I_M e^{j\theta_i} = I_M \underline{/\theta_i}$$

From Eq. (7.21) we see that $\theta_v = \theta_i$ and thus the current and voltage for this circuit are *in phase*.

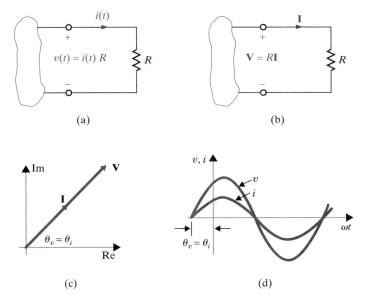

Figure 7.6
Voltage–current relationships for a resistor.

Historically, complex numbers have been represented as points on a graph in which the
x-axis represents the real axis and the y-axis the imaginary axis. The line segment connecting
the origin with the point provides a convenient representation of the magnitude and angle when
the complex number is written in a polar form. A review of the Appendix will indicate how these
complex numbers or line segments can be added, subtracted, and so on. Since phasors are complex numbers, it is convenient to represent the phasor voltage and current graphically as line
segments. A plot of the line segments representing the phasors is called a *phasor diagram*.
This pictorial representation of phasors provides immediate information on the relative magnitude of one phasor with another, the angle between two phasors, and the relative position of
one phasor with respect to another (i.e., leading or lagging). A phasor diagram and the sinusoidal waveforms for the resistor are shown in Figs. 7.6c and d, respectively. A phasor diagram will be drawn for each of the other circuit elements in the remainder of this section.

LEARNING Example 7.6

If the voltage $v(t) = 24 \cos(377t + 75°)$ V is applied to a 6-Ω
resistor as shown in Fig. 7.6a, we wish to determine the resultant
current.

SOLUTION   Since the phasor voltage is

$$\mathbf{V} = 24 \underline{/75°} \text{ V}$$

the phasor current from Eq. (7.22) is

$$\mathbf{I} = \frac{24 \underline{/75°}}{6} = 4 \underline{/75°} \text{ A}$$

which in the time domain is

$$i(t) = 4 \cos(377t + 75°) \text{ A}$$

E7.5 The current in a 4-Ω resistor is known to be $\mathbf{I} = 12\ \underline{/60°}$ A. Express the voltage across the resistor as a time function if the frequency of the current is 4 kHz.

ANSWER

$v(t) = 48\cos(8000\pi t + 60°)$ V.

The voltage–current relationship for an inductor, as shown in Fig. 7.7a, is

$$v(t) = L\frac{di(t)}{dt} \qquad\qquad 7.23$$

Substituting the complex voltage and current into this equation yields

$$V_M e^{j(\omega t + \theta_v)} = L\frac{d}{dt} I_M e^{j(\omega t + \theta_i)}$$

which reduces to

$$V_M e^{j\theta_v} = j\omega L I_M e^{j\theta_i} \qquad\qquad 7.24$$

LEARNING Hint

The derivative process yields a frequency-dependent impedance.

Equation (7.24) in phasor notation is

$$\mathbf{V} = j\omega L\mathbf{I} \qquad\qquad 7.25$$

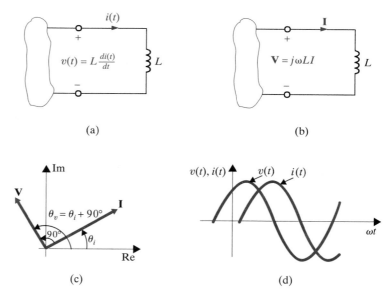

(a) (b) (c) (d)

Figure 7.7
Voltage–current relationships for an inductor.

Note that the differential equation in the time domain (7.23) has been converted to an algebraic equation with complex coefficients in the frequency domain. This relationship is shown in Fig. 7.7b. Since the imaginary operator $j = 1e^{j90°} = 1\,\underline{/90°} = \sqrt{-1}$, Eq. (7.24) can be written as

$$V_M e^{j\theta_v} = \omega L I_M e^{j(\theta_i + 90°)} \qquad\qquad \textbf{7.26}$$

Therefore, the voltage and current are *90° out of phase*, and in particular the voltage leads the current by 90° or the current lags the voltage by 90°. The phasor diagram and the sinusoidal waveforms for the inductor circuit are shown in Figs. 7.7c and d, respectively.

LEARNING Hint

The voltage leads the current or the current lags the voltage.

LEARNING Example 7.7

The voltage $v(t) = 12 \cos(377t + 20°)$ V is applied to a 20-mH inductor as shown in Fig. 7.7a. Find the resultant current.

SOLUTION   The phasor current is

$$\mathbf{I} = \frac{\mathbf{V}}{j\omega L} = \frac{12\,\underline{/20°}}{\omega L\,\underline{/90°}}$$

$$= \frac{12\,\underline{/20°}}{(377)(20 \times 10^{-3})\,\underline{/90°}}$$

$$= 1.59\,\underline{/-70°}\ \text{A}$$

or

$$i(t) = 1.59 \cos(377t - 70°)\ \text{A}$$

LEARNING Hint

Applying $\mathbf{V} = j\omega L\mathbf{I}$

$$\frac{x_1\,\underline{/\theta_1}}{x_2\,\underline{/\theta_2}} = \frac{x_1}{x_2}\,\underline{/\theta_1 - \theta_2}$$

LEARNING EXTENSION

E7.6   The current in a 0.05-H inductor is $\mathbf{I} = 4\,\underline{/-30°}$ A. If the frequency of the current is 60 Hz, determine the voltage across the inductor.

ANSWER
$v_L(t) = 75.4 \cos(377t + 60°)$ V.

The voltage–current relationship for our last passive element, the capacitor, as shown in Fig. 7.8a, is

$$i(t) = C\,\frac{dv(t)}{dt} \qquad\qquad \textbf{7.27}$$

LEARNING Hint

The current leads the voltage or the voltage lags the current.

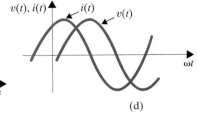

(a)   (b)   (c)   (d)

Figure 7.8   *Voltage–current relationships for a capacitor.*

Once again employing the complex voltage and current, we obtain

$$I_M e^{j(\omega t + \theta_i)} = C \frac{d}{dt} V_M e^{j(\omega t + \theta_v)}$$

which reduces to

$$I_M e^{j\theta_i} = j\omega C V_M e^{j\theta_v} \qquad \textbf{7.28}$$

In phasor notation this equation becomes

$$\mathbf{I} = j\omega C \mathbf{V} \qquad \textbf{7.29}$$

Equation (7.27), a differential equation in the time domain, has been transformed into Eq. (7.29), an algebraic equation with complex coefficients in the frequency domain. The phasor relationship is shown in Fig. 7.8b. Substituting $j = 1e^{j90°}$ into Eq. (7.28) yields

$$I_M e^{j\theta_i} = \omega C V_M e^{j(\theta_v + 90°)} \qquad \textbf{7.30}$$

Note that the voltage and current are *90° out of phase*. Equation (7.30) states that the current leads the voltage by 90° or the voltage lags the current by 90°. The phasor diagram and the sinusoidal waveforms for the capacitor circuit are shown in Figs. 7.8c and d, respectively.

LEARNING Example 7.8

The voltage $v(t) = 100 \cos(314t + 15°)$ V is applied to a 100-μF capacitor as shown in Fig. 7.8a. Find the current.

SOLUTION   The resultant phasor current is

$$\mathbf{I} = j\omega C \left(100 \underline{/15°}\right)$$
$$= (314)\left(100 \times 10^{-6} \underline{/90°}\right)\left(100 \underline{/15°}\right)$$
$$= 3.14 \underline{/105°} \text{ A}$$

Therefore, the current written as a time function is

$$i(t) = 3.14 \cos(314t + 105°) \text{ A}$$

E7.7   The current in a 150-μF capacitor is $\mathbf{I} = 3.6 \underline{/-145°}$ A. If the frequency of the current is 60 Hz, determine the voltage across the capacitor.

ANSWER
$v_C(t) = 63.66 \cos(377t - 235°)$ V

7.5 Impedance and Admittance

We have examined each of the circuit elements in the frequency domain on an individual basis. We now wish to treat these passive circuit elements in a more general fashion. We now define the two-terminal input *impedance* **Z**, also referred to as the driving point impedance, in exactly the same manner in which we defined resistance earlier. Later we will examine another type of impedance, called transfer impedance.

Impedance is defined as the ratio of the phasor voltage **V** to the phasor current **I**:

$$\mathbf{Z} = \frac{\mathbf{V}}{\mathbf{I}} \qquad\qquad 7.31$$

at the two terminals of the element related to one another by the passive sign convention, as illustrated in Fig. 7.9. Since **V** and **I** are complex, the impedance **Z** is complex and

$$\mathbf{Z} = \frac{V_M \underline{/\theta_v}}{I_M \underline{/\theta_i}} = \frac{V_M}{I_M} \underline{/\theta_v - \theta_i} = Z \underline{/\theta_z} \qquad\qquad 7.32$$

Figure 7.9
General impedance relationship.

Since **Z** is the ratio of **V** to **I**, the units of **Z** are ohms. Thus, impedance in an ac circuit is analogous to resistance in a dc circuit. In rectangular form, impedance is expressed as

$$\mathbf{Z}(\omega) = R(\omega) + jX(\omega) \qquad\qquad 7.33$$

where $R(\omega)$ is the real, or resistive, component and $X(\omega)$ is the imaginary, or reactive, component. In general, we simply refer to R as the resistance and X as the reactance. It is important to note that R and X are real functions of ω and therefore $\mathbf{Z}(\omega)$ is frequency dependent. Equation (7.33) clearly indicates that **Z** is a complex number; however, it is not a phasor, since phasors denote sinusoidal functions.

Equations (7.32) and (7.33) indicate that

$$Z \underline{/\theta_z} = R + jX \qquad\qquad 7.34$$

Therefore,

$$Z = \sqrt{R^2 + X^2}$$
$$\theta_z = \tan^{-1}\frac{X}{R}$$

7.35

where

$$R = Z\cos\theta_z$$
$$X = Z\sin\theta_z$$

For the individual passive elements the impedance is as shown in Table 7.2. However, just as it was advantageous to know how to determine the equivalent resistance in dc circuits, we want to learn how to determine the equivalent impedance in ac circuits.

Table 7.2 Passive element impedance

Passive element	Impedance
R	$\mathbf{Z} = R$
L	$\mathbf{Z} = j\omega L = jX_L = \omega L\,\underline{/90°}, X_L = \omega L$
C	$\mathbf{Z} = \dfrac{1}{j\omega C} = jX_C = -\dfrac{1}{\omega C}\,\underline{/90°}, X_C = -\dfrac{1}{\omega C}$

KCL and KVL are both valid in the frequency domain and we can use this fact, as was done in Chapter 2 for resistors, to show that impedances can be combined using the same rules that we established for resistor combinations. That is, if $\mathbf{Z}_1, \mathbf{Z}_2, \mathbf{Z}_3, \ldots, \mathbf{Z}_n$ are connected in series, the equivalent impedance $\mathbf{Z}_s$ is

$$\mathbf{Z}_s = \mathbf{Z}_1 + \mathbf{Z}_2 + \mathbf{Z}_3 + \cdots + \mathbf{Z}_n$$

7.36

and if $\mathbf{Z}_1, \mathbf{Z}_2, \mathbf{Z}_3, \ldots, \mathbf{Z}_n$ are connected in parallel, the equivalent impedance is given by

$$\frac{1}{\mathbf{Z}_p} = \frac{1}{\mathbf{Z}_1} + \frac{1}{\mathbf{Z}_2} + \frac{1}{\mathbf{Z}_3} + \cdots + \frac{1}{\mathbf{Z}_n}$$

7.37

LEARNING Example 7.9

Determine the equivalent impedance of the network shown in Fig. 7.10 if the frequency is $f = 60$ Hz. Then compute the current $i(t)$ if the voltage source is $v(t) = 50\cos(\omega t + 30°)$ V. Finally, calculate the equivalent impedance if the frequency is $f = 400$ Hz.

SOLUTION The impedances of the individual elements at 60 Hz are

$$\mathbf{Z}_R = 25\ \Omega$$

$$\mathbf{Z}_L = j\omega L = j(2\pi \times 60)(20 \times 10^{-3}) = j7.54\ \Omega$$

$$\mathbf{Z}_C = \frac{-j}{\omega C} = \frac{-j}{(2\pi \times 60)(50 \times 10^{-6})} = -j53.05\ \Omega$$

Since the elements are in series,

$$\mathbf{Z} = \mathbf{Z}_R + \mathbf{Z}_L + \mathbf{Z}_C$$

$$= 25 - j45.51\ \Omega$$

The current in the circuit is given by

$$\mathbf{I} = \frac{\mathbf{V}}{\mathbf{Z}} = \frac{50\ \underline{/30^\circ}}{25 - j45.51} = \frac{50\ \underline{/30^\circ}}{51.93\ \underline{/-61.22^\circ}} = 0.96\ \underline{/91.22^\circ}\ \text{A}$$

or in the time domain, $i(t) = 0.96 \cos(377t + 91.22^\circ)$ A.
    If the frequency is 400 Hz, the impedance of each element is

$$\mathbf{Z}_R = 25\ \Omega$$

$$\mathbf{Z}_L = j\omega L = j50.27\ \Omega$$

$$\mathbf{Z}_C = \frac{-j}{\omega C} = -j7.96\ \Omega$$

The total impedance is then

$$\mathbf{Z} = 25 + j42.31 = 49.14\ \underline{/59.42^\circ}\ \Omega$$

It is important to note that at the frequency $f = 60$ Hz, the reactance of the circuit is capacitive; that is, if the impedance is written as $R + jX$, $X < 0$; however, at $f = 400$ Hz the reactance is inductive since $X > 0$.

LEARNING Hint

Technique

1. Express $v(t)$ as a phasor and determine the impedance of each passive element.
2. Combine impedances and solve for the phasor **I**.
3. Convert the phasor **I** to $i(t)$.

Figure 7.10
Series ac circuit.

LEARNING EXTENSION

E7.8 Find the current $i(t)$ in the network in Fig. E7.8.

Figure E7.8

ANSWER
$i(t) = 3.88 \cos(377t - 39.2^\circ)$ A.

Another quantity that is very useful in the analysis of ac circuits is the two-terminal input *admittance*, which is the reciprocal of impedance; that is,

$$\mathbf{Y} = \frac{1}{\mathbf{Z}} = \frac{\mathbf{I}}{\mathbf{V}}$$ 7.38

The units of $\mathbf{Y}$ are siemens, and this quantity is analogous to conductance in resistive dc circuits. Since $\mathbf{Z}$ is a complex number, $\mathbf{Y}$ is also a complex number.

$$\mathbf{Y} = Y_M \underline{/\theta_y}$$ 7.39

which is written in rectangular form as

$$\mathbf{Y} = G + jB$$ 7.40

where G and B are called *conductance* and *susceptance*, respectively. Because of the relationship between $\mathbf{Y}$ and $\mathbf{Z}$, we can express the components of one quantity as a function of the components of the other. From the expression

$$G + jB = \frac{1}{R + jX}$$ 7.41

we can show that

$$G = \frac{R}{R^2 + X^2}, \qquad B = \frac{-X}{R^2 + X^2}$$ 7.42

and in a similar manner, we can show that

$$R = \frac{G}{G^2 + B^2}, \qquad X = \frac{-B}{G^2 + B^2}$$ 7.43

LEARNING by Doing

D 7.2 Find $\mathbf{Y}_p$

ANSWER
$\mathbf{Y}_p = 0.1 + j1$ S.

D 7.3 Find $\mathbf{Y}_s$

ANSWER
$\mathbf{Y}_s = 0.05 - j0.05$ S.

It is very important to note that in general R and G are *not* reciprocals of one another. The same is true for X and B. The purely resistive case is an exception. In the purely reactive case the quantities are negative reciprocals of one another.

The admittance of the individual passive elements are

$$\mathbf{Y}_R = \frac{1}{R} = G$$

$$\mathbf{Y}_L = \frac{1}{j\omega L} = -\frac{1}{\omega L} \underline{/90°}$$ 7.44

$$\mathbf{Y}_C = j\omega C = \omega C \underline{/90°}$$

Once again, since KCL and KVL are valid in the frequency domain, we can show, using the same approach outlined in Chapter 2 for conductance in resistive circuits, that the rules for combining admittances are the same as those for combining conductances; that is, if $\mathbf{Y}_1, \mathbf{Y}_2, \mathbf{Y}_3, \ldots, \mathbf{Y}_n$ are connected in parallel, the equivalent admittance is

$$\mathbf{Y}_p = \mathbf{Y}_1 + \mathbf{Y}_2 + \cdots + \mathbf{Y}_n$$ 7.45

and if $\mathbf{Y}_1, \mathbf{Y}_2, \ldots, \mathbf{Y}_n$ are connected in series, the equivalent admittance is

$$\frac{1}{\mathbf{Y}_S} = \frac{1}{\mathbf{Y}_1} + \frac{1}{\mathbf{Y}_2} + \cdots + \frac{1}{\mathbf{Y}_n}$$ 7.46

LEARNING Example 7.10

Calculate the equivalent admittance $\mathbf{Y}_p$ for the network in Fig. 7.11 and use it to determine the current $\mathbf{I}$ if $\mathbf{V}_S = 60\,\underline{/45°}$ V.

Figure 7.11
An example parallel circuit.

SOLUTION   From Fig. 7.11 we note that

$$\mathbf{Y}_R = \frac{1}{\mathbf{Z}_R} = \frac{1}{2} \, \text{S}$$

$$\mathbf{Y}_L = \frac{1}{\mathbf{Z}_L} = \frac{-j}{4} \, \text{S}$$

Therefore,

$$\mathbf{Y}_p = \frac{1}{2} - j\frac{1}{4} \, \text{S}$$

and hence,

$$\mathbf{I} = \mathbf{Y}_p \mathbf{V}_S$$

$$= \left(\frac{1}{2} - j\frac{1}{4} \right)(60\,\underline{/45°})$$

$$= 33.5\,\underline{/18.43°} \, \text{A}$$

> **LEARNING Hint**
>
> Admittances add in parallel.

LEARNING EXTENSION

E7.9   Find the current $\mathbf{I}$ in the network in Fig. E7.9.

ANSWER $\mathbf{I} = 9.01\,\underline{/53.7°}$ A.

$\mathbf{V} = 10\,\underline{/20°}$ V $\mathbf{Z}_R = 2\,\Omega$ $\mathbf{Z}_L = j2\,\Omega$ $\mathbf{Z}_C = -j1\,\Omega$ $\mathbf{Z}_R = 4\,\Omega$

Figure E7.9

As a prelude to our analysis of more general ac circuits, let us examine the techniques for computing the impedance or admittance of circuits in which numerous passive elements are interconnected. The following example illustrates that our technique is analogous to our earlier computations of equivalent resistance.

LEARNING Example 7.11

Consider the network shown in Fig. 7.12a. The impedance of each element is given in the figure. We wish to calculate the equivalent impedance of the network $\mathbf{Z}_{eq}$ at terminals A-B.

SOLUTION   The equivalent impedance $\mathbf{Z}_{eq}$ could be calculated in a variety of ways; we could use only impedances, or only

> **LEARNING Hint**
>
> **Technique:**
> 1. Add the admittances of elements in parallel.
> 2. Add the impedances of elements in series.
> 3. Convert back and forth between admittance and impedance in order to combine neighboring elements.

(continued)

admittances, or a combination of the two. We will use the latter. We begin by noting that the circuit in Fig. 7.12a can be represented by the circuit in Fig. 7.12b.

Note that

$$\mathbf{Y}_4 = \mathbf{Y}_L + \mathbf{Y}_C$$
$$= \frac{1}{j4} + \frac{1}{-j2}$$
$$= j\frac{1}{4} \text{ S}$$

Therefore,

$$\mathbf{Z}_4 = -j4 \ \Omega$$

Now

$$\mathbf{Z}_{34} = \mathbf{Z}_3 + \mathbf{Z}_4$$
$$= (4 + j2) + (-j4)$$
$$= 4 - j2 \ \Omega$$

and hence,

$$\mathbf{Y}_{34} = \frac{1}{\mathbf{Z}_{34}}$$
$$= \frac{1}{4 - j2}$$
$$= 0.20 + j0.10 \text{ S}$$

Since

$$\mathbf{Z}_2 = 2 + j6 - j2$$
$$= 2 + j4 \ \Omega$$

then

$$\mathbf{Y}_2 = \frac{1}{2 + j4}$$
$$= 0.10 - j0.20 \text{ S}$$
$$\mathbf{Y}_{234} = \mathbf{Y}_2 + \mathbf{Y}_{34}$$
$$= 0.30 - j0.10 \text{ S}$$

The reader should note carefully our approach—we are adding impedances in series and adding admittances in parallel.

From $\mathbf{Y}_{234}$ we can compute $\mathbf{Z}_{234}$ as

$$\mathbf{Z}_{234} = \frac{1}{\mathbf{Y}_{234}}$$
$$= \frac{1}{0.30 - j0.10}$$
$$= 3 + j1 \ \Omega$$

Now

$$\mathbf{Y}_1 = \mathbf{Y}_R + \mathbf{Y}_C$$
$$= \frac{1}{1} + \frac{1}{-j2}$$
$$= 1 + j\frac{1}{2} \text{ S}$$

Figure 7.12
Example circuit for determining equivalent impedance in two steps.

and then

$$\mathbf{Z}_1 = \cfrac{1}{1 + j\cfrac{1}{2}}$$

$$= 0.8 - j0.4 \ \Omega$$

Therefore,

$$\mathbf{Z}_{eq} = \mathbf{Z}_1 + \mathbf{Z}_{234}$$

$$= 0.8 - j0.4 + 3 + j1$$

$$= 3.8 + j0.6 \ \Omega$$

LEARNING EXTENSION

E7.10 Compute the impedance $\mathbf{Z}_T$ in the network in Fig. E7.10.

ANSWER

$\mathbf{Z}_T = 3.38 + j1.08 \ \Omega.$

Figure E7.10

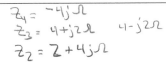

$Z_4 = -4j\ \Omega$
$Z_3 = 4+j2\ \Omega \qquad 4-j2\ \Omega$
$Z_2 = 2+4j\ \Omega$

7.6 Phasor Diagrams

Impedance and admittance are functions of frequency, and therefore their values change as the frequency changes. These changes in $\mathbf{Z}$ and $\mathbf{Y}$ have a resultant effect on the current–voltage relationships in a network. This impact of changes in frequency on circuit parameters can be easily seen via a phasor diagram. The following examples will serve to illustrate these points.

LEARNING Example 7.12

Let us sketch the phasor diagram for the network shown in Fig. 7.13.

Figure 7.13   Example parallel circuit.

SOLUTION   The pertinent variables are labeled on the figure. For convenience in forming a phasor diagram, we select $\mathbf{V}$ as a reference phasor and arbitrarily assign it a $0°$ phase angle. We will, therefore, measure all currents with respect to this phasor. We suffer no loss of generality by assigning $\mathbf{V}$ a $0°$ phase angle, since if it is actually $30°$, for example, we will simply rotate the entire phasor diagram by $30°$ because all the currents are measured with respect to this phasor.

At the upper node in the circuit KCL is

$$\mathbf{I}_S = \mathbf{I}_R + \mathbf{I}_L + \mathbf{I}_C = \frac{\mathbf{V}}{R} + \frac{\mathbf{V}}{j\omega L} + \frac{\mathbf{V}}{1/j\omega C}$$

(continued)

Since $\mathbf{V} = V_M\,\underline{/0°}$, then

$$\mathbf{I}_S = \frac{V_M\,\underline{/0°}}{R} + \frac{V_M\,\underline{/-90°}}{\omega L} + V_M\omega C\,\underline{/90°}$$

The phasor diagram that illustrates the phase relationship between $\mathbf{V}$, $\mathbf{I}_R$, $\mathbf{I}_L$, and $\mathbf{I}_C$ is shown in Fig. 7.14a. For small values of ω such that the magnitude of $\mathbf{I}_L$ is greater than that of $\mathbf{I}_C$, the phasor diagram for the currents is shown in Fig. 7.14b. In the case of

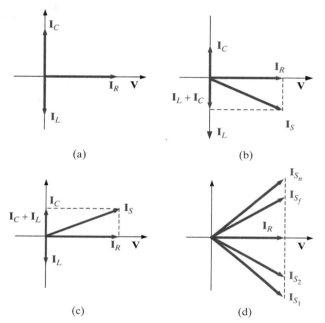

(a)   (b)

(c)   (d)

Figure 7.14   Phasor diagrams for the circuit in Fig. 7.13.

large values of ω—that is, those for which $\mathbf{I}_C$ is greater than $\mathbf{I}_L$—the phasor diagram for the currents is shown in Fig. 7.14c. Note that as ω increases, the phasor $\mathbf{I}_S$ moves from $\mathbf{I}_{S_1}$ to $\mathbf{I}_{S_n}$ along a locus of points specified by the dashed line shown in Fig. 7.14d.

Note that $\mathbf{I}_S$ is in phase with $\mathbf{V}$ when $\mathbf{I}_C = \mathbf{I}_L$ or, in other words, when $\omega L = 1/\omega C$. Hence, the node voltage $\mathbf{V}$ is in phase with the current source $\mathbf{I}_S$ when

$$\omega = \frac{1}{\sqrt{LC}}$$

This can also be seen from the KCL equation

$$\mathbf{I} = \left[\frac{1}{R} + j\left(\omega C - \frac{1}{\omega L}\right)\right]\mathbf{V}$$

LEARNING by Doing

D 7.4 Find the frequency at which $v(t)$ and $i(t)$ are in phase.

ANSWER $f = 5.03 \times 10^3$ Hz.

LEARNING Hint

From a graphical standpoint, phasors can be manipulated like vectors.

LEARNING Example 7.13

Let us determine the phasor diagram for the series circuit shown in Fig. 7.15a.

SOLUTION   KVL for this circuit is of the form

$$\mathbf{V}_S = \mathbf{V}_R + \mathbf{V}_L + \mathbf{V}_C$$

$$= \mathbf{I}R + \omega L\mathbf{I}\,\underline{/90°} + \frac{\mathbf{I}}{\omega C}\,\underline{/-90°}$$

If we select $\mathbf{I}$ as a reference phasor so that $\mathbf{I} = I_M\,\underline{/0°}$, then if $\omega L I_M > I_M/\omega C$, the phasor diagram will be of the form shown in Fig. 7.15b. Specifically, if $\omega = 377$ rad/s (i.e., $f = 60$ Hz), then $\omega L = 6$ and $1/\omega C = 2$. Under these conditions the phasor

diagram is as shown in Fig. 7.15c. If, however, we select $\mathbf{V}_S$ as reference with, for example,

$$v_S(t) = 12\sqrt{2}\cos(377t + 90°)\ \text{V}$$

then

$$\mathbf{I} = \frac{\mathbf{V}}{\mathbf{Z}} = \frac{12\sqrt{2}\,\underline{/90°}}{4 + j6 - j2}$$

$$= \frac{12\sqrt{2}\,\underline{/90°}}{4\sqrt{2}\,\underline{/45°}}$$

$$= 3\,\underline{/45°}\ \text{A}$$

and the entire phasor diagram, as shown in Figs. 7.15b and c, is rotated 45°, as shown in Fig. 7.15d.

(a)

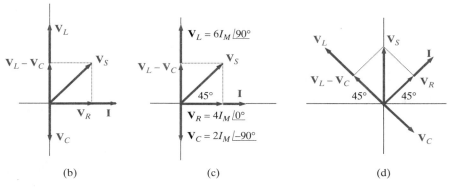

(b)   (c)   (d)

Figure 7.15
Series circuit and certain specific phasor diagrams (plots are not drawn to scale).

LEARNING EXTENSION

E7.11 Draw a phasor diagram illustrating all currents and voltages for the network in Fig. E7.11.

ANSWER

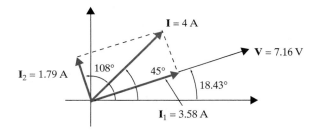

Figure E7.11

7.7 Basic Analysis Using Kirchhoff's Laws

We have shown that Kirchhoff's laws apply in the frequency domain, and therefore they can be used to compute steady-state voltages and currents in ac circuits. This approach involves expressing these voltages and currents as phasors, and once this is done, the ac steady-state analysis employing phasor equations is performed in an identical fashion to that used in the dc analysis of resistive circuits. Complex number algebra is the tool that is used for the mathematical manipulation of the phasor equations, which, of course, have complex coefficients. We will begin by illustrating that the techniques we have applied in the solution of dc resistive circuits are valid in ac circuit analysis also—the only difference being that in steady-state ac circuit analysis the algebraic phasor equations have complex coefficients.

Problem-Solving Strategy
AC Steady-State Analysis

▶ For relatively simple circuits (e.g., those with a single source), use

 ▶ Ohm's law for ac analysis, i.e., $\mathbf{V} = \mathbf{IZ}$

 ▶ The rules for combining $\mathbf{Z}_s$ and $\mathbf{Y}_p$

 ▶ KCL and KVL

 ▶ Current and voltage division

▶ For more complicated circuits with multiple sources, use

 ▶ Nodal analysis

 ▶ Loop or mesh analysis

 ▶ Superposition

 ▶ Source exchange

 ▶ Thévenin's and Norton's theorems

 ▶ MATLAB

 ▶ PSPICE

At this point, it is important for the reader to understand that in our manipulation of algebraic phasor equations with complex coefficients we will, for the sake of simplicity, normally carry only two digits to the right of the decimal point. In doing so, we will introduce round-off errors in our calculations. Nowhere are these errors more evident than when two or more approaches are used to solve the same problem, as is done in the following example.

LEARNING Example 7.14

We wish to calculate all the voltages and currents in the circuit shown in Fig. 7.16a.

SOLUTION   Our approach will be as follows. We will calculate the total impedance seen by the source $\mathbf{V}_S$. Then we will use this to determine $\mathbf{I}_1$. Knowing $\mathbf{I}_1$, we can compute $\mathbf{V}_1$ using KVL. Knowing $\mathbf{V}_1$, we can compute $\mathbf{I}_2$ and $\mathbf{I}_3$, and so on.

LEARNING Hint

Technique

1. Compute $\mathbf{I}_1$.

2. Determine $\mathbf{V}_1 = \mathbf{V}_S - \mathbf{I}_1\mathbf{Z}_1$.

Then $\mathbf{I}_2 = \dfrac{\mathbf{V}_1}{\mathbf{Z}_2}$ and $\mathbf{I}_3 = \dfrac{\mathbf{V}_1}{\mathbf{Z}_3}$

Current and voltage division are also applicable.

The total impedance seen by the source $\mathbf{V}_S$ is

$$\mathbf{Z}_{eq} = 4 + \frac{(j6)(8 - j4)}{j6 + 8 - j4}$$

$$= 4 + \frac{24 + j48}{8 + j2}$$

$$= 4 + 4.24 + j4.94$$

$$= 9.61 \,\underline{/30.94°}\ \Omega$$

Then

$$\mathbf{I}_1 = \frac{\mathbf{V}_S}{\mathbf{Z}_{eq}} = \frac{24\,\underline{/60°}}{9.61\,\underline{/30.94°}}$$

$$= 2.5\,\underline{/29.06°}\ \text{A}$$

$\mathbf{V}_1$ can be determined using KVL:

$$\mathbf{V}_1 = \mathbf{V}_S - 4\mathbf{I}_1$$

$$= 24\,\underline{/60°} - 10\,\underline{/29.06°}$$

$$= 3.26 + j15.93$$

$$= 16.26\,\underline{/78.43°}\ \text{V}$$

Note that $\mathbf{V}_1$ could also be computed via voltage division:

$$\mathbf{V}_1 = \frac{\mathbf{V}_S\,\dfrac{(j6)(8 - j4)}{j6 + 8 - j4}}{4 + \dfrac{(j6)(8 - j4)}{j6 + 8 - 4}}\ \text{V}$$

which from our previous calculation is

$$\mathbf{V}_1 = \frac{(24\,\underline{/60°})(6.51\,\underline{/49.36°})}{9.61\,\underline{/30.94°}}$$

$$= 16.26\,\underline{/78.42°}\ \text{V}$$

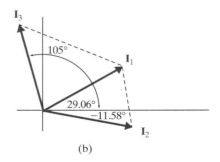

(a)   (b)

Figure 7.16
(a) Example ac circuit, (b) phasor diagram for the currents (plots are not drawn to scale).

(continued)

Knowing $\mathbf{V}_1$, we can calculate both $\mathbf{I}_2$ and $\mathbf{I}_3$:

$$\mathbf{I}_2 = \frac{\mathbf{V}_1}{j6} = \frac{16.26 \,\underline{/78.43^\circ}}{6 \,\underline{/90^\circ}}$$

$$= 2.71 \,\underline{/-11.58^\circ} \text{ A}$$

and

$$\mathbf{I}_3 = \frac{\mathbf{V}_1}{8 - j4}$$

$$= 1.82 \,\underline{/105^\circ} \text{ A}$$

Note that $\mathbf{I}_2$ and $\mathbf{I}_3$ could have been calculated by current division. For example, $\mathbf{I}_2$ could be determined by

$$\mathbf{I}_2 = \frac{\mathbf{I}_1(8 - j4)}{8 - j4 + j6}$$

$$= \frac{(2.5 \,\underline{/29.06^\circ})(8.94 \,\underline{/-26.57^\circ})}{8 + j2}$$

$$= 2.71 \,\underline{/-11.55^\circ} \text{ A}$$

Finally, $\mathbf{V}_2$ can be computed as

$$\mathbf{V}_2 = \mathbf{I}_3(-j4)$$

$$= 7.28 \,\underline{/15^\circ} \text{ V}$$

This value could also have been computed by voltage division. The phasor diagram for the currents $\mathbf{I}_1$, $\mathbf{I}_2$, and $\mathbf{I}_3$ is shown in Fig. 7.16b and is an illustration of KCL.

Finally, the reader is encouraged to work the problem in reverse; that is, given $\mathbf{V}_2$, find $\mathbf{V}_S$. Note that if $\mathbf{V}_2$ is known, $\mathbf{I}_3$ can be computed immediately using the capacitor impedance. Then $\mathbf{V}_2 + \mathbf{I}_3(8)$ yields $\mathbf{V}_1$. Knowing $\mathbf{V}_1$ we can find $\mathbf{I}_2$. Then $\mathbf{I}_2 + \mathbf{I}_3 = \mathbf{I}_1$, and so on. Note that this analysis, which is the subject of Learning Extensions Exercise E7.12, involves simply a repeated application of Ohm's law, KCL, and KVL.

LEARNING EXTENSION

E7.12 In the network in Fig. E7.12, $\mathbf{V}_o$ is known to be $8 \,\underline{/45^\circ}$ V. Compute $\mathbf{V}_S$.

ANSWER
$\mathbf{V}_S = 17.89 \,\underline{/-18.43^\circ}$ V.

Figure E7.12

7.8 Analysis Techniques

In this section we revisit the circuit analysis methods that were successfully applied earlier to dc circuits and illustrate their applicability to ac steady-state analysis. The vehicle we employ to present these techniques is examples in which all the theorems, together with nodal analysis and loop analysis, are used to obtain a solution.

LEARNING Example 7.15

Let us determine the current I_o in the network in Fig. 7.17a using nodal analysis, loop analysis, superposition, source exchange, Thévenin's theorem, and Norton's theorem.

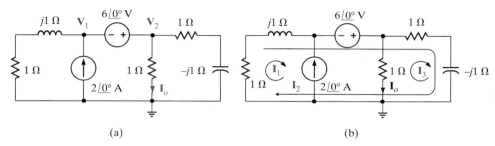

Figure 7.17
Circuits used in Example 7.15 for node and loop analysis.

SOLUTION

(1) Nodal Analysis   We begin with a nodal analysis of the network. The KCL equation for the supernode that includes the voltage source is

$$\frac{V_1}{1+j} - 2\underline{/0^\circ} + \frac{V_2}{1} + \frac{V_2}{1-j} = 0$$

and the associated KVL constraint equation is

$$V_1 + 6\underline{/0^\circ} = V_2$$

Solving for V_1 in the second equation and using this value in the first equation yields

$$\frac{V_2 - 6\underline{/0^\circ}}{1+j} - 2\underline{/0^\circ} + V_2 + \frac{V_2}{1-j} = 0$$

or

$$V_2\left[\frac{1}{1+j} + 1 + \frac{1}{1-j}\right] = \frac{6+2+2j}{1+j}$$

Solving for V_2, we obtain

$$V_2 = \left(\frac{4+j}{1+j}\right) V$$

Therefore,

$$I_o = \frac{4+j}{1+j} = \left(\frac{5}{2} - \frac{3}{2}j\right) A$$

(2) Loop Analysis   The network in Fig. 7.17b is used to perform a loop analysis. Note that one loop current is selected that passes through the independent current source. The three loop equations are

$$I_1 = -2\underline{/0^\circ}$$

$$1(I_1 + I_2) + j1(I_1 + I_2) - 6\underline{/0^\circ} + 1(I_2 + I_3) - j1(I_2 + I_3) = 0$$

$$1I_3 + 1(I_2 + I_3) - j1(I_2 + I_3) = 0$$

Combining the first two equations yields

$$I_2(2) + I_3(1 - j) = 8 + 2j$$

The third loop equation can be simplified to the form

$$I_2(1 - j) + I_3(2 - j) = 0$$

Solving this last equation for I_2 and substituting the value into the previous equation yields

$$I_3\left[\frac{-4+2j}{1-j} + 1 - j\right] = 8 + 2j$$

(continued)

Figure 7.18
Circuits used in Example 7.15 for a superposition analysis.

(c)

or

$$\mathbf{I}_3 = \frac{-10 + 6j}{4}$$

and finally

$$\mathbf{I}_o = -\mathbf{I}_3 = \left(\frac{5}{2} - \frac{3}{2}j\right) \text{A}$$

and hence,

$$\mathbf{I}_0'' = \frac{6}{4}(1 - j) \text{ A}$$

Then

$$\mathbf{I}_o = \mathbf{I}_o' + \mathbf{I}_o'' = 1 + \frac{6}{4}(1 - j) = \left(\frac{5}{2} - \frac{3}{2}j\right) \text{A}$$

LEARNING Hint

In applying superposition in this case, each source is applied independently and the results are added to obtain the solution.

(3) Superposition In using superposition, we apply one independent source at a time. The network in which the current source acts alone is shown in Fig. 7.18a. By combining the two parallel impedances on each end of the network, we obtain the circuit in Fig. 7.18b, where

$$\mathbf{Z}' = \frac{(1 + j)(1 - j)}{(1 + j) + (1 - j)} = 1 \ \Omega$$

Therefore, using current division

$$\mathbf{I}_o' = 1 \ \underline{/0°} \text{ A}$$

The circuit in which the voltage source acts alone is shown in Fig. 7.18c. The voltage $\mathbf{V}_1''$ obtained using voltage division is

$$\mathbf{V}_1'' = \frac{(6 \underline{/0°}) \left[\dfrac{1(1 - j)}{1 + 1 - j}\right]}{1 + j + \left[\dfrac{1(1 - j)}{1 + 1 - j}\right]}$$

$$= \frac{6(1 - j)}{4} \text{ V}$$

LEARNING Hint

In source exchange, a voltage source in series with an impedance can be exchanged for a current source in parallel with the impedance, and vice versa. Repeated application systematically reduces the number of circuit elements.

(4) Source Exchange As a first step in the source exchange approach, we exchange the current source and parallel impedance for a voltage source in series with the impedance, as shown in Fig. 7.19a.

Adding the two voltage sources and transforming them and the series impedance into a current source in parallel with that impedance are shown in Fig. 7.19b. Combining the two impedances that are in parallel with the 1-Ω resistor produces the network in Fig. 7.19c, where

$$\mathbf{Z} = \frac{(1 + j)(1 - j)}{1 + j + 1 - j} = 1 \ \Omega$$

Therefore, using current division,

$$\mathbf{I}_o = \left(\frac{8 + 2j}{1 + j}\right)\left(\frac{1}{2}\right) = \frac{4 + j}{1 + j}$$

$$= \left(\frac{5}{2} - \frac{3}{2}j\right) \text{A}$$

(a)

(b)

(c)

Figure 7.19
Circuits used in Example 7.15 for a sourrce exchange analysis.

LEARNING Hint

In this Thévenin analysis,

1. Remove the 1-Ω load and find the voltage across the open terminals, $\mathbf{V}_{oc}$.
2. Determine the impedance $\mathbf{Z}_{Th}$ at the open terminals with all sources made zero.
3. Construct the following circuit and determine $\mathbf{I}_o$.

(5) Thévenin Analysis   In applying Thévenin's theorem to the circuit in Fig. 7.17a, we first find the open-circuit voltage, $\mathbf{V}_{oc}$, as shown in Fig. 7.20a. To simplify the analysis, we perform a source exchange on the left end of the network, which results in the circuit in Fig. 7.20b. Now using voltage division,

$$\mathbf{V}_{oc} = [6 + 2(1 + j)]\left[\frac{1 - j}{1 - j + 1 + j}\right]$$

or

$$\mathbf{V}_{oc} = (5 - 3j) \text{ V}$$

The Thévenin equivalent impedance, $\mathbf{Z}_{Th}$, obtained at the open-circuit terminals when the current source is replaced with an open circuit and the voltage source is replaced with a short circuit is shown in Fig. 7.20c and calculated to be

$$\mathbf{Z}_{Th} = \frac{(1 + j)(1 - j)}{1 + j + 1 - j} = 1 \ \Omega$$

Connecting the Thévenin equivalent circuit to the 1-Ω resistor containing $\mathbf{I}_o$ in the original network yields the circuit in Fig. 7.20d. The current $\mathbf{I}_o$ is then

$$\mathbf{I}_o = \left(\frac{5}{2} - \frac{3}{2}j\right) \text{ A}$$

LEARNING Hint

In this Norton analysis,

1. Remove the 1-Ω load and find the current $\mathbf{I}_{sc}$ through the short-circuited terminals.
2. Determine the impedance $\mathbf{Z}_{Th}$ at the open load terminals with all sources made zero.
3. Construct the following circuit and determine $\mathbf{I}_o$.

(6) Norton Analysis   Finally, in applying Norton's theorem to the circuit in Fig. 7.17a, we calculate the short-circuit current, $\mathbf{I}_{sc}$, using the network in Fig. 7.21a. Note that because of the short circuit, the voltage source is directly across the impedance in the left-most branch. Therefore,

$$\mathbf{I}_1 = \frac{6\ \underline{/0^\circ}}{1 + j}$$

(continued)

Then using KCL,

$$\mathbf{I}_{sc} = \mathbf{I}_1 + 2\,\underline{/0°} = 2 + \frac{6}{1+j}$$

$$= \left(\frac{8+2j}{1+j}\right) \text{A}$$

The Thévenin equivalent impedance, $\mathbf{Z}_{Th}$, is known to be $1\,\Omega$ and, therefore, connecting the Norton equivalent to the 1-Ω resistor con-

taining $\mathbf{I}_o$ yields the network in Fig. 7.21b. Using current division, we find that

$$\mathbf{I}_o = \frac{1}{2}\left(\frac{8+2j}{1+j}\right)$$

$$= \left(\frac{5}{2} - \frac{3}{2}j\right) \text{A}$$

(a)

(b)

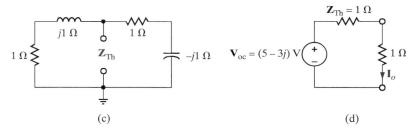

(c)   (d)

Figure 7.20
Circuits used in Example 7.15 for a Thévenin analysis.

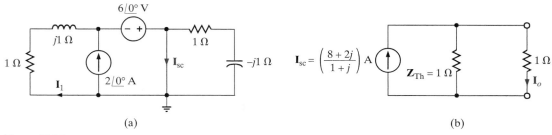

(a)   (b)

Figure 7.21
Circuits used in Example 7.15 for a Norton analysis.

Let us now consider an example containing a dependent source.

LEARNING Example 7.16

Let us determine the voltage $\mathbf{V}_o$ in the circuit in Fig. 7.22a. In this example we will use node equations, loop equations, Thévenin's theorem, and Norton's theorem. We will omit the techniques of superposition and source transformation. Why?

SOLUTION

LEARNING Hint

How does the presence of a dependent source affect superposition and source exchange?

(1) Nodal Analysis   To perform a nodal analysis, we label the node voltages and identify the supernode as shown in Fig. 7.22b. The constraint equation for the supernode is

$$\mathbf{V}_3 + 12\,\underline{/0^\circ} = \mathbf{V}_1$$

and the KCL equations for the nodes of the network are

$$\frac{\mathbf{V}_1 - \mathbf{V}_2}{-j1} + \frac{\mathbf{V}_3 - \mathbf{V}_2}{1} - 4\,\underline{/0^\circ} + \frac{\mathbf{V}_3 - \mathbf{V}_o}{1} + \frac{\mathbf{V}_3}{j1} = 0$$

$$\frac{\mathbf{V}_2 - \mathbf{V}_1}{-j1} + \frac{\mathbf{V}_2 - \mathbf{V}_3}{1} - 2\left(\frac{\mathbf{V}_3 - \mathbf{V}_o}{1}\right) = 0$$

$$4\,\underline{/0^\circ} + \frac{\mathbf{V}_o - \mathbf{V}_3}{1} + \frac{\mathbf{V}_o}{1} = 0$$

At this point we can solve the foregoing equations using a matrix analysis or, for example, substitute the first and last equations into the remaining two equations, which yields

$$3\mathbf{V}_o - (1 + j)\mathbf{V}_2 = -(4 + j12)$$

$$-(4 + j2)\mathbf{V}_o + (1 + j)\mathbf{V}_2 = 12 + j16$$

Solving these equations for $\mathbf{V}_o$ yields

$$\mathbf{V}_o = \frac{-(8 + j4)}{1 + j2}$$

$$= +4\,\underline{/143.13^\circ}\ \text{V}$$

(2) Loop Analysis   The mesh currents for the network are defined in Fig. 7.22c. The constraint equations for the circuit are

$$\mathbf{I}_2 = -4\,\underline{/0^\circ}$$

$$\mathbf{I}_x = \mathbf{I}_4 - \mathbf{I}_2 = \mathbf{I}_4 + 4\,\underline{/0^\circ}$$

$$\mathbf{I}_3 = 2\mathbf{I}_x = 2\mathbf{I}_4 + 8\,\underline{/0^\circ}$$

The KVL equations for mesh 1 and mesh 4 are

$$-j1\mathbf{I}_1 + 1(\mathbf{I}_1 - \mathbf{I}_3) = -12\,\underline{/0^\circ}$$

$$j1(\mathbf{I}_4 - \mathbf{I}_3) + 1(\mathbf{I}_4 - \mathbf{I}_2) + 1\mathbf{I}_4 = 0$$

Note that if the constraint equations are substituted into the second KVL equation, the only unknown in the equation is $\mathbf{I}_4$. This substitution yields

$$\mathbf{I}_4 = +4\,\underline{/143.13^\circ}\ \text{A}$$

and hence,

$$\mathbf{V}_o = +4\,\underline{/143.13^\circ}\ \text{V}$$

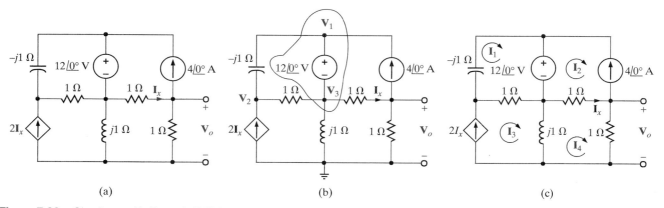

(a)   (b)   (c)

Figure 7.22   *Circuits used in Example 7.16 for nodal and loop analysis.*

(continued)

(3) Thévenin's Theorem   In applying Thévenin's theorem, we will find the open-circuit voltage and then determine the Thévenin equivalent impedance using a test source at the open-circuit terminals. We could determine the Thévenin equivalent impedance by calculating the short-circuit current; however, we will determine this current when we apply Norton's theorem.

The open-circuit voltage is determined from the network in Fig. 7.23a. Note that $\mathbf{I}'_x = 4\,\underline{/0°}$ A and since $2\mathbf{I}'_x$ flows through the inductor, the open-circuit voltage $\mathbf{V}_{oc}$ is

$$\mathbf{V}_{oc} = -1\left(4\,\underline{/0°}\right) + j1\left(2\mathbf{I}'_x\right)$$
$$= -4 + j8 \text{ V}$$

To determine the Thévenin equivalent impedance, we turn off the independent sources, apply a test voltage source to the output terminals, and compute the current leaving the test source. As shown in Fig. 7.23b, since $\mathbf{I}''_x$ flows in the test source, KCL requires that the current in the inductor be $\mathbf{I}''_x$ also. KVL around the mesh containing the test source indicates that

$$j1\mathbf{I}''_x - 1\mathbf{I}''_x - \mathbf{V}_{\text{test}} = 0$$

Therefore,

$$\mathbf{I}''_x = \frac{-\mathbf{V}_{\text{test}}}{1 - j}$$

Then

$$\mathbf{Z}_{\text{Th}} = \frac{\mathbf{V}_{\text{test}}}{-\mathbf{I}''_x}$$
$$= 1 - j\ \Omega$$

If the Thévenin equivalent network is now connected to the load, as shown in Fig. 7.23c, the output voltage $\mathbf{V}_o$ is found to be

$$\mathbf{V}_o = \frac{-4 + 8j}{2 - j1}\ (1)$$
$$= +4\,\underline{/143.13°} \text{ V}$$

(4) Norton's Theorem   In using Norton's theorem, we will find the short-circuit current from the network in Fig. 7.24a. Once again, using the supernode, the constraint and KCL equations are

$$\mathbf{V}_3 + 12\,\underline{/0°} = \mathbf{V}_1$$

$$\frac{\mathbf{V}_2 - \mathbf{V}_1}{-j1} + \frac{\mathbf{V}_2 - \mathbf{V}_3}{1} - 2\mathbf{I}'''_x = 0$$

$$\frac{\mathbf{V}_1 - \mathbf{V}_2}{-j1} + \frac{\mathbf{V}_3 - \mathbf{V}_2}{1} - 4\,\underline{/0°} + \frac{\mathbf{V}_3}{j1} + \mathbf{I}'''_x = 0$$

$$\mathbf{I}'''_x = \frac{\mathbf{V}_3}{1}$$

Substituting the first and last equations into the remaining equations yields

$$(1 + j)\mathbf{V}_2 - (3 + j)\mathbf{I}'''_x = j12$$
$$-(1 + j)\mathbf{V}_2 + (2)\mathbf{I}'''_x = 4 - j12$$

Solving these equations for $\mathbf{I}'''_x$ yields

$$\mathbf{I}'''_x = \frac{-4}{1 + j}\ \text{A}$$

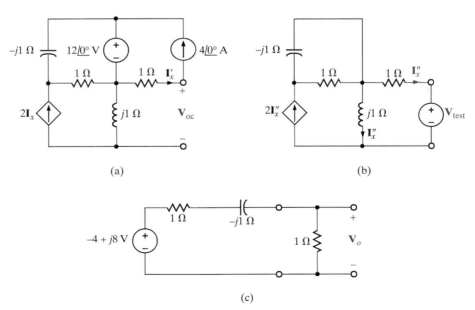

(a)   (b)

(c)

Figure 7.23
Circuits used in Example 7.16 when applying Thévenin's theorem.

The KCL equation at the right-most node in the network in Fig. 7.24a is

$$\mathbf{I}_x''' = 4\,\underline{/0°} + \mathbf{I}_{sc}$$

Solving for $\mathbf{I}_{sc}$, we obtain

$$\mathbf{I}_{sc} = \frac{-(8 + j4)}{1 + j}\,\text{A}$$

The Thévenin equivalent impedance was found earlier to be

$$\mathbf{Z}_{Th} = 1 - j\,\Omega$$

Using the Norton equivalent network, the original network is reduced to that shown in Fig. 7.24b. The voltage $\mathbf{V}_o$ is then

$$\mathbf{V}_o = \frac{-(8 + j4)}{1 + j}\left[\frac{(1)(1 - j)}{1 + 1 - j}\right]$$

$$= -4\left[\frac{3 - j}{3 + j}\right]$$

$$= +4\,\underline{/-143.13°}\,\text{V}$$

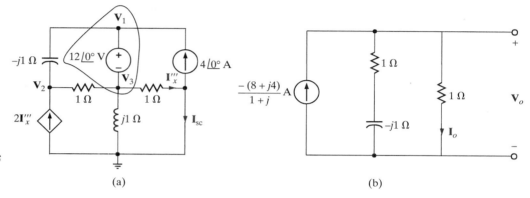

Figure 7.24
Circuits used in Example 7.16 when applying Norton's theorem.

(a)   (b)

LEARNING EXTENSIONS

E7.13 Use nodal analysis to find $\mathbf{V}_o$ in the network in Fig. E7.13.

Figure E7.13

ANSWER
$\mathbf{V}_o = 2.12\,\underline{/75°}\,\text{V}.$

E7.14 Use (a) mesh equations and (b) Thévenin's theorem to find $\mathbf{V}_o$ in the network in Fig. E7.14.

Figure E7.14

ANSWER
$\mathbf{V}_o = 10.88\,\underline{/36°}\,\text{V}.$

E7.15 Use (a) superposition, (b) source transformation, and (c) Norton's theorem to find $\mathbf{V}_o$ in the network in Fig. E7.15.   **ANSWER** $\mathbf{V}_o = 12\,\underline{/90°}$ V.

Figure E7.15

MATLAB Analysis

When dealing with large networks it is impractical to determine the currents and voltages within the network without the help of a mathematical software or CAD package. Thus, we will once again demonstrate how to bring this computing power to bear when solving more complicated ac networks.

Earlier we used MATLAB to solve a set of simultaneous equations, which yielded the node voltages or loop currents in dc circuits. We now apply this technique to ac circuits. In the ac case where the number-crunching involves complex numbers, we use j to represent the imaginary part of a complex number (unless it has been previously defined as something else), and the complex number $x + jy$ is expressed in MATLAB as `x+j*y`. Although we use j in defining a complex number, MATLAB will list the complex number using i.

Complex sources are expressed in rectangular form, and we use the fact that 360° equals 2 pi radians. For example, a source $\mathbf{V} = 10\,\underline{/45°}$ will be entered into MATLAB data as

$$\mathbf{V} = 10\,\underline{/45°} = \texttt{x+j*y}$$

where the real and imaginary components are

$$\texttt{x = 10*cos(45*pi/180)}$$
$$\texttt{= 7.07}$$

and

$$\texttt{y = 10*sin(45*pi/180)}$$
$$\texttt{= 7.07}$$

When using MATLAB to determine the node voltages in ac circuits, we enter the Y matrix, the I vector, and then the solution equation

$$\texttt{V = inv(Y)*I}$$

as was done in the dc case. The following example will serve to illustrate the use of MATLAB in the solution of ac circuits.

LEARNING Example 7.17

Consider the network in Fig. 7.25. We wish to find all the node voltages in this network. The five simultaneous equations describing the node voltages are

$$V_1 = 12\,\underline{/30°}$$

$$\frac{V_2 - V_1}{1} + \frac{V_2 - V_5}{j2} + \frac{V_2 - V_3}{-j1} = 0$$

$$\frac{V_3 - V_2}{-j1} + \frac{V_3 - V_5}{2} + \frac{V_3}{1} = 0$$

$$\frac{V_4 - V_1}{2} + \frac{V_4 - V_5}{1} + \frac{V_4}{-j1} = 0$$

$$\frac{V_5 - V_2}{j2} + \frac{V_5 - V_3}{2} + \frac{V_5}{-j1} + \frac{V_5 - V_4}{1} = 2\,\underline{/45°}$$

Figure 7.25   *Circuit used in Example 7.17.*

Expressing the equations in matrix form, we obtain

$$\begin{Bmatrix} 1 & 0 & 0 & 0 & 0 \\ -1 & 1+j0.5 & -j1 & 0 & j0.5 \\ 0 & -j1 & 1.5+j1 & 0 & -0.5 \\ -0.5 & 0 & 0 & 1.5+j1 & -1 \\ 0 & j0.5 & -0.5 & -1 & 1.5+j0.5 \end{Bmatrix} \begin{Bmatrix} V_1 \\ V_2 \\ V_3 \\ V_4 \\ V_5 \end{Bmatrix}$$

$$= \begin{Bmatrix} 12\,\underline{/30°} \\ 0 \\ 0 \\ 0 \\ 2\,\underline{/45°} \end{Bmatrix}$$

The following MATLAB data consist of the conversion of the sources to rectangular form, the coefficient matrix Y, the vector I, the solution equation V = inv(Y)*I, and the solution vector V.

```
>> x = 12*cos(30*pi/180)
x =
    10.3923
>> y = 12*sin(30*pi/180)
y =
    6.0000
>> v1 = x+j*y
v1 =
    10.3923 + 6.0000i
>> I2 = 2*cos(45*pi/180)
+ j*2*sin(45*pi/180)
I2 =
    1.4142 + 1.4142i
>> Y = [1 0 0 0 0; -1 1+j*0.5 -j*1 0
j*0.5; 0 -j*1 1.5+j*1 0 -0.5; -0.5 0 0
1.5+j*1 -1; 0 j*0.5 -0.5 -1 1.5+j*0.5]
Y =
Columns 1 through 4

 1.0000        0                 0                0
-1.0000 1.0000 +0.5000i          0 -1.0000i        0
      0        0 -1.0000i  1.5000 +1.0000i          0
-0.5000        0                 0          1.5000 +1.0000i
      0        0 +0.5000i -0.5000          -1.0000

Column 5
      0
      0  +0.5000i
-0.5000
-1.0000
 1.5000  +0.5000i
>> I = [v1 ; 0; 0; 0; I2]
I =
    10.3923  +6.0000i
         0
         0
         0
     1.4142  +1.4142i
>> V = inv(Y) *I
V =
    10.3923  +6.0000i
     7.0766  +2.1580i
     1.4038  +2.5561i
     3.7661  -2.9621i
     3.4151  -3.6771i
```

7.9 *AC PSPICE Analysis Using Schematic Capture*

INTRODUCTION In this chapter we found that an ac steady-state analysis is facilitated by the use of phasors. PSPICE can perform ac steady-state simulations, outputting magnitude and phase data for any voltage or current phasors of interest. Additionally, PSPICE can perform an AC SWEEP in which the frequency of the sinusoidal sources is varied over a user-defined range. In this case, the simulation results are the magnitude and phase of every node voltage and branch current as a function of frequency.

We will introduce five new *Schematics*/PSPICE topics in this section: defining AC sources, simulating at a single frequency, simulating over a frequency range, using the PROBE feature to create plots and, finally, saving and printing these plots. *Schematics* fundamentals such as getting parts, wiring, and changing part names and values were already discussed in Chapter 4. As in Chapter 4, we will use the following font conventions. Uppercase text refers to programs and utilities within PSPICE such as the AC SWEEP feature and PROBE graphing utility. All boldface text, whether upper or lowercase, denotes keyboard or mouse entries. For example, when placing a resistor into a circuit schematic, one must specify the resistor **VALUE** using the keyboard. The case of the boldface text matches that used in PSPICE.

DEFINING AC SOURCES Figure 7.26 shows the circuit we will simulate at a frequency of 60 Hz. We will continue to follow the flowchart shown in Fig. 4.20 in performing this simulation. Inductor and capacitor parts are in the ANALOG library and are called L and C, respectively. The AC source, VAC, is in the SOURCE library. Figure 7.27 shows the resulting *Schematics* diagram after wiring and editing the part's names and values.

Figure 7.26
Circuit for ac simulation.

Figure 7.27
The Schematics diagram for the circuit in Figure 7.26.

To set up the AC source for simulation, double-click on the source symbol to open its ATTRIBUTES box, which is shown, after editing, in Fig. 7.28. As discussed in Chapter 6, we deselect the fields **Include Non-changeable Attributes** and **Include System-defined Attributes**. Each line in the ATTRIBUTES box is called an attribute of the ac source. Each attribute has a name and a value. The **DC** attribute is the dc value of the source for dc analyses. The **ACMAG** and **ACPHASE** attributes set the magnitude and phase of the phasor representing Vin for ac analyses. Each of these attributes defaults to zero. The value of the **ACMAG** attribute was set to 4 V when we created the schematic in Fig. 7.27. To set the **ACPHASE** attribute to 10°, click on the **ACPHASE** attribute line, enter 10 in the **Value** field, press **Save Attr** and **OK**. When the ATTRIBUTE box looks like that shown in Fig. 7.28, the source is ready for simulation.

Figure 7.28
Setting the ac source phase angle.

SINGLE FREQUENCY AC SIMULATIONS   Next, we must specify the frequency for simulation. This is done by selecting **Setup** from the **Analysis** menu. The SETUP box in Fig. 7.29 should appear. If we double-click on the text **AC Sweep**, the AC SWEEP AND NOISE ANALYSIS window in Fig. 7.30 will open. All of the fields in Fig. 7.30 have been set for our 60-Hz simulation.

Figure 7.29
The ANALYSIS SETUP window.

AC Sweep and Noise Analysis ☒

AC Sweep Type
- ⦿ Linear
- ◯ Octave
- ◯ Decade

Sweep Parameters

Total Pts.: `1`

Start Freq.: `60`

End Freq.: `60`

Noise Analysis

☐ Noise Enabled

Output Voltage:

I/V

Interval:

[OK] [Cancel]

Figure 7.30
Setting the frequency range for a single frequency simulation.

Since the simulation will be performed at only one frequency, 60 Hz, graphing the simulation results is not an attractive option. Instead, we will write the magnitude and phase of the phasors **Vout** and **I** to the output file using the VPRINT1 and IPRINT parts, from the SPECIAL library, which have been added to the circuit diagram as shown in Fig. 7.31. The VPRINT1 part acts as a voltmeter, measuring the voltage at any single node with respect to the ground node. There is also a VPRINT2 part, which measures the voltage between any two nonreference nodes. Similarly, the IPRINT part acts as an ammeter and must be placed in series with the branch current of interest. By convention, current in the IPRINT part is assumed to exit from its negatively marked terminal. To find the clockwise loop current, as defined in Fig. 7.26, the IPRINT part has been flipped. The FLIP command is in the EDIT menu.

Figure 7.31
A Schematics diagram ready for single-frequency ac simulation.

After placing the VPRINT1 part, double-click on it to open its ATTRIBUTES box, shown in Fig. 7.32. The VPRINT1 part can be configured to meter the node voltage in any kind of simulation: dc, ac, or transient. Since an ac analysis was specified in the SETUP window in Fig. 7.29, the values of the **AC**, **MAG**, and **PHASE** attributes are set to **Y**, where Y stands for YES. This process is repeated for the IPRINT part. When we return to *Schematics*, the simulation is ready to run.

Figure 7.32
Setting the VPRINT1 measurements for ac magnitude and phase.

When an AC SWEEP is performed, PSPICE, unless instructed otherwise, will attempt to plot the results using the PROBE plotting program. To turn off this feature, select **Probe Setup** in the **Analysis** menu. When the PROBE SETUP window shown in Fig. 7.33 appears, select **Do Not Auto-Run Probe** and **OK**.

Figure 7.33
The PROBE SETUP window.

The circuit is simulated by selecting **Simulate** from the **Analysis** menu. Since the results are in the output file, select **Examine Output** from the **Analysis** menu to view the data. At the bottom of the file, we find the results as seen in Fig. 7.34: **Vout** $= 2.651\ \underline{/-38.54°}$ V and **I** $= 2.998\ \underline{/51.46°}$ mA.

Figure 7.34 *Magnitude and phase data for* **Vout** *and* **I** *are at the bottom of the output file.*

VARIABLE FREQUENCY AC SIMULATIONS

To sweep the frequency over a range, 1 Hz to 10 MHz, for example, return to the AC SWEEP AND NOISE ANALYSIS box shown in Fig. 7.30. Change the fields to those shown in Fig. 7.35. Since the frequency range is so large, we have chosen a log axis for frequency with 50 data points in each decade. We can now plot the data using the PROBE utility. This procedure requires two steps. First, we remove the VPRINT1 and IPRINT parts in Fig. 7.31. Second, we return to the PROBE SETUP window shown in Fig. 7.33, and select **Automatically Run Probe After Simulation** and **OK**.

Figure 7.35
Setting the frequency range for a swept frequency simulation.

CREATING PLOTS IN PROBE   When the PSPICE simulation is finished, the PROBE window shown in Fig. 7.36 will open. Actually, there are three windows here: the main display window, the output window, and the simulation status window. The latter two can be toggled off and on in the **View** menu. We will focus on the main display window, where the frequency is displayed on a log axis, as requested. To plot the magnitude and phase of **Vout**, select **Add** from the **Trace** menu. The ADD TRACES window shown in Fig. 7.37 will appear. To display the magnitude of **Vout**, we select **V(Vout)** from the left column. When a voltage or current is selected, the magnitude of the phasor will be plotted. Now the PROBE window should look like that shown in Fig. 7.38.

Before adding the phase to the plot, we note that **Vout** spans a small range—that is, 0 to 4 V. Since the phase change could span a much greater range, we will plot the phase on a second y-axis. From the **Plot** menu, select **Add Y Axis**. To add the phase to the plot, select **Add** from the **Trace** menu. On the right side of the ADD TRACES window in Fig. 7.37, scroll down to the entry, **P()**. Click on that, and then click on **V(Vout)** in the left column. The TRACE EXPRESSION line at the bottom of the window will contain the expression **P(V(Vout))**—the phase of **Vout**. Figure 7.39 shows the PROBE plot for both magnitude and phase of **Vout**.

To plot the current, **I**, on a new plot, we select **New** from the **Window** menu. Then, add the traces for the magnitude and phase of the current through R1 (PSPICE calls it **I(R1)**) using the process described above for plotting the magnitude and phase of **Vout**. The results are shown in Fig. 7.40.

The procedures for saving and printing PROBE plots, as well as the techniques for plot manipulation and data extraction, are described in Chapter 6.

Figure 7.36   *The PROBE window.*

Figure 7.37 *The Add Traces window.*

Figure 7.38 *The magnitude of **Vout**.*

Figure 7.39   *The magnitude and phase of* **Vout**.

Figure 7.40   *The magnitude and phase of the current,* **I**.

LEARNING Example 7.18

Using the PSPICE *Schematics* editor, draw the circuit in Fig. 7.41, and use the PROBE utility to create plots for the magnitude and phase of $\mathbf{V}_{out}$ and $\mathbf{I}$. At what frequency does maximum $|\mathbf{I}|$ occur? What are the phasors $\mathbf{V}_{out}$ and $\mathbf{I}$ at that frequency?

SOLUTION The *Schematics* diagram for the simulation is shown in Fig. 7.42, where an AC SWEEP has been set up for the frequency range 10 Hz to 10 MHz at 100 data points per decade. Plots for $\mathbf{V}_{out}$ and $\mathbf{I}$ magnitudes and phases are given in Figs. 7.43a and b, respectively. From Fig. 7.43b we see that the maximum inductor current magnitude occurs at 31.42 kHz. At that frequency, the phasors of interest are $\mathbf{I} = 5.94 \underline{/16.06°}$ A and $\mathbf{V}_{out} = 299.5 \underline{/-68.15°}$ V.

Figure 7.41 Circuit for Example 7.18.

Figure 7.42
PSPICE Schematics diagram
for Example 7.18.

(a)

Figure 7.43 Simulation results for Example 7.18, (a) **Vout** and (b) **I**.

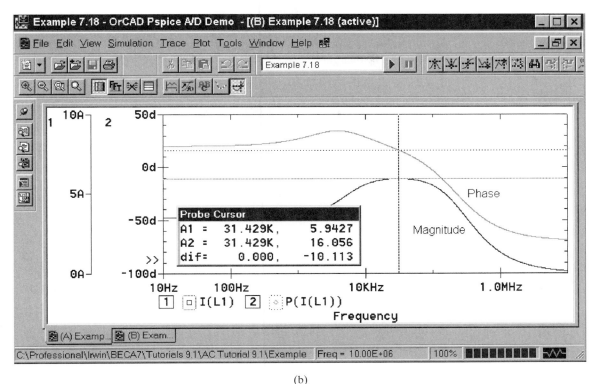

(b)

Figure 7.43 (continued)

LEARNING Check

Summary

▶ **The sinusoidal function definition** The sinusoidal function $x(t) = X_M \sin(\omega t + \theta)$ has an amplitude of X_M, a radian frequency of ω, a period of $2\pi/\omega$, and a phase angle of θ.

▶ **The phase lead and phase lag definitions** If $x_1(t) = X_{M_1} \sin(\omega t + \theta)$ and $x_2(t) = X_{M_2} \sin(\omega t + \phi)$, $x_1(t)$ leads $x_2(t)$ by $\theta - \phi$ radians and $x_2(t)$ lags $x_1(t)$ by $\theta - \phi$ radians.

▶ **The phasor definition** The sinusoidal voltage $v(t) = V_M \cos(\omega t + \theta)$ can be written in exponential form as $v(t) = \text{Re}\left[V_M e^{j(\omega t + \theta)}\right]$ and in phasor form as $\mathbf{V} = V_M \underline{/\theta}$.

▶ **The phase relationship in θ_v and θ_i for elements R, L, and C** If θ_v and θ_i represent the phase angles of the voltage across and the current through a circuit element, then $\theta_i = \theta_v$ if the element is a resistor, θ_i lags θ_v by 90° if the element is an inductor, θ_i leads θ_v by 90° if the element is a capacitor.

▶ **The impedances of R, L, and C** Impedance, $\mathbf{Z}$, is defined as the ratio of the phasor voltage, $\mathbf{V}$, to the phasor current, $\mathbf{I}$, where $\mathbf{Z} = R$ for a resistor, $\mathbf{Z} = j\omega L$ for an inductor, and $\mathbf{Z} = 1/j\omega C$ for a capacitor.

▶ **The phasor diagrams** Phasor diagrams can be used to display the magnitude and phase relationships of various voltages and currents in a network.

▶ **Frequency domain analysis:**

1. Represent all voltages, $v_i(t)$, and all currents, $i_j(t)$, as phasors and represent all passive elements by their impedance or admittance.

2. Solve for the unknown phasors in the frequency (ω) domain.

3. Transform the now-known phasors back to the time domain.

▶ **Solution techniques for ac steady-state problems:**

Ohm's law
KCL and KVL
PSPICE
MATLAB
Nodal and loop analysis
Superposition and source exchange
Thévenin's Theorem
Norton's Theorem

Problems

SECTION 7.1

7.1 Given $i(t) = 5\cos(400t - 120°)$ A, determine the period of the current and the frequency in hertz.

7.2 Determine the relative position of the two sine waves.

$$v_1(t) = 12\sin(377t - 45°)$$
$$v_2(t) = 6\sin(377t + 675°)$$

7.3 Given the following currents

$$i_1(t) = 4\sin(377t - 10°) \text{ A}$$
$$i_2(t) = -2\cos(377t - 195°) \text{ A}$$
$$i_3(t) = -1\sin(377t - 250°) \text{ A}$$

compute the phase angle between each pair of currents.

7.4 Determine the phase angles by which $v_1(t)$ leads $i_1(t)$ and $v_1(t)$ leads $i_2(t)$, where

$$v_1(t) = 4\sin(377t + 25°) \text{ V}$$
$$i_1(t) = 0.05\cos(377t - 10°) \text{ A}$$
$$i_2(t) = -0.1\sin(377t - 75°) \text{ A}$$

SECTIONS 7.2, 7.3, 7.4, 7.5

7.5 Calculate the current in the resistor in Fig. P7.5 if the voltage input is

(a) $v_1(t) = 10\cos(377t + 180°)$ V.

(b) $v_2(t) = 12\sin(377t + 45°)$ V.

Give the answers in both the time and frequency domains.

Figure P7.5

7.6 Calculate the current in the capacitor shown in Fig. P7.6 if the voltage input is

(a) $v_1(t) = 16\cos(377t - 22°)$ V.

(b) $v_2(t) = 8\sin(377t + 64°)$ V.

Give the answers in both the time and frequency domains.

Figure P7.6

7.7 Calculate the current in the inductor shown in Fig. P7.7 if the voltage input is

(a) $v_1(t) = 24 \cos(377t + 12°)$ V.

(b) $v_2(t) = 18 \sin(377t + 48°)$ V.

Give the answers in both the time and frequency domains.

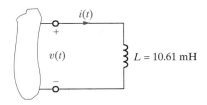

Figure P7.5

7.8 Find the frequency-domain impedance, **Z**, for the network in Fig. P7.8.

Figure P7.8

7.9 Find the frequency-domain impedance, **Z**, as shown in Fig. P7.9.

Figure P7.9

7.10 Find the frequency-domain impedance, **Z**, for the circuit in Fig. P7.10.

Figure P7.10

7.11 Find the frequency-domain impedance, **Z**, as shown in Fig. P7.11.

Figure P7.11

7.12 Find the frequency-domain impedance, **Z**, as shown in Fig. P7.12.

Figure P7.12

7.13 In the network in Fig. P7.13, find $\mathbf{Z}(j\omega)$ at a frequency of 60 Hz.

Figure P7.13

7.14 Calculate the equivalent impedance at terminals A-B in the circuit shown in Fig. P7.14.

Figure P7.14

7.15 Find $\mathbf{Z}$ in the network in Fig. P7.15.

Figure P7.15

7.16 Find $\mathbf{Z}$ in the network in Fig. P7.16.

Figure P7.16

7.17 Find $\mathbf{Z}$ in the network in Fig. P7.17.

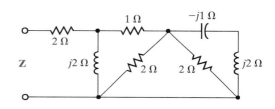

Figure P7.17

7.18 The impedance of the network in Fig. P7.18 is found to be purely real at $f = 60$ Hz. What is the value of C?

Figure P7.18

7.19 In the circuit shown in Fig. P7.19, determine the value of the inductance such that the current is in phase with the source voltage.

Figure P7.19

7.20 Draw the frequency-domain circuit and calculate $i(t)$ for the circuit shown in Fig. P7.20 if
$$v_s(t) = 10\cos(377t + 30°)\,\text{V}.$$

Figure P7.20

7.21 Draw the frequency-domain circuit and calculate $v(t)$ for the circuit shown in Fig. P7.21 if
$$i_s(t) = 20\cos(377t + 120°)\,\text{A}.$$

Figure P7.21

SECTION 7.6

7.22 The voltages $v_R(t)$, $v_L(t)$, and $v_C(t)$ in the circuit shown in Fig. P7.20 can be drawn as phasors in a phasor diagram. Show that $v_R(t) + v_L(t) + v_C(t) = v_S(t)$.

7.23 The currents $i_R(t)$, $i_L(t)$, and $i_C(t)$ in the circuit shown in Fig. P7.21 can be drawn as phasors in a phasor diagram. Show that $i_R(t) + i_L(t) + i_C(t) = i_S(t)$.

7.24 The voltages $v_R(t)$ and $v_L(t)$ in the circuit shown in Fig. P7.24 can be drawn as phasors in a phasor diagram. Use a phasor diagram to show that $v_R(t) + v_L(t) = v_S(t)$.

Figure P7.24

7.25 The currents $i_R(t)$ and $i_C(t)$ in the circuit shown in Fig. P7.25 can be drawn as phasors in a phasor diagram. Use the diagram to show that $i_R(t) + i_C(t) = i_S(t)$.

Figure P7.25

7.27 In the circuit shown in Fig. P7.27, determine the frequency at which $i(t)$ is in phase with $v_s(t)$.

Figure P7.27

7.26 The currents $i_L(t)$ and $i_C(t)$ of the inductor and capacitor in the circuit shown in Fig. P7.26 can be drawn as phasors in a phasor diagram. Show that $i_L(t) + i_C(t) = i(t)$.

Figure P7.26

SECTION 7.7

7.28 Find the frequency-domain voltage $\mathbf{V}_o$, as shown in Fig. P7.28.

Figure P7.28

7.29 Find the frequency-domain voltage $\mathbf{V}_o$, as shown in Fig. P7.29.

Figure P7.29

7.30 Find the frequency-domain current, $\mathbf{I}_o$, as shown in Fig. P7.30.

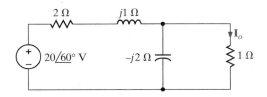

Figure P7.30

7.31 Find the frequency-domain voltage $\mathbf{V}_o$, as shown in Fig. P7.31.

Figure P7.31

7.32 Draw the frequency-domain network and calculate $v_o(t)$ in the circuit shown in Fig. P7.32 if $i_S(t)$ is $100 \cos(5000t + 8.13°)$ mA. Also, using a phasor diagram, show that $i_L(t) + i_R(t) = i_S(t)$.

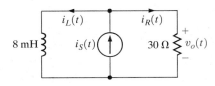

Figure P7.32

7.33 Draw the frequency-domain network and calculate $v_o(t)$ in the circuit shown in Fig. P7.33 if $i_S(t)$ is $300 \sin(10^4 t - 45°)$ mA. Also, using a phasor diagram, show that $i_1(t) + i_2(t) = i_S(t)$.

Figure P7.33

7.34 Find $\mathbf{I}_o$ in the network in Fig. P7.34.

Figure P7.34

7.35 Find $\mathbf{I}_o$ in the network in Fig. P7.35.

Figure P7.35

7.36 In the circuit in Fig. P7.36, if $\mathbf{V}_o = 4\,\underline{/45°}$ V, find $\mathbf{I}_1$.

Figure P7.36

7.37 Find $\mathbf{V}_S$ in the network in Fig. P7.37, if $\mathbf{V}_1 = 4\,\underline{/0°}$ V.

Figure P7.37

7.38 Find $\mathbf{V}_S$ in the network in Fig. P7.38 if $\mathbf{I}_1 = 2\,\underline{/0°}$ A.

Figure P7.38

7.39 Find $\mathbf{V}_S$ in the network in Fig. P7.39 if $\mathbf{I}_o = 2\,\underline{/0°}$ A.

Figure P7.39

7.40 Find $\mathbf{I}_S$ in the network in Fig. P7.40 if $\mathbf{V}_1 = 8\,\underline{/0°}$ V.

Figure P7.40

7.41 If $\mathbf{V}_1 = 4\,\underline{/0°}$ V, find $\mathbf{I}_o$ in Fig. P7.41.

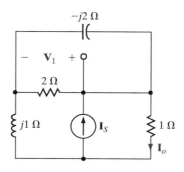

Figure P7.41

7.42 In the network in Fig. P7.42, $\mathbf{V}_o$ is known to be $4\,\underline{/45°}$ V. Find $\mathbf{Z}$.

Figure P7.42

7.44 Determine $\mathbf{I}_o$ in the network in Fig. P7.44 if $\mathbf{I}_S = 12\,\underline{/0°}$ A.

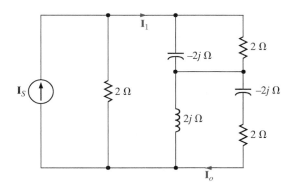

Figure P7.44

7.43 In the network in Fig. P7.43, $\mathbf{V}_1 = 2\,\underline{/45°}$ V. Find $\mathbf{Z}$.

Figure P7.43

SECTION 7.8

7.45 Using nodal analysis, find $\mathbf{I}_o$ in the circuit in Fig. P7.45.

Figure P7.45

7.46 Use nodal analysis to find $\mathbf{V}_o$ in the circuit in Fig. P7.46.

Figure P7.46

7.47 Find $\mathbf{V}_o$ in the network in Fig. P7.47 using nodal analysis.

Figure P7.47

7.48 Use nodal analysis to determine $\mathbf{I}_o$ in the network in Fig. P7.48. In addition, solve the problem using MATLAB.

Figure P7.48

7.49 Find $\mathbf{V}_o$ in the network in Fig. P7.49.

Figure P7.49

7.50 Find the voltage across the inductor in the circuit shown in Fig. P7.50 using nodal analysis.

Figure P7.50

7.51 Use mesh analysis to find $\mathbf{V}_o$ in the circuit shown in Fig. P7.51.

Figure P7.51

7.52 Use mesh analysis to find $\mathbf{V}_o$ in the circuit shown in Fig. P7.52.

Figure P7.52

7.53 Using both loop analysis and MATLAB, find $\mathbf{I}_o$ in the network in Fig. P7.53.

Figure P7.53

7.54 Find $\mathbf{V}_o$ in the network in Fig. P7.54.

Figure P7.54

7.55 Find $\mathbf{V}_o$ in the network in Fig. P7.55.

Figure P7.55

7.56 Use superposition to find $\mathbf{V}_o$ in the network in Fig. P7.56.

Figure P7.56

7.57 Find $\mathbf{V}_o$ in the network in Fig. P7.57 using superposition.

Figure P7.57

7.58 Using superposition, find $\mathbf{V}_o$ in the circuit in Fig. P7.58.

Figure P7.58

7.59 Use both superposition and MATLAB to determine $\mathbf{V}_o$ in the circuit in Fig. P7.59.

Figure P7.59

7.60 Use source exchange to determine $\mathbf{V}_o$ in the network in Fig. P7.60.

Figure P7.60

7.61 Use source exchange to find the current $\mathbf{I}_o$ in the network in Fig. P7.61.

Figure P7.61

7.62 Use source transformation to determine $\mathbf{I}_o$ in the network in Fig. P7.45.

7.63 Use Thévenin's theorem to find $\mathbf{V}_o$ in the circuit in Fig. P7.63.

Figure P7.63

7.64 Using Thévenin's theorem, find $\mathbf{V}_o$ in the network in Fig. P7.51.

7.65 Use Thévenin's theorem to find $\mathbf{V}_o$ in the circuit in Fig. P7.52.

7.66 Solve Problem 7.49 using Thévenin's theorem.

7.67 Apply Thévenin's theorem twice to find $\mathbf{V}_o$ in the circuit in Fig. P7.67.

Figure P7.67

7.68 Find $\mathbf{V}_o$ in the network in Fig. P7.68 using Thévenin's theorem.

Figure P7.68

7.69 Find the Thévenin's equivalent for the network in Fig. P7.69 at the terminals A-B.

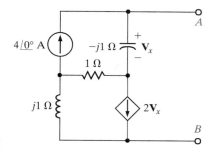

Figure P7.69

7.70 Find $\mathbf{V}_x$ in the circuit in Fig. P7.70 using Norton's theorem.

Figure P7.70

7.71 Find $\mathbf{I}_o$ in the network in Fig. P7.71 using Norton's theorem.

Figure P7.71

7.72 Apply both Norton's theorem and MATLAB to find $\mathbf{V}_o$ in the network in Fig. P7.72.

Figure P7.72

7.73 Find $\mathbf{V}_o$ using Norton's theorem for the circuit in Fig. P7.73.

Figure P7.73

7.74 Use Norton's theorem to find $\mathbf{V}_o$ in the network in Fig. P7.74.

Figure P7.74

7.75 Use MATLAB to find the node voltages in the network in Fig. P7.75.

Figure P7.75

7.76 Using the PSPICE *Schematics* editor, draw the circuit in Fig. P7.76. At what frequency are the magnitudes of $i_C(t)$ and $i_L(t)$ equal?

7.77 Using the PSPICE *Schematics* editor, draw the circuit in Fig. P7.77. At what frequency are the phases of $i_1(t)$ and $v_x(t)$ equal?

Figure P7.76

Figure P7.77

Typical Problems Found on the FE Exam

7FE-1 Find V_o in the network in Fig. 7PFE-1.

Figure 7PFE-1

7FE-3 Find V_o in the network in Fig. 7PFE-3.

Figure 7PFE-3

7FE-2 Find V_o in the circuit in Fig. 7PFE-2.

Figure 7PFE-2

7FE-4 Determine the midband (where the coupling capacitors can be ignored) gain of the single-stage transistor amplifier shown in Fig. 7PFE-4.

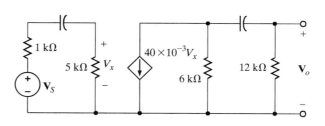

Figure 7PFE-4

Magnetically Coupled Networks

8

High-voltage power lines crisscross the country to supply electric power to us in every area of our lives. Many of the electric utilities are tied together to form a power grid. As indicated earlier, in order to deliver power in an efficient manner, it is transmitted at high voltage. Five-hundred-kilovolt lines are typical transmission facilities. However, the power that is supplied to our homes, for example, is typically 208/120 V rms. Thus, it is natural for us to question how the high voltage supplied through the transmission lines is stepped down to this level. Hence, one of the topics that we will address in this chapter is the transformer, which is the primary component in this application.

The recent advances in integrated circuits have caused designers to avoid the use of transformers (and inductors too) in many situations. For example, to avoid the use of a transformer, we may use a large number of other components such as resistors, capacitors, op-amps, and the like. However, although the trade-off in number of components may seem unreasonable, it is not. It is very easy to fabricate such components as resistors and capacitors, and they require much less space than a single inductor.

Nevertheless, transformers remain an important electrical component. In addition to power systems, where the transformers play a fundamental role, there are other applications. For example, transformers are used to reject high-frequency noise in audio and industrial control systems, and they are built into special wall plugs that are used to step down the voltage for the purpose of recharging batteries for calculators and hand tools.

8.1 Mutual Inductance

Figure 8.1
Magnetic flux φ linking an N-turn coil.

As we introduce this subject, we feel compelled to remind the reader, once again, that in our analyses we assume that we are dealing with "ideal" elements. For example, we ignore the resistance of the coil used to make an inductor and any stray capacitance that might exist. This approach is especially important in our discussion of mutual inductance since an exact analysis of this topic is quite involved. As is our practice, we will treat the subject in a straightforward manner and ignore issues, beyond the scope of this book, that only serve to complicate the presentation.

To begin our discussion of mutual inductance, we recall two important laws: Ampère's law and Faraday's law. Ampère's law predicts that the flow of electric current will create a magnetic field. If that field links an electric circuit, and that field (or flux linkage) is time-varying, Faraday's law predicts the creation of a voltage within the linked circuit. Although this occurs to some extent in all circuits, the effect is magnified in coils, because the circuit geometry amplifies the linkage effect. With these ideas in mind, consider an ideal situation in which a magnetic field, represented by flux φ, links N turns of a coil as shown in Fig. 8.1. The flux linkage is then

$$\lambda = N\phi \qquad \text{webers} \qquad \textbf{8.1}$$

If the flux φ, and thus the flux linkage λ, is created by a current i, and furthermore if the field exists in a constant permeable medium, it can be shown that

$$\lambda = Ki \qquad \textbf{8.2}$$

This constant of proportionality K is defined as inductance L; that is,

$$K = L \triangleq \frac{\lambda}{i} \qquad \text{webers/amperes or henrys} \qquad \textbf{8.3}$$

If λ varies with time, then according to Faraday's law a voltage v is induced such that

$$v = \frac{d\lambda}{dt} = L\frac{di}{dt} \qquad \textbf{8.4}$$

Figure 8.2
An ideal inductor.

If the current that causes the induced voltage flows in the same circuit, then this defining equation represents the ideal inductor, as shown in Fig. 8.2.

However, what if this field in circuit i is created by a remote current in some circuit j, that may or may not be conductively coupled to circuit i? To consider this situation, we expand our concept of flux linkage in the following manner:

$$\lambda_i = \lambda_{ii} + \lambda_{ij} \qquad \textbf{8.5}$$

where λ_i = the total flux linking circuit i

λ_{ii} = the flux linking circuit i caused by the current i

λ_{ij} = the flux linking circuit i caused by the current j

In the event that there are n circuits involved, then

$$\lambda_i = \sum_{j=1}^{n} \lambda_{ij} \qquad \textbf{8.6}$$

where λ_i = the total flux linking circuit i

λ_{ij} = the flux linking circuit i caused by the current in circuit j—that is, i_j

If the field exists in a constant permeable medium, it can be shown that

$$\lambda_i = \sum_{j=1}^{n} L_{ij} i_j \qquad \textbf{8.7}$$

where $L_{ij} = L_{ji}$ = the "mutual" inductance between circuit i and circuit j

L_{ii} = the "self" inductance of circuit i

When only two coils are involved, the relationship reduces to

$$\lambda_1 = L_{11}i_1 + L_{12}i_2$$
$$\lambda_2 = L_{21}i_1 + L_{22}i_2$$

8.8

Differentiating these two equations, we obtain

$$\frac{d\lambda_1}{dt} = v_1 = L_{11}\frac{di_1}{dt} + L_{12}\frac{di_2}{dt}$$

$$\frac{d\lambda_2}{dt} = v_2 = L_{21}\frac{di_1}{dt} + L_{22}\frac{di_2}{dt}$$

8.9

If the medium through which the magnetic flux passes is linear, then $L_{12} = L_{21} = M$, and for convenience we define $L_1 = L_{11}$ and $L_2 = L_{22}$. Thus the defining equations are reduced to

$$v_1(t) = L_1\frac{di_1(t)}{dt} + M\frac{di_2(t)}{dt}$$

$$v_2(t) = M\frac{di_1(t)}{dt} + L_2\frac{di_2(t)}{dt}$$

8.10

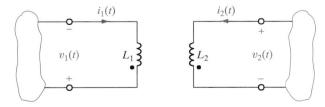

Figure 8.3
Mutually
coupled coils.

The circuit corresponding to these equations is shown in Fig. 8.3, where the "dots" are used to indicate whether λ_{11} and λ_{12} add or subtract to produce λ_1. Now, note carefully that when we applied Ohm's law, $v = Ri$, the relationship between the variables was that specified in Fig. 8.4. Furthermore, if we reversed the direction of either the current or voltage in the circuit, we changed the sign of that variable in the equation. The exact same relationship exists between Eqs. (8.10) and the circuit in Fig. 8.3, including the position of the dots. In other words, if the direction of i_1 is reversed in the network in Fig. 8.3, the sign on i_1 is changed in the equations, and if v_1 is reversed in the network, then v_1 is replaced with $-v_1$ in the circuit equations. In addition, both v_1 and v_2 are positive at the dots. Thus, if the dot on L_1 is moved from top to bottom, then the signs on both v_1 and i_1 in the circuit equations would be changed. It may be necessary to make several changes to force the circuit into the relationship that exists between Eqs. (8.10) and the network in Fig. 8.3. However, the technique is foolproof and all we have to remember is this latter relationship. The following example illustrates the approach.

Figure 8.4
The circuit relationship for Ohm's law, v = Ri.

LEARNING Example 8.1

Determine the equations for $v_1(t)$ and $v_2(t)$ in the circuit shown in Fig. 8.5.

Figure 8.5   *Circuit used in Example 8.1.*

SOLUTION   If we compare this network to that shown in Fig. 8.3, we find that to force the variables in the network in

Fig. 8.5 to the same form as those shown in Fig. 8.3, we simply reverse v_2, i_2 and i_1. Under these conditions, the signs on these three variables are changed in Eqs. (8.10) and the results are

$$v_1(t) = L_1\frac{d}{dt}\big[-i_1(t)\big] + M\frac{d}{dt}\big[-i_2(t)\big]$$

$$-v_2(t) = L_1\frac{d}{dt}\big[-i_1(t)\big] + M\frac{d}{dt}\big[-i_2(t)\big]$$

which can be written as

$$v_1(t) = -L_1\frac{di_1(t)}{dt} - M\frac{di_2(t)}{dt}$$

$$v_2(t) = L_2\frac{di_2(t)}{dt} + M\frac{di_1(t)}{dt}$$

LEARNING EXTENSION

E8.1 Write the equations for $v_1(t)$ and $v_2(t)$ in the circuit in Fig. E8.1.

ANSWER

$$v_1(t) = L_1 \frac{di_1(t)}{dt} + M \frac{di_2(t)}{dt}$$

$$v_2(t) = -L_2 \frac{di_2(t)}{dt} - M \frac{di_1(t)}{dt}$$

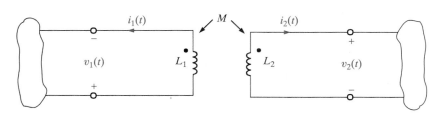

Figure E8.1

Assume that the coupled circuit in Fig. 8.3 is excited with a sinusoidal source. The voltages will be of the form $V_1 e^{j\omega t}$ and $V_2 e^{j\omega t}$, and the currents will be of the form $I_1 e^{j\omega t}$ and $I_2 e^{j\omega t}$, where V_1, V_2, I_1, and I_2 are phasors. Substituting these voltages and currents into Eqs. (8.10), we obtain

$$V_1 = j\omega L_1 I_1 + j\omega M I_2$$
$$V_2 = j\omega L_2 I_2 + j\omega M I_1$$

8.11

The model of the coupled circuit in the frequency domain is identical to that in the time domain except for the way the elements and variables are labeled. The sign on the mutual terms is handled in the same manner as is done in the time domain.

LEARNING Example 8.2

The two mutually coupled coils in Fig. 8.6a can be interconnected in four possible ways. We wish to determine the equivalent inductance of each of the four possible interconnections.

SOLUTION Case 1 is shown in Fig. 8.6b. In this case

$$V = j\omega L_1 I + j\omega M I + j\omega L_2 I + j\omega M I$$

$$= j\omega L_{eq} I$$

where $L_{eq} = L_1 + L_2 + 2M$.

Case 2 is shown in Fig. 8.6c. Using KVL, we obtain

$$V = j\omega L_1 I - j\omega M I + j\omega L_2 I - j\omega M I$$

$$= j\omega L_{eq} I$$

where $L_{eq} = L_1 + L_2 - 2M$.

Case 3 is shown in Fig. 8.6d and redrawn in Fig. 8.6e. The two KVL equations are

$$V = j\omega L_1 I_1 + j\omega M I_2$$
$$V = j\omega M I_1 + j\omega L_2 I_2$$

Solving these equations for I_1 and I_2 yields

$$I_1 = \frac{V(L_2 - M)}{j\omega(L_1 L_2 - M^2)}$$

$$I_2 = \frac{V(L_1 - M)}{j\omega(L_1 L_2 - M^2)}$$

Using KCL gives us

$$I = I_1 + I_2 = \frac{V(L_1 + L_2 - 2M)}{j\omega(L_1 L_2 - M^2)} = \frac{V}{j\omega L_{eq}}$$

where

$$L_{eq} = \frac{L_1 L_2 - M^2}{L_1 + L_2 - 2M}$$

Case 4 is shown in Fig. 8.6f. The voltage equations in this case will be the same as those in case 3 except that the signs of the mutual terms will be negative. Therefore,

$$L_{eq} = \frac{L_1 L_2 - M^2}{L_1 + L_2 + 2M}$$

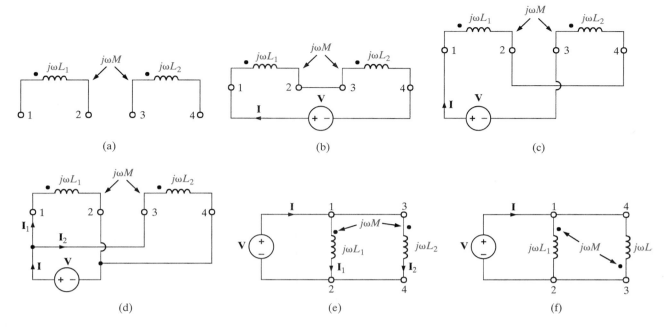

Figure 8.6 *Circuits used in Example 8.2.*

LEARNING Example 8.3

We wish to determine the output voltage $\mathbf{V}_o$ in the circuit in Fig. 8.7.

Figure 8.7 *Example of a magnetically coupled circuit.*

SOLUTION The two KVL equations for the network are

$$(2 + j4)\mathbf{I}_1 - j2\mathbf{I}_2 = 24\,\underline{/30°}$$

$$-j2\mathbf{I}_1 + (2 + j6 - j2)\mathbf{I}_2 = 0$$

Solving the equations yields

$$\mathbf{I}_2 = 2.68\,\underline{/3.43°}\ \text{A}$$

Therefore,

$$\mathbf{V}_o = 2\mathbf{I}_2$$

$$= 5.36\,\underline{/3.43°}\ \text{V}$$

Let us now consider a more complicated example involving mutual inductance.

LEARNING Example 8.4

Consider the circuit in Fig. 8.8. We wish to write the mesh equations for this network.

SOLUTION Because of the multiple currents that are present in the coupled inductors, we must be very careful in writing the circuit equations.

The mesh equations for the phasor network are

$$\mathbf{I}_1 R_1 + j\omega L_1(\mathbf{I}_1 - \mathbf{I}_2) + j\omega M(\mathbf{I}_2 - \mathbf{I}_3) + \frac{1}{j\omega C_1}(\mathbf{I}_1 - \mathbf{I}_2) = \mathbf{V}$$

$$\frac{1}{j\omega C_1}(\mathbf{I}_2 - \mathbf{I}_1) + j\omega L_1(\mathbf{I}_2 - \mathbf{I}_1) + j\omega M(\mathbf{I}_3 - \mathbf{I}_2) + R_2 \mathbf{I}_2$$
$$+ j\omega L_2(\mathbf{I}_2 - \mathbf{I}_3) + j\omega M(\mathbf{I}_2 - \mathbf{I}_1) + R_3(\mathbf{I}_2 - \mathbf{I}_3) = 0$$

$$R_3(\mathbf{I}_3 - \mathbf{I}_2) + j\omega L_2(\mathbf{I}_3 - \mathbf{I}_2) + j\omega M(\mathbf{I}_2 - \mathbf{I}_1)$$
$$+ \frac{1}{j\omega C_2}\mathbf{I}_3 + R_4 \mathbf{I}_3 = 0$$

which can be rewritten in the form

(continued)

$$\left(R_1 + j\omega L_1 + \frac{1}{j\omega C_1}\right)\mathbf{I}_1 - \left(j\omega L_1 + \frac{1}{j\omega C_1} - j\omega M\right)\mathbf{I}_2$$
$$- j\omega M\mathbf{I}_3 = \mathbf{V}$$

$$-\left(j\omega L_1 + \frac{1}{j\omega C_1} - j\omega M\right)\mathbf{I}_1$$
$$+\left(\frac{1}{j\omega C_1} + j\omega L_1 + R_2 + j\omega L_2 + R_3 - j2\omega M\right)\mathbf{I}_2$$
$$-(j\omega L_2 + R_3 - j\omega M)\mathbf{I}_3 = 0$$

$$-j\omega M\mathbf{I}_1 - (R_3 + j\omega L_2 - j\omega M)\mathbf{I}_2$$
$$+\left(R_3 + j\omega L_2 + \frac{1}{j\omega C_2} + R_4\right)\mathbf{I}_3 = 0$$

Note the symmetrical form of these equations.

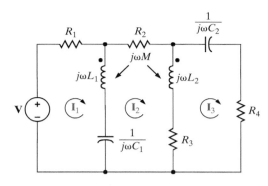

Figure 8.8 *Example of a magnetically coupled circuit.*

LEARNING EXTENSIONS

E8.2 Find the currents $\mathbf{I}_1$ and $\mathbf{I}_2$ and the output voltage $\mathbf{V}_o$ in the network in Fig. E8.2.

Figure E8.2

ANSWER
$\mathbf{I}_1 = +4.29 \underline{/137.2°}$ A,
$\mathbf{I}_2 = 0.96 \underline{/-16.26°}$ A,
$\mathbf{V}_o = 3.84 \underline{/-106.26°}$ V.

E8.3 Write the KVL equations in standard form for the network in Fig. E8.3.

Figure E8.3

ANSWER
$$(R_1 + j\omega L_1 + R_2)\mathbf{I}_1$$
$$- (R_2 + j\omega M)\mathbf{I}_2 = -\mathbf{V}_1,$$
$$-(R_2 + j\omega M)\mathbf{I}_1$$
$$+ (R_2 + j\omega L_2 + R_3)\mathbf{I}_2$$
$$= \mathbf{V}_1.$$

LEARNING Example 8.5

Given the network in Fig. 8.9 with the parameters $\mathbf{Z}_S = 3 + j1\ \Omega$, $j\omega L_1 = j2\ \Omega$, $j\omega L_2 = j2\ \Omega$, $j\omega M = j1\ \Omega$, and $\mathbf{Z}_L = 1 - j1\ \Omega$, determine the impedance seen by the source $\mathbf{V}_S$.

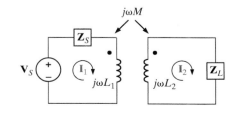

Figure 8.9
*Circuit employed in
Example 8.5.*

SOLUTION The mesh equations for the network are

$$\mathbf{V}_S = (\mathbf{Z}_S + j\omega L_1)\mathbf{I}_1 - j\omega M\mathbf{I}_2$$
$$0 = -j\omega M\mathbf{I}_1 + (j\omega L_2 + \mathbf{Z}_L)\mathbf{I}_2$$

If we now define $\mathbf{Z}_{11} = \mathbf{Z}_S + j\omega L_1$ and $\mathbf{Z}_{22} = j\omega L_2 + \mathbf{Z}_L$, then the second equation yields

$$\mathbf{I}_2 = \frac{j\omega M}{\mathbf{Z}_{22}}\mathbf{I}_1$$

If this secondary mesh equation is substituted into the primary mesh equation, we obtain

$$V_S = Z_{11}I_1 + \frac{\omega^2 M^2}{Z_{22}} I_1$$

and therefore

$$\frac{V_S}{I_1} = Z_{11} + \frac{\omega^2 M^2}{Z_{22}}$$

which is the impedance seen by V_S. Note that the mutual term is squared, and therefore the impedance is independent of the location of the dots.

Using the values of the circuit parameters, we find that

$$\frac{V_S}{I_1} = (3 + j1 + j2) + \frac{1}{j2 + 1 - j1}$$

$$= 3 + j3 + 0.5 - j0.5$$

$$= 3.5 + j2.5 \ \Omega$$

LEARNING EXTENSION

E8.4   Find the impedance seen by the source in the circuit in Fig. E8.4.

ANSWER
$Z_S = 2.25\ \underline{/20.9°}\ \Omega.$

Figure E8.4

8.2 Energy Analysis

We now perform an energy analysis on a pair of mutually coupled inductors, which will yield some interesting relationships for the circuit elements. Our analysis will involve the performance of an experiment on the network shown in Fig. 8.10. Before beginning the experiment, we set all voltages and currents in the circuit equal to zero. Once the circuit is quiescent, we begin by letting the current $i_1(t)$ increase from zero to some value I_1 with the right-side terminals open-circuited. Since the right-side terminals are open, $i_2(t) = 0$, and therefore the power entering these terminals is zero. The instantaneous power entering the left-side terminals is

$$p(t) = v_1(t)i_1(t) = \left[L_1 \frac{di_1(t)}{dt} \right] i_1(t)$$

The energy stored within the coupled circuit at t_1 when $i_1(t) = I_1$ is then

$$\int_0^{t_1} v_1(t)i_1(t)\ dt = \int_0^{I_1} L_1 i_1(t)\ di_1(t) = \tfrac{1}{2} L_1 I_1^2$$

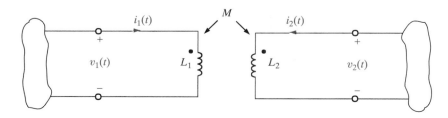

Figure 8.10
Magnetically coupled circuit.

Continuing our experiment, starting at time t_1, we let the current $i_2(t)$ increase from zero to some value I_2 at time t_2 while holding $i_1(t)$ constant at I_1. The energy delivered through the right-side terminals is

$$\int_{t_1}^{t_2} v_2(t)i_2(t)\, dt = \int_0^{I_2} L_2 i_2(t)\, di_2(t) = \tfrac{1}{2}L_2 I_2^2$$

However, during the interval t_1 to t_2 the voltage $v_1(t)$ is

$$v_1(t) = L_1 \frac{di_1(t)}{dt} + M\frac{di_2(t)}{dt}$$

Since $i_1(t)$ is a constant I_1, the energy delivered through the left-side terminals is

$$\int_{t_1}^{t_2} v_1(t)i_1(t)\, dt = \int_{t_1}^{t_2} M\frac{di_2(t)}{dt} I_1\, dt = MI_1 \int_0^{I_2} di_2(t)$$

$$= MI_1 I_2$$

Therefore, the total energy stored in the network for $t > t_2$ is

$$w = \tfrac{1}{2}L_1 I_1^2 + \tfrac{1}{2}L_2 I_2^2 + MI_1 I_2 \qquad \textbf{8.12}$$

We could, of course, repeat our entire experiment with either the dot on L_1 or L_2, but not both, reversed, and in this case the sign on the mutual inductance term would be negative, producing

$$w = \tfrac{1}{2}L_1 I_1^2 + \tfrac{1}{2}L_2 I_2^2 - MI_1 I_2$$

It is very important for the reader to realize that in our derivation of the preceding equation, by means of the experiment, the values I_1 and I_2 could have been any values at any *time*; therefore, the energy stored in the magnetically coupled inductors at any instant of time is given by the expression

$$w(t) = \tfrac{1}{2}L_1\big[i_1(t)\big]^2 + \tfrac{1}{2}L_2\big[i_2(t)\big]^2 \pm Mi_1(t)i_2(t) \qquad \textbf{8.13}$$

The two coupled inductors represent a passive network, and therefore, the energy stored within this network must be nonnegative for any values of the inductances and currents.

The equation for the instantaneous energy stored in the magnetic circuit can be written as

$$w(t) = \tfrac{1}{2}L_1 i_1^2 + \tfrac{1}{2}L_2 i_2^2 \pm Mi_1 i_2$$

Adding and subtracting the term $\tfrac{1}{2}(M^2/L_2)i_1^2$ and rearranging the equation yields

$$w(t) = \frac{1}{2}\left(L_1 - \frac{M^2}{L_2}\right)i_1^2 + \frac{1}{2}L_2\left(i_2 + \frac{M}{L_2}i_1\right)^2$$

From this expression we recognize that the instantaneous energy stored will be nonnegative if

$$M \le \sqrt{L_1 L_2} \qquad \textbf{8.14}$$

Note that this equation specifies an upper limit on the value of the mutual inductance. We define the coefficient of coupling between the two inductors L_1 and L_2 as

$$k = \frac{M}{\sqrt{L_1 L_2}} \qquad \textbf{8.15}$$

and we note from Eq. (8.14) that its range of values is

$$0 \le k \le 1 \qquad \textbf{8.16}$$

This coefficient is an indication of how much flux in one coil is linked with the other coil; that is, if all the flux in one coil reaches the other coil, then we have 100% coupling and $k = 1$. For large values of k (i.e., $k > 0.5$), the inductors are said to be tightly coupled, and for small values of k (i.e., $k \leq 0.5$), the coils are said to be loosely coupled. If there is no coupling, $k = 0$. The previous equations indicate that the value for the mutual inductance is confined to the range

$$0 \leq M \leq \sqrt{L_1 L_2} \qquad \qquad \textbf{8.17}$$

and that the upper limit is the geometric mean of the inductances L_1 and L_2.

LEARNING Example 8.6

The coupled circuit in Fig. 8.11a has a coefficient of coupling of 1 (i.e., $k = 1$). We wish to determine the energy stored in the mutually coupled inductors at time $t = 5$ ms. $L_1 = 2.653$ mH and $L_2 = 10.61$ mH.

SOLUTION From the data the mutual inductance is

$$M = \sqrt{L_1 L_2} = 5.31 \text{ mH}$$

The frequency domain equivalent circuit is shown in Fig. 8.11b, where the impedance values for X_{L_1}, X_{L_2}, and X_M are 1, 4, and 2, respectively. The mesh equations for the network are then

$$(2 + j1)\mathbf{I}_1 - j2\mathbf{I}_2 = 24 \underline{/0°}$$

$$-j2\mathbf{I}_1 + (4 + j4)\mathbf{I}_2 = 0$$

Solving these equations for the two mesh currents yields

$$\mathbf{I}_1 = 9.41 \underline{/-11.31°} \text{ A} \quad \text{and} \quad \mathbf{I}_2 = 3.33 \underline{/+33.69°} \text{ A}$$

and therefore,

$$i_1(t) = 9.41 \cos(377t - 11.31°) \text{ A}$$

$$i_2(t) = 3.33 \cos(377t + 33.69°) \text{ A}$$

At $t = 5$ ms, $377t = 1.885$ rad or $108°$, and therefore,

$$i_1(t = 5 \text{ ms}) = 9.41 \cos(108° - 11.31°) = -1.10 \text{ A}$$

$$i_2(t = 5 \text{ ms}) = 3.33 \cos(108° + 33.69°) = -2.61 \text{ A}$$

Therefore, the energy stored in the coupled inductors at $t = 5$ ms is

$$w(t)|_{t=0.005 \, s} = \tfrac{1}{2}(2.653)(10^{-3})(-1.10)^2 + \tfrac{1}{2}(10.61)(10^{-3})(-2.61)^2$$

$$-(5.31)(10^{-3})(-1.10)(-2.61)$$

$$= (1.61)(10^{-3}) + (36.14)(10^{-3}) - (15.25)(10^{-3})$$

$$= 22.5 \text{ mJ}$$

(a)   (b)

Figure 8.11 *Example of a magnetically coupled circuit drawn in the time and frequency domains.*

LEARNING EXTENSION

E8.5 The network in Fig. E8.5 operates at 60 Hz. Compute the energy stored in the mutually coupled inductors at time $t = 10$ ms.

ANSWER
$w(10 \text{ ms}) = 39$ mJ.

Figure E8.5

8.3 The Ideal Transformer

Consider the situation illustrated in Fig. 8.12, showing two coils of wire wound around a single closed magnetic core. Assume a core flux ϕ, which links all the turns of both coils. In the ideal case we also neglect wire resistance. Let us now examine the coupling equations under the condition that the same flux goes through each winding and so,

$$v_1(t) = N_1 \frac{d\phi}{dt}$$

Figure 8.12
Transformer employing a magnetic core.

and

$$v_2(t) = N_2 \frac{d\phi}{dt}$$

and therefore,

$$\frac{v_1}{v_2} = \frac{N_1}{N_2} \frac{\dfrac{d\phi}{dt}}{\dfrac{d\phi}{dt}} = \frac{N_1}{N_2} \qquad \textbf{8.18}$$

Ampère's law requires that

$$\oint H \cdot dl = i_{\text{enclosed}} = N_1 i_1 + N_2 i_2 \qquad \textbf{8.19}$$

where H is the magnetic field intensity and the integral is over the closed path traveled by the flux around the transformer core. If $H = 0$, which is the case for an ideal magnetic core with infinite permeability, then

$$N_1 i_1 + N_2 i_2 = 0 \qquad \textbf{8.20}$$

or

$$\frac{i_1}{i_2} = -\frac{N_2}{N_1} \qquad \textbf{8.21}$$

Note that if we divide Eq. (8.20) by N_1 and multiply it by v_1, we obtain

$$v_1 i_1 + \frac{N_2}{N_1} v_1 i_2 = 0$$

However, since $v_1 = (N_1/N_2)v_2$,

$$v_1 i_1 + v_2 i_2 = 0$$

and hence the total power into the device is zero, which means that an ideal transformer is lossless.

The symbol we employ for the ideal transformer is shown in Fig. 8.13a, and the corresponding equations are

$$\frac{v_1}{v_2} = \frac{N_1}{N_2}$$

8.22

$$N_1 i_1 + N_2 i_2 = 0$$

The normal power flow through a transformer occurs from an input current (i_1) on the primary to an output current (i_2) on the secondary. This situation is shown in Fig. 8.13b and the corresponding equations are

$$\frac{v_1}{v_2} = \frac{N_1}{N_2}$$

8.23

$$N_1 i_1 = N_2 i_2$$

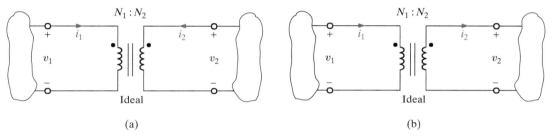

Figure 8.13 *Symbol for an ideal transformer: (a) primary and secondary currents into the dots; (b) primary current into, and secondary current out of, the dots.*

Note that although the voltage, current, and impedance levels change through a transformer, the power levels do not. The vertical lines between the coils, shown in the figures, represent the magnetic core. Although practical transformers do not use dots, per se, they use markings specified by the National Electrical Manufacturers Association (NEMA) that are conceptually equivalent to the dots.

Thus, our model for the ideal transformer is specified by the circuit in Fig. 8.13a and the corresponding Eqs. (8.22), or alternatively by the circuit in Fig. 8.13b, together with Eqs. (8.23). Therefore, it is important to note carefully that our model specifies the equations as well as the relationship among the voltages, currents, and the position of the dots. In other words, the equations are valid only for the corresponding circuit diagram. Thus, in a direct analogy to our discussion of the mutual inductance equations and their corresponding circuit, if we change the direction of the current or voltage or the position of the dots, we must make a corresponding change in the equations. The following material will clarify this critical issue.

Consider now the circuit shown in Fig. 8.14. If we compare this circuit to that shown in Fig. 8.13b, we find that the direction of both the currents and voltages are the same. Hence the equations for the network are

$$\frac{\mathbf{V}_1}{\mathbf{V}_2} = \frac{N_1}{N_2}$$

Figure 8.14
Ideal transformer circuit used to illustrate input impedance.

and

$$\frac{\mathbf{I}_1}{\mathbf{I}_2} = \frac{N_2}{N_1}$$

These equations can be written as

$$\mathbf{V}_1 = \frac{N_1}{N_2}\mathbf{V}_2$$

$$\mathbf{I}_1 = \frac{N_2}{N_1}\mathbf{I}_2$$

8.24

Also note that

$$\mathbf{Z}_L = \frac{\mathbf{V}_2}{\mathbf{I}_2}$$

and therefore the input impedance is

$$\mathbf{Z}_1 = \frac{\mathbf{V}_1}{\mathbf{I}_1} = \left(\frac{N_1}{N_2}\right)^2 \mathbf{Z}_L$$

8.25

where $\mathbf{Z}_L$ is reflected into the primary side by the turns ratio.
If we now define the turns ratio as

$$n = \frac{N_2}{N_1}$$

8.26

then the defining equations for the *ideal transformer* in this configurarion are

$$\mathbf{V}_1 = \frac{\mathbf{V}_2}{n}$$

$$\mathbf{I}_1 = n\mathbf{I}_2$$

$$\mathbf{Z}_1 = \frac{\mathbf{Z}_L}{n^2}$$

8.27

Care must be exercised in using these equations because the signs on the voltages and currents are dependent on the assigned references and their relationship to the dots.

LEARNING Example 8.7

Given the circuit shown in Fig. 8.15, we wish to determine all indicated voltages and currents.

Figure 8.15   *Ideal transformer circuit.*

SOLUTION   Because of the relationships between the dots and the currents and voltages, the transformer equations are

$$\mathbf{V}_1 = -\frac{\mathbf{V}_2}{n} \quad \text{and} \quad \mathbf{I}_1 = -n\mathbf{I}_2$$

where $n = \frac{1}{4}$. The reflected impedance at the input to the transformer is

$$\mathbf{Z}_1 = 4^2\mathbf{Z}_L = 16(2 + j1) = 32 + j16 \ \Omega$$

Therefore, the current in the source is

$$\mathbf{I}_1 = \frac{120\,/0^\circ}{18 - j4 + 32 + j16} = 2.33\,/{-13.5^\circ}\ \text{A}$$

The voltage across the input to the transformer is then

$$\mathbf{V}_1 = \mathbf{I}_1\mathbf{Z}_1$$
$$= (2.33\,/{-13.5^\circ})(32 + j16)$$
$$= 83.49\,/13.07^\circ\ \text{V}$$

Hence, $\mathbf{V}_2$ is

$$\mathbf{V}_2 = -n\mathbf{V}_1$$
$$= -\tfrac{1}{4}(83.49\,/13.07^\circ)$$
$$= 20.87\,/193.07^\circ\ \text{V}$$

The current $\mathbf{I}_2$ is

$$\mathbf{I}_2 = -\frac{\mathbf{I}_1}{n}$$
$$= -4(2.33\,/{-13.5^\circ})$$
$$= 9.33\,/166.50^\circ\ \text{A}$$

LEARNING EXTENSIONS

E8.6   Compute the current $\mathbf{I}_1$ in the network in Fig. E8.6.

Figure E8.6

ANSWER
$\mathbf{I}_1 = 3.07\,/39.81^\circ$ A.

E8.7   Find $\mathbf{V}_o$ in the network in Fig. E8.6.

ANSWER
$\mathbf{V}_o = 3.07\,/39.81^\circ$ V.

Another technique for simplifying the analysis of circuits containing an ideal transformer involves the use of either Thévenin's or Norton's theorem to obtain an equivalent circuit that replaces the transformer and either the primary or secondary circuit. This technique usually requires more effort, however, than the approach presented thus far. Let us demonstrate this

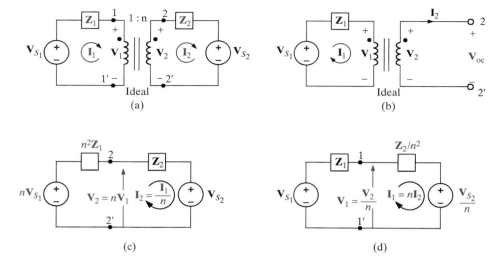

Figure 8.16

Circuit containing an ideal transformer and some of its equivalent networks.

approach by employing Thévenin's theorem to derive an equivalent circuit for the transformer and primary circuit of the network shown in Fig. 8.16a. The equations for the transformer in view of the direction of the currents and voltages and the position of the dots are

$$\mathbf{I}_1 = n\mathbf{I}_2$$

$$\mathbf{V}_1 = \frac{\mathbf{V}_2}{n}$$

Forming a Thévenin equivalent at the secondary terminals $2 = 2'$, as shown in Fig. 8.16b, we note that $\mathbf{I}_2 = 0$ and therefore $\mathbf{I}_1 = 0$. Hence

$$\mathbf{V}_{oc} = \mathbf{V}_2 = n\mathbf{V}_1 = n\mathbf{V}_{S_1}$$

The Thévenin equivalent impedance obtained by looking into the open-circuit terminals with $\mathbf{V}_{S_1}$ replaced by a short circuit is $\mathbf{Z}_1$, which when reflected into the secondary by the turns ratio is

$$\mathbf{Z}_{Th} = n^2\mathbf{Z}_1$$

Therefore, one of the resulting equivalent circuits for the network in Fig. 8.16a is as shown in Fig. 8.16c. In a similar manner we can show that replacing the transformer and its secondary circuit by an equivalent circuit results in the network shown in Fig. 8.16d.

It can be shown in general that when developing an equivalent circuit for the transformer and its primary circuit, each primary voltage is multiplied by n, each primary current is divided by n, and each primary impedance is multiplied by n^2. Similarly, when developing an equivalent circuit for the transformer and its secondary circuit, each secondary voltage is divided by n, each secondary current is multiplied by n, and each secondary impedance is divided by n^2. Powers are the same, whether calculated on the primary or secondary side.

Recall from our previous analysis that if either dot on the transformer is reversed, then n is replaced by $-n$ in the equivalent circuits. In addition, note that the development of these equivalent circuits is predicated on the assumption that removing the transformer will divide the network into two parts; that is, there are no connections between the primary and secondary other than through the transformer. If any external connections exist, the equivalent circuit technique cannot in general be used. Finally, we point out that if the primary or secondary circuits are more complicated than those shown in Fig. 8.16a, Thévenin's theorem may be applied to reduce the network to that shown in Fig. 8.16a. Also, we can simply reflect the complicated circuit component by component from one side of the transformer to the other.

LEARNING Example 8.8

Given the circuit in Fig. 8.17a, we wish to draw the two networks obtained by replacing the transformer and the primary, and the transformer and the secondary, with equivalent circuits.

SOLUTION   Due to the relationship between the assigned currents and voltages and the location of the dots, the network

containing an equivalent circuit for the primary and the network containing an equivalent circuit for the secondary are shown in Figs. 8.17b and c, respectively. The reader should note carefully the polarity of the voltage sources in the equivalent networks.

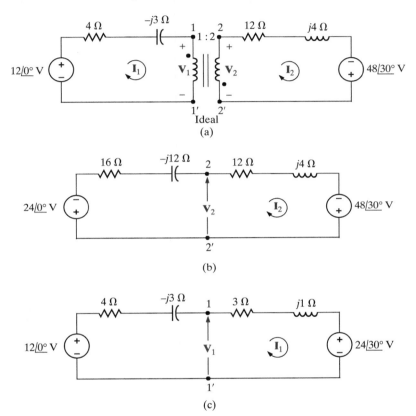

Figure 8.17
Example circuit and two equivalent circuits.

LEARNING Example 8.9

Let us determine the output voltage V_o in the circuit in Fig. 8.18a.

SOLUTION   We begin our attack by forming a Thévenin equivalent for the primary circuit. From Fig. 8.18b we can show that the open-circuit voltage is

$$V_{oc} = \frac{24\,\underline{/0^\circ}}{4 - j4}(-j4) - 4\,\underline{/-90^\circ}$$

$$= 12 - j8 = 14.42\,\underline{/-33.69^\circ}\text{ V}$$

The Thévenin equivalent impedance looking into the open-circuit terminals with the voltage sources replaced by short circuits is

$$Z_{Th} = \frac{(4)(-j4)}{4 - j4} + 2$$

$$= 4 - j2\ \Omega$$

The circuit in Fig. 8.18a thus reduces to that shown in Fig. 8.18c. Forming an equivalent circuit for the transformer and primary results in the network shown in Fig. 8.18d. Therefore, the voltage V_o is

$$V_o = \frac{-28.84\,\underline{/-33.69^\circ}}{20 - j5}\quad(2)$$

$$= 2.80\,\underline{/160.35^\circ}\text{ V}$$

(continued)

(a)

(b)   (c)

(d)

Figure 8.18   *Example network and other circuits used to derive an equivalent network.*

● ● ●

LEARNING EXTENSIONS

E8.8   Given the network in Fig. E8.8, form an equivalent circuit for the transformer and secondary, and use the resultant network to compute $\mathbf{I}_1$.

ANSWER
$\mathbf{I}_1 = 13.12 \underline{/38.66°}$ A.

Figure E8.8

E8.9   Given the network in Fig. E8.9, form an equivalent circuit for the transformer and primary, and use the resultant network to find $\mathbf{V}_o$.

ANSWER
$\mathbf{V}_o = 3.12 \underline{/38.66°}$ V.

Figure E8.9

LEARNING Example 8.10

Determine $\mathbf{I}_1$, $\mathbf{I}_2$, $\mathbf{V}_1$, and $\mathbf{V}_2$ in the network in Fig. 8.19.

Figure 8.19

SOLUTION   The nodal equations at nodes 1 and 2 are

$$\frac{10 - \mathbf{V}_1}{2} = \frac{\mathbf{V}_1 - \mathbf{V}_2}{2} + \mathbf{I}_1$$

$$\mathbf{I}_2 + \frac{\mathbf{V}_1 - \mathbf{V}_2}{2} = \frac{\mathbf{V}_2}{2j}$$

The transformer relationships are $\mathbf{V}_2 = 2\mathbf{V}_1$ and $\mathbf{I}_1 = 2\mathbf{I}_2$. The first nodal equation yields $\mathbf{I}_1 = 5$ A and therefore $\mathbf{I}_2 = 2.5$ A. The second nodal equation, together with the constraint equations specified by the transformer, yields $\mathbf{V}_1 = \sqrt{5} \underline{/63°}$ V and $\mathbf{V}_2 = 2\sqrt{5} \underline{/63°}$ V.

LEARNING EXTENSION

E8.10   Determine $\mathbf{I}_1$, $\mathbf{I}_2$, $\mathbf{V}_1$, and $\mathbf{V}_2$ in the network in Fig. E8.10.

Figure E8.10

ANSWER

$\mathbf{I}_1 = 4.48 \underline{/-26.57°}$ A,
$\mathbf{I}_2 = 2.24 \underline{/153.43°}$ A

$\mathbf{V}_1 = 2.24 \underline{/63.43°}$ V,
$\mathbf{V}_2 = 4.48 \underline{/-116.57°}$ V.

8.4 Safety Considerations

Transistors are used extensively in modern electronic equipment to provide a low-voltage power supply. As examples, a common voltage level in computer systems is 5 V dc, portable radios use 9 V dc, and military and airplane equipment operates at 28 V dc. When transformers are used to connect these low-voltage transistor circuits to the power line, there is generally less danger of shock within the system because the transformer provides electrical isolation from the line voltage. However, from a safety standpoint, a transformer, although helpful in many situations, is not an absolute solution. We must always be vigilant when working with any electrical equipment to minimize the dangers of electrical shock.

In power electronics equipment or power systems, the danger is severe. The problem in these cases is that of high voltage from a low-impedance source, and we must constantly remember that the line voltage in our homes can be lethal.

Consider now the following example, which illustrates a hidden danger that could surprise even the experienced professional, with devastating consequences.

LEARNING Example 8.11

Two adjacent homes, *A* and *B*, are fed from different transformers, as shown in Fig. 8.20a. A surge on the line feeding house *B* has caused the circuit breaker *X-Y* to open. House *B* is now left without power. In an attempt to help his neighbor, the resident of house *A* volunteers to connect a long extension cord between a wall plug in house *A* and a wall plug in house *B*, as shown in Fig. 8.20b. Later, the line technician from the utility company comes to reconnect the circuit breaker. Is the line technician in any danger in this situation?

SOLUTION Unaware of the extension cord connection, the line technician believes that there is no voltage between points *X* and *Z*. However, because of the electrical connection between the two homes, 7200 V exist between the two points, and the line technician could be seriously injured or even killed if he comes in contact with this high voltage.

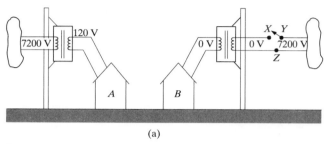

(a)   (b)

Figure 8.20 *Diagrams used in Example 8.11.*

Learning by Application

The following examples demonstrate a couple of applications for transformers. We first revisit the problem of transmitting power over long distances and illustrate the important role the transformer plays in that application. We then examine the use of a transformer in distributing power to our homes.

LEARNING Example 8.12

Consider the problem of transporting 24 MW over a distance of 100 miles (160.9 km) using a two-conductor line. Determine the requisite conductor radius to achieve a transmission efficiency of 95%, considering only the line resistance, if the line operates at (a) 240 V or (b) 240 kV. Assume that conductor resistivity is $\rho = 8 \times 10^{-8}$ Ω-m.

SOLUTION
(a) At 240 V:

$$I = \frac{P}{V} = \frac{24\ M}{240} = 100\ kA$$

If
$$\eta = 95\%,$$
$$P_{loss} = 0.05(24\ M) = 1.2\ MW = I^2R$$

Therefore,
$$R = \frac{P_{loss}}{I^2} = \frac{1.2\ M}{(100k)^2} = 1.2 \times 10^{-4}Ω$$

Since
$$R = \frac{\rho l}{A} = \frac{(8 \times 10^{-8})(2 \times 160.9 \times 10^3)}{A}$$
$$A = \frac{0.25744}{1.2 \times 10^{-4}} = 214.5\ m^2 = \pi r^2$$

Therefore,
$$r = 8.624\ m$$

(a huge conductor and totally impractical!)

(b) At 240 kV:
$$I = 100\ A$$

and
$$R = \frac{1.2 \times 10^6}{(100)^2} = 120\ Ω$$

$$A = \frac{0.25744}{120} = 2.145 \times 10^{-4}\ m$$

and
$$r = 0.8264\ cm$$

(which is a very practical value!)

The point of this example is that practical transmission of bulk electrical energy requires operation at high voltage. What is needed is an economical device that can efficiently convert one voltage level to another. Such a device is, as we have shown, the power transformer.

LEARNING Example 8.13

The local transformer in Fig. 8.21 provides the last voltage step-down in a power distribution system. A common sight on utility poles in residential areas, it is a single-phase transformer that typically has a 13.8-kV line to neutral on its primary coil, and a center tap secondary coil provides both 120 V and 240 V to service several residences. Let us find the turns ratio necessary to produce the 240-V secondary voltage. Assuming that the transformer provides 200-A service to each of 10 houses, let us determine the minimum power rating for the transformer and the maximum current in the primary.

SOLUTION The turns ratio is given by

$$n = \frac{V_2}{V_1} = \frac{240}{13,800} = \frac{1}{57.5}$$

If I_H is the maximum current per household, then the maximum primary current is

$$I_1 = nI_2 = n(10I_H) = 34.78 \text{ A}$$

The maximum power delivered to the primary is then

$$S_1 = V_1I_1 = (13,800)(34.78) = 480 \text{ kVA}.$$

Therefore, the transformer must have a power rating of at least 480 kVA.

Figure 8.21 *Local transformer subcircuit with center tap.*

Learning by Design

The following simple example illustrates a technique for employing a transformer in a configuration that will extend the life of a set of Christmas tree lights.

LEARNING Example 8.14

The bulbs in a set of Christmas tree lights normally operate at 120 V rms. However, they last much longer if they are instead connected to 108 V rms. Using a 120-V–12-V transformer, let us design an autotransformer that will provide 108 V rms to the bulbs.

SOLUTION The two-winding transformers we have presented thus far provide electrical isolation between primary and secondary windings, as shown in Fig. 8.22a. It is possible, however, to interconnect primary and secondary windings serially,

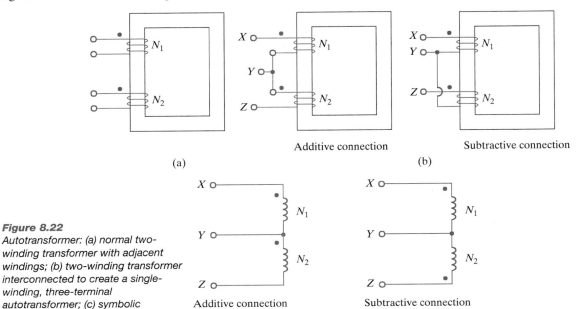

Additive connection   Subtractive connection

(a)   (b)

Figure 8.22
Autotransformer: (a) normal two-winding transformer with adjacent windings; (b) two-winding transformer interconnected to create a single-winding, three-terminal autotransformer; (c) symbolic representation of (b).

Additive connection   Subtractive connection

(c)

(continued)

creating a three-terminal device, known as an autotransformer, as shown in Fig. 8.22b and represented in Fig. 8.22c. As we shall see, this arrangement offers certain practical advantages over the isolated case. Note that the three-terminal arrangement is essentially one continuous winding with an internal tap.

To reduce the voltage from 120 V to 108 V, the two coils must be connected such that their voltages are in opposition to each other, corresponding to a subtractive connection (in Fig. 8.22b), as shown in Fig. 8.23. In this arrangement, the voltage across both coils is

$$V_o = V_1 - V_2 = 120 - 12 = 108 \text{ V rms}$$

and the lights are simply connected across both coils.

Figure 8.23 *Autotransformer for low-voltage Christmas tree lights.*

LEARNING Check

Summary

▶ **Mutual inductance** Mutual inductance occurs when inductors are placed in close proximity to one another and share a common magnetic flux.

▶ **The dot convention for mutual inductance** The dot convention governs the sign of the induced voltage in one coil based on the current direction in another.

▶ **The relationship between the mutual inductance and self-inductance of two coils** An energy analysis indicates that $M = k\sqrt{L_1 L_2}$, where k, the coefficient of coupling, has a value between 0 and 1.

▶ **The ideal transformer** An ideal transformer has infinite core permeability and winding conductance. The voltage and current can be transformed between the primary

and secondary ends based on the ratio of the number of winding turns between the primary and secondary.

▶ **The dot convention for an ideal transformer** The dot convention for ideal transformers, like that for mutual inductance, specifies the manner in which a current in one winding induces a voltage in another winding.

▶ **Equivalent circuits involving ideal transformers** Based on the location of the circuits' unknowns, either the primary or secondary can be reflected to the other side of the transformer to form a single circuit containing the desired unknown. The reflected voltages, currents, and impedances are a function of the dot convention and turns ratio.

Problems

SECTION 8.1

8.1 Given the network in Fig. P8.1,
  (a) find the equations for $v_a(t)$ and $v_b(t)$.
  (b) find the equations for $v_C(t)$ and $v_d(t)$.

8.2 Given the network in Fig. P8.2,
  (a) write the equations for $v_a(t)$ and $v_b(t)$.
  (b) write the equations for $v_C(t)$ and $v_d(t)$.

Figure P8.1

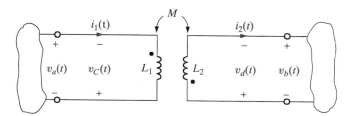

Figure P8.2

8.3 Given the network in Fig. P8.3,
  (a) write the equations for $v_a(t)$ and $v_b(t)$.
  (b) write the equations for $v_C(t)$ and $v_d(t)$.

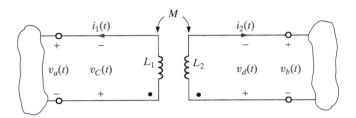

Figure P8.3

8.4 Given the network in Fig. P8.4,
  (a) write the equations for $v_a(t)$ and $v_b(t)$.
  (b) write the equations for $v_C(t)$ and $v_d(t)$.

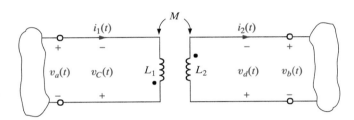

Figure P8.4

8.5 Find the voltage gain $\mathbf{V}_o/\mathbf{V}_S$ of the network shown in Fig. P8.5.

Figure P8.5

8.6 Find $\mathbf{V}_o$ in the network in Fig. P8.6.

Figure P8.6

8.7 Find $\mathbf{V}_o$ in the network in Fig. P8.7.

Figure P8.7

8.8 Given the network in Fig. P8.8, find $\mathbf{V}_o$.

Figure P8.8

8.9 Find $\mathbf{V}_o$ in the circuit in Fig. P8.9.

Figure P8.9

8.10 Find $\mathbf{V}_o$ in the network in Fig. P8.10.

Figure P8.10

8.11 Find $\mathbf{V}_o$ in the network in Fig. P8.11.

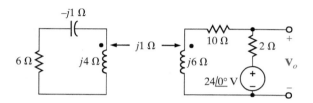

Figure P8.11

8.12 Find $\mathbf{V}_o$ in the network in Fig. P8.12.

Figure P8.12

8.13 Find $\mathbf{V}_o$ in the circuit in Fig. P8.13.

Figure P8.13

8.14 Find $\mathbf{I}_o$ in the circuit in Fig. P8.14.

Figure P8.14

8.15 Write the mesh equations for the network in Fig. P8.15.

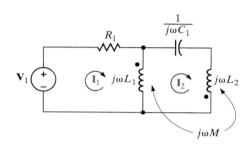

Figure P8.15

8.16 Write the mesh equations for the network shown in Fig. P8.16.

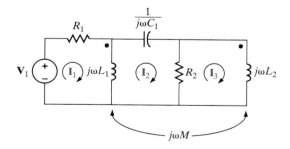

Figure P8.16

8.17 Write the mesh equations for the network shown in Fig. P8.17.

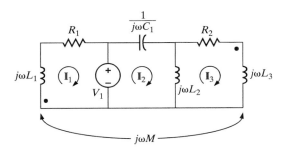

Figure P8.17

8.18 Write the mesh equations for the network shown in Fig. P8.18.

Figure P8.18

8.19 Write the mesh equations for the network in Fig. P8.19.

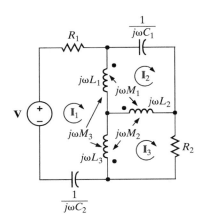

Figure P8.19

8.20 Find $\mathbf{V}_o$ in the network in Fig. P8.20.

Figure P8.20

8.21 Find $\mathbf{V}_o$ in the network in Fig. P8.21.

Figure P8.21

8.22 Find $\mathbf{V}_o$ in the network in Fig. P8.22.

Figure P8.22

8.23 Find $\mathbf{V}_o$ in the network in Fig. P8.23.

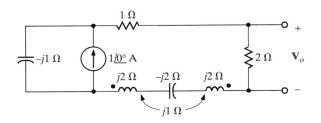

Figure P8.23

8.24 Find $\mathbf{V}_o$ in the network in Fig. P8.24.

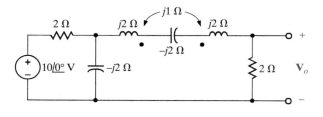

Figure P8.24

8.25 Find $\mathbf{V}_o$ in the network in Fig. P8.25.

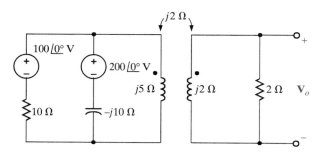

Figure P8.25

8.26 Find $\mathbf{V}_o$ in the network in Fig. P8.26.

Figure P8.26

8.27 Find $\mathbf{V}_o$ in the network in Fig. P8.27.

Figure P8.27

8.28 Find $\mathbf{V}_o$ in the network in Fig. P8.28.

Figure P8.28

8.29 Find $\mathbf{V}_o$ in the network in Fig. P8.29.

Figure P8.29

8.30 Find $\mathbf{I}_o$ in the circuit in Fig. P8.30.

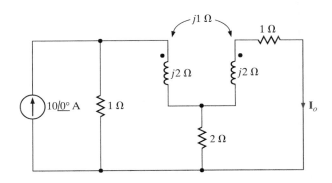

Figure P8.30

8.31 Find $\mathbf{V}_o$ in the circuit in Fig. P8.31.

Figure P8.31

8.32 Find $\mathbf{I}_o$ in the circuit in Fig. P8.32.

Figure P8.32

8.33 Find $\mathbf{V}_o$ in the network in Fig. P8.33.

Figure P8.33

8.34 Find $\mathbf{V}_o$ in the network in Fig. P8.34.

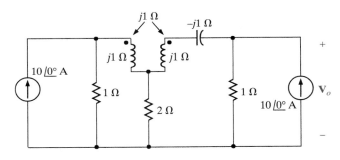

Figure P8.34

8.35 Determine the impedance seen by the source in the network shown in Fig. P8.35.

Figure P8.35

8.36 Determine the impedance seen by the source in the network shown in Fig. P8.36.

Figure P8.36

8.37 Determine the input impedance $\mathbf{Z}_{in}$ of the circuit in Fig. P8.37.

Figure P8.37

8.38 Determine the input impedance $\mathbf{Z}_{in}$ in the network in Fig. P8.38.

Figure P8.38

8.39 Given the network shown in Fig. P8.39, determine the value of the capacitor C that will cause the impedance seen by the $24\,\underline{/0°}$ V voltage source to be purely resistive. $f = 60$ Hz.

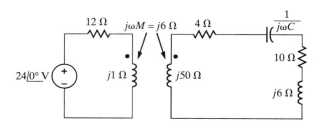

Figure P8.39

8.40 Analyze the network in Fig. P8.40 and determine whether a value of X_C can be found such that the output voltage is equal to twice the input voltage.

Figure P8.40

SECTION 8.2

8.41 Two coils in a network are positioned such that there is 100% coupling between them. If the inductance of one coil is 10 mH and the mutual inductance is 6 mH, compute the inductance of the other coil.

8.42 The currents in the magnetically coupled inductors shown in Fig. P8.42 are known to be $i_1(t) = 8\cos(377t - 20°)$ mA and $i_2(t) = 4\cos(377t - 50°)$ mA. The inductor values are $L_1 = 2$ H, $L_2 = 1$ H, and $k = 0.6$. Determine $v_1(t)$ and $v_2(t)$.

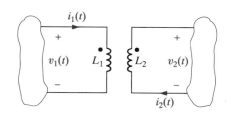

Figure P8.42

8.43 Determine the energy stored in the coupled inductors in Problem 8.42 at $t = 1$ ms.

8.44 If $L_1 = L_2 = 4$ H+ and $k = 0.8$, find $i_1(t)$ and $i_2(t)$ in the circuit in Fig. P8.44.

Figure P8.44

SECTION 8.3

8.46 Find all currents and voltages in the network in Fig. P8.46.

Figure P8.46

8.47 Determine I_1, I_2, V_1, and V_2 in the network in Fig. P8.47.

Figure P8.47

8.45 Determine the energy stored in the coupled inductors in the network in Fig. P8.44 at $t = 2$ ms.

8.48 Determine V_o in the circuit in Fig. P8.48.

Figure P8.48

8.49 Determine I_1, I_2, V_1, and V_2 in the network in Fig. P8.49.

Figure P8.49

8.50 Determine I_1, I_2, V_1, and V_2 in the network in Fig. P8.50.

Figure P8.50

8.51 Determine I_1, I_2, V_1, and V_2 in the network in Fig. P8.51.

Figure P8.51

8.52 Determine the input impedance seen by the source in the circuit in Fig. P8.52.

Figure P8.52

8.53 Determine the input impedance seen by the source in the circuit in Fig. P8.53.

Figure P8.53

8.54 Determine the input impedance seen by the source in the network shown in Fig. P8.54.

Figure P8.54

8.55 Determine the input impedance seen by the source in the network shown in Fig. P8.55.

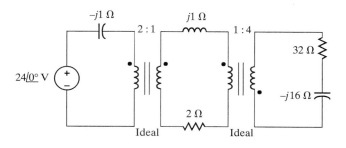

Figure P8.55

8.56 The output stage of an amplifier in an old radio is to be matched to the impedance of a speaker, as shown in Fig. P8.56. If the impedance of the speaker is 8 Ω and the amplifier requires a load impedance of 3.2 kΩ, determine the turns ratio of the ideal transformer.

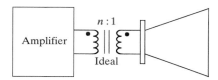

Figure P8.56

8.57 Given that $V_o = 48\underline{/30°}$ V in the circuit shown in Fig. P8.57 determine V_S.

Figure P8.57

8.58 If the voltage source V_S in the circuit of Problem 8.57 is $50\underline{/0°}$ V, determine V_o.

8.59 Determine $\mathbf{V}_S$ in the circuit in Fig. P8.59.

Figure P8.59

8.60 Determine $\mathbf{I}_S$ in the circuit in Fig. P8.60.

Figure P8.60

8.61 Find the current $\mathbf{I}$ in the network in Fig. P8.61.

Figure P8.61

8.62 Find the voltage $\mathbf{V}_o$ in the network in Fig. P8.62.

Figure P8.62

8.63 Find $\mathbf{V}_o$ in the network in Fig. P8.63.

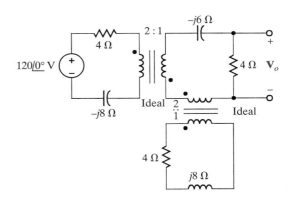

Figure P8.63

8.64 Find $\mathbf{V}_o$ in the circuit in Fig. P8.64.

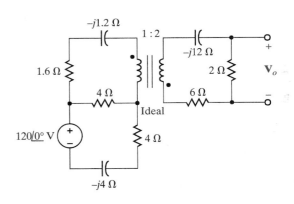

Figure P8.64

8.65 Find $\mathbf{V}_o$ in the circuit in Fig. P8.65.

Figure P8.65

8.66 In the circuit in Fig. P8.66, if $\mathbf{I}_x = 6\underline{/45°}$ A, find $\mathbf{V}_o$.

8.67 In the network in Fig. P8.67, if $\mathbf{I}_1 = 4\underline{/0°}$ A, find $\mathbf{V}_S$.

Figure P8.66

Figure P8.67

Typical Problems Found on the FE Exam

8FE-1 In the network in Fig. 8PFE-1, find the impedance seen by the source.

8FE-3 In the circuit in Fig. 8FE-3, select the turns ratio of the ideal transformer that will match the output of the transistor amplifier to the speaker represented by the 16-Ω load.

Figure 8PFE-1

Figure 8PFE-3

8FE-2 In the circuit in Fig. 8PFE-2, select the value of the transformer's turns ratio $n = N_2/N_1$ to achieve impedance matching for maximum power transfer. Using this value of n, calculate the power absorbed by the 3-Ω resistor.

Figure 8PFE-2

Steady-State Power Analysis

Do you understand the electrical wiring circuit in your home that is used to power all the appliances and lights? Do you know the hazards associated with working with electrical power circuits? These are just two of the topics that will be examined in this chapter.

In the preceding chapters we have been concerned primarily with determining the voltage or current at some point in a network. Of equal importance to us in many situations is the power that is supplied or absorbed by some element. Typically, electrical and electronic devices have peak power or maximum instantaneous power ratings that cannot be exceeded without damaging the devices.

In electrical and electronic systems, power comes in all sizes. The power absorbed by some device on an integrated circuit chip may be in picowatts, whereas the power supplied by a large generating station may be in gigawatts. Note that the range between these two examples is phenomenally large $\left(10^{21}\right)$.

In our previous work we defined instantaneous power to be the product of voltage and current. Average power, obtained by averaging the instantaneous power, is the average rate at which energy is absorbed or supplied. In the dc case, where both current and voltage are constant, the instantaneous power is equal to the average power. However, as we will demonstrate, this is not the case when the currents and voltages are sinusoidal functions of time.

In this chapter we explore the many ramifications of power in ac circuits. We examine instantaneous power, average power, maximum power transfer, average power for periodic nonsinusoidal waveforms, the power factor, complex power, and power measurement.

Some very practical safety considerations will be introduced and discussed through a number of examples.

Finally, a wide variety of application-oriented examples is presented. The origin of many of these examples is the common household wiring circuit.

9.1 Instantaneous Power

Figure 9.1
Simple ac network.

By employing the sign convention adopted in the earlier chapters, we can compute the instantaneous power supplied or absorbed by any device as the product of the instantaneous voltage across the device and the instantaneous current through it.

Consider the circuit shown in Fig. 9.1. In general, the steady-state voltage and current for the network can be written as

$$v(t) = V_M \cos(\omega t + \theta_v) \qquad\qquad 9.1$$

$$i(t) = I_M \cos(\omega t + \theta_i) \qquad\qquad 9.2$$

The instantaneous power is then

$$
\begin{aligned}
p(t) &= v(t)i(t) \\
&= V_M I_M \cos(\omega t + \theta_v)\cos(\omega t + \theta_i) \qquad\qquad 9.3
\end{aligned}
$$

Employing the following trigonometric identity,

$$\cos\phi_1 \cos\phi_2 = \tfrac{1}{2}\big[\cos(\phi_1 - \phi_2) + \cos(\phi_1 + \phi_2)\big] \qquad\qquad 9.4$$

we find that the instantaneous power can be written as

$$p(t) = \frac{V_M I_M}{2}\big[\cos(\theta_v - \theta_i) + \cos(2\omega t + \theta_v + \theta_i)\big] \qquad\qquad 9.5$$

Note that the instantaneous power consists of two terms. The first term is a constant (i.e., it is time independent), and the second term is a cosine wave of twice the excitation frequency. We will examine this equation in more detail in Section 9.2.

LEARNING Example 9.1

The circuit in Fig. 9.1 has the following parameters: $v(t) = 4\cos(\omega t + 60°)$ V and $\mathbf{Z} = 2\,\underline{/30°}\ \Omega$. We wish to determine equations for the current and the instantaneous power as a function of time and plot these functions with the voltage on a single graph for comparison.

A plot of this function, together with plots of the voltage and current, is shown in Fig. 9.2. As can be seen in the figure, the instantaneous power has an average value, and the frequency is twice that of the voltage or current.

SOLUTION Since

$$
\begin{aligned}
\mathbf{I} &= \frac{4\,\underline{/60°}}{2\,\underline{/30°}} \\
&= 2\,\underline{/30°}\ \text{A}
\end{aligned}
$$

then

$$i(t) = 2\cos(\omega t + 30°)\ \text{A}$$

From Eq. (9.5),

$$
\begin{aligned}
p(t) &= 4\big[\cos(30°) + \cos(2\omega t + 90°)\big] \\
&= 3.46 + 4\cos(2\omega t + 90°)\ \text{W}
\end{aligned}
$$

LEARNING Hint

Note that p(t) contains a dc term and a cosine wave with twice the frequency of v(t) and i(t).

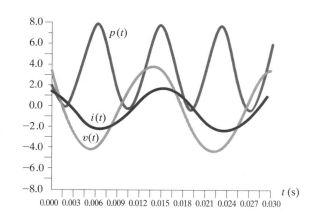

Figure 9.2 Plots of $v(t), i(t)$, and $p(t)$ for the circuit in Example 9.1 using $f = 60$ Hz.

9.2 Average Power

The average value of any periodic waveform (e.g., a sinusoidal function) can be computed by integrating the function over a complete period and dividing this result by the period. Therefore, if the voltage and current are given by Eqs. (9.1) and (9.2), respectively, the average power is

$$P = \frac{1}{T} \int_{t_0}^{t_0+T} p(t) \, dt$$

$$= \frac{1}{T} \int_{t_0}^{t_0+T} V_M I_M \cos(\omega t + \theta_v) \cos(\omega t + \theta_i) \, dt \qquad \textbf{9.6}$$

where t_0 is arbitrary, $T = 2\pi/\omega$ is the period of the voltage or current, and P is measured in watts. Actually, we may average the waveform over any integral number of periods so that Eq. (9.6) can also be written as

$$P = \frac{1}{nT} \int_{t_0}^{t_0+nT} V_M I_M \cos(\omega t + \theta_v) \cos(\omega t + \theta_i) \, dt \qquad \textbf{9.7}$$

where n is a positive integer.

Employing Eq. (9.5) for the expression in (9.6), we obtain

$$P = \frac{1}{T} \int_{t_0}^{t_0+T} \frac{V_M I_M}{2} \left[\cos(\theta_v - \theta_i) + \cos(2\omega t + \theta_v + \theta_i) \right] dt \qquad \textbf{9.8}$$

We could, of course, plod through the indicated integration; however, with a little forethought we can determine the result by inspection. The first term is independent of t, and therefore a constant in the integration. Integrating the constant over the period and dividing by the period simply results in the original constant. The second term is a cosine wave. It is well known that the average value of a cosine wave over one complete period or an integral number of periods is zero, and therefore the second term in Eq. (9.8) vanishes. In view of this discussion, Eq. (9.8) reduces to

LEARNING Hint

A frequently used equation for calculating the average power

$$P = \tfrac{1}{2} V_M I_M \cos(\theta_v - \theta_i) \qquad \textbf{9.9}$$

Note that since $\cos(-\theta) = \cos(\theta)$, the argument for the cosine function can be either $\theta_v - \theta_i$ or $\theta_i - \theta_v$. In addition, note that $\theta_v - \theta_i$ is the angle of the circuit impedance as shown in Fig. 9.1. Therefore, *for a purely resistive circuit,*

$$P = \tfrac{1}{2} V_M I_M \qquad \textbf{9.10}$$

and *for a purely reactive circuit,*

$$P = \tfrac{1}{2} V_M I_M \cos(90°)$$
$$= 0$$

Because purely reactive impedances absorb no average power, they are often called *lossless elements*. The purely reactive network operates in a mode in which it stores energy over one part of the period and releases it over another.

LEARNING by Doing

D 9.1   Find the average power for the following **V** and **I**:

(1) $\mathbf{V} = 2\,\underline{/30°}$ V,
     $\mathbf{I} = 2\,\underline{/30°}$ A

(2) $\mathbf{V} = 2\,\underline{/30°}$ V,
     $\mathbf{I} = 2\,\underline{/120°}$ A

(3) $\mathbf{V} = 2\,\underline{/30°}$ V,
     $\mathbf{I} = 2\,\underline{/60°}$ A

ANSWER   *(1) 2 W*
(2) 0 W (3) 1.732 W

LEARNING Example 9.2

We wish to determine the average power absorbed by the impedance shown in Fig. 9.3.

Figure 9.3
Example RL circuit.

SOLUTION From the figure we note that

$$I = \frac{V}{Z} = \frac{V_M \underline{/\theta_v}}{2 + j2} = \frac{10\underline{/60°}}{2.83\underline{/45°}} = 3.53\underline{/15°} \text{ A}$$

Therefore,

$$I_M = 3.53 \text{ A} \quad \text{and} \quad \theta_i = 15°$$

Hence,

$$P = \tfrac{1}{2}V_M I_M \cos(\theta_v - \theta_i)$$
$$= \tfrac{1}{2}(10)(3.53)\cos(60° - 15°)$$
$$= 12.5 \text{ W}$$

Since the inductor absorbs no power, we can employ Eq. (9.10) provided that V_M in that equation is the voltage across the resistor. Using voltage division, we obtain

$$V_R = \frac{(10\underline{/60°})(2)}{2 + j2} = 7.07\underline{/15°} \text{ V}$$

and therefore,

$$P = \tfrac{1}{2}(7.07)(3.53)$$
$$= 12.5 \text{ W}$$

In addition, using Ohm's law, we could also employ the expressions

$$P = \frac{1}{2}\frac{V_M^2}{R}$$

or

$$P = \tfrac{1}{2}I_M^2 R$$

where once again we must be careful that the V_M and I_M in these equations refer to the voltage across the resistor and the current through it, respectively.

LEARNING Example 9.3

For the circuit shown in Fig. 9.4, we wish to determine both the total average power absorbed and the total average power supplied.

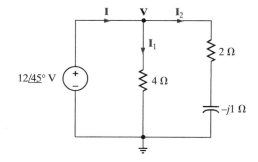

Figure 9.4
Example circuit for illustrating a power balance.

SOLUTION From the figure we note that

$$I_1 = \frac{12\underline{/45°}}{4} = 3\underline{/45°} \text{ A}$$

$$I_2 = \frac{12\underline{/45°}}{2 - j1} = \frac{12\underline{/45°}}{2.24\underline{/-26.57°}} = 5.36\underline{/71.57°} \text{ A}$$

and therefore,

$$I = I_1 + I_2$$

$$= 3\underline{/45°} + 5.36\underline{/71.57°}$$
$$= 8.15\underline{/62.10°} \text{ A}$$

The average power absorbed in the 4-Ω resistor is

$$P_4 = \tfrac{1}{2}V_M I_M = \tfrac{1}{2}(12)(3) = 18 \text{ W}$$

The average power absorbed in the 2-Ω resistor is

$$P_2 = \tfrac{1}{2}I_M^2 R = \tfrac{1}{2}(5.34)^2(2) = 28.7 \text{ W}$$

Therefore, the total average power absorbed is

$$P_A = 18 + 28.7 = 46.7 \text{ W}$$

Note that we could have calculated the power absorbed in the 2-Ω resistor using $\tfrac{1}{2}V_M^2/R$ if we had first calculated the voltage across the 2-Ω resistor.

The total average power supplied by the source is

$$P_S = \tfrac{1}{2}V_M I_M \cos(\theta_v - \theta_i)$$
$$= \tfrac{1}{2}(12)(8.15)\cos(45° - 62.10°)$$
$$= 46.7 \text{ W}$$

Thus, the total average power supplied is, of course, equal to the total average power absorbed.

LEARNING EXTENSIONS

E9.1   Find the average power absorbed by each resistor in the network in Fig. E9.1.

ANSWER $P_{2\Omega} = 7.20$ W,
$P_{4\Omega} = 7.20$ W.

Figure E9.1

E9.2   Given the network in Fig. E9.2, find the average power absorbed by each passive circuit element and the total average power supplied by the current source.

ANSWER $P_{3\Omega} = 56.60$ W,
$P_{4\Omega} = 33.96$ W, $P_L = 0$,
$P_{CS} = 90.50$ W.

Figure E9.2

When determining average power, if more than one source is present in a network, we can use any of our network analysis techniques to find the necessary voltage and/or current to compute the power. However, we must remember that in general we cannot apply superposition to power.

LEARNING Hint

Superposition is not applicable to power. Why?

LEARNING Example 9.4

Consider the network shown in Fig. 9.5. We wish to determine the total average power absorbed and supplied by each element.

Figure 9.5   *Example RL circuit with two sources.*

SOLUTION   From the figure we note that

$$\mathbf{I}_2 = \frac{12 \, \underline{/30°}}{2} = 6 \, \underline{/30°} \text{ A}$$

and

$$\mathbf{I}_3 = \frac{12 \, \underline{/30°} - 6 \, \underline{/0°}}{j1} = \frac{4.39 + j6}{j1} = 7.44 \, \underline{/-36.21°} \text{ A}$$

The power absorbed by the 2-Ω resistor is

$$P_2 = \tfrac{1}{2} V_M I_M = \tfrac{1}{2}(12)(6) = 36 \text{ W}$$

According to the direction of $\mathbf{I}_3$ the $6 \, \underline{/0°}$-V source is absorbing power. The power it absorbs is given by

$$P_{6\underline{/0°}} = \tfrac{1}{2} V_M I_M \cos(\theta_v - \theta_i)$$
$$= \tfrac{1}{2}(6)(7.44) \cos[0° - (-36.21°)]$$
$$= 18 \text{ W}$$

LEARNING Hint

Under the following condition

if $P = \mathbf{IV}$ is positive, power is being absorbed.
If $P = \mathbf{IV}$ is negative, power is being generated.

(continued)

At this point an obvious question arises: How do we know whether the $6\,\underline{/0°}$-V source is supplying power to the remainder of the network or absorbing it? The answer to this question is actually straightforward. If we employ our passive sign convention that was adopted in the earlier chapters—that is, if the current reference direction enters the positive terminal of the source and the answer is positive—the source is absorbing power. If the answer is negative, the source is supplying power to the remainder of the circuit. A generator sign convention could have been used, and under this condition the interpretation of the sign of the answer would be reversed. Note that once the sign convention is adopted and used, the sign for average power will be negative only if the angle difference is greater than 90° (i.e., $\left|\theta_v - \theta_i\right| > 90°$).

To obtain the power supplied to the network, we compute $\mathbf{I}_1$ as

$$\mathbf{I}_1 = \mathbf{I}_2 + \mathbf{I}_3$$
$$= 6\,\underline{/30°} + 7.44\,\underline{/-36.21°}$$
$$= 11.29\,\underline{/-7.10°}\ \text{A}$$

Therefore, the power supplied by the $12\,\underline{/30°}$-V source using the generator sign convention is

$$P_S = \tfrac{1}{2}(12)(11.29)\cos(30° + 7.10°)$$
$$= 54\ \text{W}$$

and hence the power absorbed is equal to the power supplied.

LEARNING EXTENSIONS

E9.3 Determine the total average power absorbed and supplied by each element in the network in Fig. E9.3.

ANSWER $P_{CS} = -69.4$ W, $P_{VS} = 19.8$ W, $P_{4\Omega} = 49.6$ W, $P_C = 0$.

Figure E9.3

E9.4 Given the network in Fig. E9.4, determine the total average power absorbed or supplied by each element.

ANSWER $P_{24\,\underline{/0°}} = -55.4$ W, $P_{12\,\underline{/0°}} = 5.5$ W, $P_{2\Omega} = 22.2$ W, $P_{4\Omega} = 27.7$ W, $P_L = 0$.

Figure E9.4

9.3 Maximum Average Power Transfer

In our study of resistive networks we addressed the problem of maximum power transfer to a resistive load. We showed that if the network excluding the load was represented by a Thévenin equivalent circuit, maximum power transfer would result if the value of the load resistor was equal to the Thévenin equivalent resistance $\left(\text{i.e., } R_L = R_{\text{Th}}\right)$. We will now reexamine this issue within the present context to determine the load impedance for the network shown in Fig. 9.6 that will result in maximum average power being absorbed by the load impedance $\mathbf{Z}_L$.

The equation for average power at the load is

$$P_L = \tfrac{1}{2} V_L I_L \cos\left(\theta_{v_L} - \theta_{i_L}\right) \qquad \textbf{9.11}$$

The phasor current and voltage at the load are given by the expressions

$$\mathbf{I}_L = \frac{\mathbf{V}_{\text{oc}}}{\mathbf{Z}_{\text{Th}} + \mathbf{Z}_L} \qquad \textbf{9.12}$$

$$\mathbf{V}_L = \frac{\mathbf{V}_{\text{oc}}\mathbf{Z}_L}{\mathbf{Z}_{\text{Th}} + \mathbf{Z}_L} \qquad \textbf{9.13}$$

where

$$\mathbf{Z}_{\text{Th}} = R_{\text{Th}} + jX_{\text{Th}} \qquad \textbf{9.14}$$

and

$$\mathbf{Z}_L = R_L + jX_L \qquad \textbf{9.15}$$

The magnitude of the phasor current and voltage are given by the expressions

$$I_L = \frac{V_{\text{oc}}}{\left[\left(R_{\text{Th}} + R_L\right)^2 + \left(X_{\text{Th}} + X_L\right)^2\right]^{1/2}} \qquad \textbf{9.16}$$

$$V_L = \frac{V_{\text{oc}}\left(R_L^2 + X_L^2\right)^{1/2}}{\left[\left(R_{\text{Th}} + R_L\right)^2 + \left(X_{\text{Th}} + X_L\right)^2\right]^{1/2}} \qquad \textbf{9.17}$$

The phase angles for the phasor current and voltage are contained in the quantity $\left(\theta_{v_L} - \theta_{i_L}\right)$. Note also that $\theta_{v_L} - \theta_{i_L} = \theta_{\mathbf{Z}_L}$ and, in addition,

$$\cos\theta_{\mathbf{Z}_L} = \frac{R_L}{\left(R_L^2 + X_L^2\right)^{1/2}} \qquad \textbf{9.18}$$

Substituting Eqs. (9.16) to (9.18) into Eq. (9.11) yields

$$P_L = \frac{1}{2}\frac{V_{\text{oc}}^2 R_L}{\left(R_{\text{Th}} + R_L\right)^2 + \left(X_{\text{Th}} + X_L\right)^2} \qquad \textbf{9.19}$$

which could, of course, be obtained directly from Eq. (9.16) using $P_L = \tfrac{1}{2} I_L^2 R_L$. Once again, a little forethought will save us some work. From the standpoint of maximizing P_L, V_{oc} is a constant. The quantity $\left(X_{\text{Th}} + X_L\right)$ absorbs no power, and therefore any nonzero value of this quantity only serves to reduce P_L. Hence, we can eliminate this term by selecting $X_L = -X_{\text{Th}}$. Our problem then reduces to maximizing

$$P_L = \frac{1}{2}\frac{V_{\text{oc}}^2 R_L}{\left(R_L + R_{\text{Th}}\right)^2} \qquad \textbf{9.20}$$

However, this is the same quantity we maximized in the purely resistive case by selecting $R_L = R_{\text{Th}}$. Therefore, for maximum average power transfer to the load shown in Fig. 9.6, $\mathbf{Z}_L$ should be chosen so that

$$\mathbf{Z}_L = R_L + jX_L = R_{\text{Th}} - jX_{\text{Th}} = \mathbf{Z}_{\text{Th}}^* \qquad \textbf{9.21}$$

Finally, if the load impedance is purely resistive $\left(\text{i.e., } X_L = 0\right)$, the condition for maximum average power transfer can be derived via the expression

$$\frac{dP_L}{dR_L} = 0$$

Figure 9.6
Circuit used to examine maximum power transfer.

LEARNING Hint

This impedance-matching concept is an important issue in the design of high-speed computer chips and motherboards. For today's high-speed chips with internal clocks running at about 1 GHz and motherboards with a bus speed above 100 MHz, impedance matching is necessary in order to obtain the required speed for signal propagation. Although this high-speed transmission line is based on a distributed circuit (discussed later in electrical engineering courses), the impedance-matching technique for the transmission line is the same as that of the lumped parameter circuit for maximum average power transfer.

LEARNING by Doing

D 9.2 Determine $\mathbf{Z}_L$ for maximum average power transfer to the load and the maximum average power transferred to the load.

ANSWER
$\mathbf{Z}_L = 1 - j1\ \Omega$
Maximum average power transfer $= \tfrac{1}{2}$ W.

where P_L is the expression in Eq. (9.19) with $X_L = 0$. *The value of R_L that maximizes P_L under the condition $X_L = 0$ is*

$$R_L = \sqrt{R_{Th}^2 + X_{Th}^2}$$ 9.22

Problem-Solving Strategy
Maximum Average Power Transfer

1. Remove the load $\mathbf{Z}_L$ and find the Thévenin equivalent for the remainder of the circuit.
2. Construct the circuit shown in Fig. 9.6.
3. Select $\mathbf{Z}_L = \mathbf{Z}_{Th}^* = R_{Th} - jX_{Th}$, and then $\mathbf{I}_L = \mathbf{V}_{oc}/2\,R_{Th}$ and the maximum average power transfer $= \dfrac{1}{2}\mathbf{I}_L^2 R_{Th} = \mathbf{V}_{oc}^2/8\,R_{Th}$.

LEARNING Example 9.5

Given the circuit in Fig. 9.7a, we wish to find the value of $\mathbf{Z}_L$ for maximum average power transfer. In addition, we wish to find the value of the maximum average power delivered to the load.

SOLUTION To solve the problem, we form a Thévenin equivalent at the load. The circuit in Fig. 9.7b is used to compute the open-circuit voltage

$$\mathbf{V}_{oc} = \frac{4\,\underline{/0^\circ}\,(2)}{6 + j1}\,(4) = 5.26\,\underline{/-9.46^\circ}\,\text{V}$$

The Thévenin equivalent impedance can be derived from the circuit in Fig. 9.7c. As shown in the figure,

$$\mathbf{Z}_{Th} = \frac{4(2 + j1)}{6 + j1} = 1.41 + j0.43\,\Omega$$

Therefore, $\mathbf{Z}_L$ for maximum average power transfer is

$$\mathbf{Z}_L = 1.41 - j0.43\,\Omega$$

With $\mathbf{Z}_L$ as given previously, the current in the load is

$$\mathbf{I} = \frac{5.26\,\underline{/-9.46^\circ}}{2.82} = 1.87\,\underline{/-9.46^\circ}\,\text{A}$$

Therefore, the maximum average power transferred to the load is

$$P_L = \tfrac{1}{2}I_M^2 R_L = \tfrac{1}{2}(1.87)^2(1.41) = 2.47\,\text{W}$$

LEARNING Hint

In this Thévenin analysis,
1. Remove $\mathbf{Z}_L$ and find the voltage across the open terminals, $\mathbf{V}_{oc}$.
2. Determine the impedance $\mathbf{Z}_{Th}$ at the open terminals with all independent sources made zero.
3. Construct the following circuit and determine $\mathbf{I}$ and P_L.

(a) (b) (c)

Figure 9.7 Circuits for illustrating maximum average power transfer.

LEARNING Example 9.6

For the circuit shown in Fig. 9.8a, we wish to find the value of $\mathbf{Z}_L$ for maximum average power transfer. In addition, let us determine the value of the maximum average power delivered to the load.

SOLUTION   We will first reduce the circuit, with the exception of the load, to a Thévenin equivalent circuit. The open-circuit voltage can be computed from Fig. 9.8b. The equations for the circuit are

$$\mathbf{V}'_x + 4 = (2 + j4)\mathbf{I}_1$$

$$\mathbf{V}'_x = -2\mathbf{I}_1$$

Solving for $\mathbf{I}_1$, we obtain

$$\mathbf{I}_1 = \frac{1\,\underline{/-45°}}{\sqrt{2}}$$

The open-circuit voltage is then

$$\begin{aligned}
\mathbf{V}_{oc} &= 2\mathbf{I}_1 - 4\,\underline{/0°} \\
&= \sqrt{2}\,\underline{/-45°} - 4\,\underline{/0°} \\
&= -3 - j1 \\
&= -3.16\,\underline{/-161.57°}\ \text{V}
\end{aligned}$$

The short-circuit current can be derived from Fig. 9.8c. The equations for this circuit are

$$\mathbf{V}''_x + 4 = (2 + j4)\mathbf{I} - 2\mathbf{I}_{sc}$$

$$-4 = -2\mathbf{I} + (2 - j2)\mathbf{I}_{sc}$$

$$\mathbf{V}''_x = -2\big(\mathbf{I} - \mathbf{I}_{sc}\big)$$

Solving these equations for $\mathbf{I}_{sc}$ yields

$$\mathbf{I}_{sc} = -(1 + j2)\ \text{A}$$

The Thévenin equivalent impedance is then

$$\mathbf{Z}_{Th} = \frac{\mathbf{V}_{oc}}{\mathbf{I}_{sc}} = \frac{3 + j1}{1 + j2} = 1 - j1\ \Omega$$

Therefore, for maximum average power transfer the load impedance should be

$$\mathbf{Z}_L = 1 + j1\ \Omega$$

The current in this load $\mathbf{Z}_L$ is then

$$\mathbf{I}_L = \frac{\mathbf{V}_{oc}}{\mathbf{Z}_{Th} + \mathbf{Z}_L} = \frac{-3 - j1}{2} = 1.58\,\underline{/-161.57°}\ \text{A}$$

Hence, the maximum average power transferred to the load is

$$P_L = \tfrac{1}{2}(1.58)^2(1)$$

$$= 1.25\ \text{W}$$

LEARNING Hint

When there is a dependent source, both $\mathbf{V}_{oc}$ and $\mathbf{I}_{sc}$ must be found and $\mathbf{Z}_{Th}$ computed from the equation

$$\mathbf{Z}_{Th} = \frac{\mathbf{V}_{oc}}{\mathbf{I}_{sc}}$$

Figure 9.8   *Circuits for illustrating maximum average power transfer.*

E9.5 Given the network in Fig. E9.5, find $\mathbf{Z}_L$ for maximum average power transfer and the maximum average power transferred to the load.

ANSWER $\mathbf{Z}_L = 1 + j1\ \Omega$,
$P_L = 45$ W.

Figure E9.5

E9.6 Find $\mathbf{Z}_L$ for maximum average power transfer and the maximum average power transferred to the load in the network in Fig. E9.6.

ANSWER $\mathbf{Z}_L = 2 - j2\ \Omega$,
$P_L = 45$ W.

Figure E9.6

9.4 Effective or rms Values

In the preceding sections of this chapter, we have shown that the average power absorbed by a resistive load is directly dependent on the type, or types, of sources that are delivering power to the load. For example, if the source was dc, the average power absorbed was I^2R, and if the source was sinusoidal, the average power was $\frac{1}{2}I_M^2 R$. Although these two types of waveforms are extremely important, they are by no means the only waveforms we will encounter in circuit analysis. Therefore, a technique by which we can compare the *effectiveness* of different sources in delivering power to a resistive load would be quite useful.

To accomplish this comparison, we define what is called the *effective value of a periodic waveform*, representing either voltage or current. Although either quantity could be used, we will employ current in the definition. Hence, we define the effective value of a periodic current as a constant or dc value, which as current would deliver the same average power to a resistor R. Let us call the constant current I_{eff}. Then the average power delivered to a resistor as a result of this current is

$$P = I_{\text{eff}}^2 R$$

Similarly, the average power delivered to a resistor by a periodic current $i(t)$ is

$$P = \frac{1}{T}\int_{t_0}^{t_0+T} i^2(t)R\,dt$$

Equating these two expressions, we find that

$$I_{\text{eff}} = \sqrt{\frac{1}{T}\int_{t_0}^{t_0+T} i^2(t)\,dt}$$

9.23

Note that this effective value is found by first determining the *square* of the current, then computing the average or *mean* value, and finally taking the square *root*. Thus in "reading" the mathematical Eq. (9.23), we are determining the root mean square, which we abbreviate as rms, and therefore I_{eff} is called I_{rms}.

Since dc is a constant, the rms value of dc is simply the constant value. Let us now determine the rms value of other waveforms. The most important waveform is the sinusoid, and therefore, we address this particular one in the following example.

LEARNING Example 9.7

We wish to compute the rms value of the waveform $i(t) = I_M \cos(\omega t - \theta)$, which has a period of $T = 2\pi/\omega$.

SOLUTION Substituting these expressions into Eq. (9.23) yields

$$I_{rms} = \left[\frac{1}{T} \int_0^T I_M^2 \cos^2(\omega t - \theta)\, dt \right]^{1/2}$$

Using the trigonometric identity

$$\cos^2 \phi = \tfrac{1}{2} + \tfrac{1}{2} \cos 2\phi$$

we find that the preceding equation can be expressed as

$$I_{rms} = I_M \left\{ \frac{\omega}{2\pi} \int_0^{2\pi/\omega} \left[\tfrac{1}{2} + \tfrac{1}{2} \cos(2\omega t - 2\theta) \right] dt \right\}^{1/2}$$

Since we know that the average or mean value of a cosine wave is zero,

$$I_{rms} = I_M \left(\frac{\omega}{2\pi} \int_0^{2\pi/\omega} \frac{1}{2}\, dt \right)^{1/2}$$

$$= I_M \left[\frac{\omega}{2\pi} \left(\frac{t}{2} \right) \Big|_0^{2\pi/\omega} \right]^{1/2} = \frac{I_M}{\sqrt{2}} \qquad \textbf{9.24}$$

Therefore, the rms value of a sinusoid is equal to the maximum value divided by the $\sqrt{2}$. Hence, a sinusoidal current with a maximum value of I_M delivers the same average power to a resistor R as a dc current with a value of $I_M/\sqrt{2}$. Recall that earlier a phasor **X** was defined as $X_M \underline{/\theta}$ for a sinusoidal wave of the form $X_M \cos(\omega t + \theta)$. This phasor can also be represented as $X_M/\sqrt{2} \underline{/\theta}$ if the units are given in rms. For example, $120 \underline{/30°}$ V rms is equivalent to $170 \underline{/30°}$ V.

On using the rms values for voltage and current, the average power can be written, in general, as

$$P = V_{rms} I_{rms} \cos(\theta_v - \theta_i) \qquad \textbf{9.25}$$

The power absorbed by a resistor R is

$$P = I_{rms}^2 R = \frac{V_{rms}^2}{R} \qquad \textbf{9.26}$$

In dealing with voltages and currents in numerous electrical applications, it is important to know whether the values quoted are maximum, average, rms, or what. For example, the normal 120-V ac electrical outlets have an rms value of 120 V, an average value of 0 V, and a maximum value of $120\sqrt{2}$ V.

LEARNING Example 9.8

We wish to compute the rms value of the voltage waveform shown in Fig. 9.9.

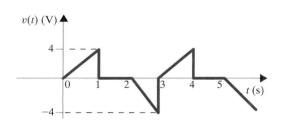

$v(t)$ (V)

Figure 9.9 *Waveform used to illustrate rms values.*

SOLUTION The waveform is periodic with period $T = 3$ s. The equation for the voltage in the time frame $0 \le t \le 3$ s is

$$v(t) = \begin{bmatrix} 4t \text{ V} & 0 < t \le 1 \text{ s} \\ 0 \text{ V} & 1 < t \le 2 \text{ s} \\ -4t + 8 \text{ V} & 2 < t \le 3 \text{ s} \end{bmatrix}$$

The rms value is

$$V_{rms} = \left\{ \frac{1}{3} \left[\int_0^1 (4t)^2\, dt + \int_1^2 (0)^2\, dt + \int_2^3 (8 - 4t)^2\, dt \right] \right\}^{1/2}$$

$$= \left[\frac{1}{3} \left(\frac{16t^3}{3} \Big|_0^1 + \left(64t - \frac{64t^2}{2} + \frac{16t^3}{3} \right) \Big|_2^3 \right) \right]^{1/2}$$

$$= 1.89 \text{ V}$$

LEARNING Example 9.9

Determine the rms value of the current waveform in Fig. 9.10 and use this value to compute the average power delivered to a 2-Ω resistor through which this current is flowing.

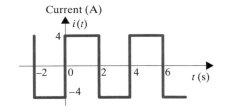

Figure 9.10
Waveform used to
illustrate rms values.

SOLUTION The current waveform is periodic with a period of $T = 4$ s. The rms value is

$$I_{rms} = \left\{ \frac{1}{4} \left[\int_0^2 (4)^2 \, dt + \int_2^4 (-4)^2 \, dt \right] \right\}^{1/2}$$

$$= \left[\frac{1}{4} \left(16t \Big|_0^2 + 16t \Big|_2^4 \right) \right]^{1/2}$$

$$= 4 \text{ A}$$

The average power delivered to a 2-Ω resistor with this current is

$$P = I_{rms}^2 R = (4)^2(2) = 32 \text{ W}$$

LEARNING EXTENSIONS

E9.7 Compute the rms value of the voltage waveform shown in Fig. E9.7.

ANSWER $V_{rms} = 1.633$ V.

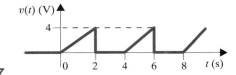

Figure E9.7

E9.8 The current waveform in Fig. E9.8 is flowing through a 4-Ω resistor. Compute the average power delivered to the resistor.

ANSWER $P = 32$ W.

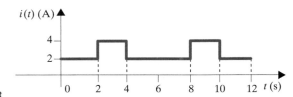

Figure E9.8

E9.9 The current waveform in Fig. E9.9 is flowing through a 10-Ω resistor. Determine the average power delivered to the resistor.

ANSWER $P = 80$ W.

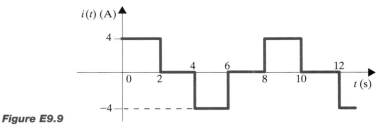

Figure E9.9

9.5 The Power Factor

The power factor is a very important quantity. Its importance stems in part from the economic impact it has on industrial users of large amounts of power. In this section we carefully define this term and then illustrate its significance via some practical examples.

In Section 9.4 we showed that a load operating in the ac steady state is delivered an average power of

$$P = V_{rms} I_{rms} \cos(\theta_v - \theta_i)$$

We will now further define the terms in this important equation. The product $V_{rms} I_{rms}$ is referred to as the *apparent power*. Although the term $\cos(\theta_v - \theta_i)$ is a dimensionless quantity, and the units of P are watts, apparent power is normally stated in volt-amperes (VA) or kilovolt-amperes (kVA) to distinguish it from average power.

We now define the *power factor* (pf) as the ratio of the average power to the apparent power; that is,

$$pf = \frac{P}{V_{rms} I_{rms}} = \cos(\theta_v - \theta_i) \qquad \textbf{9.27}$$

where

$$\cos(\theta_v - \theta_i) = \cos\theta_{\mathbf{Z}_L} \qquad \textbf{9.28}$$

The angle $\theta_v - \theta_i = \theta_{\mathbf{Z}_L}$ is the phase angle of the load impedance and is often referred to as the *power factor angle*. The two extreme positions for this angle correspond to a purely resistive load where $\theta_{\mathbf{Z}_L} = 0$ and the pf is 1, and the purely reactive load where $\theta_{\mathbf{Z}_L} = \pm 90°$ and the pf is 0. It is, of course, possible to have a unity pf for a load containing R, L, and C elements if the values of the circuit elements are such that a zero phase angle is obtained at the particular operating frequency.

There is, of course, a whole range of power factor angles between $\pm 90°$ and $0°$. If the load is an equivalent RC combination, then the pf angle lies between the limits $-90° < \theta_{\mathbf{Z}_L} < 0°$. On the other hand, if the load is an equivalent RL combination, then the pf angle lies between the limits $0 < \theta_{\mathbf{Z}_L} < 90°$. Obviously, confusion in identifying the type of load could result, due to the fact that $\cos\theta_{\mathbf{Z}_L} = \cos(-\theta_{\mathbf{Z}_L})$. To circumvent this problem, the pf is said to be either *leading* or *lagging*, where these two terms *refer to the phase of the current with respect to the voltage*. Since the current leads the voltage in an RC load, the load has a leading pf. In a similar manner, an RL load has a lagging pf; therefore, load impedances of $\mathbf{Z}_L = 1 - j1$ and $\mathbf{Z}_L = 2 + j1$ have power factors of $\cos(-45°) = 0.707$ leading and $\cos(26.57°) = 0.894$ lagging, respectively.

LEARNING by Doing

D 9.3 If a load impedance is $10\underline{/30°}$ Ω, find the power factor.

ANSWER
pf $= \cos 30° = \sqrt{3}/2.$

LEARNING Example 9.10

An industrial load consumes 88 kW at a pf of 0.707 lagging from a 480-V rms line. The transmission line resistance from the power company's transformer to the plant is 0.08 Ω. Let us determine the power that must be supplied by the power company (a) under present conditions and (b) if the pf is somehow changed to 0.90 lagging. (It is economically advantageous to have a power factor as close to one as possible.)

SOLUTION (a) The equivalent circuit for these conditions is shown in Fig. 9.11. Using Eq. (9.27), we obtain the magnitude of the rms current into the plant:

$$I_{rms} = \frac{P_L}{(pf)(V_{rms})}$$

$$= \frac{(88)(10^3)}{(0.707)(480)}$$

$$= 259.3 \text{ A rms}$$

Figure 9.11
Example circuit for examining changes in power factor.

(continued)

LEARNING Hint

Technique
1. Given P_L, pf, and V_{rms}, determine I_{rms}.
2. Then $P_S = P_L + I^2_{rms} R_{line}$, where R_{line} is the line resistance.

The power that must be supplied by the power company is

$$P_S = P_L + (0.08)I^2_{rms}$$
$$= 88,000 + (0.08)(259.3)^2$$
$$= 93.38 \text{ kW}$$

(b) Suppose now that the pf is somehow changed to 0.90 lagging but the voltage remains constant at 480 V. The rms load current for this condition is

$$I_{rms} = \frac{P_L}{(pf)(V_{rms})}$$

$$= \frac{(88)(10^3)}{(0.90)(480)}$$
$$= 203.7 \text{ A rms}$$

Under these conditions the power company must generate

$$P_S = P_L + (0.08)I^2_{rms}$$
$$= 88,000 + (0.08)(203.7)^2$$
$$= 91.32 \text{ kW}$$

Note carefully the difference between the two cases. A simple change in the pf of the load from 0.707 lagging to 0.90 lagging has had an interesting effect. Note that in the first case the power company must generate 93.38 kW in order to supply the plant with 88 kW of power because the low power factor means that the line losses will be high—5.38 kW. However, in the second case the power company need only generate 91.32 kW in order to supply the plant with its required power, and the corresponding line losses are only 3.32 kW.

The example clearly indicates the economic impact of the load's power factor. A low power factor at the load means that the utility generators must be capable of carrying more current at constant voltage, and they must also supply power for higher $I^2_{rms} R_{line}$ losses than would be required if the load's power factor were high. Since line losses represent energy expended in heat and benefit no one, the utility will insist that a plant maintain a high pf, typically 0.90 lagging, and adjust its rate schedule to penalize plants that do not conform to this requirement. In the next section we will demonstrate a simple and economical technique for achieving this power factor correction.

LEARNING EXTENSION

E9.10 An industrial load consumes 100 kW at 0.707 pf lagging. The 60-Hz line voltage at the load is $480 \underline{/0°}$ V rms. The transmission-line resistance between the power company transformer and the load is 0.1 Ω. Determine the power savings that could be obtained if the pf is changed to 0.94 lagging.

ANSWER Power saved is 3.771 kW.

9.6 Complex Power

Figure 9.12
Circuit used to explain power relationships.

In our study of ac steady-state power, it is convenient to introduce another quantity, which is commonly called *complex power*. To develop the relationship between this quantity and others we have presented in the preceding sections, consider the circuit shown in Fig. 9.12.

The complex power is defined to be

$$\mathbf{S} = \mathbf{V}_{rms}\mathbf{I}^*_{rms} \qquad \textbf{9.29}$$

where $\mathbf{I}^*_{rms}$ refers to the complex conjugate of $\mathbf{I}_{rms}$; that is, if $\mathbf{I}_{rms} = I_{rms}\underline{/\theta_i} = I_R + jI_I$, then $\mathbf{I}^*_{rms} = I_{rms}\underline{/-\theta_i} = I_R - jI_I$. Complex power is then

$$\mathbf{S} = V_{rms}\underline{/\theta_v}\, I_{rms}\underline{/-\theta_i} = V_{rms}I_{rms}\underline{/\theta_v - \theta_i} \qquad \textbf{9.30}$$

or

$$\mathbf{S} = V_{rms}I_{rms}\cos(\theta_v - \theta_i) + jV_{rms}I_{rms}\sin(\theta_v - \theta_i) \qquad \textbf{9.31}$$

where, of course, $\theta_v - \theta_i = \theta_{\mathbf{Z}}$. We note from Eq. (9.31) that the real part of the complex power is simply the *real* or *average power*. The imaginary part of **S** we call the *reactive* or *quadrature power*. Therefore, complex power can be expressed in the form

$$\mathbf{S} = P + jQ \qquad \textbf{9.32}$$

where

$$P = \mathrm{Re}(\mathbf{S}) = V_{rms}I_{rms}\cos(\theta_v - \theta_i) \qquad \textbf{9.33}$$

$$Q = \mathrm{Im}(\mathbf{S}) = V_{rms}I_{rms}\sin(\theta_v - \theta_i) \qquad \textbf{9.34}$$

As shown in Eq. (9.31), the magnitude of the complex power is what we have called the *apparent power*, and the phase angle for complex power is simply the power factor angle. Complex power, like apparent power, is measured in volt-amperes, real power is measured in watts, and to distinguish Q from the other quantities, which in fact have the same dimensions, it is measured in volt-amperes reactive, or var.

In addition to the relationships expressed previously, we note that since

$$\cos(\theta_v - \theta_i) = \frac{\mathrm{Re}(\mathbf{Z})}{Z}$$

$$\sin(\theta_v - \theta_i) = \frac{\mathrm{Im}(\mathbf{Z})}{Z}$$

and

$$I_{rms} = \frac{V_{rms}}{Z}$$

Eqs. (9.33) and (9.34) can be written as

$$P = I_{rms}^2 \mathrm{Re}(\mathbf{Z}) \qquad \textbf{9.35}$$

$$Q = I_{rms}^2 \mathrm{Im}(\mathbf{Z}) \qquad \textbf{9.36}$$

and therefore, Eq. (9.32) can be expressed as

$$\mathbf{S} = I_{rms}^2 \mathbf{Z} \qquad \textbf{9.37}$$

The diagrams in Fig. 9.13 serve to explain further the relationships among the various quantities of power. As shown in Fig. 9.13a, the phasor current can be split into two components: one that is in phase with $\mathbf{V}_{rms}$ and one that is 90° out of phase with $\mathbf{V}_{rms}$. Equations (9.33)

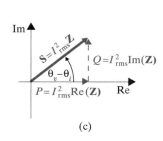

Figure 9.13
Diagram for illustrating power relationships.

and (9.34) illustrate that the in-phase component produces the real power, and the 90° component, called the *quadrature component*, produces the reactive or quadrature power. In addition, Eqs. (9.32) to (9.34) indicate that

$$\tan(\theta_v - \theta_i) = \frac{Q}{P} \qquad\qquad 9.38$$

which relates the pf angle to P and Q in what is called the *power triangle*.

The relationships among S, P, and Q can be expressed via the diagrams shown in Figs. 9.13b and c. In Fig. 9.13b we note the following conditions. If Q is positive, the load is inductive, the power factor is lagging, and the complex number S lies in the first quadrant. If Q is negative, the load is capacitive, the power factor is leading, and the complex number S lies in the fourth quadrant. If Q is zero, the load is resistive, the power factor is unity, and the complex number S lies along the positive real axis. Figure 9.13c illustrates the relationships expressed by Eqs. (9.35) to (9.37) for an inductive load.

Finally, it is important to state that complex power, like energy, is conserved; that is, the total complex power delivered to any number of individual loads is equal to the sum of the complex powers delivered to each individual load, regardless of how the loads are interconnected.

Problem-Solving Strategy
P or S

If $v(t)$ and $i(t)$ are known and we wish to find P given an impedance $\mathbf{Z}\underline{/\theta} = R + jX$, two viable approaches are as follows:

1. Determine $\mathbf{V}$ and $\mathbf{I}$ and then calculate

$$P = VI\cos\theta \quad \text{or} \quad P = VI\cos(\underline{/\mathbf{V}} - \underline{/\mathbf{I}})$$

2. Use $\mathbf{I}$ to calculate the real part of S; that is,

$$P = R_e(\mathbf{S}) = I^2 R$$

The latter method may be easier to calculate than the former. However, if the imaginary part of the impedance, X, is not zero, then

$$P \neq \frac{V^2}{R}$$

which is a common mistake. Furthermore, the P and Q portions of S are directly related to $\mathbf{Z}\underline{/\theta}$ and provide a convenient way in which to relate power, current, and impedance. That is,

$$\tan\theta = \frac{Q}{P}$$

$$\mathbf{S} = I^2\mathbf{Z}$$

The following example illustrates the usefulness of S.

LEARNING Example 9.11

A load operates at 20 kW, 0.8 pf lagging. The load voltage is $220\underline{/0°}$ V rms at 60 Hz. The impedance of the line is $0.09 + j0.3\ \Omega$. We wish to determine the voltage and power factor at the input to the line.

SOLUTION The circuit diagram for this problem is shown in Fig. 9.14. As illustrated in Fig. 9.13,

$$S = \frac{P}{\cos\theta} = \frac{P}{\text{pf}} = \frac{20{,}000}{0.8} = 25{,}000 \text{ VA}$$

Therefore, at the load

$$\mathbf{S}_L = 25{,}000 \underline{/\theta} = 25{,}000 \underline{/36.87°} = 20{,}000 + j15{,}000 \text{ VA}$$

Since $\mathbf{S}_L = \mathbf{V}_L \mathbf{I}_L^*$

$$\mathbf{I}_L = \left[\frac{25{,}000 \underline{/36.87°}}{220 \underline{/0°}} \right]^*$$

$$= 113.64 \underline{/-36.87°} \text{ A rms}$$

The complex power losses in the line are

$$\mathbf{S}_{\text{line}} = I_L^2 \mathbf{Z}_{\text{line}}$$

$$= (113.64)^2 (0.09 + j0.3)$$

$$= 1162.26 + j3874.21 \text{ VA}$$

As stated earlier, complex power is conserved and, therefore, the complex power at the generator is

$$\mathbf{S}_S = \mathbf{S}_L + \mathbf{S}_{\text{line}}$$

$$= 21{,}162.26 + j18{,}874.21$$

$$= 28{,}356.25 \underline{/41.73°} \text{ VA}$$

Figure 9.14
Example circuit for power analysis.

Hence, the generator voltage is

$$V_L = \frac{|\mathbf{S}_L|}{I_L} = \frac{28{,}356.25}{113.64}$$

$$= 249.53 \text{ V rms}$$

and the generator power factor is

$$\cos(41.73°) = 0.75 \text{ lagging}$$

We could have solved this problem using KVL. For example, we calculated the load current as

$$\mathbf{I}_L = 113.64 \underline{/-36.87°} \text{ A rms}$$

Hence, the voltage drop in the transmission line is

$$\mathbf{V}_{\text{line}} = (113.64 \underline{/-36.87°})(0.09 + j0.3)$$

$$= 35.59 \underline{/36.43°} \text{ V rms}$$

Therefore, the generator voltage is

$$\mathbf{V}_S = 220 \underline{/0°} + 35.59 \underline{/36.43°}$$

$$= 249.53 \underline{/4.86°} \text{ V rms}$$

Hence, the generator voltage is 249.53 V rms. In addition,

$$\theta_v - \theta_i = 4.86° - (-36.87°) = 41.73°$$

and therefore,

$$\text{pf} = \cos(41.73°) = 0.75 \text{ lagging}$$

LEARNING Hint

1. Use the given P_L, $\cos\theta$, and V_L rms to obtain $\mathbf{S}_L$ and $\mathbf{I}_L$ based on Eqs. (9.33) and (9.29), respectively.
2. Use $\mathbf{I}_L$ and $\mathbf{Z}_{\text{line}}$ to obtain $\mathbf{S}_{\text{line}}$ using Eq. (9.37).
3. Use $\mathbf{S}_s = \mathbf{S}_{\text{line}} + \mathbf{S}_L$.
4. $\mathbf{V}_s = \mathbf{S}_s / \mathbf{I}_L^*$ yields V_S and θ_v. Since $\mathbf{V}_s = V_s \underline{/\theta_v}$ and θ_i is the phase of $\mathbf{I}_L$, $\text{pf} = \cos(\theta_v - \theta_i)$.

LEARNING Example 9.12

Two networks A and B are connected by two conductors having a net impedance of $\mathbf{Z} = 0 + j1 \ \Omega$, as shown in Fig. 9.15. The voltages at the terminals of the networks are $\mathbf{V}_A = 120 \underline{/30°}$ V rms and $\mathbf{V}_B = 120 \underline{/0°}$ V rms. We wish to determine the average power flow between the networks and identify which is the source and which is the load.

Figure 9.15 Network used in Example 9.12.

SOLUTION As shown in Fig. 9.15,

$$\mathbf{I} = \frac{\mathbf{V}_A - \mathbf{V}_B}{\mathbf{Z}}$$

$$= \frac{120 \underline{/30°} - 120 \underline{/0°}}{j1}$$

$$= 62.12 \underline{/15°} \text{ A rms}$$

The power delivered by network A is

$$P_A = |\mathbf{V}_A||\mathbf{I}| \cos(\theta_{\mathbf{V}_A} - \theta_{\mathbf{I}})$$

$$= (120)(62.12) \cos(30° - 15°)$$

$$= 7200.4 \text{ W}$$

(continued)

The power absorbed by network B is

$$P_B = |\mathbf{V}_B||\mathbf{I}| \cos(\theta_{\mathbf{V}_B} - \theta_{\mathbf{I}})$$

$$= (120)(62.12) \cos(0° - 15°)$$

$$= 7200.4 \text{ W}$$

If the power flow had actually been from network B to network A, the resultant signs on P_A and P_B would have been negative.

E9.11 An industrial load requires 40 kW at 0.84 pf lagging. The load voltage is 220 $\underline{/0°}$ V rms at 60 Hz. The transmission-line impedance is $0.1 + j0.25$ Ω. Determine the real and reactive power losses in the line and the real and reactive power required at the input to the transmission line.

ANSWER $P_{\text{line}} = 4.685$ kW, $Q_{\text{line}} = 11.713$ kvar, $P_S = 44.685$ kW, $Q_S = 37.55$ kvar.

E9.12 A load requires 60 kW at 0.85 pf lagging. The 60-Hz line voltage at the load is 220 $\underline{/0°}$ V rms. If the transmission-line impedance is $0.12 + j0.18$ Ω, determine the line voltage and power factor at the input.

ANSWER $\mathbf{V}_{\text{in}} = 284.6 \underline{/5.8°}$ V rms, $\text{pf}_{\text{in}} = 0.792$ lagging.

9.7 Power Factor Correction

Industrial plants that require large amounts of power have a wide variety of loads. However, by nature the loads normally have a lagging power factor. In view of the results obtained in Example 9.10, we are naturally led to ask whether there is any convenient technique for raising the power factor of a load. Since a typical load may be a bank of induction motors or other expensive machinery, the technique for raising the pf should be an economical one to be feasible.

To answer the question we pose, let us examine the circuit shown in Fig. 9.16. The circuit illustrates a typical industrial load. In parallel with this load we have placed a capacitor. The original complex power for the load $\mathbf{Z}_L$, which we will denote as $\mathbf{S}_{\text{old}}$, is

$$\mathbf{S}_{\text{old}} = P_{\text{old}} + jQ_{\text{old}} = |\mathbf{S}_{\text{old}}| \underline{/\theta_{\text{old}}}$$

The complex power for the capacitor is

$$\mathbf{S}_{\text{cap}} = 0 + jQ_{\text{cap}} = |\mathbf{S}_{\text{cap}}| \underline{/-90°}$$

The new complex power that results from adding a capacitor is

$$\mathbf{S}_{\text{old}} + \mathbf{S}_{\text{cap}} = \mathbf{S}_{\text{new}} = P_{\text{old}} + jQ_{\text{new}} = |\mathbf{S}_{\text{new}}| \underline{/\theta_{\text{new}}} \qquad \textbf{9.39}$$

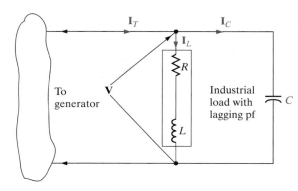

Figure 9.16
Circuit for power factor correction.

where θ_{new} is specified by the required power factor. The difference between the new and old complex powers is caused by the addition of the capacitor and since the capacitor is purely reactive,

$$\mathbf{S}_{cap} = +jQ_{cap} = -j\omega C V_{rms}^2 \qquad \textbf{9.40}$$

Equation (9.40) can be used to find the required value of C in order to achieve the new specified power factor. The procedure we have described is illustrated in Fig. 9.17, where $\theta_{old} = \theta_{v_L} - \theta_{i_L}$ and $\theta_{new} = \theta_{v_T} - \theta_{i_T}$. Hence, we can obtain a particular power factor for the total load simply by judiciously selecting a capacitor and placing it in parallel with the original load. In general, we want the power factor to be large, and therefore the power factor angle must be small $\left(\text{i.e., the larger the desired power factor, the smaller the angle } \theta_{v_T} - \theta_{i_T}\right)$.

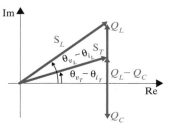

Figure 9.17
Illustration of the technique for power factor correction.

LEARNING Example 9.13

Plastic kayaks are manufactured using a process called roto-molding, which is diagrammed in Fig. 9.18. Molten plastic is injected into a mold, which is then spun on the long axis of the kayak until the plastic cools, resulting in a hollow one-piece craft. Suppose that the induction motors used to spin the molds consume 50 kW at a pf of 0.8 lagging from a 220 $\underline{/0°}$-V rms, 60-Hz line. We wish to raise the pf to 0.95 lagging by placing a bank of capacitors in parallel with the load.

Figure 9.18 *Rotomolding manufacturing process.*

SOLUTION The circuit diagram for this problem is shown in Fig. 9.19. $P_L = 50$ kW and since $\cos^{-1}0.8 = 36.87°$, $\theta_{old} = 36.87°$. Therefore,

$$Q_{old} = P_{old} \tan \theta_{old} = (50)(10^3)(0.75) = 37.5 \text{ kvar}$$

Hence,

$$\mathbf{S}_{old} = P_{old} + jQ_{old} = 50,000 + j37,500$$

and

$$\mathbf{S}_{cap} = 0 + jQ_{cap}$$

Since the required power factor is 0.95,

$$\theta_{new} = \cos^{-1}(pf_{new}) = \cos^{-1}(0.95)$$
$$= 18.19°$$

Then

$$Q_{new} = P_{old} \tan \theta_{new}$$
$$= 50,000 \tan(18.19°)$$
$$= 16,430 \text{ var}$$

Hence

$$Q_{new} - Q_{old} = Q_{cap} = -\omega C V^2 \text{ rms}$$
$$16,430 - 37,500 = -\omega C V^2 \text{ rms}$$

Solving the equation for C yields

$$C = \frac{21,070}{(377)(220)^2}$$
$$= 1155 \text{ μF}$$

By using a capacitor of this magnitude in parallel with the industrial load, we create, from the utility's perspective, a load pf of 0.95 lagging. However, the parameters of the actual load remain unchanged. Under these conditions, the current supplied by the utility to the kayak manufacturer is less and therefore they can use smaller conductors for the same amount of power. Or, if the conductor size is fixed, the line losses will be less since these losses are a function of the square of the current.

LEARNING Hint

Technique

1. Find Q_{old} from P_L and θ_{old}, or the equivalent pf_{old}.
2. Find θ_{new} from the desired pf_{new}.
3. Determine $Q_{new} = P_{old} \tan \theta_{new}$.
4. $Q_{new} - Q_{old} = Q_{cap} = -\omega C V^2$ rms.

Figure 9.19
Example circuit for power factor correction.

E9.13 Compute the value of the capacitor necessary to change the power factor in Extension Exercise E9.10 to 0.95 lagging.   **ANSWER** $C = 773\ \mu F$.

9.8 Single-Phase Three-Wire Circuits

LEARNING by Doing

D 9.5 In Fig. 9.20b, if
$V = 120\ \underline{/0°}$ V and
$Z_L = 100 + j1\ \Omega$, find
I_{nN}.

ANSWER $I_{nN} = 0$.

The single-phase three-wire ac circuit shown in Figure 9.20 is an important topic because it is the typical ac power network found in households. Note that the voltage sources are equal; that is, $\mathbf{V}_{an} = \mathbf{V}_{nb} = \mathbf{V}$. Thus, the magnitudes are equal and the phases are equal (single phase). The line-to-line voltage $\mathbf{V}_{ab} = 2\mathbf{V}_{an} = 2\mathbf{V}_{nb} = 2\mathbf{V}$. Within a household, lights and small appliances are connected from one line to *neutral n*, and large appliances such as hot water heaters and air conditioners are connected line to line. Lights operate at about 120 V rms and large appliances operate at approximately 240 V rms.

Let us now attach two identical loads to the single-phase three-wire voltage system using perfect conductors as shown in Fig. 9.20b. From the figure we note that

$$\mathbf{I}_{aA} = \frac{\mathbf{V}}{\mathbf{Z}_L}$$

LEARNING by Doing

D 9.6 In Fig. 9.20d, if

$\mathbf{V} = 120\ \underline{/0°}$ V
$\mathbf{Z}_L = 100 + j1\ \Omega$
$\mathbf{Z}_{LL} = 10 + j0.1\ \Omega$
$\mathbf{Z}_{\text{line}} = 0$
$\mathbf{Z}_n = 0$

find $\mathbf{I}_a$.

ANSWER $13.2\ \underline{/0.57°}$ A.

and

$$\mathbf{I}_{bB} = -\frac{\mathbf{V}}{\mathbf{Z}_L}$$

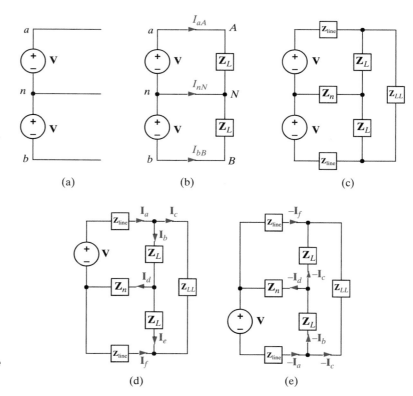

Figure 9.20
Single-phase three-wire system.

KCL at point N is

$$\mathbf{I}_{nN} = -\left(\mathbf{I}_{aA} + \mathbf{I}_{bB}\right)$$

$$= -\left(\frac{\mathbf{V}}{\mathbf{Z}_L} - \frac{\mathbf{V}}{\mathbf{Z}_L}\right)$$

$$= 0$$

Note that there is no current in the neutral wire, and therefore it could be removed without affecting the remainder of the system; that is, all the voltages and currents would be unchanged. One is naturally led to wonder just how far the simplicity exhibited by this system will extend. For example, what would happen if each line had a line impedance, if the neutral conductor had an impedance associated with it, and if there were a load tied from line to line? To explore these questions, consider the circuit in Fig. 9.20c. Although we could examine this circuit using many of the techniques we have employed in previous chapters, the symmetry of the network suggests that perhaps superposition may lead us to some conclusions without having to resort to a brute-force assault. Employing superposition, we consider the two circuits in Figs. 9.20d and e. The currents in Fig. 9.20d are labeled arbitrarily. Because of the symmetrical relationship between Figs. 9.20d and e, the currents in Fig. 9.20e correspond directly to those in Fig. 9.20d. If we add the two *phasor* currents in each branch, we find that the neutral current is again zero. A neutral current of zero is a direct result of the symmetrical nature of the network. If either the line impedances $\mathbf{Z}_{\text{line}}$ or the load impedances $\mathbf{Z}_L$ are unequal, the neutral current will be nonzero. We will make direct use of these concepts when we study three-phase networks in Chapter 10.

LEARNING Example 9.14

A three-wire single-phase household circuit is shown in Fig. 9.21a. The use of the lights, stereo, and range for a 24-hour period is demonstrated in Fig. 9.21b. Let us calculate the energy use over the 24 hours in kilowatt-hours. Assuming that this represents a typical day and that our utility rate is \$0.08/kWh, let us also estimate the power bill for a 30-day month.

SOLUTION   Applying nodal analysis to Fig. 9.21a yields

$$\mathbf{I}_{aA} = \mathbf{I}_L + \mathbf{I}_R$$

$$\mathbf{I}_{bB} = -\mathbf{I}_S - \mathbf{I}_R$$

$$\mathbf{I}_{nN} = \mathbf{I}_S - \mathbf{I}_L$$

The current magnitudes for each load can be found from the corresponding power levels as follows.

$$I_L = \frac{P_L}{V_{an}} = \frac{120}{120} = 1 \text{ A rms}$$

$$I_S = \frac{P_S}{V_{nb}} = \frac{24}{120} = 0.2 \text{ A rms}$$

$$I_R = \frac{P_R}{V_{ab}} = \frac{7200}{240} = 30 \text{ A rms}$$

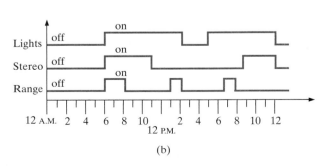

(a)   (b)

Figure 9.21   *Household three-wire network and appliance usage.*

(continued)

The energy used is simply the integral of the power delivered by the two sources over the 24-hour period. Since the voltage magnitudes are constants, we can express the energy delivered by the sources as

$$E_{an} = V_{an} \int I_{aA} \, dt$$

$$E_{nb} = V_{nb} \int -I_{bB} \, dt$$

The integrals of I_{aA} and I_{bB} can be determined graphically from Fig. 9.21b.

$$\int_{12\text{A.M.}}^{12\text{A.M.}} I_{aA} \, dt = 4I_R + 15I_L = 135$$

$$\int_{12\text{A.M.}}^{12\text{A.M.}} -I_{bB} \, dt = 8I_S + 4I_R = 121.6$$

Therefore, the daily energy for each source and the total energy is

$$E_{an} = 16.2 \text{ kWh}$$

$$E_{nb} = 14.6 \text{ kWh}$$

$$E_{\text{total}} = 30.8 \text{ kWh}$$

Over a 30-day month, a \$0.08/kWh utility rate results in a power bill of

$$\text{Cost} = (30.8)(30)(0.08) = \$73.92$$

9.9 Safety Considerations

Although this book is concerned primarily with the theory of circuit analysis, we recognize that most students, by this point in their study, will have begun to relate the theory to the electrical devices and systems that they encounter in the world around them. Thus, it seems advisable to depart briefly from the theoretical and spend some time discussing the very practical and important subject of safety. Electrical safety is a very broad and diverse topic that would require several volumes for a comprehensive treatment. Instead, we will limit our discussion to a few introductory concepts and illustrate them with examples.

It would be difficult to imagine that anyone in our society could have reached adolescence without having experienced some form of electrical shock. Whether that shock was from a harmless electrostatic discharge or from accidental contact with an energized electrical circuit, the response was probably the same—an immediate and involuntary muscular reaction. In either case, the cause of the reaction is current flowing through the body. The severity of the shock depends on several factors, the most important of which are the magnitude, the duration, and the pathway of the current through the body.

The effect of electrical shock varies widely from person to person. Figure 9.22 shows the general reactions that occur as a result of 60-Hz ac current flow through the body from hand to hand, with the heart in the conduction pathway. Observe that there is an intermediate range of current, from about 0.1 to 0.2 A, which is most likely to be fatal. Current levels in this range are apt to produce ventricular fibrillation, a disruption of the orderly contractions of the heart muscle. Recovery of the heartbeat generally does not occur without immediate medical intervention. Current levels above that fatal range tend to cause the heart muscle to contract severely, and if the shock is removed soon enough, the heart may resume beating on its own.

The voltage required to produce a given current depends on the quality of the contact to the body and the impedance of the body between the points of contact. The electrostatic voltage such as might be produced by sliding across a car seat on a dry winter day may be on the order of 20,000 to 40,000 V, and the current surge on touching the door handle, on the order of 40 A. However, the pathway for the current flow is mainly over the body surface, and its duration is for only a few microseconds. Although that shock could be disastrous for some electronic components, it causes nothing more than mild discomfort and aggravation to a human being.

Electrical appliances found about the home typically require 120 or 240 V rms for operation. Although the voltage level is small compared with that of the electrostatic shock, the potential for harm to the individual and to property is much greater. Accidental contact is more apt to result in current flow either from hand to hand or from hand to foot—either of which will subject the heart to shock. Moreover, the relatively slowly changing (low frequency)

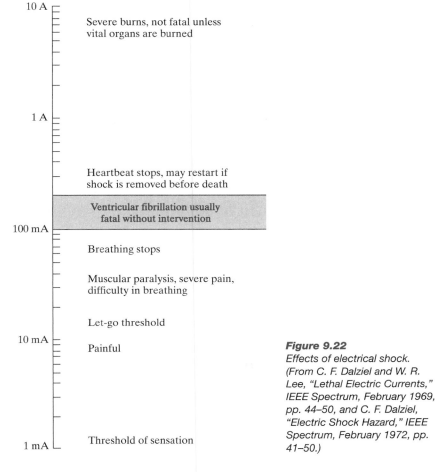

Figure 9.22
Effects of electrical shock.
(From C. F. Dalziel and W. R.
Lee, "Lethal Electric Currents,"
IEEE Spectrum, February 1969,
pp. 44–50, and C. F. Dalziel,
"Electric Shock Hazard," IEEE
Spectrum, February 1972, pp.
41–50.)

60-Hz current tends to penetrate more deeply into the body as opposed to remaining on the surface as a rapidly changing (high frequency) current would tend to do. In addition, the energy source has the capability of sustaining a current flow without depletion. Thus, subsequent discussion will concentrate primarily on hazards associated with the 60-Hz ac power system.

The single-phase three-wire system introduced earlier is commonly, although not exclusively, used for electrical power distribution in residences. Two important aspects of this, or any, system that relate to safety were not mentioned earlier: circuit fusing and grounding.

Each branch circuit, regardless of the type of load it serves, is protected from excessive current flow by circuit breakers or fuses. Receptacle circuits are generally limited to 20 amps and lighting circuits to 15 amps. Clearly, these cannot protect persons from lethal shock. The primary purpose of these current-limiting devices is to protect equipment.

The neutral conductor of the power system is connected to ground (earth) at a multitude of points throughout the system and, in particular, at the service entrance to the residence. The connection to earth may be by way of a driven ground rod or by contact to a cold water pipe of a buried metallic water system. The 120-V branch circuits radiating from the distribution panel (fuse box) generally consist of three conductors rather than only two, as was shown in Fig. 9.20. The third conductor is the ground wire, as shown in Fig. 9.23.

The ground conductor may appear to be redundant, since it plays no role in the normal operation of a load that might be connected to the receptacle. Its role is illustrated by the following example.

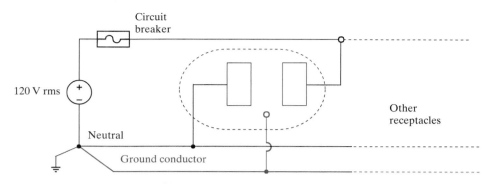

Figure 9.23
A household receptacle.

LEARNING Example 9.15

Joe has a workshop in his basement where he uses a variety of power tools such as drills, saws, and sanders. The basement floor is concrete, and being below ground level, it is usually damp. Damp concrete is a relatively good conductor. Unknown to Joe, the insulation on a wire in his electric drill has been nicked and the wire is in contact with (or shorted to) the metal case of the drill, as shown in Fig. 9.24. Is Joe in any danger when using the drill?

SOLUTION Without the ground conductor connected to the metal case of the tool, Joe would receive a severe, perhaps fatal, shock when he attempts to use the drill. The voltage between his hand and his feet would be 120 V, and the current through his body would be limited by the resistance of his body and of the concrete floor. Typically, the circuit breakers would not operate. However, if the ground conductor is present and properly connected to the drill case, the case remains at ground potential, the

120-V source becomes shorted to ground, the circuit breaker operates, and Joe lives to drill another hole.

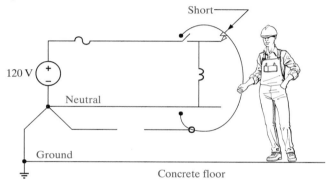

Figure 9.24   *Faulty circuit, when the case of the tool is not grounded through the power cord.*

It was mentioned earlier that the circuit breaker or fuse cannot provide effective protection against shock. There is, however, a special type of device called a ground-fault interrupter (GFI) that can provide protection for personnel. This device detects current flow outside the normal circuit. Consider the circuit of Fig. 9.24. In the normal safe operating condition, the current in the neutral conductor must be the same as that in the line conductor. If at any time the current in the line does not equal the current in the neutral, then a secondary path has somehow been established, creating an unsafe condition. This secondary path is called a fault. For example, the fault path in Fig. 9.24 is through Joe and the concrete floor. The GFI detects this fault and opens the circuit in response. Its principle of operation is illustrated by the following example.

LEARNING Example 9.16

Let us describe the operation of a GFI.

SOLUTION Consider the action of the magnetic circuit in Fig. 9.25. Under normal operating conditions, i_1 and i_2 are equal, and if the coils in the neutral and line conductors are identical, as we learned in basic physics, the magnetic flux in the core will be zero. Consequently, no voltage will be induced in the sensing coil.

Figure 9.25   *Ground-fault interrupter circuit.*

If a fault should occur at the load, current will flow in the ground conductor and perhaps in the earth; thus, i_1 and i_2 will no longer be equal, the magnetic flux will not be zero, and a voltage will be induced in the sensing coil. That voltage can be used to activate a circuit breaker. This is the essence of the GFI device.

Ground-fault interrupters are available in the form of circuit breakers and also as receptacles. They are now required in branch circuits that serve outlets in areas such as bathrooms, basements, garages, and outdoor sites. The devices will operate at ground-fault currents on the order of a few milliamperes. Unfortunately, the GFI is a relatively new device and electrical code requirements are generally not retroactive. Thus few older residences have them.

Requirements for the installation and maintenance of electrical systems are meticulously defined by various codes that have been established to provide protection of personnel and property. Installation, alteration, or repair of electrical devices and systems should be undertaken only by qualified persons. The subject matter that we study in circuit analysis does not provide that qualification.

The following examples illustrate the potential hazards that can be encountered in a variety of everyday situations. We begin by revisiting a situation described in a previous example.

LEARNING Example 9.17

Suppose that a man is working on the roof of a mobile home with a hand drill. It is early in the day, the man is barefoot, and dew covers the mobile home. The ground prong on the electrical plug of the drill has been removed. Will the man be shocked if the "hot" electrical line shorts to the case of the drill?

SOLUTION To analyze this problem, we must construct a model that adequately represents the situation described. In his book *Medical Instrumentation* (Boston: Houghton Mifflin, 1978), John G. Webster suggests the following values for resistance of the human body: $R_{skin}(dry) = 15 \text{ k}\Omega$, $R_{skin}(wet) = 150 \text{ }\Omega$, $R_{limb}(arm \text{ or } leg) = 100 \text{ }\Omega$, and $R_{trunk} = 200 \text{ }\Omega$.

The network model is shown in Fig. 9.26. Note that since the ground line is open-circuited, a closed path exists from the hot wire through the short, the human body, the mobile home, and the ground. For the conditions stated previously, we assume that the surface contact resistances R_{sc_1} and R_{sc_2} are 150 Ω each. The body resistance, R_{body}, consisting of arm, trunk, and leg, is 400 Ω. The mobile home resistance is assumed to be zero, and the ground resistance, R_{gnd}, from the mobile home ground to the actual source ground is assumed to be 1 Ω. Therefore, the magnitude of the current through the body from hand to foot would be

$$\mathbf{I}_{body} = \frac{120}{R_{sc_1} + R_{body} + R_{sc_2} + R_{gnd}}$$
$$= \frac{120}{701}$$
$$= 171 \text{ mA}$$

A current of this magnitude can easily cause heart failure.

It is important to note that additional protection would be provided if the circuit breaker were a ground-fault interrupter.

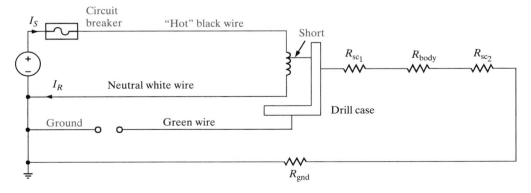

Figure 9.26 *Model for Example 9.17.*

LEARNING Example 9.18

Two boys are playing basketball in their backyard. To cool off, they decide to jump into their pool. The pool has a vinyl lining, so the water is electrically insulated from the earth. Unknown to the boys, there is a ground fault in one of the pool lights. One boy jumps in and while standing in the pool with water up to his chest, reaches up to pull in the other boy, who is holding onto a grounded hand rail, as shown in Fig. 9.27a. What is the impact of this action?

SOLUTION   The action in Fig. 9.27a is modeled as shown in Fig. 9.27b. Note that since a ground fault has occurred, there exists a current path through the two boys. Assuming that the

fault, pool, and railing resistances are approximately zero, the magnitude of the current through the two boys would be

$$I = \frac{120}{(3R_{\text{arm}}) + 3(3R_{\text{wet contact}}) + R_{\text{trunk}}}$$

$$= \frac{120}{950}$$

$$= 126 \text{ mA}$$

This current level would cause severe shock in both boys. The boy outside the pool could experience heart failure.

(a)

Figure 9.27
Diagrams used in Example 9.18.

(b)

LEARNING Example 9.19

A patient in a medical laboratory has a muscle stimulator attached to her left forearm. Her heart rate is being monitored by an EKG machine with two differential electrodes over the heart and the ground electrode attached to her right ankle. This activity is illustrated in Fig. 9.28a. The stimulator acts as a current

source that drives 150 mA through the muscle from the active electrode to the passive electrode. If the laboratory technician mistakenly decides to connect the passive electrode of the stimulator to the ground electrode of the EKG system to achieve a common ground, is there any risk?

(a)

Figure 9.28   *Diagrams used in Example 9.19.*

SOLUTION   When the passive electrode of the stimulator is connected to the ground electrode of the EKG system, the equivalent network in Fig. 9.28b illustrates the two paths for the stimulator current: one through half an arm and the other through half an arm and the body. Using current division, the body current is

$$\mathbf{I}_{body} = \frac{(150)(10^{-3})(50)}{50 + 50 + 200 + 100}$$

$$= 19 \text{ mA}$$

Therefore, a dangerously high level of current will flow from the stimulator through the body to the EKG ground.

LEARNING Example 9.20

A cardiac care patient with a pacing electrode has ignored the hospital rules and is listening to a cheap stereo. The stereo has an amplified 60-Hz hum that is very annoying. The patient decides to dismantle the stereo partially in an attempt to eliminate the hum. In the process, while he is holding one of the speaker wires, the other touches the pacing electrode. What are the risks in this situation?

SOLUTION   Let us suppose that the patient's skin is damp and that the 60-Hz voltage across the speaker wires is only 10 mV. Then the circuit model in this case would be shown in Fig. 9.29. The current through the heart would be

$$I = \frac{(10)(10^{-3})}{150 + 100 + 200}$$

$$= 22.2 \text{ μA}$$

It is known that 10 μA delivered directly to the heart is potentially lethal.

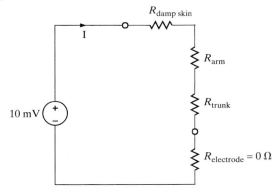

Figure 9.29   *Circuit model for Example 9.20.*

LEARNING Example 9.21

While maneuvering in a muddy area, a crane operator accidentally touched a high-voltage line with the boom of the crane, as illustrated in Fig. 9.30a. The line potential was 7200 V. The neutral conductor was grounded at the pole. When the crane operator realized what had happened, he jumped from the crane and ran in the direction of the pole, which was approximately 10 m away. He was electrocuted as he ran. Can we explain this very tragic accident?

SOLUTION   The conditions depicted in Fig. 9.30a can be modeled as shown in Fig. 9.30b. The crane was at 7200 V with respect to earth. Therefore, a gradient of 720 V/m existed along the earth between the crane and the power pole. This earth between crane and the pole is modeled as a resistance. If the man's stride was about 1 m, the difference in potential between his feet was approximately 720 V. A man standing in the same area with his feet together was unharmed.

(continued)

Figure 9.30 *Illustrations used in Example 9.21.*

The examples of this section have been provided in an attempt to illustrate some of the potential dangers that exist when working or playing around electric power. In the worst case, failure to prevent an electrical accident can result in death. However, even nonlethal electrical contacts can cause such things as burns or falls. Therefore, we must always be alert to ensure not only our own safety, but also that of others who work and play with us.

LEARNING Hint

Safety guidelines

The following guidelines will help us minimize the chances of injury.

1. Avoid working on energized electrical systems.
2. Always assume that an electrical system is energized unless you can absolutely verify that it is not.
3. Never make repairs or alterations that are not in compliance with the provisions of the prevailing code.
4. Do not work on potentially hazardous electrical systems alone.
5. If another person is "frozen" to an energized electrical circuit, deenergize the circuit, if possible. If that cannot be done, use nonconductive material such as dry wooden boards, sticks, belts, and articles of clothing to separate the body from the contact. Act quickly but take care to protect yourself.
6. When handling long metallic equipment, such as ladders, antennas, and so on, outdoors, be continuously aware of overhead power lines and avoid any possibility of contact with them.

LEARNING EXTENSION

E9.14 A woman is driving her car in a violent rainstorm. While she is waiting at an intersection, a power line falls on her car and makes contact. The power line voltage is 7200 V.

(a) Assuming that the resistance of the car is negligible, what is the potential current through her body if, while holding the door handle with a dry hand, she steps out onto the wet ground?

(b) If she remained in the car, what would happen?

ANSWER (a) $I = 463$ mA, extremely dangerous; (b) she should be safe.

Safety when working with electric power must always be a primary consideration. Regardless of how efficient or expedient an electrical network is for a particular application, it is worthless if it is also hazardous to human life.

The safety device shown in Fig. 9.31, which is also used for troubleshooting, is a proximity-type sensor that will indicate whether a circuit is energized by simply touching the conductor

Figure 9.31
*A modern safety or
troubleshooting device
(courtesy of Fluke Corporation).*

on the outside of the insulation. This device is typically carried by all electricians and is help-ful when working on electric circuits.

In addition to the numerous deaths that occur each year due to electrical accidents, fire damage that results from improper use of electrical wiring and distribution equipment amounts to millions of dollars per year.

To prevent the loss of life and damage to property, very detailed procedures and specifica-tions have been established for the construction and operation of electrical systems to ensure their safe operation. The *National Electrical Code* ANSI C1 (ANSI—American National Stan-dards Institute) is the primary guide. There are other codes, however: for example, the National Electric Safety Code, ANSI C2, which deals with safety requirements for public utilities. The Underwriters' Laboratory (UL) tests all types of devices and systems to ensure that they are safe for use by the general public. We find the UL label on all types of electrical equipment that is used in the home, such as appliances and extension cords.

Electric energy plays a very central role in our lives. It is extremely important to our gen-eral health and well-being. However, if not properly used, it can be lethal.

Learning by Application

The following application-oriented examples illustrate a practical use of the material studied in this chapter.

LEARNING Example 9.22

An electric lawn mower requires 12 A rms at 120 V rms but will operate down to 110 V rms without damage. At 110 V rms, the current draw is 13.1 A rms, as shown in Fig. 9.32. Give the maximum length extension cord that can be used with a 120-V rms power source if the extension cord is made from

1. 16-gauge wire (4 mΩ/ft)
2. 14-gauge wire (2.5 mΩ/ft)

Figure 9.32   *Circuit model in Example 9.22.*

(continued)

SOLUTION The voltage drop across the extension cord is

$$V_{cord} = (2)(13.1)R_{cord} = 10 \text{ V rms}$$

or

$$R_{cord} = 0.382 \ \Omega$$

If ℓ_{cord} is the length of the extension cord, then for 16-gauge wire

we find

$$\ell_{cord} = \frac{R_{cord}}{0.004} = 95.5 \text{ feet}$$

and for 14-gauge wire

$$\ell_{cord} = \frac{R_{cord}}{0.0025} = 152.8 \text{ feet}$$

LEARNING Example 9.23

While sitting in a house reading a book, we notice that every time the air conditioner comes on, the lights momentarily dim. Let us investigate this phenomenon using the single-phase three-wire circuit shown in Fig. 9.33a and some typical current requirements for a 10,000-Btu/h air conditioner, assuming a line resistance of 0.5 Ω.

LEARNING Hint

Technique

1. Find the resistance of the lightbulb.
2. Use a large current source to represent the transient current of the air conditioner and a small current source to represent the steady-state current.
3. Find the voltage drop across the lightbulb during both the transient and steady-state operations.

SOLUTION The 60-W lightbulb can be roughly modeled by its equivalent resistance:

$$P_{bulb} = \frac{V_{an}^2}{R_{bulb}}$$

or

$$R_{bulb} = 240 \ \Omega$$

When the air conditioner unit first turns on, the current requirement is 40 A, as shown in Fig. 9.33b. As the compressor motor comes up to speed, the current requirement drops quickly to a steady-state value of 10 A, as shown in Fig. 9.33c. We will compare the voltage across the light fixture, V_{AN}, both at turn-on and in steady state.

Figure 9.33 Networks used in Example 9.23.

Using superposition, let us first find that portion of V_{AN} caused by the voltage sources. The appropriate circuit is shown in Fig. 9.33d. Using voltage division, we find that

$$V_{AN1} = V_{AN}\left(\frac{R_{bulb}}{R_{bulb} + 2R_L}\right)$$

or

$$V_{AN1} = 119.50 \text{ V rms}$$

Figure 9.33e will yield the contribution to V_{AN} caused by the 10-A steady-state current. Using current division to calculate the current through the lightbulb, we find that

$$V_{AN2} = -\left\{I_{AB}\left(\frac{R_L}{R_{bulb} + 2R_L}\right)\right\}R_{bulb}$$

or

$$V_{AN2} = -4.98 \text{ V rms}$$

Therefore, the steady-state value of V_{AN} is

$$V_{AN} = V_{AN1} + V_{AN2} = 114.52 \text{ V rms}$$

At start-up, our expression for V_{AN2} can be used with $I_{AB} = 40 \text{ A}$, which yields $V_{AN2} = -19.92$ V rms. The resulting value for V_{AN} is

$$V_{AN} = V_{AN1} + V_{AN2} = 119.50 - 19.92 = 99.58 \text{ V rms}$$

The voltage delivered to the light fixture at start-up is 13% lower than the steady-state value, resulting in a momentary dimming of the lights.

Learning by Design

The following design example is presented in the context of a single-phase three-wire house-hold circuit and serves as a good introduction to the material in the next chapter.

LEARNING Example 9.24

A light-duty commercial single-phase three-wire 60-Hz circuit serves lighting, heating, and motor loads, as shown in Fig. 9.34a. Lighting and heating loads are essentially pure resistance and, hence, unity power factor (pf), whereas motor loads have lagging pf.

We wish to design a balanced configuration for the network and determine its economic viability using the following procedure.

(a) Compute the phase and neutral currents, and the complex power and pf for each source.

(b) Now move the heating load (panel H) to phase b, as shown in Fig. 9.34b. This is called "balancing" the load. Repeat the analysis of (a).

(c) Assume that the phase and neutral conductor resistances are each $0.05 \ \Omega$ and have negligible effect on the results of (a). Evaluate the system line losses for (a) and (b). If the loads in question operate 24 hours per day and 365 days per year, at $0.08/kWh, how much energy (and money) is saved by operating in the balanced mode?

SOLUTION   (a) The magnitudes of the rms currents are

$$I_L = I_H = \frac{P}{V} = \frac{5000}{120} = 41.67 \text{ A}$$

and

$$I_m = \frac{10,000}{240} = 41.67 \text{ A}$$

(a)

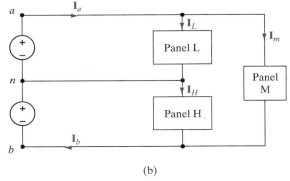

(b)

Figure 9.34   *Single-phase three-wire power distribution system.*

(continued)

In addition,
$$\theta_m = \cos^{-1}(0.8) = -36.9°$$

Therefore,
$$\mathbf{I}_a = \mathbf{I}_L + \mathbf{I}_H + \mathbf{I}_m$$
$$= 41.67\underline{/0°} + 41.67\underline{/0°} + 41.67\underline{/-36.9°}$$
$$= 119.4\underline{/-12.1°} \text{ A}$$

The currents in the neutral and phase b lines are
$$\mathbf{I}_n = \mathbf{I}_L + \mathbf{I}_H = 83.34\underline{/0°} \text{ A}$$
$$\mathbf{I}_b = 41.67\underline{/-36.9°} \text{ A}$$

The complex power and power factor for each source are
$$\mathbf{S}_a = \mathbf{V}_{an}\mathbf{I}_a^* = (120\underline{/0°})(119.4\underline{/+12.1°}) = 14 + j3 \text{ kVA}$$
$$\text{pf}_a = \cos(12.1°) = 0.9778 \text{ lagging}$$

and in a similar manner
$$\mathbf{S}_b = \mathbf{V}_{bn}\mathbf{I}_b^* = 4 + j3 \text{ kVA}$$
$$\text{pf}_b = 0.8 \text{ lagging}$$

(b) Under the balanced condition
$$\mathbf{I}_a = \mathbf{I}_L + \mathbf{I}_m = 41.67\underline{/0°} + 41.67\underline{/-36.9°}$$
$$= 79.06\underline{/-18.4°} \text{ A}$$

and
$$\mathbf{I}_b = 79.06\underline{/-18.4°} \text{ A}$$
$$\mathbf{I}_n = 0$$

Therefore,
$$\mathbf{S}_a = \mathbf{V}_{na}\mathbf{I}_a^* = 9 + j3 \text{ kVA}$$
$$\text{pf}_a = 0.9487 \text{ lagging}$$

and
$$\mathbf{S}_b = \mathbf{V}_{bn}\mathbf{I}_b^* = 9 + j3 \text{ kVA}$$
$$\text{pf}_b = 0.9487 \text{ lagging}$$

(c) The power loss in the lines in kW is
$$P_{\text{loss}} = R_a I_a^2 + R_b I_b^2 + R_n I_c^2$$
$$= 0.05(I_a^2 + I_b^2 + I_n^2)/1000$$

The total energy loss for a year is
$$W_{\text{loss}} = (24)(365)P_{\text{loss}} = 8760\, P_{\text{loss}}$$

and the annual cost is
$$\text{Cost} = \$0.08\, W_{\text{loss}}$$

A comparison of the unbalanced and balanced cases is shown in the following table.

	Unbalanced Case	Balanced Case
$P_{\text{loss}}(\text{kW})$	1.147	0.625
$W_{\text{loss}}(\text{kW-hr})$	10,034	5,475
Cost($)	804	438

Therefore, the annual savings obtained using the balanced configuration is

Annual energy savings = $10,048 - 5,475 = 4,573$ kWh

Annual cost savings = $804 - 438 = \$366$

LEARNING Check

Summary

▶ **Instantaneous power** If the current and voltage are sinusoidal functions of time, the instantaneous power is equal to a time-independent average value plus a sinusoidal term that has a frequency twice that of the voltage or current.

▶ **Average power** $P = \frac{1}{2}VI\cos(\theta_v - \theta_i) = \frac{1}{2}VI\cos\theta$, where θ is the phase of the impedance.

▶ **Resistive load** $P = \frac{1}{2}I^2R = \frac{1}{2}VI$ since $\mathbf{V}$ and $\mathbf{I}$ are in phase.

▶ **Reactive load** $P = \frac{1}{2}VI\cos(\pm 90°) = 0$

▶ **Maximum average power transfer** To obtain the maximum average power transfer to a load, the load impedance should be chosen equal to the complex conjugate of the Thévenin equivalent impedance representing the remainder of the network.

▶ **rms or effective value of a periodic waveform** The effective, or rms, value of a periodic waveform was introduced as a means of measuring the effectiveness of a source in delivering power to a resistive load. The effective value of a periodic waveform is found by determining the root-mean-square value of the waveform. The rms value of a sinusoidal function is equal to the maximum value of the sinusoid divided by $\sqrt{2}$.

▶ **Power factor**   Apparent power is defined as the product $V_{rms}I_{rms}$. The power factor is defined as the ratio of the average power to the apparent power and is said to be leading when the phase of the current leads the voltage, and lagging when the phase of the current lags the voltage. The power factor of a load with a lagging power factor can be corrected by placing a capacitor in parallel with the load.

▶ **Complex power**   The complex power, **S**, is defined as the product $\mathbf{V}_{rms}\mathbf{I}_{rms}^*$. The complex power **S** can be written

as $\mathbf{S} = P + jQ$, where P is the real or average power and Q is the imaginary or quadrature power.

$$\mathbf{S} = I^2\mathbf{Z} = I^2R + jI^2X$$

▶ **The single-phase three-wire circuit**   The single-phase three-wire circuit is the one commonly used in households. Large appliances are connected line to line and small appliances and lights are connected line to neutral.

▶ **Safety**   Safety must be a primary concern in the design and use of any electrical circuit. The National Electric Code is the primary guide for the construction and operation of electrical systems.

Problems

SECTION 9.1

9.1 Determine the equations for the current and the instantaneous power in the network in Fig. P9.1.

Figure P9.1

9.2 Determine the equations for the voltage and the instantaneous power in the network in Fig. P9.2.

Figure P9.2

SECTION 9.2

9.3 Find the average power absorbed by the network shown in Fig. P9.3.

Figure P9.3

9.4 Given the network in Fig. P9.4, find the power supplied and the average power absorbed by each element.

Figure P9.4

9.5 Given the circuit in Fig. P9.5, find the average power supplied and the average power absorbed by each element.

Figure P9.5

9.6 Compute the average power absorbed by the elements to the right of the dashed line in the network shown in Fig. P9.6.

Figure P9.6

9.7 Given the network in Fig. P9.7, determine which elements are supplying power, which ones are absorbing power, and how much power is being supplied and absorbed.

Figure P9.7

9.8 Given the circuit in Fig. P9.8, find the average power absorbed by the network.

Figure P9.8

9.9 Given the network in Fig. P9.9, find the average power supplied to the circuit.

Figure P9.9

9.10 Given the circuit in Fig. P9.10, determine the amount of average power supplied to the network.

Figure P9.10

9.11 Determine the average power absorbed by the 4-Ω resistor in the network shown in Fig. P9.11.

Figure P9.11

9.12 Given the network in Fig. P9.12, show that the power supplied by the sources is equal to the power absorbed by the passive elements.

Figure P9.12

9.13 Calculate the average power absorbed by the 1-Ω resistor in the network shown in Fig. P9.13.

Figure P9.13

9.14 Determine the average power absorbed by the 4-Ω resistor in the network shown in Fig. P9.14.

Figure P9.14

9.15 Find the average power absorbed by each element in the network shown in Fig. P9.15.

Figure P9.15

9.16 Determine the average power supplied to the network shown in Fig. P9.16.

Figure P9.16

9.17 Determine the average power supplied to the network in Fig. P9.17.

Figure P9.17

9.18 Find the average power absorbed by the 2-Ω resistor in the circuit shown in Fig. P9.18.

Figure P9.18

9.19 Given the network in Fig. P9.19, determine which elements are supplying power, which ones are absorbing power, and how much power is being supplied and absorbed.

Figure P9.19

SECTION 9.3

9.20 Determine the impedance $\mathbf{Z}_L$ for maximum average power transfer and the value of the maximum power transferred to the $\mathbf{Z}_L$ for the circuit shown in Fig. P9.20.

Figure P9.20

9.21 Determine the impedance $\mathbf{Z}_L$ for maximum average power transfer and the value of the maximum average power transferred to $\mathbf{Z}_L$ for the circuit shown in Fig. P9.21.

Figure P9.21

9.22 Determine the impedance $\mathbf{Z}_L$ for maximum average power transfer and the value of the maximum average power absorbed by the load in the network shown in Fig. P9.22.

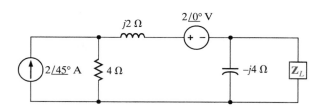

Figure P9.22

9.23 Determine the impedance $\mathbf{Z}_L$ for maximum average power transfer and the value of the maximum average power absorbed by the load in the network shown in Fig. P9.23.

Figure P9.23

9.24 Determine the impedance $\mathbf{Z}_L$ for maximum average power transfer and the value of the maximum average power transferred to $\mathbf{Z}_L$ for the circuit shown in Fig. P9.24.

Figure P9.24

9.25 In the network in Fig. P9.25, find $\mathbf{Z}_L$ for maximum average power transfer and the maximum average power transferred.

Figure P9.25

9.26 Repeat Problem 9.24 for the network in Fig. P9.26.

Figure P9.26

9.27 Repeat Problem 9.24 for the network in Fig. P9.27.

Figure P9.27

9.28 Repeat Problem 9.24 for the network in Fig. P9.28.

Figure P9.28

9.29 Repeat Problem 9.24 for the network in Fig. P9.29.

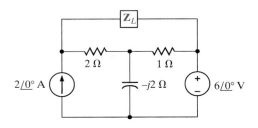

Figure P9.29

9.30 Determine the impedance $\mathbf{Z}_L$ for maximum average power transfer and the value of the maximum average power absorbed by the load in the network shown in Fig. P9.30.

Figure P9.30

9.31 Determine the impedance $\mathbf{Z}_L$ for maximum average power transfer and the value of the maximum average power absorbed by the load in the network shown in Fig. P9.31.

Figure P9.31

9.32 Find the impedance $\mathbf{Z}_L$ for maximum average power transfer and the value of the maximum average power transferred to $\mathbf{Z}_L$ for the circuit shown in Fig. P9.32.

Figure P9.32

9.33 Determine the impedance $\mathbf{Z}_L$ for maximum average power transfer and the value of the maximum average power absorbed by the load in the network shown in Fig. P9.33.

Figure P9.33

SECTION 9.4

9.34 Compute the rms value of the voltage given by the waveform shown in Fig. P9.34.

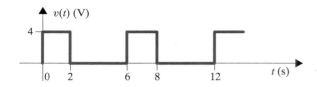

Figure P9.34

9.35 Find the rms value of the waveform shown in Fig. P9.35.

Figure P9.35

9.36 Calculate the rms value of the waveform shown in Fig. P9.36.

Figure P9.36

9.37 Calculate the rms value of the waveform shown in Fig. P9.37.

Figure P9.37

9.38 The current waveform in Fig. P9.38 is flowing through a 5-Ω resistor. Find the average power absorbed by the resistor.

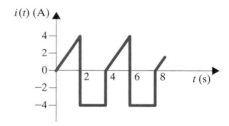

Figure P9.38

9.39 Calculate the rms value of the waveform shown in Fig. P9.39.

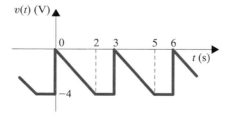

Figure P9.39

9.40 Compute the rms value of the waveform in Fig. P9.40.

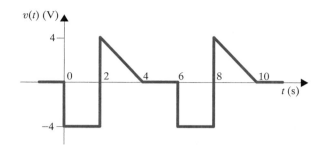

Figure P9.40

9.41 Calculate the rms value of the waveform shown in Fig. P9.41.

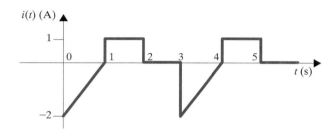

Figure P9.41

9.42 Calculate the rms value of the waveform shown in Fig. P9.42.

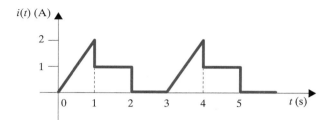

Figure P9.42

SECTION 9.5

9.43 An industrial load consumes 100 kW at 0.8 pf lagging. If an ammeter in the transmission line indicates that the load current is 284 A rms, find the load voltage.

9.44 An industrial load that consumes 80 kW is supplied by the power company through a transmission line with 0.1-Ω resistance, with 84 kW. If the voltage at the load is 440 V rms, find the power factor at the load.

9.45 The power company supplies 80 kW to an industrial load. The load draws 220 A rms from the transmission line. If the load voltage is 440 V rms and the load power factor is 0.8 lagging, find the losses in the transmission line.

9.46 An industrial plant with an inductive load consumes 10 kW of power from a 220-V rms line. If the load power factor is 0.8, what is the angle by which the load voltage leads the load current?

9.47 The power company must generate 100 kW to supply an industrial load with 94 kW through a transmission line with 0.09-Ω resistance. If the load power factor is 0.83 lagging, find the load voltage.

9.48 An industrial load operates at 30 kW, 0.8 pf lagging. The load voltage is 220 $\underline{/0°}$ V rms. The real and reactive power losses in the transmission-line feeder are 1.8 kW and 2.4 kvar, respectively. Find the impedance of the transmission line and the input voltage to the line.

9.49 A transmission line with impedance 0.08 + j0.25 Ω is used to deliver power to a load. The load is inductive and the load voltage is 220 $\underline{/0°}$ V rms at 60 Hz. If the load requires 12 kW and the real power loss in the line is 560 W, determine the power factor angle of the load.

SECTION 9.6

9.50 Find the source voltage in the network shown in Fig. P9.50.

Figure P9.50

9.51 Given the network in Fig. P9.51, compute the input source voltage and the input power factor.

Figure P9.51

9.52 Given the network in Fig. P9.52, determine the input voltage $\mathbf{V}_S$.

Figure P9.52

9.53 Find the input source voltage and the power factor of the source for the network shown in Fig. P9.53.

Figure P9.53

9.54 Use Kirchhoff's laws to compute the source voltage of the network shown in Fig. P9.54.

Figure P9.54

9.55 Given the network in Fig. P9.55 determine the input voltage $\mathbf{V}_S$.

Figure P9.55

SECTION 9.7

9.56 What value of capacitance must be placed in parallel with the 18-kW load in Problem 9.51 to raise the power factor of this load to 0.9 lagging?

9.57 An industrial load is supplied through a transmission line that has a line impedance of $0.1 + j0.2 \; \Omega$. The 60-Hz line voltage at the load is $480 \underline{/0°}$ V rms. The load consumes 124 kW at 0.75 pf lagging. What value of capacitance when placed in parallel with the load will change the power factor to 0.9 lagging?

9.58 A plant consumes 60 kW at a power factor of 0.5 lagging from a 220-V rms 60-Hz line. Determine the value of the capacitor that when placed in parallel with the load will change the load power factor to 0.9 lagging.

9.59 A small plant has a bank of induction motors that consume 64 kW at a pf of 0.68 lagging. The 60-Hz line voltage across the motors is $220 \underline{/0°}$ V rms. The local power company has told the plant to raise the pf to 0.92 lagging. What value of capacitance is required?

9.60 The 60-Hz line voltage for a 60-kW, 0.76-pf lagging industrial load is $440 \underline{/0°}$ V rms. Find the value of capacitance that when placed in parallel with the load will raise the power factor to 0.9 lagging.

9.61 A particular load has a pf of 0.8 lagging. The power delivered to the load is 40 kW from a 220-V rms 60-Hz line. What value of capacitance placed in parallel with the load will raise the pf to 0.9 lagging?

9.62 An industrial load consumes 44 kW at 0.82 pf lagging from a $220 \underline{/0°}$ V-rms, 60-Hz line. A bank of capacitors totaling 900 μF is available. If these capacitors are placed in parallel with the load, what is the new power factor of the total load?

9.63 A bank of induction motors consumes 36 kW at 0.78 pf lagging from a 60-Hz, $220 \underline{/0°}$ V-rms line. If 500 μF of capacitors is placed in parallel with the load, find the new power factor of the combined load.

SECTION 9.8

9.64 A single-phase three-wire 60-Hz circuit serves three loads, as shown in Fig. P9.64. Determine $\mathbf{I}_{aA}$, $\mathbf{I}_{nN}$, $\mathbf{I}_c$, and the energy use over a 24-hour period in kilowatt-hours.

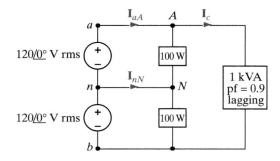

Figure P9.64

9.65 A man and his son are flying a kite. The kite becomes entangled in a 7200-V power line close to a power pole. The man crawls up the pole to remove the kite. While trying to remove the kite, the man accidentally touches the 7200-V line. Assuming the power pole is well grounded, what is the potential current through the man's body?

9.66 A number of 120-V household fixtures are to be used to provide lighting for a large room. The total lighting load is 8 kW. The National Electric Code requires that no circuit breaker be larger than 20 A with a 25% safety margin. Determine the number of identical branch circuits needed for this requirement.

9.67 A 5.1-kW household range is designed to operate on a 240-V rms sinusoidal voltage, as shown in Fig. P9.67a. However, the electrician has mistakenly connected the range to 120 V rms, as shown in Fig. P9.67b. What is the effect of this error?

Figure P9.67

9.68 To test a light socket, a woman, while standing on cushions that insulate her from the ground, sticks her finger into the socket, as shown in Fig. P9.68. The tip of her finger makes contact with one side of the line, and the side of her finger makes contact with the other side of the line. Assuming that any portion of a limb has a resistance of 95 Ω, is there any current in the body? Is there any current in the vicinity of the heart?

Figure P9.68

9.69 An inexperienced mechanic is installing a 12-V battery in a car. The negative terminal has been connected. He is currently tightening the bolts on the positive terminal. With a tight grip on the wrench, he turns it so that the gold ring on his finger makes contact with the frame of the car. This situation is modeled in Fig. P9.69, where we assume that the resistance of the wrench is negligible and the resistance of the contact is as follows:

$$R_1 = R_{\text{bolt to wrench}} = 0.012 \ \Omega$$

$$R_2 = R_{\text{wrench to ring}} = 0.012 \ \Omega$$

$$R_3 = R_{\text{ring}} = 0.012 \ \Omega$$

$$R_4 = R_{\text{ring to frame}} = 0.012 \ \Omega$$

What power is quickly dissipated in the gold ring, and what is the impact of this power dissipation?

Figure P9.69

Typical Problems Found on the FE Exam

9FE-1 An industrial load consumes 120 kW at 0.707 pf lagging and is connected to a 480 $\underline{/0°}$ V rms 60-Hz line. Determine the value of the capacitor that, when connected in parallel with the load, will raise the power factor to 0.95 lagging.

9FE-2 Determine the average and rms values of the following waveform.

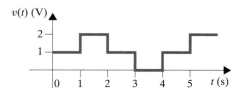

Figure 9PFE-2

9FE-3 Find the impedance $\mathbf{Z}_L$ in the network in Fig. 9PFE-3 for maximum average power transfer.

Figure 9PFE-3

9FE-4 An rms-reading voltmeter is connected to the output of the op-amp shown in Fig. 9PFE-4. Determine the meter reading.

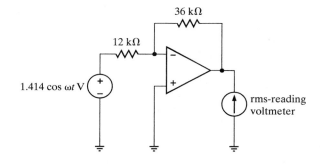

Figure 9PFE-4

9FE-5 Determine the average power delivered to the resistor in Fig. 9PFE-5a if the current waveform is shown in Fig. 9PFE-5b. k

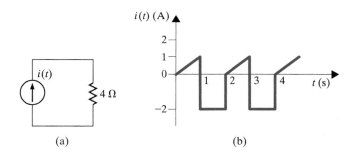

Figure 9PFE-5

10 | *Polyphase Circuits*

LEARNING Goals

While traveling throughout our city or across the countryside, we often encounter power transmission lines overhead. If we look carefully, we find that there are typically three large conductors. Depending on the configuration, there may be one or two smaller conductors that are placed above the power conductors and used for lightning protection. This three-phase power distribution line is employed to transmit power from the generating station to the user. This three-phase circuit is important because we can show that (1) power distributed in this manner is steady, rather than pulsating, as is the case in single-phase distribution and, therefore, there is less wear and tear on mechanical conversion equipment; and (2) by using a three-phase set of voltages we can deliver the same amount of power as that delivered in the single-phase case with fewer conductors and therefore much less material.

10.1 Three-Phase Circuits

In this chapter we add a new dimension to our study of ac steady-state circuits. Up to this point we have dealt with what we refer to as single-phase circuits. Now we extend our analysis techniques to polyphase circuits or, more specifically, three-phase circuits (that is, circuits containing three voltage sources that are one-third of a cycle apart in time).

There are a number of important reasons why we study three-phase circuits. It is more advantageous and economical to generate and transmit electric power in the polyphase mode than with single-phase systems. As a result, most electric power is transmitted in polyphase circuits. In the United States the power system frequency is 60 Hz, whereas in other parts of the world 50 Hz is common.

Power transmission is most efficiently accomplished at very high voltage. Since this voltage can be extremely high in comparison to the level at which it is normally used (e.g., in the household), there is a need to raise and lower the voltage. This can be accomplished in ac systems using transformers, which we studied in Chapter 8.

As the name implies, three-phase circuits are those in which the forcing function is a three-phase system of voltages. If the three sinusoidal voltages have the same magnitude and frequency and each voltage is 120° out of phase with the other two, the voltages are said to be *balanced*. If the loads are such that the currents produced by the voltages are also balanced, the entire circuit is referred to as a *balanced three-phase circuit*.

A balanced set of three-phase voltages can be represented in the frequency domain as shown in Fig. 10.1a, where we have assumed that their magnitudes are 120 V rms. From the figure we note that

$$\mathbf{V}_{an} = 120 \underline{/0°} \text{ V rms}$$
$$\mathbf{V}_{bn} = 120 \underline{/-120°} \text{ V rms}$$
$$\mathbf{V}_{cn} = 120 \underline{/-240°} \text{ V rms}$$
$$= 120 \underline{/120°} \text{ V rms}$$

10.1

Note that our double-subscript notation is exactly the same as that employed in the earlier chapters; that is, $\mathbf{V}_{an}$ means the voltage at point a with respect to the point n. We will also employ the double-subscript notation for currents; that is, $\mathbf{I}_{an}$ is used to represent the current from a to n. However, we must be very careful in this case to describe the precise path, since in a circuit there will be more than one path between the two points. For example, in the case of a single loop the two possible currents in the two paths will be 180° out of phase with each other.

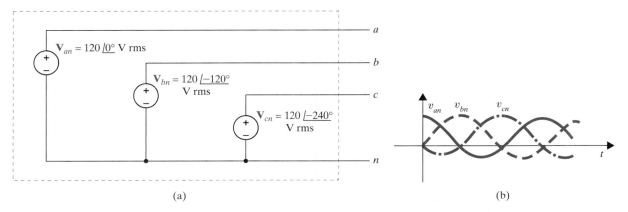

(a)   (b)

Figure 10.1   *Balanced three-phase voltages.*

The preceding phasor voltages can be expressed in the time domain as

$$v_{an}(t) = 120\sqrt{2}\cos\omega t \text{ V}$$
$$v_{bn}(t) = 120\sqrt{2}\cos(\omega t - 120°) \text{ V} \qquad \textbf{10.2}$$
$$v_{cn}(t) = 120\sqrt{2}\cos(\omega t - 240°) \text{ V}$$

These time functions are shown in Fig. 10.1b.

Finally, let us examine the instantaneous power generated by a three-phase system. Assume that the voltages in Fig. 10.1 are

$$v_{an}(t) = V_m\cos\omega t \text{ V}$$
$$v_{bn}(t) = V_m\cos(\omega t - 120°) \text{ V} \qquad \textbf{10.3}$$
$$v_{cn}(t) = V_m\cos(\omega t - 240°) \text{ V}$$

If the load is balanced, the currents produced by the sources are

$$i_a(t) = I_m\cos(\omega t - \theta) \text{ A}$$
$$i_b(t) = I_m\cos(\omega t - \theta - 120°) \text{ A} \qquad \textbf{10.4}$$
$$i_c(t) = I_m\cos(\omega t - \theta - 240°) \text{ A}$$

The instantaneous power produced by the system is

$$p(t) = p_a(t) + p_b(t) + p_c(t)$$
$$= V_m I_m[\cos\omega t \cos(\omega t - \theta) + \cos(\omega t - 120°)\cos(\omega t - \theta - 120°)$$
$$+ \cos(\omega t - 240°)\cos(\omega t - \theta - 240°)] \qquad \textbf{10.5}$$

Using the trigonometric identity

$$\cos\alpha\cos\beta = \tfrac{1}{2}[\cos(\alpha - \beta) + \cos(\alpha + \beta)] \qquad \textbf{10.6}$$

Eq. (10.5) becomes

$$p(t) = \frac{V_m I_m}{2}\Big[\cos\theta + \cos(2\omega t - \theta) + \cos\theta$$
$$+ \cos(2\omega t - \theta - 240°) + \cos\theta + \cos(2\omega t - \theta - 480°)\Big] \qquad \textbf{10.7}$$

which can be written as

$$p(t) = \frac{V_m I_m}{2}\Big[3\cos\theta + \cos(2\omega t - \theta)$$
$$+ \cos(2\omega t - \theta - 120°) + \cos(2\omega t - \theta + 120°)\Big] \qquad \textbf{10.8}$$

There exists a trigonometric identity that allows us to simplify the preceding expression. The identity, which we will prove later using phasors, is

$$\cos\phi + \cos(\phi - 120°) + \cos(\phi + 120°) = 0 \qquad \textbf{10.9}$$

If we employ this identity, the expression for the power becomes

$$p(t) = 3\frac{V_m I_m}{2}\cos\theta \text{ W} \qquad \textbf{10.10}$$

Note that this equation indicates that the instantaneous power is always constant in time, rather than pulsating as in the single-phase case. Therefore, power delivery from a three-phase voltage source is very smooth, which is another important reason why power is generated in three-phase form.

10.2 Three-Phase Connections

By far the most important polyphase voltage source is the balanced three-phase source. This source, as illustrated by Fig. 10.2, has the following properties. The phase voltages—that is, the voltage from each line a, b, and c to the neutral n—are given by

$$\mathbf{V}_{an} = V_p \underline{/0°}$$
$$\mathbf{V}_{bn} = V_p \underline{/-120°} \qquad \qquad \textbf{10.11}$$
$$\mathbf{V}_{cn} = V_p \underline{/+120°}$$

Figure 10.2
Balanced three-phase voltage source.

Figure 10.3
Phasor diagram for a balanced three-phase voltage source.

The phasor diagram for these voltages is shown in Fig. 10.3. The phase sequence of this set is said to be *abc* (called positive phase sequence), meaning that $\mathbf{V}_{bn}$ *lags* $\mathbf{V}_{an}$ by 120°.

We will standardize our notation so that we always label the voltages $\mathbf{V}_{an}$, $\mathbf{V}_{bn}$, and $\mathbf{V}_{cn}$ and observe them in the order *abc*. Furthermore, we will normally assume with no loss of generality that $\underline{/\mathbf{V}_{an}} = 0°$.

An important property of the balanced voltage set is that

$$\mathbf{V}_{an} + \mathbf{V}_{bn} + \mathbf{V}_{cn} = 0 \qquad \qquad \textbf{10.12}$$

This property can easily be seen by resolving the voltage phasors into components along the real and imaginary axes. It can also be demonstrated via Eq. (10.9).

From the standpoint of the user who connects a load to the balanced three-phase voltage source, it is not important how the voltages are generated. It is important to note, however, that if the load currents generated by connecting a load to the power source shown in Fig. 10.2 are also *balanced*, there are two possible equivalent configurations for the load. The equivalent load can be considered as being connected in either a *wye* (Y) or a *delta* (Δ) configuration. The balanced wye configuration is shown in Fig. 10.4a and equivalently in Fig. 10.4b. The delta configuration is shown in Fig. 10.5a and equivalently in Fig. 10.5b. Note that in the case of the delta connection, there is no neutral line. The actual function of the neutral line in the wye connection will be examined and it will be shown that in a balanced system the neutral line carries no current and, for purposes of analysis, may be omitted.

The wye and delta connections each have their advantages. In the wye case, we have access to two voltages, the line-to-line and line-to-neutral, and it provides a convenient place to connect to ground for system protection—that is, it limits the magnitude of surge voltages. In the delta case, this configuration stays in balance better when serving unbalanced loads, and it is capable of trapping the third harmonic.

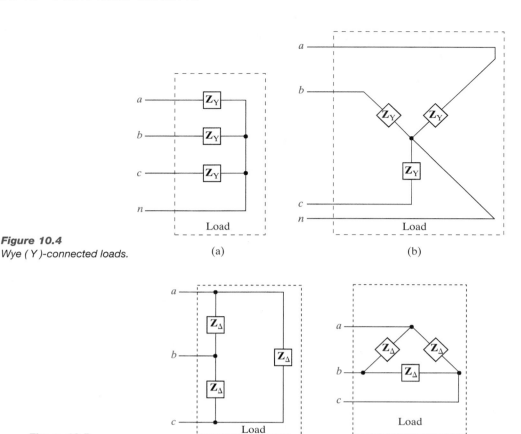

Figure 10.4
Wye (Y)-connected loads.

(a) (b)

Figure 10.5
Delta (Δ)-connected loads.

(a) (b)

10.3 Source/Load Connections

Since the source and the load can each be connected in either Y or Δ, three-phase balanced circuits can be connected Y–Y, Y–Δ, Δ–Y, or Δ–Δ. Our approach to the analysis of all of these circuits will be "Think Y" and, therefore, we will analyze the Y–Y connection first.

BALANCED WYE–WYE CONNECTION Suppose now that the source and load are both connected in a wye, as shown in Fig. 10.6. The phase voltages with positive phase sequence are

$$\mathbf{V}_{an} = V_p \underline{/0°}$$
$$\mathbf{V}_{bn} = V_p \underline{/-120°} \qquad \textbf{10.13}$$
$$\mathbf{V}_{cn} = V_p \underline{/+120°}$$

where V_p, the phase voltage, is the magnitude of the phasor voltage from the neutral to any line. The *line-to-line* or, simply, *line voltages* can be calculated using KVL; for example,

$$\mathbf{V}_{ab} = \mathbf{V}_{an} - \mathbf{V}_{bn}$$
$$= V_p \underline{/0°} - V_p \underline{/-120°}$$
$$= V_p - V_p \left[-\frac{1}{2} - j\frac{\sqrt{3}}{2} \right]$$

Figure 10.6
Balanced three-phase wye–wye connection.

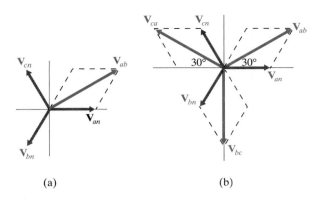

(a)   (b)

Figure 10.7
*Phasor representation of phase
and line voltages in a balanced
wye–wye system.*

$$= V_p\left[\frac{3}{2} + j\frac{\sqrt{3}}{2}\right]$$

$$= \sqrt{3}\, V_p\,\underline{/30°}$$

The phasor addition is shown in Fig. 10.7a. In a similar manner, we obtain the set of line-to-line voltages as

$$\mathbf{V}_{ab} = \sqrt{3}\, V_p\,\underline{/30°}$$
$$\mathbf{V}_{bc} = \sqrt{3}\, V_p\,\underline{/-90°} \qquad \textbf{10.14}$$
$$\mathbf{V}_{ca} = \sqrt{3}\, V_p\,\underline{/-210°}$$

LEARNING Hint

Conversion rules

$\underline{/\mathbf{V}_{ab}} = \underline{/\mathbf{V}_{an}} + 30°$
$V_{ab} = \sqrt{3}\, V_{an}$

All the line voltages together with the phase voltages are shown in Fig. 10.7b. We will denote the magnitude of the line voltages as V_L, and therefore, for a balanced system,

$$V_L = \sqrt{3}\, V_p \qquad \textbf{10.15}$$

Hence, in a wye-connected system, the line voltage is equal to $\sqrt{3}$ times the phase voltage.
   As shown in Fig. 10.6, the line current for the a phase is

$$\mathbf{I}_a = \frac{\mathbf{V}_{an}}{\mathbf{Z}_Y} = \frac{V_p\,\underline{/0°}}{\mathbf{Z}_Y} \qquad \textbf{10.16}$$

where $\mathbf{I}_b$ and $\mathbf{I}_c$ have the same magnitude but lag $\mathbf{I}_a$ by 120° and 240°, respectively.
   The neutral current $\mathbf{I}_n$ is then

$$\mathbf{I}_n = (\mathbf{I}_a + \mathbf{I}_b + \mathbf{I}_c) = 0 \qquad \textbf{10.17}$$

Since there is no current in the neutral, this conductor could contain any impedance or it could be an open or a short circuit, without changing the results found previously.
   As illustrated by the wye–wye connection in Fig. 10.6, the current in the line connecting the source to the load is the same as the phase current flowing through the impedance $\mathbf{Z}_Y$. Therefore, in a *wye–wye connection*,

$$I_L = I_Y \qquad \textbf{10.18}$$

where I_L is the magnitude of the line current and I_Y is the magnitude of the current in a wye-connected load.

LEARNING by Doing

D 10.1 If
$\mathbf{V}_{an} = 120\,\underline{/30°}$ V rms,
determine $\mathbf{V}_{ab}$.

ANSWER
$\mathbf{V}_{ab} = 208\,\underline{/60°}$ V rms

D 10.2 If
$\mathbf{V}_{ab} = 199.2\,\underline{/45°}$ V rms,
determine $\mathbf{V}_{an}$.

ANSWER
$\mathbf{V}_{an} = 115\,\underline{/15°}$ V rms

It is important to note that although we have a three-phase system composed of three sources and three loads, we can analyze a single phase and use the phase sequence to obtain the voltages and currents in the other phases. This is, of course, a direct result of the balanced condition. We may even have impedances present in the lines; however, as long as the system remains balanced, we need analyze only one phase. If the line impedances in lines *a*, *b*, and *c* are equal, the system will be balanced. Recall that the balance of the system is unaffected by whatever appears in the neutral line, and since the neutral line impedance is arbitrary, we assume that it is zero (i.e., a short circuit).

LEARNING Example 10.1

An *abc*-sequence three-phase voltage source connected in a balanced wye has a line voltage of $\mathbf{V}_{ab} = 208\,\underline{/-30°}$ V rms. We wish to determine the phase voltages.

SOLUTION The magnitude of the phase voltage is given by the expression

$$V_p = \frac{208}{\sqrt{3}}$$

$$= 120 \text{ V rms}$$

The phase relationships between the line and phase voltages are shown in Fig. 10.7. From this figure we note that

$$\mathbf{V}_{an} = 120\,\underline{/-60°} \text{ V rms}$$
$$\mathbf{V}_{bn} = 120\,\underline{/-180°} \text{ V rms}$$
$$\mathbf{V}_{cn} = 120\,\underline{/+60°} \text{ V rms}$$

The magnitudes of these voltages are quite common, and one often hears that the electric service in a building, for example, is three-phase 208/120 V.

LEARNING Hint

The phase of

$$V_{an} = \underline{/V_{an}} = \underline{/V_{ab}} - 30°$$

LEARNING Example 10.2

A three-phase wye-connected load is supplied by an *abc*-sequence balanced three-phase wye-connected source with a phase voltage of 120 V rms. If the line impedance and load impedance per phase are $1 + j1\ \Omega$ and $20 + j10\ \Omega$, respectively, we wish to determine the value of the line currents and the load voltages.

SOLUTION The phase voltages are

$$\mathbf{V}_{an} = 120\,\underline{/0°} \text{ V rms}$$
$$\mathbf{V}_{bn} = 120\,\underline{/-120°} \text{ V rms}$$
$$\mathbf{V}_{cn} = 120\,\underline{/+120°} \text{ V rms}$$

The per phase circuit diagram is shown in Fig. 10.8. The line current for the *a* phase is

$$\mathbf{I}_{aA} = \frac{120\,\underline{/0°}}{21 + j11}$$

$$= 5.06\,\underline{/-27.65°} \text{ A rms}$$

The load voltage for the *a* phase, which we call $\mathbf{V}_{AN}$ is

$$\mathbf{V}_{AN} = \left(5.06\,\underline{/-27.65°}\right)(20 + j10)$$

$$= 113.15\,\underline{/-1.08°} \text{ V rms}$$

The corresponding line currents and load voltages for the *b* and *c* phases are

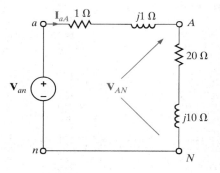

Figure 10.8
Per phase circuit diagram for the problem in Example 10.2.

$$\mathbf{I}_{bB} = 5.06\,\underline{/-147.65°} \text{ A rms} \qquad \mathbf{V}_{BN} = 113.15\,\underline{/-121.08°} \text{ V rms}$$
$$\mathbf{I}_{cC} = 5.06\,\underline{/-267.65°} \text{ A rms} \qquad \mathbf{V}_{CN} = 113.15\,\underline{/-241.08°} \text{ V rms}$$

To reemphasize and clarify our terminology, phase voltage, V_p, is the magnitude of the phasor voltage from the neutral to any line, while line voltage, V_L, is the magnitude of the phasor voltage between any two lines. Thus, the values of V_L and V_p will depend on the point at which they are calculated in the system.

LEARNING Hint

$$\underline{/I_{bB}} = \underline{/I_{aA}} - 120°$$
$$\underline{/I_{cC}} = \underline{/I_{aA}} + 120°$$

E10.1   The voltage for the *a* phase of an *abc*-phase-sequence balanced wye-connected source is $\mathbf{V}_{an} = 120 \underline{/90°}$ V rms. Determine the line voltages for this source.

ANSWER
$\mathbf{V}_{ab} = 208 \underline{/120°}$ V rms,
$\mathbf{V}_{bc} = 208 \underline{/0°}$ V rms,
$\mathbf{V}_{ca} = 208 \underline{/-120°}$ V rms.

E10.2   An *abc*-phase-sequence three-phase voltage source connected in a balanced wye has a line voltage of $\mathbf{V}_{ab} = 208 \underline{/0°}$ V rms. Determine the phase voltages of the source.

ANSWER
$\mathbf{V}_{an} = 120 \underline{/-30°}$ V rms,
$\mathbf{V}_{bn} = 120 \underline{/-150°}$ V rms,
$\mathbf{V}_{cn} = 120 \underline{/-270°}$ V rms.

E10.3   A three-phase wye-connected load is supplied by an *abc*-sequence balanced three-phase wye-connected source through a transmission line with an impedance of $1 + j1$ Ω per phase. The load impedance is $8 + j3$ Ω per phase. If the load voltage for the *a* phase is $104.02 \underline{/26.6°}$ V rms (i.e., $V_p = 104.02$ V rms at the load end), determine the phase voltages of the source.

ANSWER
$\mathbf{V}_{an} = 120 \underline{/30°}$ V rms,
$\mathbf{V}_{bn} = 120 \underline{/-90°}$ V rms,
$\mathbf{V}_{cn} = 120 \underline{/-210°}$ V rms.

The previous analysis indicates that we can simply treat a three-phase balanced circuit on a per phase basis and use the phase relationship to determine all voltages and currents. Let us now examine the situations in which either the source or the load is connected in Δ.

DELTA-CONNECTED SOURCE   Consider the delta-connected source shown in Fig. 10.9a. Note that the sources are connected line to line. We found earlier that the relationship between line-to-line and line-to-neutral voltages was given by Eq. (10.14) and illustrated in Fig. 10.7 for an *abc*-phase sequence of voltages. Therefore, if the delta sources are

$$\mathbf{V}_{ab} = V_L \underline{/0°}$$
$$\mathbf{V}_{bc} = V_L \underline{/-120°}$$
$$\mathbf{V}_{ca} = V_L \underline{/+120°}$$

10.19

(a)

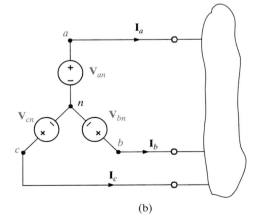

(b)

Figure 10.9   *Sources connected in delta and wye.*

where V_L is the magnitude of the phase voltage. The equivalent wye sources shown in Fig. 10.9b are

$$V_{an} = \frac{V_L}{\sqrt{3}} \underline{/-30°} = V_p \underline{/-30°}$$

$$V_{bn} = \frac{V_L}{\sqrt{3}} \underline{/-150°} = V_p \underline{/-150°}$$ **10.20**

$$V_{cn} = \frac{V_L}{\sqrt{3}} \underline{/-270°} = V_p \underline{/+90°}$$

where V_p is the magnitude of the phase voltage of an equivalent wye-connected source. Therefore, if we encounter a network containing a delta-connected source, we can easily convert the source from delta to wye so that all the techniques we have discussed previously can be applied in an analysis.

Problem-Solving Strategy
Three-Phase Balanced AC Power Circuits

1. Convert the source/load connection to a wye–wye connection if either the source, load, or both are connected in delta since the wye–wye connection can be easily used to obtain the unknown phasors.

2. Only the unknown phasors for the *a*-phase of the circuit need be determined since the three-phase system is balanced.

3. Finally, convert the now known phasors to the corresponding phasors in the original system.

LEARNING Example 10.3

Consider the network shown in Fig. 10.10a. We wish to determine the line currents and the magnitude of the line voltage at the load.

SOLUTION   The single-phase diagram for the network is shown in Fig. 10.10b. The line current I_{aA} is

$$I_{aA} = \frac{(208/\sqrt{3}) \underline{/-30°}}{12.1 + j4.2}$$

$$= 9.38 \underline{/-49.14°} \text{ A rms}$$

and thus $I_{bB} = 9.38 \underline{/-169.14°}$ V rms and $I_{cC} = 9.38 \underline{/70.86°}$ V rms. The voltage V_{AN} is then

$$V_{AN} = (9.38 \underline{/-49.14°})(12 + j4)$$

$$= 118.65 \underline{/-30.71°} \text{ V rms}$$

Figure 10.10   *Delta–wye network and an equivalent single-phase (a-phase) diagram.*

Therefore, the magnitude of the line voltage at the load is

$$V_L = \sqrt{3}\,(118.65)$$

$$= 205.51 \text{ V rms}$$

The phase voltage at the source is $V_p = 208/\sqrt{3} = 120$ V rms, while the phase voltage at the load is $V_p = 205.51/\sqrt{3} = 118.65$ V rms. Clearly, we must be careful with our notation and specify where the phase or line voltage is taken.

LEARNING EXTENSION

E10.4 Consider the network shown in Fig. E10.4. Compute the magnitude of the line voltages at the load.

ANSWER
$V_L = 205.2$ V rms.

Figure E10.4

DELTA-CONNECTED LOAD

Consider now the Δ-connected load shown in Fig. 10.11. Note that in this connection the line-to-line voltage is the voltage across each load impedance.

If the phase voltages of the source are

$$\mathbf{V}_{an} = V_p\,/0°$$

$$\mathbf{V}_{bn} = V_p\,/{-120°} \qquad \textbf{10.21}$$

$$\mathbf{V}_{cn} = V_p\,/{+120°}$$

then the line voltages are

$$\mathbf{V}_{ab} = \sqrt{3}\,V_p\,/30° = V_L\,/30° = \mathbf{V}_{AB}$$

$$\mathbf{V}_{bc} = \sqrt{3}\,V_p\,/{-90°} = V_L\,/{-90°} = \mathbf{V}_{BC} \qquad \textbf{10.22}$$

$$\mathbf{V}_{ca} = \sqrt{3}\,V_p\,/{-210°} = V_L\,/{-210°} = \mathbf{V}_{CA}$$

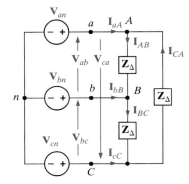

Figure 10.11
Balanced three-phase wye–delta system.

LEARNING by Doing

D 10.6 A three-phase balanced delta-connected load has a phase current of $\mathbf{I}_{AB} = 10\,\underline{/15°}$ A rms. Find the line current $\mathbf{I}_{aA}$.

ANSWER
$\mathbf{I}_{aA} = 17.32\,\underline{/-15°}$ A rms.

D 10.7 For a balanced delta-connected load, $\mathbf{Z}_\Delta = 30 + j6\ \Omega$, determine the corresponding wye-connected load.

ANSWER
$\mathbf{Z}_Y = 10 + j2\ \Omega$.

D 10.8 For a balanced wye-connected load, $\mathbf{Z}_Y = 4 + j1\ \Omega$, determine the corresponding delta connected load $\mathbf{Z}_\Delta$.

ANSWER
$\mathbf{Z}_\Delta = 12 + j3\ \Omega$.

where V_L is the magnitude of the line voltage at both the delta-connected load and at the source since there is no line impedance present in the network.

From Fig. 10.11 we note that if $\mathbf{Z}_\Delta = Z_\Delta\,\underline{/\theta}$, the phase currents at the load are

$$\mathbf{I}_{AB} = \frac{\mathbf{V}_{AB}}{\mathbf{Z}_\Delta} \qquad \text{10.23}$$

where $\mathbf{I}_{BC}$ and $\mathbf{I}_{CA}$ have the same magnitude but lag $\mathbf{I}_{AB}$ by 120° and 240°, respectively. KCL can now be employed in conjunction with the phase currents to determine the line currents. For example,

$$\mathbf{I}_{aA} = \mathbf{I}_{AB} + \mathbf{I}_{AC}$$
$$= \mathbf{I}_{AB} - \mathbf{I}_{CA}$$

However, it is perhaps easier to simply convert the balanced Δ-connected load to a balanced Y-connected load using the Δ–Y transformation. This conversion is possible since the wye–delta and delta–wye transformations outlined in Chapter 2 are also valid for impedance in the frequency domain. In the balanced case, the transformation equations reduce to

$$\mathbf{Z}_Y = \frac{1}{3}\,\mathbf{Z}_\Delta$$

and then the line current $\mathbf{I}_{aA}$ is simply

$$\mathbf{I}_{aA} = \frac{\mathbf{V}_{an}}{\mathbf{Z}_Y}$$

Finally, using the same approach as that employed earlier to determine the relationship between the line voltages and phase voltages in a Y–Y connection, we can show that the relationship between the *magnitudes* of the phase currents in the Δ-connected load and the line currents is

$$I_L = \sqrt{3}\,I_\Delta \qquad \text{10.24}$$

LEARNING Example 10.4

A balanced delta-connected load contains a 10-Ω resistor in series with a 20-mH inductor in each phase. The voltage source is an *abc*-sequence three-phase 60-Hz, balanced wye with a voltage $\mathbf{V}_{an} = 120\,\underline{/30°}$ V rms. We wish to determine all Δ currents and line currents.

SOLUTION The impedance per phase in the delta load is $\mathbf{Z}_\Delta = 10 + j7.54\ \Omega$. The line voltage $\mathbf{V}_{ab} = 120\sqrt{3}\,\underline{/60°}$ V rms. Since there is no line impedance, $\mathbf{V}_{AB} = \mathbf{V}_{ab} = 120\sqrt{3}\,\underline{/60°}$ V rms. Hence,

$$\mathbf{I}_{AB} = \frac{120\sqrt{3}\,\underline{/60°}}{10 + j7.54}$$

$$= 16.60\,\underline{/+22.98°}\ \text{A rms}$$

If $\mathbf{Z}_\Delta = 10 + j7.54\ \Omega$, then

$$\mathbf{Z}_Y = \frac{1}{3}\,\mathbf{Z}_\Delta$$
$$= 3.33 + j2.51\ \Omega$$

Then the line current

$$\mathbf{I}_{aA} = \frac{\mathbf{V}_{an}}{\mathbf{Z}_Y} = \frac{120\,\underline{/30°}}{3.33 + j2.51}$$

$$= \frac{120\,\underline{/30°}}{4.17\,\underline{/37.01°}}$$

$$= 28.78\,\underline{/-7.01°}\ \text{A rms}$$

Therefore, the remaining phase and line currents are

$$\mathbf{I}_{BC} = 16.60\,\underline{/-97.02°}\ \text{A rms} \quad \mathbf{I}_{bB} = 28.78\,\underline{/-127.01°}\ \text{A rms}$$
$$\mathbf{I}_{CA} = 16.60\,\underline{/+142.98°}\ \text{A rms} \quad \mathbf{I}_{cC} = 28.78\,\underline{/+112.99°}\ \text{A rms}$$

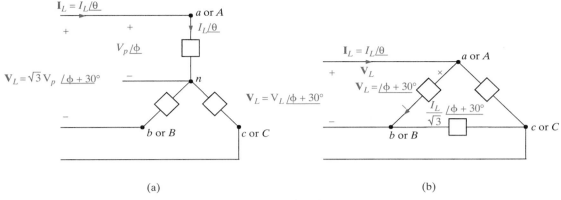

Figure 10.12 *Voltage and current relationships for Y and Δ configurations.*

In summary, the relationship between the line voltage and phase voltage and the line current and phase current for both the Y and Δ configurations are shown in Fig. 10.12. The currents and voltages are shown for one phase. The two remaining phases have the same magnitude but lag by 120° and 240°, respectively.

Careful observation of Table 10.1 indicates that the following rules apply when solving problems in balanced three-phase systems:

▶ The phase of the voltages and currents in a Δ connection is 30° ahead of those in a Y connection.

▶ The magnitude of the line voltage or, equivalently, the Δ-connection phase voltage, is $\sqrt{3}$ times that of the Y-connection phase voltage.

▶ The magnitude of the line current or, equivalently, the Y-connection phase current, is $\sqrt{3}$ times that of the Δ-connection phase current.

▶ The load impedance in the Y connection is one-third of that in the Δ-connection and the phase is identical.

Table 10.1 The voltage, current, and impedance relationships for Y and Δ configurations

	Y	**Δ**
Line voltage $(\mathbf{V}_{ab}$ or $\mathbf{V}_{AB})$	$\sqrt{3}\,V_p\,\underline{/\phi + 30°}$ $= V_L\,\underline{/\phi + 30°}$	$V_L\,\underline{/\phi + 30°}$
Line current $\mathbf{I}_{aA}$	$I_L\,\underline{/\theta}$	$I_L\,\underline{/\theta}$
Phase voltage	$V_p\,\underline{/\phi}\;(\mathbf{V}_{an}$ or $\mathbf{V}_{AN})$	$\sqrt{3}\,V_p\,\underline{/\phi + 30°}$
Phase current	$I_L\,\underline{/\theta}$	$\dfrac{I_L}{\sqrt{3}}\,\underline{/\theta + 30°}$
Load impedance	$\mathbf{Z}_Y\,\underline{/\phi - \theta}$	$3\,\mathbf{Z}_Y\,\underline{/\phi - \theta}$

LEARNING EXTENSION

E10.5 An *abc*-sequence three-phase voltage source connected in a balanced wye supplies power to a balanced delta-connected load. The line current for the *a* phase is $\mathbf{I}_{aA} = 12\,\underline{/40°}$ A rms. Find the phase currents in the delta-connected load.

ANSWER
$\mathbf{I}_{AB} = 6.93\,\underline{/70°}$ A rms,
$\mathbf{I}_{BC} = 6.93\,\underline{/-50°}$ A rms,
$\mathbf{I}_{CA} = 6.93\,\underline{/-170°}$ A rms.

10.4 Power Relationships

Whether the load is connected in a wye or a delta, the real and reactive power per phase is

$$P_p = V_p I_p \cos \theta$$
$$Q_p = V_p I_p \sin \theta$$

10.25

where θ is the angle between the phase voltage and the line current. For a Y-connected system, $I_p = I_L$ and $V_p = V_L/\sqrt{3}$, and for a Δ-connected system, $I_p = I_L/\sqrt{3}$ and $V_p = V_L$. Therefore,

$$P_p = \frac{V_L I_L}{\sqrt{3}} \cos \theta$$

10.26

$$Q_p = \frac{V_L I_L}{\sqrt{3}} \sin \theta$$

The total real and reactive power for all three phases is then

$$P_T = \sqrt{3} \, V_L I_L \cos \theta$$
$$Q_T = \sqrt{3} \, V_L I_L \sin \theta$$

10.27

and, therefore, the magnitude of the complex power (apparent power) is

$$S_T = \sqrt{P_T^2 + Q_T^2}$$
$$= \sqrt{3} \, V_L I_L$$

and

$$\underline{/S_T} = \theta$$

LEARNING Example 10.5

A three-phase balanced wye–delta system has a line voltage of 208 V rms. The total real power absorbed by the load is 1200 W. If the power factor angle of the load is 20° lagging, we wish to determine the magnitude of the line current and the value of the load impedance per phase in the delta.

SOLUTION The line current can be obtained from Eq. (10.26). Since the real power per phase is 400 W,

$$400 = \frac{208 I_L}{\sqrt{3}} \cos 20°$$

$$I_L = 3.54 \text{ A rms}$$

The magnitude of the current in each leg of the delta-connected load is

$$I_\Delta = \frac{I_L}{\sqrt{3}}$$

$$= 2.05 \text{ A rms}$$

Therefore, the magnitude of the delta impedance in each phase of the load is

$$|\mathbf{Z}_\Delta| = \frac{V_L}{I_\Delta}$$

$$= \frac{208}{2.05}$$

$$= 101.46 \ \Omega$$

Since the power factor angle is 20° lagging, the load impedance is

$$\mathbf{Z}_\Delta = 101.46 \underline{/20°}$$

$$= 95.34 + j34.70 \ \Omega$$

LEARNING by Doing

D 10.9 A three-phase balanced delta-connected load has a phase voltage of 208 V rms and a phase current of 10 A rms. Determine the complex power if the pf is 0.9.

ANSWER
2080 $\underline{/25.8°}$ VA.

LEARNING Example 10.6

For the circuit in Example 10.2 we wish to determine the real and reactive power per phase at the load and the total real power, reactive power, and complex power at the source.

SOLUTION From the data in Example 10.2 the complex power per phase at the load is

$$\mathbf{S}_{\text{load}} = \mathbf{VI}^*$$
$$= (113.15 \underline{/-1.08°})(5.06 \underline{/27.65°})$$
$$= 572.54 \underline{/26.57°}$$
$$= 512.07 + j256.09 \text{ VA}$$

Therefore, the real and reactive power per phase at the load are 512.07 W and 256.09 var, respectively.

The complex power per phase at the source is

$$\mathbf{S}_{\text{source}} = \mathbf{VI}^*$$
$$= (120 \underline{/0°})(5.06 \underline{/27.65°})$$
$$= 607.2 \underline{/27.65°}$$
$$= 537.86 + j281.78 \text{ VA}$$

Therefore, total real power, reactive power, and apparent power at the source are 1613.6 W, 845.2 var, and 1821.6 VA, respectively.

LEARNING Example 10.7

A balanced three-phase source serves three loads as follows:

Load 1: 24 kW at 0.6 lagging power factor

Load 2: 10 kW at unity power factor

Load 3: 12 kVA at 0.8 leading power factor

If the line voltage at the loads is 208 V rms at 60 Hz, we wish to determine the line current and the combined power factor of the loads.

SOLUTION From the data we find that

$$\mathbf{S}_1 = 24,000 + j32,000$$
$$\mathbf{S}_2 = 10,000 + j0$$
$$\mathbf{S}_3 = 12,000 \underline{/-36.9°} = 9600 - j7200$$

Therefore,

$$\mathbf{S}_{\text{load}} = 43,600 + j24,800$$

$$= 50,160 \underline{/29.63°} \text{ VA}$$
$$I_L = \frac{|\mathbf{S}_{\text{load}}|}{\sqrt{3} \, V_L}$$
$$= \frac{50,160}{208\sqrt{3}}$$
$$I_L = 139.23 \text{ A rms}$$

and the combined power factor is

$$\text{pf}_{\text{load}} = \cos 29.63°$$
$$= 0.869 \text{ lagging}$$

LEARNING Hint

The sum of three complex powers
$\mathbf{S}_{\text{load}} = \mathbf{S}_1 + \mathbf{S}_2 + \mathbf{S}_3$

LEARNING Example 10.8

Given the three-phase system in Example 10.7, let us determine the line voltage and power factor at the source if the line impedance is $\mathbf{Z}_{\text{line}} = 0.05 + j0.02 \ \Omega$.

LEARNING Hint

Recall that the complex power for all three lines is $\mathbf{S}_{\text{line}} = 3I_L^2 \mathbf{Z}_{\text{line}}$

SOLUTION The complex power absorbed by the line impedances is

$$\mathbf{S}_{\text{line}} = 3(R_{\text{line}} I_L^2 + jX_{\text{line}} I_L^2)$$
$$= 2908 + j1163 \text{ VA}$$

The complex power delivered by the source is then

$$\mathbf{S}_S = \mathbf{S}_{\text{load}} + \mathbf{S}_{\text{line}}$$
$$= 43,600 + j24,800 + 2908 + j1163$$
$$= 53,264 \underline{/29.17°} \text{ VA}$$

The line voltage at the source is then

$$V_{L_S} = \frac{S_S}{\sqrt{3} \, I_L}$$
$$= 220.87 \text{ V rms}$$

and the power factor at the source is

$$\text{pf}_S = \cos 29.17°$$
$$= 0.873 \text{ lagging}$$

LEARNING EXTENSIONS

E10.6 A three-phase balanced wye–wye system has a line voltage of 208 V rms. The total real power absorbed by the load is 12 kW at 0.8 pf lagging. Determine the per-phase impedance of the load.

ANSWER
$\mathbf{Z} = 2.88 \,\underline{/36.87°}\ \Omega$.

E10.7 For the balanced wye–wye system described in Extension E10.3, determine the real and reactive power and the complex power at both the source and the load.

ANSWER
$\mathbf{S}_{load} = 1186.77 + j444.66$ VA;
$\mathbf{S}_{source} = 1335.65 + j593.55$ VA;

E10.8 A 480-V rms line feeds two balanced three-phase loads. If the two loads are rated as follows,

Load 1: 5 kVA at 0.8 pf lagging

Load 2: 10 kVA at 0.9 pf lagging

determine the magnitude of the line current from the 480-V rms source.

ANSWER
$I_L = 17.97$ A rms.

10.5 Power Factor Correction

In Section 9.7 we illustrated a simple technique for raising the power factor of a load. The method involved judiciously selecting a capacitor and placing it in parallel with the load. In a balanced three-phase system, power factor correction is performed in exactly the same manner. It is important to note, however, that the $\mathbf{S}_{cap}$ specified in Eq. (9.40) is provided by three capacitors, and in addition, V_{rms} in the equation is the voltage across each capacitor. The following example illustrates the technique.

LEARNING Example 10.9

In the balanced three-phase system shown in Fig. 10.13, the line voltage is 34.5 kV rms at 60 Hz. We wish to find the values of the capacitors C such that the total load has a power factor of 0.94 leading.

SOLUTION Following the development outlined in Section 9.7 for single-phase power factor correction, we obtain

$$\mathbf{S}_{old} = 24 \,\underline{/\cos^{-1} 0.78}\ \text{MVA}$$

$$= 18.72 + j15.02 \text{ MVA}$$

and

$$\theta_{new} = -\cos^{-1} 0.94$$

$$= -19.95°$$

Therefore,

$$\mathbf{S}_{new} = 18.72 + j18.72 \tan(-19.95°)$$

$$= 18.72 - j6.80 \text{ MVA}$$

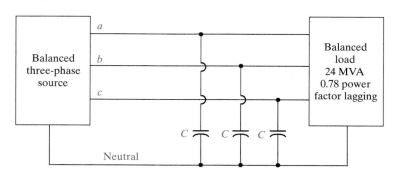

Figure 10.13
Network used in Example 10.9.

and

$$\mathbf{S}_{cap} = \mathbf{S}_{new} - \mathbf{S}_{old}$$
$$= -j21.82 \text{ MVA}$$

However,

$$-j\omega C \, V_{rms}^2 = -j21.82 \text{ MVA}$$

and since the line voltage is 34.5 kV rms, then

$$(377)\left(\frac{34.5k}{\sqrt{3}}\right)^2 C = \frac{21.82}{3} \text{ M}$$

Hence,

$$C = 48.6 \text{ μF}$$

LEARNING Hint

The reactive power to be supplied by C is derived from the expression

$$jQ_{cap} = -j\omega C V_{rms}^2$$

The phase voltage for the Y connection is

$$V_Y = \frac{34.5k}{\sqrt{3}}$$

LEARNING EXTENSION

E10.9   Find C in Example 10.9 such that the load has a power factor of 0.90 lagging.

ANSWER $C = 13.26$ μF.

Finally, recall that our entire discussion in this chapter has focused on balanced systems. It is extremely important, however, to point out that in the unbalanced three-phase system the problem is much more complicated because of the mutual inductive coupling between phases in power apparatus.

Learning by Application

The first of the following two examples illustrates the manner in which power flow is measured when utilities are interconnected, answering the question of who is supplying power to whom. The second example demonstrates the actual method in which capacitors are specified by the manufacturer for power factor correction.

LEARNING Example 10.10

Two balanced three-phase systems, X and Y, are interconnected with lines having impedance $\mathbf{Z}_{line} = 1 + j2 \; \Omega$. The line voltages are $\mathbf{V}_{ab} = 12 \underline{/0°}$ kV rms and $\mathbf{V}_{AB} = 12 \underline{/5°}$ kV rms, as shown in Fig. 10.14a. We wish to determine which system is the source, which is the load, and the average power supplied by the source and absorbed by the load.

SOLUTION   When we draw the per phase circuit for the system as shown in Fig. 10.14b, the analysis will be essentially the same as that of Example 9.13.

The network in Fig. 10.14b indicates that

$$\mathbf{I}_{aA} = \frac{\mathbf{V}_{an} - \mathbf{V}_{AN}}{\mathbf{Z}_{line}}$$

$$= \frac{\dfrac{12{,}000}{\sqrt{3}} \underline{/-30°} - \dfrac{12{,}000}{\sqrt{3}} \underline{/-25°}}{\sqrt{5} \underline{/63.43°}}$$

$$= 270.30 \underline{/-180.93°} \text{ A rms}$$

The average power absorbed by system Y is

$$P_Y = \sqrt{3} \, V_{AB} I_{aA} \cos(\theta_{\mathbf{V}_{an}} - \theta_{\mathbf{I}_{aA}})$$
$$= \sqrt{3} \, (12{,}000)(270.30) \cos(-25° + 180.93°)$$
$$= -5.130 \text{ MW}$$

Note that system Y is not the load, but rather the source and supplies 5.130 MW.

System X absorbs the following average power:

$$P_X = \sqrt{3} \, V_{ab} I_{Aa} \cos(\theta_{\mathbf{V}_{an}} - \theta_{\mathbf{I}_{aA}})$$

where

$$\mathbf{I}_{Aa} = -\mathbf{I}_{aA} = 270.30 \underline{/-0.93°} \text{ A rms}$$

(continued)

Therefore,

$$P_X = \sqrt{3}\,(12{,}000)(270.30)\cos(-30° + 0.93°)$$

$$= 4.910 \text{ MW}$$

and hence system X is the load.

The difference in the power supplied by system Y and that absorbed by system X is, of course, the power absorbed by the resistance of the three lines.

Figure 10.14
Circuits used in Example 10.11: (a) original three-phase system, (b) per phase circuit.

The preceding example illustrates an interesting point. Note that the phase difference between the two ends of the power line determines the direction of the power flow. Since the numerous power companies throughout the United States are tied together to form the U.S. power grid, the phase difference across the interconnecting transmission lines reflects the manner in which power is transferred between power companies.

Capacitors for power factor correction are usually specified by the manufacturer in vars rather than in farads. Of course, the supplier must also specify the voltage at which the capacitor is designed to operate and a frequency of 60 Hz is assumed. The relationship between capacitance and the vars rating is

$$Q_R = \frac{V^2}{Z_C}$$

where Q_R is vars rating, V is the voltage rating, and Z_C is the capacitor's impedance at 60 Hz. Thus, a 500-V, 600-var capacitor has a capacitance of

$$C = \frac{Q_R}{\omega V^2} = \frac{600}{(377)(500)^2}$$

or

$$C = 6.37 \ \mu\text{F}$$

and can be used in any application where the voltage across the capacitor does not exceed the rated value of 500 V.

LEARNING Example 10.11

Table 10.2 lists the voltage and power ratings for three power factor correction capacitors. Let us determine which of them, if any, can be employed in Example 10.9.

Table 10.2 Rated voltage and vars for power factor correction capacitors

Capacitor	Rated Voltage (kV)	Rated Q (Mvars)
1	10.0	4.0
2	50.0	25.0
3	20.0	7.5

SOLUTION From Fig. 10.13 we see that the voltage across the capacitors is the line-to-neutral voltage, which is

$$V_{an} = \frac{V_{ab}}{\sqrt{3}} = \frac{34{,}500}{\sqrt{3}}$$

or

$$V_{an} = 19.9 \text{ kV}$$

Therefore, only those capacitors with rated voltages greater than or equal to 19.9 kV can be used in this application, which eliminates capacitor 1. Let us now determine the capacitance of capacitors 2 and 3. For capacitor 2,

$$C_2 = \frac{Q}{\omega V^2} = \frac{25 \times 10^6}{(377)(50{,}000)^2}$$

or

$$C_2 = 26.53 \text{ }\mu\text{F}$$

which is much smaller than the required 48.6 μF. The capacitance of capacitor 3 is

$$C_3 = \frac{Q}{\omega V^2} = \frac{7.5 \times 10^6}{(377)(20{,}000)^2}$$

or

$$C_3 = 49.7 \text{ }\mu\text{F}$$

which is within 2.5% of the required value. Obviously, capacitor 3 is the best choice.

Learning by Design

In the following example we examine the selection of both the conductor and the capacitor in a practical power factor correction problem.

LEARNING Example 10.12

Two stores, as shown in Fig. 10.15, are located at a busy intersection. The stores are fed from a balanced three-phase 60-Hz source with a line voltage of 13.8 kV rms. The power line is constructed of a #4ACSR (aluminum cable steel reinforced) conductor that is rated at 170 A rms.

A third store, shown in Fig. 10.15, wishes to locate at the intersection. Let us determine (1) if the #4ACSR conductor will permit the addition of this store, and (2) the value of the capacitors connected in wye that are required to change the overall power factor for all three stores to 0.92 lagging.

Figure 10.15
Circuit used in
Example 10.12.

(continued)

SOLUTION (1) The complex power for each of the three loads is

$$\mathbf{S}_1 = 700\,\underline{/36.9°} = 560 + j420 \text{ kVA}$$

$$\mathbf{S}_2 = 1000\,\underline{/60°} = 500 + j866 \text{ kVA}$$

$$\mathbf{S}_3 = 800\,\underline{/25.8°} = 720 + j349 \text{ kVA}$$

Therefore, the total complex power is

$$\mathbf{S}_T = \mathbf{S}_1 + \mathbf{S}_2 + \mathbf{S}_3$$

$$= 1780 + j1635$$

$$= 2417\,\underline{/42.57°} \text{ kVA}$$

Since

$$\mathbf{S}_T = \sqrt{3}\, V_L I_L$$

the line current is

$$I_L = \frac{(2417)(10^3)}{\sqrt{3}\,(13.8)(10^3)}$$

$$= 101.1 \text{ A rms}$$

Since this value is well below the rated value of 170 A rms, the conductor is sized properly and we can safely add the third store.

(2) The combined power factor for the three loads is found from the expression

$$\cos\theta = \text{pf} = \frac{1780}{2417} = 0.7365 \text{ lagging}$$

By adding capacitors we wish to change this power factor to 0.92 lagging. This new power factor corresponds to a θ_{new} of 23.07°. Therefore, the new complex power is

$$\mathbf{S}_{new} = 1780 + j1780 \tan(23.07°)$$

$$= 1780 + j758.28 \text{ kVA}$$

As illustrated in Fig. 9.17, the difference between $\mathbf{S}_{new}$ and $\mathbf{S}_T$ is that supplied by the purely reactive capacitor and, therefore,

$$\mathbf{S}_{cap} = jQ_C = \mathbf{S}_{new} - \mathbf{S}_T$$

or

$$jQ_C = j(758.28 - 1635)$$

$$= -j876.72 \text{ kVA}$$

Thus,

$$-j\omega C\, V_{rms}^2 = \frac{-j876.72\text{k}}{3}$$

and

$$377\left(\frac{13.8 \times 10^3}{\sqrt{3}}\right)^2 C = \frac{876.72}{3} \times 10^3$$

Therefore,

$$C = 12.2 \text{ μF}$$

Hence, three capacitors of this value connected in wye at the load will yield a total power factor of 0.92 lagging.

LEARNING Check

Summary

▶ An important advantage of the balanced three-phase system is that it provides very smooth power delivery.

▶ Because of the balanced condition, it is possible to analyze a circuit on a per phase basis, thereby providing a significant computational shortcut to a solution.

▶ A balanced three-phase source and load system: A balanced three-phase voltage source has three sinusoidal voltages of the same magnitude and frequency, and each voltage is 120° out of phase with the others. A positive-phase-sequence balanced voltage source is one in which $\mathbf{V}_{bn}$ lags $\mathbf{V}_{an}$ by 120° and $\mathbf{V}_{cn}$ lags $\mathbf{V}_{bn}$ by 120°.

▶ The relationships between wye- and delta-connected sources are shown in Table 10.1.

▶ The three-phase terminology is as shown in Table 10.3.

Table 10.3 Three-phase terminology

Quantity	Wye	Delta
$\mathbf{I}_a, \mathbf{I}_b, \mathbf{I}_c$	Line current (I_L)	
	Phase current (I_p)	
$\mathbf{V}_{an}, \mathbf{V}_{bn}, \mathbf{V}_{cn}$	Line-to-neutral voltage (V_p)	
	Phase voltage (V_p)	
$\mathbf{V}_{ab}, \mathbf{V}_{bc}, \mathbf{V}_{ca}$	Line-to-line, phase-to-phase, line voltage (V_L)	
		Phase voltage (V_p)
$\mathbf{I}_{ab}, \mathbf{I}_{bc}, \mathbf{I}_{ca}$		Phase current (I_p)

▶ In a balanced system the voltages and currents sum to zero.

$$\mathbf{V}_{an} + \mathbf{V}_{bn} + \mathbf{V}_{cn} = 0$$

$$\mathbf{I}_a + \mathbf{I}_b + \mathbf{I}_c = 0 \quad \text{(no current in the neutral line)}$$

and

$$\mathbf{V}_{ab} + \mathbf{V}_{bc} + \mathbf{V}_{ca} = 0$$

$$\mathbf{I}_{ab} + \mathbf{I}_{bc} + \mathbf{I}_{ca} = 0$$

▶ The steps recommended for solving balanced three-phase ac circuits are as follows:

1. If the source/load connection is not wye–wye, then transform the system to a wye–wye connection.

2. Determine the unknown phasors in the wye–wye connection and deal only with the phase a.

3. Convert the now-known phasors back to the corresponding phasors in the original connection.

▶ Power factor correction in a balanced three-phase environment is performed in the same manner as in the single-phase case. Three capacitors are put in parallel with the load to reduce the lagging phase caused by the three-phase load.

Problems

SECTION 10.2

10.1 Sketch a phasor representation of a balanced three-phase system containing both phase voltages and line voltages if $\mathbf{V}_{an} = 100\,\underline{/45°}$ V rms. Label all magnitudes and assume an *abc*-phase sequence.

10.2 A positive-sequence three-phase balanced wye voltage source has a phase voltage of $\mathbf{V}_{an} = 100\,\underline{/20°}$ V rms. Determine the line voltages of the source.

10.3 Sketch a phasor representation of an *abc*-sequence balanced three-phase Δ-connected source, including $\mathbf{V}_{ab}$, $\mathbf{V}_{bc}$, and $\mathbf{V}_{ca}$ if $\mathbf{V}_{ab} = 208\,\underline{/0°}$ V rms.

10.4 Sketch a phasor representation of a balanced three-phase system containing both phase voltages and line voltages if $\mathbf{V}_{an} = 120\,\underline{/60°}$ V rms. Label all magnitudes and assume an *abc*-phase sequence.

10.5 A positive-sequence three-phase balanced wye voltage source has a phase voltage of $\mathbf{V}_{an} = 240\,\underline{/90°}$ V rms. Determine the line voltages of the source.

10.6 Sketch a phasor representation of a balanced three-phase system containing both phase voltages and line voltages if $\mathbf{V}_{ab} = 208\,\underline{/45°}$ V rms. Label all phasors and assume an *abc*-phase sequence.

SECTION 10.3

10.7 A positive-sequence balanced three-phase wye-connected source with a phase voltage of 100 V rms supplies power to a balanced wye-connected load. The per phase load impedance is $40 + j10\ \Omega$. Determine the line currents in the circuit if $\underline{/\mathbf{V}_{an}} = 0°$.

10.8 A positive-sequence balanced three-phase wye-connected source supplies power to a balanced wye-connected load. The magnitude of the line voltages is 150 V rms. If the load impedance per phase is $36 + j12\ \Omega$, determine the line currents if $\underline{/\mathbf{V}_{an}} = 0°$.

10.9 An *abc*-sequence balanced three-phase wye-connected source supplies power to a balanced wye-connected load. The line impedance per phase is $1 + j0\ \Omega$, and the load

impedance per phase is $20 + j20\ \Omega$. If the source line voltage $\mathbf{V}_{ab}$ is $100\,\underline{/0°}$ V rms, find the line currents.

10.10 In a three-phase balanced wye–wye system, the source is an *abc*-sequence set of voltages with $\mathbf{V}_{an} = 120\,\underline{/60°}$ V rms. The per phase impedance of the load is $12 + j16\ \Omega$. If the line impedance per phase is $0.8 + j1.4\ \Omega$, find the line currents and the load voltages.

10.11 An *abc*-sequence set of voltages feeds a balanced three-phase wye–wye system. The line and load impedances are $0.6 + j1\ \Omega$ and $18 + j14\ \Omega$, respectively. If the load voltage on the *a* phase is $\mathbf{V}_{AN} = 114.47\,\underline{/18.99°}$ V rms, determine the voltages at the line input.

10.12 In a balanced three-phase wye–wye system, the source is an *abc*-sequence set of voltages. The load voltage on the *a* phase is $\mathbf{V}_{AN} = 108.58\,\underline{/79.81°}$ V rms, $\mathbf{Z}_{\text{line}} = 1 + j1.4\ \Omega$, and $\mathbf{Z}_{\text{load}} = 10 + j13\ \Omega$. Determine the input sequence of voltages.

10.13 A balanced *abc*-sequence of voltages feeds a balanced three-phase wye–wye system. The line and load impedances are $0.6 + j0.9\ \Omega$ and $8 + j12\ \Omega$, respectively. The load voltage on the *a* phase is $\mathbf{V}_{AN} = 116.63\,\underline{/10°}$ V rms. Find the line voltage $\mathbf{V}_{ab}$.

10.14 In a balanced three-phase wye–wye system, the source is an *abc*-sequence set of voltages. $\mathbf{Z}_{\text{line}} = 1 + j1.8\ \Omega$, $\mathbf{Z}_{\text{load}} = 14 + j12\ \Omega$, and the load voltage on the *a* phase is $\mathbf{V}_{AN} = 398.1\,\underline{/17.99°}$ V rms. Find the line voltage $\mathbf{V}_{ab}$.

10.15 An *abc*-phase sequence balanced three-phase source feeds a balanced load. The system is connected wye–wye and $\underline{/\mathbf{V}_{an}} = 0°$. The line impedance is $0.5 + j0.2\ \Omega$, the load impedance is $16 + j10\ \Omega$, and the total power absorbed by the load is 1836.54 W. Determine the magnitude of the source voltage $\mathbf{V}_{an}$.

10.16 In a balanced three-phase wye–wye system, the total power loss in the lines is 272.57 W. $\mathbf{V}_{AN} = 105.28\,\underline{/31.65°}$ V rms and the power factor of the load is 0.77 lagging. If the line impedance is $2 + j1\ \Omega$, determine the load impedance.

10.17 In a balanced three-phase wye–wye system, the load impedance is $8 + j4\ \Omega$. The source has phase sequence *abc* and $\mathbf{V}_{an} = 120\,\underline{/0°}$ V rms. If the load voltage is $\mathbf{V}_{AN} = 111.62\,\underline{/-1.33°}$ V rms, determine the line impedance.

10.18 In a balanced three-phase wye–wye system, the load impedance is $10 + j1\ \Omega$. The source has phase sequence *abc* and the line voltage $\mathbf{V}_{ab} = 220\,\underline{/30°}$ V rms. If the load voltage $\mathbf{V}_{AN} = 120\,\underline{/0°}$ V rms, determine the line impedance.

10.19 In a balanced three-phase wye–wye system, the load impedance is $20 + j12\ \Omega$. The source has an *abc*-phase sequence and $\mathbf{V}_{an} = 120\,\underline{/0°}$ V rms. If the load voltage is $\mathbf{V}_{AN} = 111.49\,\underline{/-0.2°}$ V rms, determine the magnitude of the line current if the load is suddenly short-circuited.

10.20 In a balanced three-phase wye–wye system, the source is an *abc*-sequence set of voltages and $\mathbf{V}_{an} = 120\,\underline{/50°}$ V rms. The load voltage on the *a* phase is $110.65\,\underline{/29.03°}$ V rms and the load impedance is $16 + j20\ \Omega$. Find the line impedance.

10.21 In a balanced three-phase wye–wye system, the source is an *abc*-sequence set of voltages and $\mathbf{V}_{an} = 120\,\underline{/40°}$ V rms. If the *a*-phase line current and line impedance are known to be $7.10\,\underline{/-10.28°}$ A rms and $0.8 + j1\ \Omega$, respectively, find the load impedance.

10.22 An *abc*-sequence set of voltages feeds a balanced three-phase wye–wye system. If $\mathbf{V}_{an} = 440\,\underline{/30°}$ V rms, $\mathbf{V}_{AN} = 413.28\,\underline{/29.78°}$ V rms, and $\mathbf{Z}_{\text{line}} = 1.2 + j1.5\ \Omega$, find the load impedance.

10.23 In a three-phase balanced system, a delta-connected source supplies power to a wye-connected load. If the line impedance is $0.2 + j0.4\ \Omega$, the load impedance $6 + j4\ \Omega$, and the source phase voltage $\mathbf{V}_{ab} = 208\,\underline{/40°}$ V rms, find the magnitude of the line voltage at the load.

10.24 Given the network in Fig. P10.24, compute the line currents and the magnitude of the phase voltage at the load.

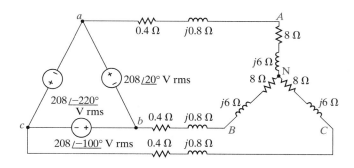

Figure P10.24

10.25 In a balanced three-phase delta–wye system, the source has an *abc*-phase sequence. The line and load impedances are $0.6 + j0.3\ \Omega$ and $12 + j7\ \Omega$, respectively. If the line current $\mathbf{I}_{aA} = 9.6\,\underline{/-20°}$ A rms, determine the phase voltages of the source.

10.26 An *abc*-phase-sequence three-phase balanced wye-connected 60-Hz source supplies a balanced delta-connected load. The phase impedance in the load consists of a 20-Ω resistor in series with a 50-mH inductor, and the phase voltage at the source is $\mathbf{V}_{an} = 120 \underline{/20°}$ V rms. If the line impedance is zero, find the line currents in the system.

10.27 An *abc*-phase-sequence three-phase balanced wye-connected source supplies a balanced delta-connected load. The impedance per phase in the delta load is $12 + j9 \ \Omega$. The line voltage at the source is $\mathbf{V}_{ab} = 120\sqrt{3} \underline{/40°}$ V rms. If the line impedance is zero, find the line currents in the balanced wye–delta system.

10.28 An *abc*-phase-sequence three-phase balanced wye-connected source supplies power to a balanced delta-connected load. The impedance per phase in the load is $14 + j12 \ \Omega$. If the source voltage for the *a* phase is $\mathbf{V}_{an} = 120 \underline{/80°}$ V rms, and the line impedance is zero, find the phase currents in the wye-connected source.

10.29 An *abc*-phase-sequence three-phase balanced wye-connected source supplies a balanced delta-connected load. The impedance per phase of the delta load is $10 + j8 \ \Omega$. If the line impedance is zero and the line current in the *a* phase is known to be $\mathbf{I}_{aA} = 28.10 \underline{/-28.66°}$ A rms, find the load voltage $\mathbf{V}_{AB}$.

10.30 An *abc*-phase-sequence three-phase balanced wye-connected source supplies power to a balanced delta-connected load. The impedance per phase of the delta load is $14 + j11 \ \Omega$. If the line impedance is zero and the line current in the *a* phase is $\mathbf{I}_{aA} = 20.22 \underline{/31.84°}$ A rms, find the voltages of the balanced source.

10.31 In a balanced three-phase wye–delta system, the source has an *abc*-phase sequence and $\mathbf{V}_{an} = 120 \underline{/0°}$ V rms. If the line current is $\mathbf{I}_{aA} = 4.8 \underline{/20°}$ A rms, find the load impedance per phase in the delta.

10.32 In a balanced three-phase wye–delta system, the source has an *abc*-phase sequence and $\mathbf{V}_{an} = 120 \underline{/40°}$ V rms. The line and load impedance are $0.5 + j0.4 \ \Omega$ and $24 + j18 \ \Omega$, respectively. Find the delta currents in the load.

10.33 In a three-phase balanced delta–delta system, the source has an *abc*-phase sequence. The line and load impedances are $0.3 + j0.2 \ \Omega$ and $9 + j6 \ \Omega$, respectively. If the load current in the delta is $\mathbf{I}_{AB} = 15 \underline{/40°}$ A rms, find the phase voltages of the source.

10.34 In a balanced three-phase delta–delta system, the source has an *abc*-phase sequence. The phase angle for the source voltage is $\underline{/\mathbf{V}_{ab}} = 40°$ and $\mathbf{I}_{ab} = 4 \underline{/15°}$ A rms. If the total power absorbed by the load is 1400 W, find the load impedance.

10.35 A three-phase load impedance consists of a balanced wye in parallel with a balanced delta. What is the equivalent wye load and what is the equivalent delta load if the phase impedance of the wye and delta are $6 + j3 \ \Omega$ and $15 + j12 \ \Omega$, respectively?

10.36 In a balanced three-phase system, the *abc*-phase-sequence source is wye connected and $\mathbf{V}_{an} = 120 \underline{/20°}$ V rms. The load consists of two balanced wyes with phase impedances of $8 + j6 \ \Omega$ and $12 + j8 \ \Omega$. If the line impedance is zero, find the line currents and the phase currents in each load.

10.37 In a balanced three-phase system, the source is a balanced wye with an *abc*-phase sequence and $\mathbf{V}_{ab} = 208 \underline{/60°}$ V rms. The load consists of a balanced wye with a phase impedance of $8 + j5 \ \Omega$ in parallel with a balanced delta with a phase impedance of $21 + j12 \ \Omega$. If the line impedance is $1.2 + j1 \ \Omega$, find the phase currents in the balanced wye load.

10.38 In a balanced three-phase system, the source is a balanced wye with an *abc*-phase sequence and $\mathbf{V}_{ab} = 208 \underline{/50°}$ V rms. The load is a balanced wye in parallel with a balanced delta. The phase impedance of the wye is $5 + j3 \ \Omega$ and the phase impedance of the delta is $18 + j12 \ \Omega$. If the line impedance is $1 + j0.8 \ \Omega$, find the line currents and the phase currents in the loads.

10.39 In a balanced three-phase system, the source, which has an *abc*-phase sequence, is connected in delta and $\mathbf{V}_{ab} = 208 \underline{/55°}$ V rms. There are two loads connected in parallel. Load 1 is connected in wye, and the phase impedance is $4 + j3 \ \Omega$. Load 2 is connected in wye, and the phase impedance is $8 + j6 \ \Omega$. Compute the delta currents in the source if the line impedance connecting the source to the loads is $0.2 + j0.1 \ \Omega$.

10.40 In a balanced three-phase system, the source has an *abc*-phase sequence and is connected in delta. There are two parallel wye-connected loads. The phase impedance of load 1 and load 2 is $4 + j4 \ \Omega$ and $10 + j4 \ \Omega$, respectively. The line impedance connecting the source to the loads is $0.3 + j0.2 \ \Omega$. If the current in the *a* phase of load 1 is $\mathbf{I}_{AN_1} = 10 \ \underline{/20°}$ A rms, find the delta currents in the source.

10.41 In a balanced three-phase system, the source has an *abc*-phase sequence and is connected in delta. There are two loads connected in parallel. The line connecting the source to the loads has an impedance of $0.2 + j0.1 \ \Omega$. Load 1 is connected in wye and the phase impedance is $4 + j2 \ \Omega$. Load 2 is connected in delta, and the phase impedance is $12 + j9 \ \Omega$. The current $\mathbf{I}_{AB}$ in the delta load is $16 \ \underline{/45°}$ A rms. Find the phase voltages of the source.

10.42 In a balanced three-phase system, the source has an *abc*-phase sequence and is connected in delta. There are two loads connected in parallel. Load 1 is connected in wye and has a phase impedance of $6 + j2 \ \Omega$. Load 2 is connected in delta and has a phase impedance of $9 + j3 \ \Omega$. The line impedance is $0.4 + j0.3 \ \Omega$. Determine the phase voltages of the source if the current in the *a* phase of load 1 is $\mathbf{I}_{AN_1} = 12 \ \underline{/30°}$ A rms.

10.43 A balanced three-phase delta-connected source supplies power to a load consisting of a balanced delta in parallel with a balanced wye. The phase impedance of the delta is $24 + j12 \ \Omega$, and the phase impedance of the wye is $12 + j8 \ \Omega$. The *abc*-phase-sequence source voltages are $\mathbf{V}_{ab} = 440 \ \underline{/60°}$ V rms, $\mathbf{V}_{bc} = 440 \ \underline{/-60°}$ V rms, and $\mathbf{V}_{ca} = 440 \ \underline{/-180°}$ V rms, and the line impedance per phase is $1 + j0.8 \ \Omega$. Find the line currents and the power absorbed by the wye-connected load.

SECTION 10.4

10.44 An *abc*-sequence wye-connected source having a phase-*a* voltage of $120 \ \underline{/0°}$ V rms is attached to a wye-connected load having an impedance of $80 \ \underline{/70°} \ \Omega$. If the line impedance is $4 \ \underline{/20°} \ \Omega$, determine the total complex power produced by the voltage sources and the real and reactive power dissipated by the load.

10.45 The magnitude of the complex power (apparent power) supplied by a three-phase balanced wye–wye system is 3600 VA. The line voltage is 208 V rms. If the line impedance is negligible and the power factor angle of the load is 25°, determine the load impedance.

10.46 A balanced three-phase wye–wye system has two parallel loads. Load 1 is rated at 3000 VA, 0.7 pf lagging, and load 2 is rated at 2000 VA, 0.75 pf leading. If the line voltage is 208 V rms, find the magnitude of the line current.

10.47 Two industrial plants represent balanced three-phase loads. The plants receive their power from a balanced three-phase source with a line voltage of 4.6 kV rms. Plant 1 is rated at 300 kVA, 0.8 pf lagging and plant 2 is rated at 350 kVA, 0.84 pf lagging. Determine the power line current.

10.48 A cluster of loads are served by a balanced three-phase source with a line voltage of 4160 V rms. Load 1 is 240 kVA at 0.8 pf lagging and load 2 is 160 kVA at 0.92 pf lagging. A third load is unknown except that it has a power factor of unity. If the line current is measured and found to be 62 A rms, find the complex power of the unknown load.

10.49 A balanced three-phase source serves two loads:

Load 1: 36 kVA at 0.8 pf lagging

Load 2: 18 kVA at 0.6 pf lagging

The line voltage at the load is 208 V rms at 60 Hz. Find the line current and the combined power factor at the load.

10.50 A balanced three-phase source serves the following loads:

Load 1:   48 kVA at 0.9 pf lagging

Load 2:   24 kVA at 0.75 pf lagging

The line voltage at the load is 208 V rms at 60 Hz. Determine the line current and the combined power factor at the load.

10.51 A small shopping center contains three stores that represent three balanced three-phase loads. The power lines to the shopping center represent a three-phase source with a line voltage of 13.8 kV rms. The three loads are

Load 1:   500 kVA at 0.8 pf lagging

Load 2:   400 kVA at 0.85 pf lagging

Load 3:   300 kVA at 0.90 pf lagging

Find the power line current.

10.52 The following loads are served by a balanced three-phase source:

Load 1:   18 kVA at 0.8 pf lagging

Load 2:   8 kVA at 0.8 pf leading

Load 3:   12 kVA at 0.75 pf lagging

The load voltage is 208 V rms at 60 Hz. If the line impedance is negligible, find the power factor of the source.

10.53 A balanced three-phase source supplies power to three loads. The loads are

Load 1:   30 kVA at 0.8 pf lagging

Load 2:   24 kW at 0.6 pf leading

Load 3:   unknown

If the line voltage and total complex power at the load are 208 V rms and $60 \underline{/0°}$ kVA, respectively, find the unknown load.

10.54 A balanced three-phase source serves the following loads:

Load 1:   18 kVA at 0.8 pf lagging

Load 2:   10 kVA at 0.7 pf leading

Load 3:   12 kW at unity pf

Load 4:   16 kVA at 0.6 pf lagging

The line voltage at the load is 208 V rms at 60 Hz, and the line impedance is $0.02 + j0.04$ Ω. Find the line voltage and power factor at the source.

10.55 A balanced three-phase source supplies power to three loads. The loads are

Load 1:   18 kW at 0.8 pf lagging

Load 2:   10 kVA at 0.6 pf leading

Load 3:   unknown

If the line voltage at the load is 208 V rms, the magnitude of the total complex power is 41.93 kVA, and the combined power factor at the load is 0.86 lagging, find the unknown load.

10.56 A balanced three-phase source supplies power to three loads. The loads are

Load 1:   20 kVA at 0.6 pf lagging

Load 2:   12 kW at 0.75 pf lagging

Load 3:   unknown

If the line voltage at the load is 208 V rms, the magnitude of the total complex power is 35.52 kVA, and the combined power factor at the load is 0.88 lagging, find the unknown load.

10.57 A standard practice for utility companies is to divide its customers into single-phase users and three-phase users. The utility must provide three-phase users, typically industries, with all three phases. However, single-phase users, residential, and light commercial are connected to only one phase. To reduce cable costs, all single-phase users in a neighborhood are connected together. This means that even if the three-phase users present perfectly balanced loads to the power grid, the single-phase loads will never be in balance, resulting in current flow in the neutral connection.

Consider the 60-Hz, *abc*-sequence network in Fig. P10.57. With a line voltage of $416\underline{/30°}$ V rms, phase *a* supplies the single-phase users on A Street, phase *b* supplies B Street, and phase *c* supplies C Street. Furthermore, the three-phase industrial load, which is connected in delta, is balanced. Find the neutral current.

Figure P10.57

SECTION 10.5

10.58 A three-phase *abc*-sequence wye-connected source with $\mathbf{V}_{an} = 220\underline{/0°}$ V rms supplies power to a wye-connected load that consumes 150 kW of power at a pf of 0.8 lagging. Three capacitors are found that each have an impedance of $-j2.0\ \Omega$, and they are connected in parallel with the previous load in a wye configuration. Determine the power factor of the combined load as seen by the source.

10.59 If the three capacitors in the network in Problem 10.58 are connected in a delta configuration, determine the power factor of the combined load as seen by the source.

10.60 Find *C* in the network in Fig. P10.60 such that the total load has a power factor of 0.9 lagging.

Figure P10.60

10.61 Find *C* in the network in Fig. P10.60 so that the total load has a power factor of 0.9 leading.

10.62 Find *C* in the network in Fig. P10.62 such that the total load has a power factor of 0.92 leading.

Figure P10.62

10.63 Find the value of *C* in Problem 10.62 such that the total load has a power factor of 0.92 lagging.

Typical Problems Found on the FE Exam

10FE-1 A wye-connected load consists of a series *RL* impedance. Measurements indicate that the rms voltage across each element is 84.85 V. If the rms line current is 6 A, find the total complex power for the three-phase load configuration.

10FE-2 A balanced three-phase delta-connected load consists of an impedance of $12 + j12\ \Omega$. If the line voltage at the load is measured to be 230 V rms, find the magnitude of the line current and the total real power absorbed by the three-phase configuration.

10FE-3 Two balanced three-phase loads are connected in parallel. One load with a phase impedance of $24 + j18\ \Omega$ is connected in delta, and the other load has a phase impedance of $6 + j4\ \Omega$ and is connected in wye. If the line-to-line voltage is 208 V rms, determine the line current.

10FE-4 The total complex power at the load of a three-phase balanced system is $24\ \underline{/30^\circ}$ kVA. Find the real power per phase.

Variable-Frequency Network Performance

In Chapter 7 we demonstrated that a network containing a capacitor and an inductor operated differently if the frequency was changed from the U.S. power grid frequency of 60 Hz to the aircraft frequency of 400 Hz. This phenomenon, although not surprising since the impedance of both these circuit elements is frequency dependent, indicates that if the frequency of the network sources is varied over some range, we can also expect the network to undergo variations in response to these changes in frequency.

Consider for a moment your stereo amplifier. The input signal contains sound waves of frequencies that range from soup to nuts; and yet, the amplifier must amplify each frequency component exactly the same amount in order to achieve perfect sound reproduction. Achieving perfect sound reproduction is a nontrivial task; and when you buy a very good amplifier, part of the price reflects the design necessary to achieve constant amplification over a wide range of frequencies.

In this chapter we examine the performance of electrical networks when excited from variable-frequency sources. Such effects are important in the analysis and design of networks such as filters, tuners, and amplifiers that have wide application in communication and control systems. Terminology and techniques for frequency-response analysis are introduced including standard plots (Bode plots) for describing network performance graphically. In particular, Bode plot construction and interpretation are discussed in detail.

The concept of resonance is introduced in reference to frequency selectivity and tuning. The various parameters used to define selectivity such as bandwidth, cutoff frequency, and quality factor are defined and discussed. Network scaling for both magnitude and phase is also presented.

Networks with special filtering properties are examined. Specifically, low-pass, high-pass, band-pass, and band-elimination filters are discussed. Techniques for designing active filters (containing op-amps) are presented.

11.1 Variable Frequency-Response Analysis

In previous chapters we investigated the response of *RLC* networks to sinusoidal inputs. In particular, we considered 60-Hz sinusoidal inputs. In this chapter we allow the frequency of excitation to become a variable and evaluate network performance as a function of frequency. To begin, let us consider the effect of varying frequency on elements with which we are already quite familiar—the resistor, inductor, and capacitor. The frequency-domain impedance of the resistor shown in Fig. 11.1a is

$$\mathbf{Z}_R = R = R\underline{/0°}$$

(a)

(b)

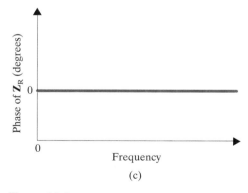

(c)

Figure 11.1
Frequency-dependent impedance of a resistor.

The magnitude and phase are constant and independent of frequency. Sketches of the magnitude and phase of $\mathbf{Z}_R$ are shown in Figs. 11.1b and c. Obviously, this is a very simple situation. For the inductor in Fig. 11.2a, the frequency-domain impedance $\mathbf{Z}_L$ is

$$\mathbf{Z}_L = j\omega L = \omega L \,\underline{/90°}$$

The phase is constant at 90° but the magnitude of $\mathbf{Z}_L$ is directly proportional to frequency. Figures 11.2b and c show sketches of the magnitude and phase of $\mathbf{Z}_L$ versus frequency. Note that at low frequencies the inductor's impedance is quite small. In fact, at dc, $\mathbf{Z}_L$ is zero and the inductor appears as a short circuit. Conversely, as frequency increases, the impedance also increases.

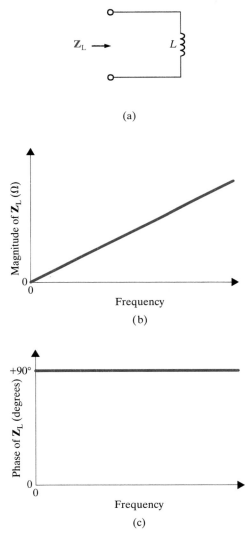

Figure 11.2
Frequency-dependent impedance of an inductor.

Next consider the capacitor of Fig. 11.3a. The impedance is

$$\mathbf{Z}_C = \frac{1}{j\omega C} = \frac{1}{\omega C} \underline{/-90^\circ}$$

Once again the phase of the impedance is constant but now the magnitude is inversely proportional to frequency, as shown in Figs. 11.3b and c. Note that the impedance approaches infinity, or an open circuit, as ω approaches zero and $\mathbf{Z}_C$ approaches zero as ω approaches infinity.

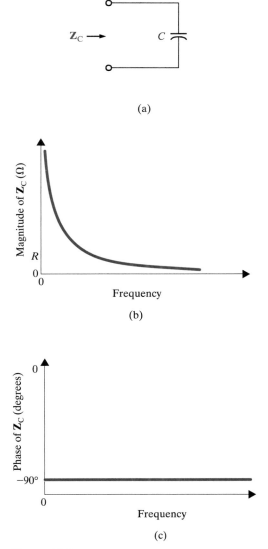

Figure 11.3
Frequency-dependent impedance of a capacitor.

Now let us investigate a more complex circuit: the *RLC* series network in Fig. 11.4a. The equivalent impedance is

$$\mathbf{Z}_{eq} = R + j\omega L + \frac{1}{j\omega C}$$

or

$$\mathbf{Z}_{eq} = \frac{(j\omega)^2 LC + j\omega RC + 1}{j\omega C}$$

Sketches of the magnitude and phase of this function are shown in Figs. 11.4b and c.

(a)

(b)

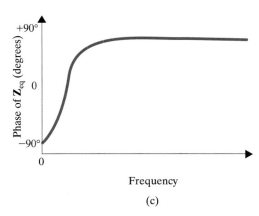

(c)

Figure 11.4

Frequency-dependent impedance of an RLC series network.

Note that at very low frequencies, the capacitor appears as an open circuit and, therefore, the impedance is very large in this range. At high frequencies, the capacitor has very little effect and the impedance is dominated by the inductor, whose impedance keeps rising with frequency.

As the circuits become more complicated, the equations become more cumbersome. In an attempt to simplify them, let us make the substitution $j\omega = s$. (This substitution has a more important meaning, which we will describe in later chapters.) With this substitution, the expression for $\mathbf{Z}_{eq}$ becomes

$$\mathbf{Z}_{eq} = \frac{s^2 LC + sRC + 1}{sC}$$

If we review the four circuits we investigated thus far, we will find that in every case the impedance is the ratio of two polynomials in s and is of the general form

$$\mathbf{Z}(s) = \frac{N(s)}{D(s)} = \frac{a_m s^m + a_{m-1} s^{m-1} + \cdots + a_1 s + a_0}{b_n s^n + b_{n-1} s^{n-1} + \cdots + b_1 s + b_0} \qquad \textbf{11.1}$$

where $N(s)$ and $D(s)$ are polynomials of order m and n, respectively. An extremely important aspect of Eq. (11.1) is that it holds not only for impedances but also for all voltages, currents, admittances, and gains in the network. The only restriction is that the values of all circuit elements (resistors, capacitors, inductors, and dependent sources) must be real numbers.

Let us now demonstrate the manner in which the voltage across an element in a series RLC network varies with frequency.

LEARNING Example 11.1

Consider the network in Fig. 11.5a. We wish to determine the variation of the output voltage as a function of frequency over the range from 0 to 1 kHz.

SOLUTION Using voltage division, the output can be expressed as

$$\mathbf{V}_o = \left(\frac{R}{R + j\omega L + \dfrac{1}{j\omega C}}\right)\mathbf{V}_S$$

or, equivalently,

$$\mathbf{V}_o = \left(\frac{j\omega CR}{(j\omega)^2 LC + j\omega CR + 1}\right)\mathbf{V}_S$$

Using the element values, the equation becomes

$$\mathbf{V}_o = \left(\frac{(j\omega)(37.95 \times 10^{-3})}{(j\omega)^2(2.53 \times 10^{-4}) + j\omega(37.95 \times 10^{-3}) + 1}\right)10\,\underline{/0°}$$

MATLAB can be effectively employed here to determine the magnitude and phase of the voltages as a function of frequency. The program required to generate the magnitude and phase plots

consists of three simple statements. If, in general, the function is of the form

$$\mathbf{V}_o(s) = \frac{a_m s^m + a_{m-1} s^{m-1} + \cdots + a_1 s + a_0}{b_n s^n + b_{n-1} s^{n-1} + \cdots + b_1 s + b_0}$$

then the plots, generated by the statements (in the following exact format)

```
>>B = [a_m  a_{m-1} .......... a_1  a_0]
>>A = [b_n  b_{n-1} .......... b_1  b_0]
>>freqs(B,A)
```

will produce the magnitude and phase characteristics as a function of frequency. The resultant characteristics are semilog plots in which the frequency is displayed on the log axis. The program for the output voltage in this example is

```
>>B = [ 37.95e-3 0];
>>A = [ 2.53e-4 37.95e-3 1];
>>freqs(B,A)
```

and the results are shown in Fig. 11.5b.

(continued)

(a)

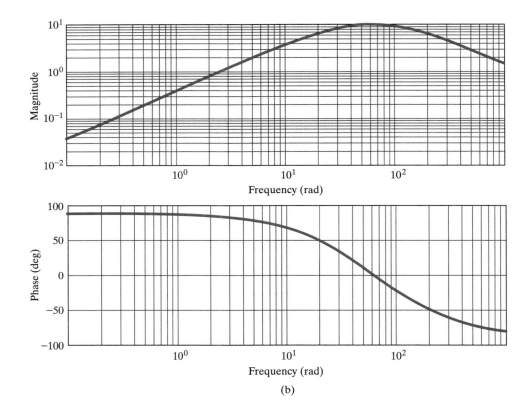

(b)

Figure 11.5
(a) Network and (b) its frequency-response simulation.

We will illustrate in subsequent sections that the use of a semilog plot is a very useful tool in deriving frequency-response information.

As an introductory application of variable frequency-response analysis and characterization, let us consider a stereo amplifier. In particular we should consider first the frequency range over which the amplifier must perform and then exactly what kind of performance we desire. The frequency range of the amplifier must exceed that of the human ear, which is roughly 50 Hz to 15,000 Hz. Accordingly, typical stereo amplifiers are designed to operate in the frequency range from 50 Hz to 20,000 Hz. Furthermore, we want to preserve the fidelity of the signal as it passes through the amplifier. Thus, the output signal should be an exact duplicate of the input

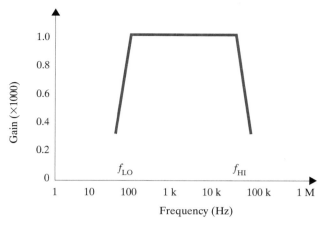

Figure 11.6 *Amplifier frequency-response requirements.*

signal times a gain factor. This requires that the gain be independent of frequency over the specified frequency range of 50 Hz to 20,000 Hz. An ideal sketch of this requirement for a gain of 1000 is shown in Fig. 11.6, where the midband region is defined as that portion of the plot where the gain is constant and is bounded by two points, which we will refer to as f_{LO} and f_{HI}. Notice once again that the frequency axis is a log axis and, thus, the frequency response is displayed on a semilog plot.

A model for the amplifier described graphically in Fig. 11.6 is shown in Fig. 11.7a with the frequency-domain equivalent circuit in Fig. 11.7b.

If the input is a steady-state sinusoid, we can use frequency-domain analysis to find the gain

$$\mathbf{G}_v(j\omega) = \frac{\mathbf{V}_o(j\omega)}{\mathbf{V}_S(j\omega)}$$

(a)

(b)

Figure 11.7 *Amplifier equivalent network.*

which with the substitution $s = j\omega$ can be expressed as

$$\mathbf{G}_v(s) = \frac{\mathbf{V}_o(s)}{\mathbf{V}_S(s)}$$

Using voltage division, we find that the gain is

$$\mathbf{G}_v(s) = \frac{\mathbf{V}_o(s)}{\mathbf{V}_S(s)} = \frac{\mathbf{V}_{\text{in}}(s)}{\mathbf{V}_S(s)} \frac{\mathbf{V}_o(s)}{\mathbf{V}_{\text{in}}(s)} = \left[\frac{R_{\text{in}}}{R_{\text{in}} + 1/sC_{\text{in}}}\right](1000)\left[\frac{1/sC_o}{R_o + 1/sC_o}\right]$$

or

$$\mathbf{G}_v(s) = \left[\frac{sC_{\text{in}}R_{\text{in}}}{1 + sC_{\text{in}}R_{\text{in}}}\right](1000)\left[\frac{1}{1 + sC_oR_o}\right]$$

Using the element values in Fig. 11.8a,

$$\mathbf{G}_v(s) = \left[\frac{s}{s + 100\pi}\right](1000)\left[\frac{40,000\pi}{s + 40,000\pi}\right]$$

where 100π and $40,000\pi$ are the radian equivalents of 50 Hz and 20,000 Hz, respectively. Since $s = j\omega$, the network function is indeed complex. An exact plot of $\mathbf{G}_v(s)$ is shown in Fig. 11.8 superimposed over the sketch of Fig. 11.6. The exact plot exhibits smooth transitions at f_{LO} and f_{HI}; otherwise the plots match fairly well.

Let us examine our expression for $\mathbf{G}_v(s)$ more closely with respect to the plot in Fig. 11.8. Assume that f is well within the midband frequency range; that is,

$$f_{\text{LO}} \ll f \ll f_{\text{HI}}$$

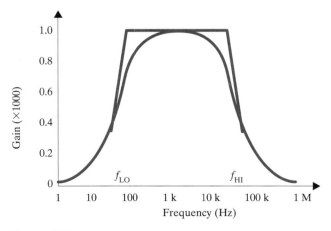

Figure 11.8
Exact and approximate amplifier gain versus frequency plots.

or

$$100\pi \ll |s| \ll 40{,}000\pi$$

Under these conditions, the network function becomes

$$\mathbf{G}_v(s) \approx \left[\frac{s}{s}\right](1000)\left[\frac{1}{1 + 0}\right]$$

or

$$\mathbf{G}_v(s) = 1000$$

Thus, well within midband, the gain is constant. However, if the frequency of excitation decreases toward f_{LO}, then $|s|$ is comparable to 100π and

$$\mathbf{G}_v(s) \approx \left[\frac{s}{s + 100\pi}\right](1000)$$

Since $R_{in}C_{in} = 1/100\pi$, we see that C_{in} causes the rolloff in gain at low frequencies. Similarly, when the frequency approaches f_{HI}, the gain rolloff is due to C_o.

Through this amplifier example, we have introduced the concept of frequency-dependent networks and have demonstrated that frequency-dependent network performance is caused by the reactive elements in a network.

NETWORK FUNCTIONS In the previous section, we introduced the term *voltage gain*, $\mathbf{G}_v(s)$. This term is actually only one of several network functions, designated generally as $\mathbf{H}(s)$, which define the ratio of response to input. Since the function describes a reaction due to an excitation at some other point in the circuit, network functions are also called *transfer functions*. Furthermore, transfer functions are not limited to voltage ratios. Since in electrical networks inputs and outputs can be either voltages or currents, there are four possible network functions, as listed in Table 11.1.

There are also *driving point functions*, which are impedances or admittances defined at a single pair of terminals. For example, the input impedance of a network is a driving point function.

Table 11.1 Network transfer functions

Input	Output	Transfer Function	Symbol
Voltage	Voltage	Voltage gain	$\mathbf{G}_v(s)$
Current	Voltage	Transimpedance	$\mathbf{Z}(s)$
Current	Current	Current gain	$\mathbf{G}_i(s)$
Voltage	Current	Transadmittance	$\mathbf{Y}(s)$

LEARNING Example 11.2

We wish to determine the transfer admittance $[\mathbf{I}_2(s)/\mathbf{V}_1(s)]$ and the voltage gain of the network shown in Fig. 11.9.

SOLUTION   The mesh equations for the network are

$$(R_1 + sL)\mathbf{I}_1(s) - sL\mathbf{I}_2(s) = \mathbf{V}_1(s)$$

$$-sL\mathbf{I}_1(s) + \left(R_2 + sL + \frac{1}{sC}\right)\mathbf{I}_2(s) = 0$$

$$\mathbf{V}_2(s) = \mathbf{I}_2(s)R_2$$

Solving the equations for $\mathbf{I}_2(s)$ yields

$$\mathbf{I}_2(s) = \frac{sL\mathbf{V}_1(s)}{(R_1 + sL)(R_2 + sL + 1/sC) - s^2L^2}$$

Therefore, the transfer admittance $[\mathbf{I}_2(s)/\mathbf{V}_1(s)]$ is

$$\mathbf{Y}_T(s) = \frac{\mathbf{I}_2(s)}{\mathbf{V}_1(s)} = \frac{LCs^2}{(R_1 + R_2)LCs^2 + (L + R_1R_2C)s + R_1}$$

and the voltage gain is

$$\mathbf{G}_v(s) = \frac{\mathbf{V}_2(s)}{\mathbf{V}_1(s)} = \frac{LCR_2s^2}{(R_1 + R_2)LCs^2 + (L + R_1R_2C)s + R_1}$$

Figure 11.9
Circuit employed in Example 11.2.

POLES AND ZEROS   As we have indicated, the network function can be expressed as the ratio of the two polynomials in s. In addition, we note that since the values of our circuit elements, or controlled sources, are real numbers, the coefficients of the two polynomials will be real. Therefore, we will express a network function in the form

$$\mathbf{H}(s) = \frac{N(s)}{D(s)} = \frac{a_m s^m + a_{m-1}s^{m-1} + \cdots + a_1 s + a_0}{b_n s^n + b_{n-1}s^{n-1} + \cdots + b_1 s + b_0} \qquad \textbf{11.2}$$

where $N(s)$ is the numerator polynomial of degree m and $D(s)$ is the denominator polynomial of degree n. Equation (11.2) can also be written in the form

$$\mathbf{H}(s) = \frac{K_0(s - z_1)(s - z_2)\cdots(s - z_m)}{(s - p_1)(s - p_2)\cdots(s - p_n)} \qquad \textbf{11.3}$$

where K_0 is a constant, $z_1, \ldots, z_m$ are the roots of $N(s)$, and $p_1, \ldots, p_n$ are the roots of $D(s)$. Note that if $s = z_1$, or $z_2, \ldots, z_m$, then $\mathbf{H}(s)$ becomes zero and hence $z_1, \ldots, z_m$ are called zeros of the transfer function. Similarly, if $s = p_1$, or $p_2, \ldots, p_n$, then $\mathbf{H}(s)$ becomes infinite and, therefore, $p_1, \ldots, p_n$ are called poles of the function. The zeros or poles may actually be complex. However, if they are complex, they must occur in conjugate pairs since the coefficients of the polynomial are real. The representation of the network function specified in Eq. (11.3) is extremely important and is generally employed to represent any linear time-invariant system.

The importance of this form stems from the fact that the dynamic properties of a system can be gleaned from an examination of the system poles.

E11.1 Find the driving point impedance at $\mathbf{V}_S(s)$ in the amplifier shown in Fig. 11.7b.

ANSWER

$$\mathbf{Z}(s) = R_{\text{in}} + \frac{1}{sC_{\text{in}}}$$

$$= \left[1 + \left(\frac{100\pi}{s} \right) \right] \text{M}\Omega$$

E11.2 Find the pole and zero locations in hertz and the value of K_0 for the amplifier network in Fig. 11.7.

ANSWER $z_1 = 0$ Hz (dc),
$p_1 = -50$ Hz,
$p_2 = -20{,}000$ Hz,
$K_0 = (4 \times 10^7)\pi$.

11.2 Sinusoidal Frequency Analysis

Although there are specific cases in which a network operates at only one frequency (e.g., a power system network), in general we are interested in the behavior of a network as a function of frequency. In a sinusoidal steady-state analysis, the network function can be expressed as

$$\mathbf{H}(j\omega) = M(\omega)e^{j\phi(\omega)} \qquad \textbf{11.4}$$

where $M(\omega) = |\mathbf{H}(j\omega)|$ and $\phi(\omega)$ is the phase. A plot of these two functions, which are commonly called the *magnitude* and *phase characteristics,* displays the manner in which the response varies with the input frequency ω. We will now illustrate the manner in which to perform a frequency-domain analysis by simply evaluating the function at various frequencies within the range of interest.

FREQUENCY RESPONSE USING A BODE PLOT If the network characteristics are plotted on a semilog scale (that is, a linear scale for the ordinate and a logarithmic scale for the abscissa), they are known as *Bode plots* (named after Hendrik W. Bode). This graph is a powerful tool in both the analysis and design of frequency-dependent systems and networks such as filters, tuners, and amplifiers. In using the graph, we plot $20 \log_{10} M(\omega)$ versus $\log_{10}(\omega)$ instead of $M(\omega)$ versus ω. The advantage of this technique is that rather than plotting the characteristic point by point, we can employ straight-line approximations to obtain the characteristic very efficiently. The ordinate for the magnitude plot is the decibel (dB). This unit was originally employed to measure the ratio of powers; that is,

$$\text{number of dB} = 10 \log_{10} \frac{P_2}{P_1} \qquad \textbf{11.5}$$

If the powers are absorbed by two equal resistors, then

$$
\begin{aligned}
\text{number of dB} &= 10 \log_{10} \frac{|\mathbf{V}_2|^2/R}{|\mathbf{V}_1|^2/R} = 10 \log_{10} \frac{|\mathbf{I}_2|^2 R}{|\mathbf{I}_1|^2 R} \\
&= 20 \log_{10} \frac{|\mathbf{V}_2|}{|\mathbf{V}_1|} = 20 \log_{10} \frac{|\mathbf{I}_2|}{|\mathbf{I}_1|}
\end{aligned}
$$

11.6

The term "dB" has become so popular that it now is used for voltage and current ratios, as illustrated in Eq. (11.6), without regard to the impedance employed in each case.

In the sinusoidal steady-state case, $\mathbf{H}(j\omega)$ in Eq. (11.3) can be expressed in general as

$$
\mathbf{H}(j\omega) = \frac{K_0(j\omega)^{\pm N}\left(1 + j\omega\tau_1\right)\left[1 + 2\zeta_3(j\omega\tau_3) + (j\omega\tau_3)^2\right]\cdots}{\left(1 + j\omega\tau_a\right)\left[1 + 2\zeta_b(j\omega\tau_b) + (j\omega\tau_b)^2\right]\cdots}
$$

11.7

Note that this equation contains the following typical factors:

1. A frequency-independent factor $K_0 > 0$
2. Poles or zeros at the origin of the form $j\omega$; that is, $(j\omega)^{+N}$ for zeros and $(j\omega)^{-N}$ for poles
3. Poles or zeros of the form $(1 + j\omega\tau)$
4. Quadratic poles or zeros of the form $1 + 2\zeta(j\omega\tau) + (j\omega\tau)^2$

Taking the logarithm of the magnitude of the function $\mathbf{H}(j\omega)$ in Eq. (11.7) yields

$$
\begin{aligned}
20 \log_{10}|\mathbf{H}(j\omega)| &= 20 \log_{10} K_0 \pm 20N \log_{10}|j\omega| \\
&\quad + 20 \log_{10}|1 + j\omega\tau_1| \\
&\quad + 20 \log_{10}\left|1 + 2\zeta_3(j\omega\tau_3) + (j\omega\tau_3)^2\right| \\
&\quad + \cdots - 20 \log_{10}|1 + j\omega\tau_a| \\
&\quad - 20 \log_{10}\left|1 + 2\zeta_b(j\omega\tau_b) + (j\omega\tau_b)^2\right|\cdots
\end{aligned}
$$

11.8

Note that we have used the fact that the log of the product of two or more terms is equal to the sum of the logs of the individual terms, the log of the quotient of two terms is equal to the difference of the logs of the individual terms, and $\log_{10} A^n = n \log_{10} A$.

The phase angle for $\mathbf{H}(j\omega)$ is

$$
\begin{aligned}
\underline{/\mathbf{H}(j\omega)} &= 0 \pm N(90°) + \tan^{-1}\omega\tau_1 + \tan^{-1}\left(\frac{2\zeta_3\omega\tau_3}{1 - \omega^2\tau_3^2}\right) \\
&\quad + \cdots - \tan^{-1}\omega\tau_a - \tan^{-1}\left(\frac{2\zeta_b\omega\tau_b}{1 - \omega^2\tau_b^2}\right)\cdots
\end{aligned}
$$

11.9

As Eqs. (11.8) and (11.9) indicate, we will simply plot each factor individually on a common graph and then sum them algebraically to obtain the total characteristic. Let us examine some of the individual terms and illustrate an efficient manner in which to plot them on the Bode diagram.

Constant Term   The term $20 \log_{10} K_0$ represents a constant magnitude with zero phase shift, as shown in Fig. 11.10a.

Figure 11.10
Magnitude and phase characteristics for a constant term and poles and zeros at the origin.

Poles or Zeros at the Origin   Poles or zeros at the origin are of the form $(j\omega)^{\pm N}$, where $+$ is used for a zero and $-$ is used for a pole. The magnitude of this function is $\pm 20 N \log_{10} \omega$, which is a straight line on semilog paper with a slope of $\pm 20 N$ dB/decade; that is, the value will change by $20 N$ each time the frequency is multiplied by 10, and the phase of this function is a constant $\pm N(90°)$. The magnitude and phase characteristics for poles and zeros at the origin are shown in Figs. 11.10b and c, respectively.

Simple Pole or Zero Linear approximations can be employed when a simple pole or zero of the form $(1 + j\omega\tau)$ is present in the network function. For $\omega\tau \ll 1$, $(1 + j\omega\tau) \approx 1$, and therefore, $20\log_{10}|(1 + j\omega\tau)| = 20\log_{10}1 = 0$ dB. Similarly, if $\omega\tau \gg 1$, then $(1 + j\omega\tau) \approx j\omega\tau$, and hence $20\log_{10}|(1 + j\omega\tau)| \approx 20\log_{10}\omega\tau$. Therefore, for $\omega\tau \ll 1$ the response is 0 dB and for $\omega\tau \gg 1$ the response has a slope that is the same as that of a simple pole or zero at the origin. The intersection of these two asymptotes, one for $\omega\tau \ll 1$ and one for $\omega\tau \gg 1$, is the point where $\omega\tau = 1$ or $\omega = 1/\tau$, which is called the *break frequency*. At this break frequency, where $\omega = 1/\tau$, $20\log_{10}|(1 + j1)| = 20\log_{10}(2)^{1/2} = 3$ dB. Therefore, the actual curve deviates from the asymptotes by 3 dB at the break frequency. It can be shown that at one-half and twice the break frequency, the deviations are 1 dB. The phase angle associated with a simple pole or zero is $\phi = \tan^{-1}\omega\tau$, which is a simple arctangent curve. Therefore, the phase shift is 45° at the break frequency and 26.6° and 63.4° at one-half and twice the break frequency, respectively. The actual magnitude curve for a pole of this form is shown in Fig. 11.11a. For a zero the magnitude curve and the asymptote for $\omega\tau \gg 1$ have a positive slope, and the phase curve extends from 0° to +90°, as shown in Fig. 11.11b. If multiple poles or zeros of the form $(1 + j\omega\tau)^N$ are present, then the slope of the high-frequency asymptote is multiplied by

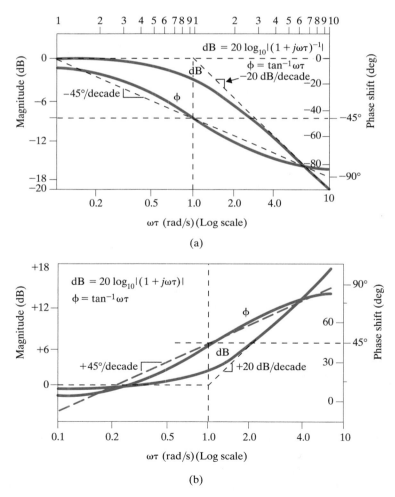

(a)

(b)

Figure 11.11
Magnitude and phase plot (a) for a simple pole, and (b) for a simple zero.

N, the deviation between the actual curve and the asymptote at the break frequency is $3N$ dB, and the phase curve extends from 0 to $N(90°)$ and is $N(45°)$ at the break frequency.

Quadratic Poles or Zeros   Quadratic poles or zeros are of the form $1 + 2\zeta(j\omega\tau) + (j\omega\tau)^2$. This term is a function not only of ω, but also of the dimensionless term ζ, which is called the *damping ratio*. If $\zeta > 1$ or $\zeta = 1$, the roots are real and unequal or real and equal, respectively, and these two cases have already been addressed. If $\zeta < 1$, the roots are complex conjugates, and it is this case that we will examine now. Following the preceding argument for a simple pole or zero, the log magnitude of the quadratic factor is 0 dB for $\omega\tau \ll 1$. For $\omega\tau \gg 1$.

$$20 \log_{10}\left|1 - (\omega\tau)^2 + 2j\zeta(\omega\tau)\right| \approx 20 \log_{10}\left|(\omega\tau)^2\right| = 40 \log_{10}|\omega\tau|$$

and therefore, for $\omega\tau \gg 1$, the slope of the log magnitude curve is $+40$ dB/decade for a quadratic zero and -40 dB/decade for a quadratic pole. Between the two extremes, $\omega\tau \ll 1$ and $\omega\tau \gg 1$, the behavior of the function is dependent on the damping ratio ζ. Figure 11.12a

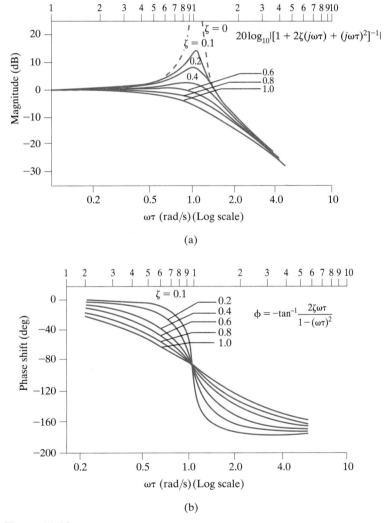

Figure 11.12
Magnitude and phase characteristics for quadratic poles.

illustrates the manner in which the log magnitude curve for a quadratic *pole* changes as a function of the damping ratio. The phase shift for the quadratic factor is $\tan^{-1} 2\zeta\omega\tau/[1 - (\omega\tau)^2]$. The phase plot for quadratic *poles* is shown in Fig. 11.12b. Note that in this case the phase changes from 0° at frequencies for which $\omega\tau \ll 1$ to $-180°$ at frequencies for which $\omega\tau \gg 1$. For quadratic zeros the magnitude and phase curves are inverted; that is, the log magnitude curve has a slope of +40 dB/decade for $\omega\tau \gg 1$, and the phase curve is 0° for $\omega\tau \ll 1$ and $+180°$ for $\omega\tau \gg 1$.

LEARNING Example 11.3

We want to generate the magnitude and phase plots for the transfer function

$$G_v(j\omega) = \frac{10(0.1j\omega + 1)}{(j\omega + 1)(0.02j\omega + 1)}$$

SOLUTION Note that this function is in standard form, since every term is of the form $(j\omega\tau + 1)$. To determine the composite magnitude and phase characteristics, we will plot the individual asymptotic terms and then add them as specified in Eqs. (11.8) and (11.9). Let us consider the magnitude plot first. Since $K_0 = 10$, $20 \log_{10} 10 = 20$ dB, which is a constant independent of frequency, as shown in Fig. 11.13a. The zero

of the transfer function contributes a term of the form $+20 \log_{10}|1 + 0.1j\omega|$, which is 0 dB for $0.1\omega \ll 1$, has a slope of +20 dB/decade for $0.1\omega \gg 1$, and has a break frequency at $\omega = 10$ rad/s. The poles have break frequencies at $\omega = 1$ and $\omega = 50$ rad/s. The pole with break frequency at $\omega = 1$ rad/s contributes a term of the form $-20 \log_{10}|1 + j\omega|$, which is 0 dB for $\omega \ll 1$ and has a slope of -20 dB/decade for $\omega \gg 1$. A similar argument can be made for the pole that has a break frequency at $\omega = 50$ rad/s. These factors are all plotted individually in Fig. 11.13a.

Consider now the individual phase curves. The term K_0 is not a function of ω and does not contribute to the phase of the transfer function. The phase curve for the zero is $+\tan^{-1} 0.1\omega$,

(a)

Figure 11.13

(a) Magnitude and phase components for the poles and zeros of the transfer function in Example 11.3; (b) Bode plot for the transfer function in Example 11.3.

Figure 11.13 (continued)

which is an arctangent curve that extends from $0°$ for $0.1\omega \ll 1$ to $+90°$ for $0.1\omega \gg 1$ and has a phase of $+45°$ at the break frequency. The phase curves for the two poles are $-\tan^{-1} \omega$ and $-\tan^{-1} 0.02\omega$. The term $-\tan^{-1} \omega$ is $0°$ for $\omega \ll 1$, $-90°$ for $\omega \gg 1$, and $-45°$ at the break frequency $\omega = 1$. The phase curve for the remaining pole is plotted in a similar fashion. All the individual phase curves are shown in Fig. 11.13a.

As specified in Eqs. (11.8) and (11.9), the composite magnitude and phase of the transfer function are obtained simply by adding the individual terms. The composite curves are plotted in Fig. 11.13b. Note that the actual magnitude curve (solid line) differs from the straight-line approximation (dashed line) by 3 dB at the break frequencies and 1 dB at one-half and twice the break frequencies.

LEARNING Example 11.4

Let us draw the Bode plot for the following transfer function:

$$\mathbf{G}_v(j\omega) = \frac{25(j\omega + 1)}{(j\omega)^2 (0.1 j\omega + 1)}$$

SOLUTION Once again all the individual terms for both magnitude and phase are plotted in Fig. 11.14a. The straight line with a slope of -40 dB/decade is generated by the double pole at the origin. This line is a plot of $-40 \log_{10} \omega$ versus ω and therefore passes through 0 dB at $\omega = 1$ rad/s. The phase for the double pole is a constant $-180°$ for all frequencies. The remainder of the terms are plotted as illustrated in Example 11.3.

The composite plots are shown in Fig. 11.14b. Once again they are obtained simply by adding the individual terms in Fig. 11.14a. Note that for frequencies for which $\omega \ll 1$, the slope of the magnitude curve is -40 dB/decade. At $\omega = 1$ rad/s, which is the break frequency of the zero, the magnitude curve changes slope to -20 dB/decade. At $\omega = 10$ rad/s, which is the break frequency of the pole, the slope of the magnitude curve changes back to -40 dB/decade.

The composite phase curve starts at $-180°$ due to the double pole at the origin. Since the first break frequency encountered is a zero, the phase curve shifts toward $-90°$. However, before the composite phase reaches $-90°$, the pole with break frequency $\omega = 10$ rad/s begins to shift the composite curve back toward $-180°$.

(continued)

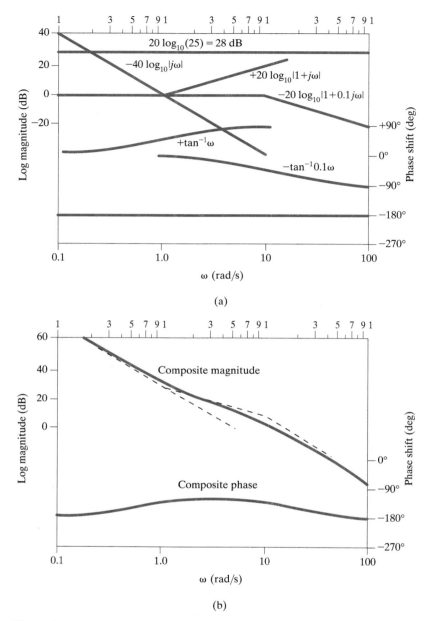

Figure 11.14
(a) Magnitude and phase components for the poles and zeros of the transfer function in Example 11.4; (b) Bode plot for the transfer function in Example 11.4.

Example 11.4 illustrates the manner in which to plot directly terms of the form $K_0/(j\omega)^N$. For terms of this form, the initial slope of $-20N$ dB/decade will intersect the 0-dB axis at a frequency of $(K_0)^{1/N}$ rad/s; that is, $-20\log_{10}|K_0/(j\omega)^N| = 0$ dB implies that $K_0/(j\omega)^N = 1$, and therefore, $\omega = (K_0)^{1/N}$ rad/s. Note that the projected slope of the magnitude curve in Example 11.4 intersects the 0-dB axis at $\omega = (25)^{1/2} = 5$ rad/s.

Similarly, it can be shown that for terms of the form $K_0(j\omega)^N$, the initial slope of $+20N$ dB/decade will intersect the 0-dB axis at a frequency of $\omega = (1/K_0)^{1/N}$ rad/s; that is, $+20 \log_{10} |K_0/(j\omega)^N| = 0$ dB implies that $K_0/(j\omega)^N = 1$, and therefore $\omega = (1/K_0)^{1/N}$ rad/s.

By applying the concepts we have just demonstrated, we can normally plot the log magnitude characteristic of a transfer function directly in one step.

LEARNING EXTENSIONS

E11.3   Sketch the magnitude characteristic of the Bode plot, labeling all critical slopes and points for the function

$$G(j\omega) = \frac{10^4(j\omega + 2)}{(j\omega + 10)(j\omega + 100)}$$

ANSWER

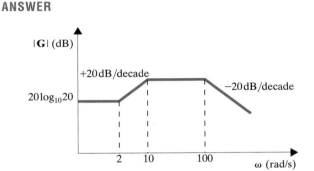

Figure E11.3

E11.4   Sketch the magnitude characteristic of the Bode plot, labeling all critical slopes and points for the function

$$G(j\omega) = \frac{100(0.02j\omega + 1)}{(j\omega)^2}$$

ANSWER

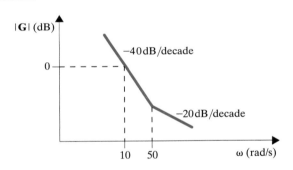

Figure E11.4

E11.5   Sketch the magnitude characteristic of the Bode plot, labeling all critical slopes and points for the function

$$G(j\omega) = \frac{10j\omega}{(j\omega + 1)(j\omega + 10)}$$

ANSWER

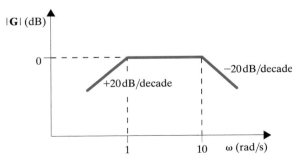

Figure E11.5

LEARNING Example 11.5

We wish to generate the Bode plot for the following transfer function:

$$\mathbf{G}_v(j\omega) = \frac{25j\omega}{(j\omega + 0.5)[(j\omega)^2 + 4j\omega + 100]}$$

SOLUTION Expressing this function in standard form, we obtain

$$\mathbf{G}_v(j\omega) = \frac{0.5j\omega}{(2j\omega + 1)[(j\omega/10)^2 + j\omega/25 + 1]}$$

The Bode plot is shown in Fig. 11.15. The initial low-frequency slope due to the zero at the origin is +20 dB/decade, and this slope intersects the 0-dB line at $\omega = 1/K_0 = 2$ rad/s. At $\omega = 0.5$ rad/s the slope changes from +20 dB/decade to 0 dB/decade due to the presence of the pole with a break frequency at $\omega = 0.5$ rad/s. The quadratic term has a center frequency of $\omega = 10$ rad/s (i.e., $\tau = 1/10$). Since

$$2\zeta\tau = \frac{1}{25}$$

and

$$\tau = 0.1$$

then

$$\zeta = 0.2$$

Plotting the curve in Fig. 11.12a with a damping ratio of $\zeta = 0.2$ at the center frequency $\omega = 10$ rad/s completes the composite magnitude curve for the transfer function.

The initial low-frequency phase curve is +90°, due to the zero at the origin. This curve and the phase curve for the simple pole and the phase curve for the quadratic term, as defined in Fig. 11.12b, are combined to yield the composite phase curve.

MATLAB could also be used to generate the magnitude and phase plots. The transfer function can be expressed in the form

$$\mathbf{G} = \frac{25j\omega}{(j\omega)^3 + 4.5(j\omega)^2 + 102j\omega + 50}$$

The MATLAB program for plotting the function is

```
>>B = [25 0];
>>A = [ 1 4.5 102 50];
>>freqs(B,A)
```

and the results are shown in Fig. 11.16.

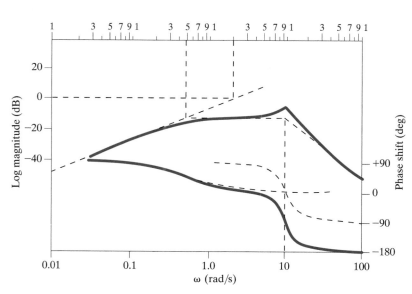

Figure 11.15
Bode plot for the transfer function in Example 11.5.

Figure 11.16 MATLAB-generated Bode plot.

E11.6 Given the following function $\mathbf{G}(j\omega)$, sketch the magnitude characteristic of the Bode plot, labeling all critical slopes and points.

$$\mathbf{G}(j\omega) = \frac{0.2(j\omega + 1)}{j\omega\left[(j\omega/12)^2 + j\omega/36 + 1\right]}$$

ANSWER

Figure E11.6

DERIVING THE TRANSFER FUNCTION FROM THE BODE PLOT

LEARNING Example 11.6

Given the asymptotic magnitude characteristic shown in Fig. 11.17, we wish to determine the transfer function $\mathbf{G}_v(j\omega)$.

SOLUTION Since the initial slope is 0 dB/decade, and the level of the characteristic is 20 dB, the factor K_0 can be obtained from the expression

$$20 \text{ dB} = 20 \log_{10} K_0$$

and hence

$$K_0 = 10$$

The −20-dB/decade slope starting at $\omega = 0.1$ rad/s indicates that the first pole has a break frequency at $\omega = 0.1$ rad/s, and

therefore one of the factors in the denominator is $(10j\omega + 1)$. The slope changes by +20 dB/decade at $\omega = 0.5$ rad/s, indicating that there is a zero present with a break frequency at $\omega = 0.5$ rad/s, and therefore the numerator has a factor of $(2j\omega + 1)$. Two additional poles are present with break frequencies at $\omega = 2$ rad/s and $\omega = 20$ rad/s. Therefore, the composite transfer function is

$$\mathbf{G}_v(j\omega) = \frac{10(2j\omega + 1)}{(10j\omega + 1)(0.5j\omega + 1)(0.05j\omega + 1)}$$

Note carefully the ramifications of this example with regard to network design.

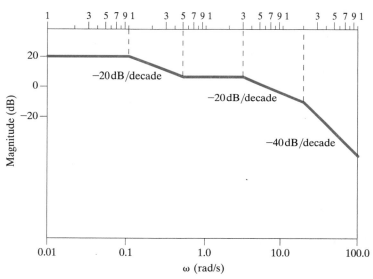

Figure 11.17
Straight-line magnitude plot employed in Example 11.6.

E11.7 Determine the transfer function $\mathbf{G}(j\omega)$ if the straight-line magnitude characteristic approximation for this function is as shown in Fig. E11.7.

ANSWER

$$\mathbf{G}(j\omega) = \frac{5\left(\dfrac{j\omega}{5} + 1\right)\left(\dfrac{j\omega}{50} + 1\right)}{j\omega\left(\dfrac{j\omega}{20} + 1\right)\left(\dfrac{j\omega}{100} + 1\right)}$$

Figure E11.7

11.3 Resonant Circuits

Two circuits with extremely important frequency characteristics are shown in Fig. 11.18. The input impedance for the series *RLC* circuit is

$$\mathbf{Z}(j\omega) = R + j\omega L + \frac{1}{j\omega C} \qquad\qquad \textbf{11.10}$$

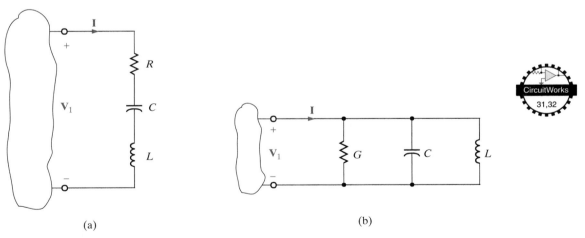

(a)

(b)

Figure 11.18 Series and parallel RLC circuits.

and the input admittance for the parallel *RLC* circuit is

$$\mathbf{Y}(j\omega) = G + j\omega C + \frac{1}{j\omega L} \qquad\qquad \textbf{11.11}$$

Note that these two equations have the same general form. The imaginary terms in both of the preceding equations will be zero if

$$\omega L = \frac{1}{\omega C}$$

The value of ω that satisfied this equation is

$$\omega_0 = \frac{1}{\sqrt{LC}} \qquad\qquad \textbf{11.12}$$

LEARNING Hint

Recall that ω_0 is the undamped natural frequency.

and at this value of ω the impedance of the series circuit becomes

$$\mathbf{Z}(j\omega_0) = R \qquad\qquad \textbf{11.13}$$

and the admittance of the parallel circuit is

$$\mathbf{Y}(j\omega_0) = G \qquad\qquad \textbf{11.14}$$

This frequency ω_0, at which the impedance of the series circuit or the admittance of the parallel circuit is purely real, is also called the *resonant frequency,* and the circuits themselves, at this frequency, are said to be *in resonance.* Resonance is a very important consideration in engineering design. For example, engineers designing the attitude control system for the Saturn vehicles had to ensure that the control system frequency did not excite the body bending (resonant) frequencies of the vehicle. Excitation of the bending frequencies would cause oscillations that, if continued unchecked, would result in a buildup of stress until the vehicle would finally break apart.

At resonance the voltage and current are in phase and, therefore, the phase angle is zero and the power factor is unity. In the series case, at resonance the impedance is a minimum and, therefore, the current is maximum for a given voltage. Figure 11.19 illustrates the frequency response of both the series and parallel *RLC* circuits. Note that at low frequencies the impedance of the series circuit is dominated by the capacitive term and the admittance of the parallel circuit is dominated by the inductive term. At high frequencies the impedance of the series circuit is dominated by the inductive term, and the admittance of the parallel circuit is dominated by the capacitive term.

Resonance can be viewed from another perspective—that of the phasor diagram. Once again we will consider the series and parallel cases together to illustrate the similarities between them. In the series case the current is common to every element, and in the parallel case the voltage is a common variable. Therefore, the current in the series circuit and the voltage in the parallel circuit are employed as references. Phasor diagrams for both circuits are shown in Fig. 11.20 for the three frequency values $\omega < \omega_0$, $\omega = \omega_0$, $\omega > \omega_0$.

In the series case when $\omega < \omega_0$, $\mathbf{V}_C > \mathbf{V}_L$, θ_Z is negative and the voltage $\mathbf{V}_1$ lags the current. If $\omega = \omega_0$, $\mathbf{V}_L = \mathbf{V}_C$, θ_Z is zero, and the voltage $\mathbf{V}_1$ is in phase with the current. If $\omega > \omega_0$, $\mathbf{V}_L > \mathbf{V}_C$, θ_Z is positive, and the voltage $\mathbf{V}_1$ leads the current. Similar statements can be made for the parallel case in Fig. 11.20b. Because of the close relationship between series and parallel resonance, as illustrated by the preceding material, we will concentrate most of our discussion on the series case in the following developments.

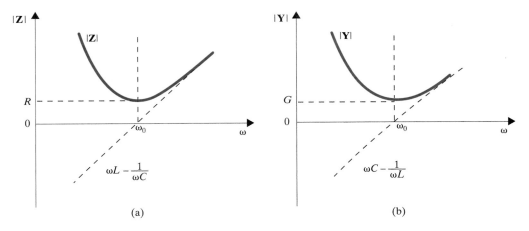

Figure 11.19
Frequency response of (a) a series and (b) a parallel RLC circuit.

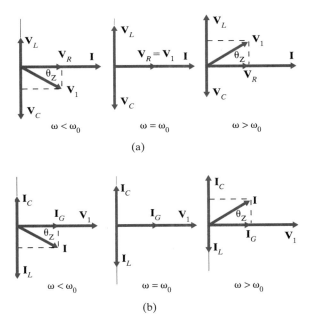

Figure 11.20
Phasor diagrams for (a) a series RLC circuit, and (b) a parallel GLC circuit.

For the series circuit we define what is commonly called the *quality factor Q* as

$$Q = \frac{\omega_0 L}{R} = \frac{1}{\omega_0 C R} = \frac{1}{R}\sqrt{\frac{L}{C}} \qquad \textbf{11.15}$$

Q is a very important factor in resonant circuits, and its ramifications will be illustrated throughout the remainder of this section.

LEARNING Example 11.7

Consider the network shown in Fig. 11.21. Let us determine the resonant frequency, the voltage across each element at resonance, and the value of the quality factor.

SOLUTION The resonant frequency is obtained from the expression

$$\omega_0 = \frac{1}{\sqrt{LC}}$$

$$= \frac{1}{\sqrt{(25)(10^{-3})(10)(10^{-6})}}$$

$$= 2000 \text{ rad/s}$$

At this resonant frequency

$$\mathbf{I} = \frac{\mathbf{V}}{\mathbf{Z}} = \frac{\mathbf{V}}{R} = 5\underline{/0°} \text{ A}$$

Therefore,

$$\mathbf{V}_R = (5\underline{/0°})(2) = 10\underline{/0°} \text{ V}$$

$$\mathbf{V}_L = j\omega_0 L\mathbf{I} = 250\underline{/90°} \text{ V}$$

$$\mathbf{V}_C = \frac{\mathbf{I}}{j\omega_0 C} = 250\underline{/-90°} \text{ V}$$

Note the magnitude of the voltages across the inductor and capacitor with respect to the input voltage. Note also that these voltages are equal and are 180° out of phase with one another. Therefore, the phasor diagram for this condition is shown in Fig. 11.20a for $\omega = \omega_0$. The quality factor Q derived from Eq. (11.15) is

$$Q = \frac{\omega_0 L}{R} = \frac{(2)(10^3)(25)(10^{-3})}{2} = 25$$

It is interesting to note that the voltages across the inductor and capacitor can be written in terms of Q as

$$|\mathbf{V}_L| = \omega_0 L|\mathbf{I}| = \frac{\omega_0 L}{R} \mathbf{V}_S = Q\mathbf{V}_S$$

and

$$|\mathbf{V}_C| = \frac{|\mathbf{I}|}{\omega_0 C} = \frac{1}{\omega_0 C R} \mathbf{V}_S = Q\mathbf{V}_S$$

This analysis indicates that for a given current there is a resonant voltage rise across the inductor and capacitor that is equal to the product of Q and the applied voltage.

Figure 11.21   Series circuit.

LEARNING Example 11.8

In an undergraduate circuits laboratory, students are asked to construct an RLC network that will demonstrate resonance at $f = 1000$ Hz given a 0.02 H inductor that has a Q of 200. One student produced the circuit shown in Fig. 11.22, where the inductor's internal resistance is represented by R.

If the capacitor chosen to demonstrate resonance was an oil-impregnated paper capacitor rated at 300 V, let us determine the network parameters and the effect of this choice of capacitor.

Figure 11.22   *RLC series resonant network.*

SOLUTION   For resonance at 1000 Hz, the student found the required capacitor value using the expression

$$\omega_0 = 2\pi f_0 = \frac{1}{\sqrt{LC}}$$

which yields

$$C = 1.27 \ \mu F$$

The student selected an oil-impregnated paper capacitor rated at 300 V. The resistor value was found using the expression for Q

$$Q = \frac{\omega_0 L}{R} = 200$$

or

$$R = 1.59 \ \Omega$$

At resonance, the current would be

$$I = \frac{V_S}{R}$$

or

$$I = 6.28 \underline{/0°} \ A$$

When constructed, the current was measured to be only

$$I \sim 1 \underline{/0°} \ mA$$

This measurement clearly indicated that the impedance seen by the source was about 10 kΩ of resistance instead of 1.59 Ω—quite a drastic difference. Suspecting that the capacitor that was selected was the source of trouble, the student calculated what the capacitor voltage should be. If operated as designed, then at resonance,

$$V_C = \frac{V_S}{R} \left(\frac{1}{j\omega C} \right) = QV_S$$

or

$$V_C = 2000 \underline{/-90°} \ V$$

which is more than six times the capacitor's rated voltage! This overvoltage had damaged the capacitor so that it did not function properly. When a new capacitor was selected and the source voltage reduced by a factor of 10, the network performed properly as a high Q circuit.

LEARNING EXTENSIONS

E11.8   Given the network in Fig. E11.8, find the value C that will place the circuit in resonance at 1800 rad/s.

ANSWER $C = 3.09 \ \mu F.$

Figure E11.8

E11.9   Given the network in E11.8, determine the Q of the network and the magnitude of the voltage across the capacitor.

ANSWER
$Q = 60, |V_C| = 600 \ V.$

The impedance of the circuit in Fig. 11.18a is given by Eq. (11.10), which can be expressed as an admittance,

$$\mathbf{Y}(j\omega) = \frac{1}{R[1 + j(1/R)(\omega L - 1/\omega C)]}$$

$$= \frac{1}{R[1 + j(\omega L/R - 1/\omega CR)]}$$

$$= \frac{1}{R[1 + jQ(\omega L/RQ - 1/\omega CRQ)]} \qquad \textbf{11.16}$$

Using the fact that $Q = \omega_0 L/R = 1/\omega_0 CR$ Eq. (11.16) becomes

$$\mathbf{Y}(j\omega) = \frac{1}{R[1 + jQ(\omega/\omega_0 - \omega_0/\omega)]} \qquad \textbf{11.17}$$

Since $\mathbf{I} = \mathbf{Y}\mathbf{V}_1$ and the voltage across the resistor is $\mathbf{V}_R = \mathbf{I}R$, then

$$\frac{\mathbf{V}_R}{\mathbf{V}_1} = \mathbf{G}_v(j\omega) = \frac{1}{1 + jQ(\omega/\omega_0 - \omega_0/\omega)} \qquad \textbf{11.18}$$

and the magnitude and phase are

$$M(\omega) = \frac{1}{[1 + Q^2(\omega/\omega_0 - \omega_0/\omega)^2]^{1/2}} \qquad \textbf{11.19}$$

and

$$\phi(\omega) = -\tan^{-1}Q\left(\frac{\omega}{\omega_0} - \frac{\omega_0}{\omega}\right) \qquad \textbf{11.20}$$

The sketches for these functions are shown in Fig. 11.23. Note that the circuit has the form of a band-pass filter. The bandwidth as shown is the difference between the two half-power frequencies. Since power is proportional to the square of the magnitude, these two frequencies may be derived by setting the magnitude $M(\omega) = 1/\sqrt{2}$; that is,

$$\left|\frac{1}{1 + jQ(\omega/\omega_0 - \omega_0/\omega)}\right| = \frac{1}{\sqrt{2}}$$

Therefore,

$$Q\left(\frac{\omega}{\omega_0} - \frac{\omega_0}{\omega}\right) = \pm 1 \qquad \textbf{11.21}$$

Solving this equation, we obtain four frequencies,

$$\omega = \pm\frac{\omega_0}{2Q} \pm \omega_0\sqrt{\left(\frac{1}{2Q}\right)^2 + 1} \qquad \textbf{11.22}$$

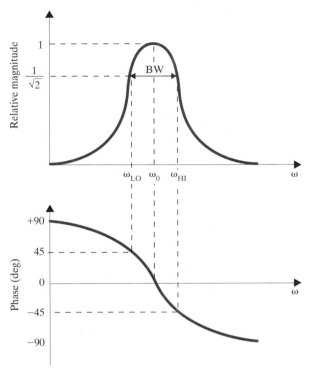

Figure 11.23
Magnitude and phase curves for Eq. (11.18).

Taking only the positive values, we obtain

$$\omega_{LO} = \omega_0 \left[-\frac{1}{2Q} + \sqrt{\left(\frac{1}{2Q}\right)^2 + 1} \right]$$ **11.23**

$$\omega_{HI} = \omega_0 \left[\frac{1}{2Q} + \sqrt{\left(\frac{1}{2Q}\right)^2 + 1} \right]$$

LEARNING Hint

Half-power frequencies and their dependence on ω_0 and Q

Subtracting these two equations yields the bandwidth as shown in Fig. 11.23:

$$BW = \omega_{HI} - \omega_{LO} = \frac{\omega_0}{Q}$$ **11.24**

and multiplying the two equations yields

$$\omega_0^2 = \omega_{LO}\omega_{HI}$$ **11.25**

LEARNING Hint

The bandwidth is the difference between the half-power frequencies and a function of ω_0 and Q.

which illustrates that the resonant frequency is the geometric mean of the two half-power frequencies. Recall that the half-power frequencies are the points at which the log-magnitude curve is down 3 dB from its maximum value. Therefore, the difference between the 3-dB frequencies, which is, of course, the bandwidth, is often called the 3-dB bandwidth.

LEARNING EXTENSION

E11.10 For the network in Fig. E11.8, compute the two half-power frequencies and the bandwidth of the network.

ANSWER
ω_{HI} = 1815 rad/s,
ω_{LO} = 1785 rad/s,
BW = 30 rad/s.

Equation (11.15) indicates the dependence of Q on R. A high-Q series circuit has a small value of R, and, as we will illustrate later, a high-Q parallel circuit has a relatively large value of R.

Equation (11.24) illustrates that the bandwidth is inversely proportional to Q. Therefore, the frequency selectivity of the circuit is determined by the value of Q. A high-Q circuit has a small bandwidth and, therefore, the circuit is very selective. The manner in which Q affects the frequency selectivity of the network is graphically illustrated in Fig. 11.24. Hence, if we pass a signal with a wide frequency range through a high-Q circuit, only the frequency components within the bandwidth of the network will not be attenuated; that is, the network acts like a band-pass filter.

Q has a more general meaning that we can explore via an energy analysis of the series circuit. Recall from Chapter 5 that an inductor stores energy in its magnetic field and a capacitor stores energy in its electric field. When a network is in resonance, there is a continuous exchange of energy between the magnetic field of the inductor and the electric field of the capacitor. During each half-cycle the energy stored in the inductor's magnetic field will vary from zero to a maximum value and back to zero again. The capacitor operates in a similar manner. The energy exchange takes place in the following way. During one quarter-cycle the capacitor absorbs energy as quickly as the inductor gives it up, and during the following one quarter-cycle the inductor absorbs energy as fast as it is released by the capacitor. Although the energy stored in each element is continuously varying, the total energy stored in the resonant circuit is constant and therefore not changing with time.

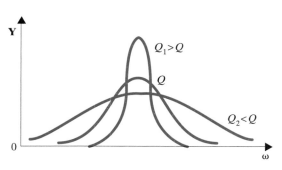

Figure 11.24
Network frequency response as a function of Q.

Q can also be defined as

$$Q = 2\pi \frac{W_S}{W_D} \qquad \textbf{11.26}$$

where W_S is the maximum energy stored at resonance and W_D is the energy dissipated per cycle. The importance of this definition of Q stems from the fact that this expression is applicable to acoustic, electrical, and mechanical systems and therefore is generally considered to be the basic definition of Q.

LEARNING Example 11.9

Given a series circuit with $R = 2\ \Omega$, $L = 2$ mH, and $C = 5\ \mu$F, we wish to determine the resonant frequency, the quality factor, and the bandwidth for the circuit. Then we will determine the change in Q and the BW if R is changed from 2 to 0.2 Ω.

SOLUTION   Using Eq. (11.12), we have

$$\omega_0 = \frac{1}{\sqrt{LC}} = \frac{1}{\left[(2)(10^{-3})(5)(10^{-6})\right]^{1/2}}$$

$$= 10^4 \text{ rad/s}$$

and therefore, the resonant frequency is $10^4/2\pi = 1592$ Hz.

The quality factor is

$$Q = \frac{\omega_0 L}{R} = \frac{(10^4)(2)(10^{-3})}{2}$$

$$= 10$$

and the bandwidth is

$$\text{BW} = \frac{\omega_0}{Q} = \frac{10^4}{10}$$

$$= 10^3 \text{ rad/s}$$

If R is changed to $R = 0.2\ \Omega$, the new value of Q is 100, and therefore the new BW is 10^2 rad/s.

LEARNING EXTENSIONS

E11.11   A series circuit is composed of $R = 2\ \Omega$, $L = 40$ mH, and $C = 100\ \mu$F. Determine the bandwidth of this circuit about its resonant frequency.

ANSWER   BW = 50 rad/s, $\omega_0 = 500$ rad/s.

E11.12   A series RLC circuit has the following properties: $R = 4\ \Omega$, $\omega_0 = 4000$ rad/s, and the BW = 100 rad/s. Determine the values of L and C.

ANSWER $L = 40$ mH, $C = 1.56\ \mu$F.

LEARNING Example 11.10

We wish to determine the parameters R, L, and C so that the circuit shown in Fig. 11.25 operates as a band-pass filter with an ω_0 of 1000 rad/s and a bandwidth of 100 rad/s.

SOLUTION   The voltage gain for the network is

$$\mathbf{G}_v(j\omega) = \frac{(R/L)j\omega}{(j\omega)^2 + (R/L)j\omega + 1/LC}$$

(continued)

Figure 11.25   *Series RLC circuit.*

Hence,

$$\omega_0 = \frac{1}{\sqrt{LC}}$$

and since $\omega_0 = 10^3$,

$$\frac{1}{LC} = 10^6$$

The bandwidth is

$$BW = \frac{\omega_0}{Q}$$

Then

$$Q = \frac{\omega_0}{BW} = \frac{1000}{100}$$
$$= 10$$

However,

$$Q = \frac{\omega_0 L}{R}$$

Therefore,

$$\frac{1000L}{R} = 10$$

Note that we have two equations in the three unknown circuit parameters R, L, and C. Hence, if we select $C = 1\ \mu F$, then

$$L = \frac{1}{10^6 C} = 1\ H$$

and

$$\frac{1000(1)}{R} = 10$$

yields

$$R = 100\ \Omega$$

Therefore, the parameters $R = 100\ \Omega$, $L = 1\ H$, and $C = 1\ \mu F$ will produce the proper filter characteristics.

In Examples 11.7 and 11.8 we found that the voltage across the capacitor or inductor in the series resonant circuit could be quite high. In fact, it was equal to Q times the magnitude of the source voltage. With this in mind, let us reexamine this network as shown in Fig. 11.26. The output voltage for the network is

$$\mathbf{V}_o = \left(\frac{1/j\omega C}{R + j\omega L + 1/j\omega C} \right) \mathbf{V}_S$$

Figure 11.26
Series resonant circuit.

which can be written as

$$\mathbf{V}_o = \frac{\mathbf{V}_S}{1 - \omega^2 LC + j\omega CR}$$

The magnitude of this voltage can be expressed as

$$|\mathbf{V}_o| = \frac{|\mathbf{V}_S|}{\sqrt{(1 - \omega^2 LC)^2 + (\omega CR)^2}} \qquad \textbf{11.27}$$

In view of the previous discussion, we might assume that the maximum value of the output voltage would occur at the resonant frequency ω_0. Let us see whether this assumption is correct. The frequency at which $|\mathbf{V}_o|$ is maximum is the nonzero value of ω, which satisfies the equation

$$\frac{d|\mathbf{V}_o|}{d\omega} = 0 \qquad \textbf{11.28}$$

If we perform the indicated operation and solve for the nonzero ω_{max}, we obtain

$$\omega_{max} = \sqrt{\frac{1}{LC} - \frac{1}{2}\left(\frac{R}{L}\right)^2} \qquad \textbf{11.29}$$

By employing the relationships $\omega_0^2 = 1/LC$ and $Q = \omega_0 L/R$, the expression for ω_{max} can be written as

$$\omega_{max} = \sqrt{\omega_0^2 - \frac{1}{2}\left(\frac{\omega_0}{Q}\right)^2}$$

$$= \omega_0\sqrt{1 - \frac{1}{2Q^2}} \qquad \textbf{11.30}$$

Clearly, $\omega_{max} \neq \omega_0$; however, ω_0 closely approximates ω_{max} if the Q is high. In addition, if we substitute Eq. (11.30) into Eq. (11.27) and use the relationships $\omega_0^2 = 1/LC$ and $\omega_0^2 C^2 R^2 = 1/Q^2$, we find that

$$|\mathbf{V}_o|_{max} = \frac{Q|\mathbf{V}_S|}{\sqrt{1 - 1/4Q^2}} \qquad \textbf{11.31}$$

Again, we see that $|\mathbf{V}_o|_{max} \approx Q|\mathbf{V}_S|$ if the network has a high Q.

LEARNING Example 11.11

Given the network in Fig. 11.26, we wish to determine ω_0 and ω_{max} for $R = 50\ \Omega$ and $R = 1\ \Omega$ if $L = 50$ mH and $C = 5\ \mu$F.

SOLUTION The network parameters yield

$$\omega_0 = \frac{1}{\sqrt{LC}}$$

$$= \frac{1}{\sqrt{(5)(10^{-2})(5)(10^{-6})}}$$

$$= 2000 \text{ rad/s}$$

(continued)

If $R = 50\ \Omega$, then

$$Q = \frac{\omega_0 L}{R}$$

$$= \frac{(2000)(0.05)}{50}$$

$$= 2$$

and

$$\omega_{max} = \omega_0\sqrt{1 - \frac{1}{2Q^2}}$$

$$= 2000\sqrt{1 - \frac{1}{8}}$$

$$= 1871\ \text{rad/s}$$

If $R = 1\ \Omega$, then $Q = 100$ and $\omega_{max} = 2000$ rad/s.

MATLAB can be used to plot the frequency response of the network transfer function for $R = 50\ \Omega$ and $R = 1\ \Omega$. The transfer function is

$$\frac{\mathbf{V}_o}{\mathbf{V}_s} = \frac{1}{2.5 \times 10^{-7}(j\omega)^2 + 2.5 \times 10^{-4}(j\omega) + 1}$$

for $R = 50\ \Omega$ and

$$\frac{\mathbf{V}_o}{\mathbf{V}_s} = \frac{1}{2.5 \times 10^{-7}(j\omega)^2 + 5 \times 10^{-6}(j\omega) + 1}$$

for $R = 1\ \Omega$. The corresponding MATLAB programs are

```
>>B = [1];
>>A = [ 2.5e-7  2.5e-4  1];
>>freqs(B, A)
```

and

```
>>B = [ 1 ];
>>A = [2.5e-7  5e-6  1]
>>freqs(B,A)
```

The magnitude and phase characteristics for the network with $R = 50\ \Omega$ and $R = 1\ \Omega$ are shown in Figs. 11.27a and b, respectively.

Note that when the Q of the network is small, the frequency response is not selective and $\omega_0 \neq \omega_{max}$. However, if the Q is large, the frequency response is very selective and $\omega_0 \simeq \omega_{max}$.

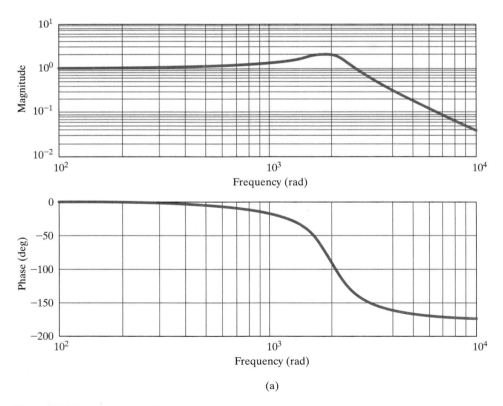

(a)

Figure 11.27
Frequency response plots for the network in Fig. 11.26 with (a) R = 50 Ω and (b) R = 1 Ω.

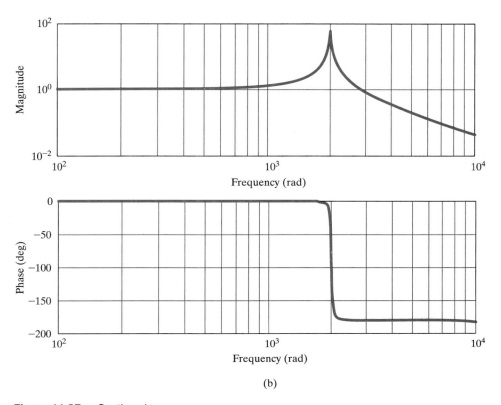

(b)

Figure 11.27 *Continued*

LEARNING Example 11.12

On July 1, 1940, the third longest bridge in the nation, the Tacoma Narrows Bridge, was opened to traffic across Puget Sound in Washington. On November 7, 1940, the structure collapsed in what has become the most celebrated structural failure of that century. A photograph of the bridge, taken as it swayed back and forth just before breaking apart, is shown in Fig. 11.28. Explaining the disaster in quantitative terms is a feat for civil engineers and structures experts, and several theories have been presented. However, the one common denominator in each explanation is that wind blowing across the bridge caused the entire structure to resonate to such an extent that the bridge tore itself apart. One can theorize that the wind, fluctuating at a frequency near the natural frequency of the bridge (0.2 Hz), drove the structure into resonance. Thus, the bridge can be roughly modeled as a second-order system. Let us design an *RLC* resonance network to demonstrate the bridge's vertical movement and investigate the effect of the wind's frequency.

SOLUTION The *RLC* network shown in Fig. 11.29 is a second-order system in which $v_{\text{in}}(t)$ is analogous to vertical deflection of the bridge's roadway (1 volt = 1 foot). The values of C, L, R_A, and R_B can be derived from the data taken at the site and from scale models, as follows:

$$\text{vertical deflection at failure} \approx 4 \text{ feet}$$
$$\text{wind speed at failure} \approx 42 \text{ mph}$$
$$\text{resonant frequency} = f_0 \approx 0.2 \text{ Hz}$$

The output voltage can be expressed as

$$\mathbf{V}_o(j\omega) = \frac{j\omega\left(\dfrac{R_B}{L}\right)\mathbf{V}_{\text{in}}(j\omega)}{-\omega^2 + j\omega\left(\dfrac{R_A + R_B}{L}\right) + \dfrac{1}{LC}}$$

(continued)

Figure 11.28 *Tacoma Narrows Bridge on the verge of collapse. (Used with permission from Special Collection Division, University of Washington Libraries. Photo by Farguharson, negative number 12.)*

Figure 11.29 *RLC resonance network for a simple Tacoma Narrows Bridge simulation.*

from which we can easily extract the following expressions:

$$\omega_0 = \frac{1}{\sqrt{LC}} = 2\pi(0.2) \text{ rad/s}$$

$$2\zeta\omega_0 = \frac{R_A + R_B}{L}$$

and

$$\frac{\mathbf{V}_o(j\omega_0)}{\mathbf{V}_{in}(j\omega_0)} = \frac{R_B}{R_A + R_B} \approx \frac{4 \text{ feet}}{42 \text{ mph}}$$

Let us choose $R_B = 1\ \Omega$ and $R_A = 9.5\ \Omega$. Having no data for the damping ratio, ζ, we will select $L = 20$ H, which yields $\zeta = 0.209$ and $Q = 2.39$, which seem reasonable for such a large structure. Given the aforementioned choices, the required capacitor value is $C = 31.66$ mF. Using these circuit values, we now simulate the effect of 42 mph winds fluctuating at 0.05 Hz, 0.1 Hz, and 0.2 Hz using an ac analysis at the three frequencies of interest.

The results are shown in Fig. 11.30. Note that at 0.05 Hz the vertical deflection (1 ft/V) is only 0.44 feet, whereas at 0.1 Hz the bridge undulates about 1.07 feet. Finally, at the bridge's resonant frequency of 0.2 Hz, the bridge is oscillating 3.77 feet—catastrophic failure.

Clearly, we have used an extremely simplistic approach to modeling something as complicated as the Tacoma Narrows

Bridge. However, we will revisit this event in Chapter 13 and examine it more closely with a more accurate model (K. Y. Billah and R. H. Scalan, "Resonance, Tacoma Narrows Bridge Failure, and Undergraduate Physics Textbooks," *American Journal of Physics* vol. 59, no. 2, pp. 118–124).

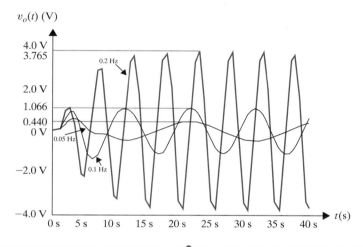

Figure 11.30
Simulated vertical deflection (1 volt = 1 foot) for the Tacoma Narrows Bridge for wind shift frequencies of 0.05, 0.1, and 0.2 Hz.

In our presentation of resonance thus far, we have focused most of our discussion on the series resonant circuit. We should recall, however, that the equations for the impedance of the series circuit and the admittance of the parallel circuit are similar. Therefore, the networks possess similar properties, as we illustrate in the following examples.

Consider the network shown in Fig. 11.31. The source current $\mathbf{I}_S$ can be expressed as

$$\mathbf{I}_S = \mathbf{I}_G + \mathbf{I}_C + \mathbf{I}_L$$

$$= \mathbf{V}_S G + j\omega C \mathbf{V}_S + \frac{\mathbf{V}_S}{j\omega L}$$

$$= \mathbf{V}_S \left[G + j\left(\omega C - \frac{1}{\omega L} \right) \right]$$

When the network is in resonance,

$$\mathbf{I}_S = G\mathbf{V}_S$$

Figure 11.31
Parallel RLC circuit.

that is, all the source current flows through the conductance G. Does this mean that there is no current in L or C? Definitely not! $\mathbf{I}_C$ and $\mathbf{I}_L$ are equal in magnitude but 180° out of phase with one another. Therefore, $\mathbf{I}_x$, as shown in Fig. 11.31, is zero. In addition, if $G = 0$, the source current is zero. What is actually taking place, however, is an energy exchange between the electric field of the capacitor and the magnetic field of the inductor. As one increases, the other decreases, and vice versa.

LEARNING Example 11.13

The network in Fig. 11.31 has the following parameters:

$$\mathbf{V}_S = 120 \underline{/0°} \text{ V}, \qquad\qquad G = 0.01 \text{ S},$$

$$C = 600 \text{ μF}, \qquad \text{and} \qquad L = 120 \text{ mH}$$

If the source operates at the resonant frequency of the network, compute all the branch currents.

SOLUTION The resonant frequency for the network is

$$\omega_0 = \frac{1}{\sqrt{LC}}$$

$$= \frac{1}{\sqrt{(120)(10^{-3})(600)(10^{-6})}}$$

$$= 117.85 \text{ rad/s}$$

At this frequency

$$\mathbf{Y}_C = j\omega_0 C = j7.07 \times 10^{-2} \text{ S}$$

and

$$\mathbf{Y}_L = -j\left(\frac{1}{\omega_0 L}\right) = -j7.07 \times 10^{-2} \text{ S}$$

The branch currents are then

$$\mathbf{I}_G = G\mathbf{V}_S = 1.2 \underline{/0°} \text{ A}$$

$$\mathbf{I}_C = \mathbf{Y}_C\mathbf{V}_S = 8.49 \underline{/90°} \text{ A}$$

$$\mathbf{I}_L = \mathbf{Y}_L\mathbf{V}_S = 8.49 \underline{/-90°} \text{ A}$$

and

$$\mathbf{I}_S = \mathbf{I}_G + \mathbf{I}_C + \mathbf{I}_L$$

$$= \mathbf{I}_G = 1.2 \underline{/0°} \text{ A}$$

As the analysis indicates, the source supplies only the losses in the resistive element. In addition, the source voltage and current are in phase and, therefore, the power factor is unity.

LEARNING Example 11.14

Given the parallel RLC circuit in Fig. 11.32,

(a) Derive the expression for the resonant frequency, the half-power frequencies, the bandwidth, and the quality factor for the transfer characteristic $\mathbf{V}_{out}/\mathbf{I}_{in}$ in terms of the circuit parameters R, L, and C.

(b) Compute the quantities in part (a) if $R = 1 \text{ k}\Omega$, $L = 10 \text{ mH}$, and $C = 100 \text{ μF}$.

SOLUTION (a) The output voltage can be written as

$$\mathbf{V}_{out} = \frac{\mathbf{I}_{in}}{\mathbf{Y}_T}$$

Figure 11.32
Circuit used in Example 11.14.

and, therefore, the magnitude of the transfer characteristic can be expressed as

$$\left|\frac{\mathbf{V}_{out}}{\mathbf{I}_{in}}\right| = \frac{1}{\sqrt{(1/R^2) + (\omega C - 1/\omega L)^2}}$$

The transfer characteristic is a maximum at the resonant frequency

$$\omega_0 = \frac{1}{\sqrt{LC}} \qquad \textbf{11.32}$$

and at this frequency

$$\left|\frac{\mathbf{V}_{out}}{\mathbf{I}_{in}}\right|_{max} = R \qquad \textbf{11.33}$$

As demonstrated earlier, at the half-power frequencies the magnitude is equal to $1/\sqrt{2}$ of its maximum value, and hence the half-power frequencies can be obtained from the expression

$$\frac{1}{\sqrt{(1/R^2) + (\omega C - 1/\omega L)^2}} = \frac{R}{\sqrt{2}}$$

Solving this equation and taking only the positive values of ω yields

$$\omega_{LO} = -\frac{1}{2RC} + \sqrt{\frac{1}{(2RC)^2} + \frac{1}{LC}} \qquad \textbf{11.34}$$

and

$$\omega_{HI} = \frac{1}{2RC} + \sqrt{\frac{1}{(2RC)^2} + \frac{1}{LC}} \qquad \textbf{11.35}$$

Subtracting these two half-power frequencies yields the bandwidth

$$BW = \omega_{HI} - \omega_{LO} \qquad \textbf{11.36}$$

$$= \frac{1}{RC}$$

Therefore, the quality factor is

$$Q = \frac{\omega_0}{BW}$$

$$= \frac{RC}{\sqrt{LC}} \qquad \textbf{11.37}$$

$$= R\sqrt{\frac{C}{L}}$$

Using Eqs. (11.32), (11.36), and (11.37), we can write Eqs. (11.34) and (11.35) as

$$\omega_{LO} = \omega_0\left[\frac{-1}{2Q} + \sqrt{\frac{1}{(2Q)^2} + 1}\right] \qquad \textbf{11.38}$$

$$\omega_{HI} = \omega_0\left[\frac{1}{2Q} + \sqrt{\frac{1}{(2Q)^2} + 1}\right] \qquad \textbf{11.39}$$

(b) Using the values given for the circuit components, we find that

$$\omega_0 = \frac{1}{\sqrt{(10^{-2})(10^{-4})}} = 10^3 \text{ rad/s}$$

The half-power frequencies are

$$\omega_{LO} = \frac{-1}{(2)(10^3)(10^{-4})} + \sqrt{\frac{1}{[(2)(10^{-1})]^2} + 10^6}$$

$$= 995 \text{ rad/s}$$

and

$$\omega_{HI} = 1005 \text{ rad/s}$$

Therefore, the bandwidth is

$$BW = \omega_{HI} = \omega_{LO} = 10 \text{ rad/s}$$

and

$$Q = 10^3\sqrt{\frac{10^{-4}}{10^{-2}}}$$

$$= 100$$

LEARNING Example 11.15

Two radio stations, WHEW and WHAT, broadcast in the same listening area: WHEW broadcasts at 100 MHz and WHAT at 98 MHz. A single-stage tuned amplifier, such as that shown in Fig. 11.33, can be used as a tuner to filter out one of the stations. However, single-stage tuned amplifiers have poor selectivity due to their wide bandwidths. To reduce the bandwidth (increase the

quality factor) of single-stage tuned amplifiers, designers employ a technique called synchronous tuning. In this process, identical tuned amplifiers are cascaded. To demonstrate this phenomenon, let us generate a Bode plot for the amplifier shown in Fig. 11.33 when it is tuned to WHEW (100 Mhz), using one, two, three, and four stages of amplification.

(continued)

SOLUTION Using the circuit for a single-stage amplifier shown in Fig. 11.33, we can cascade the stages to form a four-stage synchronously tuned amplifier. If we now plot the frequency response over the range from 90 MHz to 110 MHz, which is easily done using PSPICE, we obtain the Bode plot shown in Fig. 11.34.

From the Bode plot in Fig. 11.34 we see that increasing the number of stages does indeed decrease the bandwidth without altering the center frequency. As a result, the quality factor and selectivity increase. Accordingly, as we add stages, the gain at 98 MHz (WHAT's frequency) decreases and that station is "tuned out."

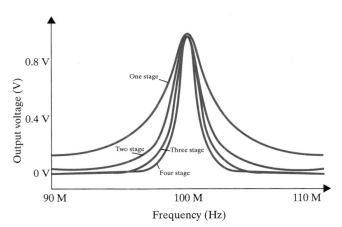

Figure 11.33
Single-stage tuned amplifier.

Figure 11.34
Bode plots for one-, two-, three-, and four-stage tuned amplifiers.

LEARNING EXTENSIONS

E11.13 A parallel RLC circuit has the following parameters: $R = 2 \text{ k}\Omega$, $L = 20 \text{ mH}$, and $C = 150 \text{ μF}$. Determine the resonant frequency, the Q, and the bandwidth of the circuit.

ANSWER $\omega_0 = 577 \text{ rad/s}$, $Q = 173$, and $BW = 3.33 \text{ rad/s}$.

E11.14 A parallel RLC circuit has the following parameters: $R = 6 \text{ k}\Omega$, $BW = 1000 \text{ rad/s}$, and $Q = 120$. Determine the values of L, C, and ω_0.

ANSWER $L = 417.5 \text{ μH}$, $C = 0.167 \text{ μF}$, and $\omega_0 = 119,760 \text{ rad/s}$.

Figure 11.35
Practical parallel resonant circuit.

In general, the resistance of the winding of an inductor cannot be neglected, and hence a more practical parallel resonant circuit is the one shown in Fig. 11.35. The input admittance of this circuit is

$$\mathbf{Y}(j\omega) = j\omega C + \frac{1}{R + j\omega L}$$

$$= j\omega C + \frac{R - j\omega L}{R^2 + \omega^2 L^2}$$

$$= \frac{R}{R^2 + \omega^2 L^2} + j\left(\omega C - \frac{\omega L}{R^2 + \omega^2 L^2}\right)$$

The frequency at which the admittance is purely real is

$$\omega_r C - \frac{\omega_r L}{R^2 + \omega_r^2 L^2} = 0$$

$$\omega_r = \sqrt{\frac{1}{LC} - \frac{R^2}{L^2}} \qquad\qquad \textbf{11.40}$$

LEARNING Example 11.16

Given the tank circuit in Fig. 11.36, let us determine ω_0 and ω_r for $R = 50 \ \Omega$ and $R = 5 \ \Omega$.

SOLUTION Using the network parameter values, we obtain

$$\omega_0 = \frac{1}{\sqrt{LC}}$$

$$= \frac{1}{\sqrt{(0.05)(5)(10^{-6})}}$$

$$= 2000 \text{ rad/s}$$

$$f_0 = 318.3 \text{ Hz}$$

Figure 11.36 *Tank circuit used in Example 11.16.*

(continued)

If $R = 50\ \Omega$, then

$$\omega_r = \sqrt{\frac{1}{LC} - \frac{R^2}{L^2}}$$

$$= \sqrt{\frac{1}{(0.05)(5)(10^{-6})} - \left(\frac{50}{0.05}\right)^2}$$

$$= 1732\ \text{rad/s}$$

$$f_r = 275.7\ \text{Hz}$$

If $R = 5\ \Omega$, then

$$\omega_r = \sqrt{\frac{1}{(0.05)(5)(10^{-6})} - \left(\frac{5}{0.05}\right)^2}$$

$$= 1997\ \text{rad/s}$$

$$f_r = 317.9\ \text{Hz}$$

Note that as $R \to 0$, $\omega_r \to \omega_0$. This fact is also illustrated in the frequency-response curves in Figs. 11.37a and b, where we have plotted $|\mathbf{V}_o|$ versus frequency for $R = 50\ \Omega$ and $R = 5\ \Omega$, respectively.

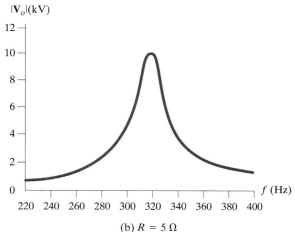

Figure 11.37 *Frequency-response curves for Example 11.16.*

Let us now try to relate some of the things we have learned about resonance to the Bode plots we presented earlier. The admittance for the series resonant circuit is

$$\mathbf{Y}(j\omega) = \frac{1}{R + j\omega L + 1/j\omega C}$$

$$= \frac{j\omega C}{(j\omega)^2 LC + j\omega CR + 1} \qquad \textbf{11.41}$$

The standard form for the quadratic factor is

$$(j\omega\tau)^2 + 2\zeta\omega\tau j + 1$$

where $\tau = 1/\omega_0$, and hence in general the quadratic factor can be written as

$$\frac{(j\omega)^2}{\omega_0^2} + \frac{2\zeta\omega}{\omega_0}j + 1 \qquad \textbf{11.42}$$

If we now compare this form of the quadratic factor with the denominator of $\mathbf{Y}(j\omega)$, we find that

$$\omega_0^2 = \frac{1}{LC}$$

$$\frac{2\zeta}{\omega_0} = CR$$

and therefore,

$$\zeta = \frac{R}{2}\sqrt{\frac{C}{L}}$$

However, from Eq. (11.15),

$$Q = \frac{1}{R}\sqrt{\frac{L}{C}}$$

and hence,

$$Q = \frac{1}{2\zeta} \qquad \text{11.43}$$

To illustrate the significance of this equation, consider the Bode plot for the function $\mathbf{Y}(j\omega)$. The plot has an initial slope of $+20$ dB/decade due to the zero at the origin. If $\zeta > 1$, the poles represented by the quadratic factor in the denominator will simply roll off the frequency response, as illustrated in Fig. 11.12a, and at high frequencies the slope of the composite characteristic will be -20 dB/decade. Note from Eq. (11.43) that if $\zeta > 1$, the Q of the circuit is very small. However, if $0 < \zeta < 1$, the frequency response will peak as shown in Fig. 11.12a, and the sharpness of the peak will be controlled by ζ. If ζ is very small, the peak of the frequency response is very narrow, the Q of the network is very large, and the circuit is very selective in filtering the input signal. Equation (11.43) and Fig. 11.24 illustrate the connections among the frequency response, the Q, and the ζ of a network.

11.4 Scaling

Throughout this book we have employed a host of examples to illustrate the concepts being discussed. In many cases the actual values of the parameters were unrealistic in a practical sense, even though they may have simplified the presentation. In this section we illustrate how to *scale* the circuits to make them more realistic.

There are two ways to scale a circuit: *magnitude* or *impedance scaling* and *frequency scaling*. To magnitude scale a circuit, we simply multiply the impedance of each element by a scale factor K_M. Therefore, a resistor R becomes $K_M R$. Multiplying the impedance of an inductor $j\omega L$ by K_M yields a new inductor $K_M L$, and multiplying the impedance of a capacitor $1/j\omega C$ by K_M yields a new capacitor C/K_M. Therefore, in magnitude scaling,

LEARNING Hint

Magnitude or impedance scaling

$$R' \rightarrow K_M R$$

$$L' \rightarrow K_M L \qquad \text{11.44}$$

$$C' \rightarrow \frac{C}{K_M}$$

since

$$\omega_0' = \frac{1}{\sqrt{L'C'}} = \frac{1}{\sqrt{K_M L C / K_M}} = \omega_0$$

and Q' is

$$Q' = \frac{\omega_0 L'}{R'} = \frac{\omega_0 K_M L}{K_M R} = Q$$

The resonant frequency, the quality factor, and therefore the bandwidth are unaffected by magnitude scaling.

In frequency scaling the scale factor is denoted as K_F. The resistor is frequency independent and, therefore, unaffected by this scaling. The new inductor L', which has the same impedance at the scaled frequency ω_1', must satisfy the equation

$$j\omega_1 L = j\omega_1' L'$$

where $\omega_1' = k_F \omega_1$. Therefore,

$$j\omega_1 L = jK_F \omega_1 L'$$

Hence, the new inductor value is

$$L' = \frac{L}{K_F}$$

Using a similar argument, we find that

$$C' = \frac{C}{K_F}$$

LEARNING Hint

Frequency scaling

Therefore, to frequency scale by a factor K_F,

$$R' \rightarrow R$$

$$L' \rightarrow \frac{L}{K_F}$$

$$C' \rightarrow \frac{C}{K_F}$$

11.45

Note that

$$\omega_0' = \frac{1}{\sqrt{(L/K_F)(C/K_F)}} = K_F \omega_0$$

and

$$Q' = \frac{K_F \omega_0 L}{RK_F} = Q$$

and therefore,

$$BW' = K_F(BW)$$

Hence, the resonant frequency and bandwidth of the circuit are affected by frequency scaling.

LEARNING Example 11.17

If the values of the circuit parameters in Fig 11.35 are $R = 2\,\Omega$, $L = 1$ H, and $C = \frac{1}{2}$F, let us determine the values of the elements if the circuit is magnitude scaled by a factor $K_M = 10^2$ and frequency scaled by a factor $K_F = 10^2$.

SOLUTION   The magnitude scaling yields

$$R' = 2K_M = 200\,\Omega$$

$$L' = (1)K_M = 100 \text{ H}$$

$$C' = \frac{1}{2}\frac{1}{K_M} = \frac{1}{200}\text{ F}$$

Applying frequency scaling to these values yields the final results:

$$R'' = 200\,\Omega$$

$$L'' = \frac{100}{K_F} = 100\ \mu\text{H}$$

$$C'' = \frac{1}{200}\frac{1}{K_F} = 0.005\ \mu\text{F}$$

E11.15   An *RLC* network has the following parameter values: $R = 10\,\Omega$, $L = 1$ H, and $C = 2$ F. Determine the values of the circuit elements if the circuit is magnitude scaled by a factor of 100 and frequency scaled by a factor of 10,000.

ANSWER $R = 1\text{ k}\Omega$, $L = 10$ mH, $C = 2\ \mu\text{F}$.

11.5 Filter Networks

PASSIVE FILTERS   A filter network is generally designed to pass signals with a specific frequency range and reject or attenuate signals whose frequency spectrum is outside this passband. The most common filters are *low-pass* filters, which pass low frequencies and reject high frequencies; *high-pass* filters, which pass high frequencies and block low frequencies; *band-pass* filters, which pass some particular band of frequencies and reject all frequencies outside the range; and *band-rejection* filters, which are specifically designed to reject a particular band of frequencies and pass all other frequencies.

The ideal frequency characteristic for a low-pass filter is shown in Fig. 11.38a. Also shown is a typical or physically realizable characteristic. Ideally, we would like the low-pass filter to pass all frequencies to some frequency ω_0 and pass no frequency above that value; however, it is not possible to design such a filter with linear circuit elements. Hence, we must be content to employ filters that we can actually build in the laboratory, and these filters have frequency characteristics that are simply not ideal.

A simple low-pass filter network is shown in Fig. 11.38b. The voltage gain for the network is

$$\mathbf{G}_v(j\omega) = \frac{1}{1 + j\omega RC} \qquad\qquad \textbf{11.46}$$

which can be written as

$$\mathbf{G}_v(j\omega) = \frac{1}{1 + j\omega\tau} \qquad \qquad \textbf{11.47}$$

where $\tau = RC$, the time constant. The amplitude characteristic is

$$M(\omega) = \frac{1}{\left[1 + (\omega\tau)^2\right]^{1/2}} \qquad \qquad \textbf{11.48}$$

and the phase characteristic is

$$\phi(\omega) = -\tan^{-1}\omega\tau \qquad \qquad \textbf{11.49}$$

Note that at the break frequency, $\omega = \frac{1}{\tau}$, the amplitude is

$$M\left(\omega = \frac{1}{\tau}\right) = \frac{1}{\sqrt{2}} \qquad \qquad \textbf{11.50}$$

The break frequency is also commonly called the *half-power frequency*. This name is derived from the fact that if the voltage or current is $1/\sqrt{2}$ of its maximum value, then the power, which is proportional to the square of the voltage or current, is one-half its maximum value.

The magnitude, in decibels, and phase curves for this simple low-pass circuit are shown in Fig. 11.38c. Note that the magnitude curve is flat for low frequencies and rolls off at high frequencies. The phase shifts from 0° at low frequencies to −90° at high frequencies.

The ideal frequency characteristic for a high-pass filter is shown in Fig. 11.39a, together with a typical characteristic that we could achieve with linear circuit components. Ideally, the high-pass filter passes all frequencies above some frequency ω_0 and no frequencies below that value.

A simple high-pass filter network is shown in Fig. 11.39b. This is the same network as shown in Fig. 11.38b, except that the output voltage is taken across the resistor. The voltage gain for this network is

$$\mathbf{G}_v(j\omega) = \frac{j\omega\tau}{1 + j\omega\tau} \qquad \qquad \textbf{11.51}$$

where once again $\tau = RC$. The magnitude of this function is

$$M(\omega) = \frac{\omega\tau}{\left[1 + (\omega\tau)^2\right]^{1/2}} \qquad \qquad \textbf{11.52}$$

and the phase is

$$\phi(\omega) = \frac{\pi}{2} - \tan^{-1}\omega\tau \qquad \qquad \textbf{11.53}$$

The half-power frequency is $\omega = 1/\tau$, and the phase at this frequency is 45°.

The magnitude and phase curves for this high-pass filter are shown in Fig. 11.39c. At low frequencies the magnitude curve has a slope of +20 dB/decade due to the term $\omega\tau$ in the numerator of Eq. (11.52). Then at the break frequency the curve begins to flatten out. The phase curve is derived from Eq. (11.53).

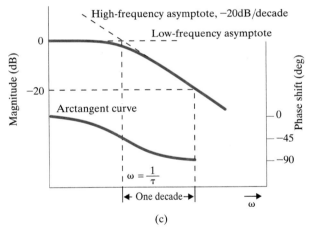

Figure 11.38
Low-pass filter circuit and its frequency characteristics.

Ideal and typical amplitude characteristics for simple band-pass and band-rejection filters are shown in Figs. 11.40a and b, respectively. Simple networks that are capable of realizing the typical characteristics of each filter are shown below the characteristics in Figs. 11.40c and d. ω_0 is the center frequency of the pass or rejection band and the frequency at which the maximum or minimum amplitude occurs. ω_{LO} and ω_{HI} are the lower and upper break frequencies or *cut-off frequencies*, where the amplitude is $1/\sqrt{2}$ of the maximum value. The width of the pass or rejection band is called *bandwidth*, and hence

$$BW = \omega_{HI} - \omega_{LO} \qquad\qquad \textbf{11.54}$$

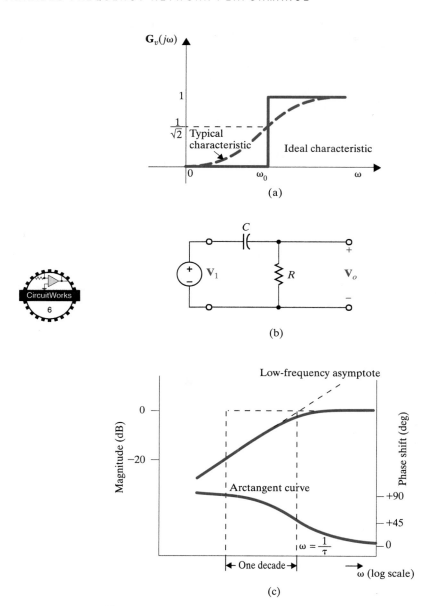

Figure 11.39

High-pass filter circuit and frequency characteristics.

To illustrate these points, let us consider the band-pass filter. The voltage transfer function is

$$\mathbf{G}_v(j\omega) = \frac{R}{R + j(\omega L - 1/\omega C)}$$

and, therefore, the amplitude characteristic is

$$M(\omega) = \frac{RC\omega}{\sqrt{(RC\omega)^2 + (\omega^2 LC - 1)^2}}$$

Figure 11.40
Band-pass and band-rejection filters and characteristics.

At low frequencies

$$M(\omega) \approx \frac{RC\omega}{1} \approx 0$$

At high frequencies

$$M(\omega) \simeq \frac{RC\omega}{\omega^2 LC} \approx \frac{R}{\omega L} \approx 0$$

In the midfrequency range $(RC\omega)^2 \gg (\omega^2 LC - 1)^2$, and thus $M(\omega) \approx 1$. Therefore, the frequency characteristic for this filter is shown in Fig. 11.40e. The center frequency is $\omega_0 = 1/\sqrt{LC}$. At the lower cutoff frequency

$$\omega^2 LC - 1 = -RC\omega$$

or

$$\omega^2 + \frac{R\omega}{L} - \omega_0^2 = 0$$

Solving this expression for ω_{LO}, we obtain

$$\omega_{LO} = \frac{-(R/L) + \sqrt{(R/L)^2 + 4\omega_0^2}}{2}$$

At the upper cutoff frequency

$$\omega^2 LC - 1 = +RC\omega$$

or

$$\omega^2 - \frac{R}{L}\omega - \omega_0^2 = 0$$

Solving this expression for ω_{HI}, we obtain

$$\omega_{HI} = \frac{+(R/L) + \sqrt{(R/L)^2 + 4\omega_0^2}}{2}$$

Therefore, the bandwidth of the filter is

$$BW = \omega_{HI} - \omega_{LO} = \frac{R}{L}$$

LEARNING Example 11.18

Consider the frequency-dependent network in Fig. 11.41. Given the following circuit parameter values: $L = 159\ \mu H$, $C = 159\ \mu F$, and $R = 10\ \Omega$, let us demonstrate that this one network can be used to produce a low-pass, high-pass, or band-pass filter.

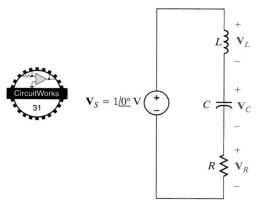

Figure 11.41   *Circuit used in Example 11.18.*

SOLUTION   The voltage gain $\mathbf{V}_R/\mathbf{V}_S$ is found by voltage division to be

$$\frac{\mathbf{V}_R}{\mathbf{V}_S} = \frac{R}{j\omega L + R + 1/(j\omega C)} = \frac{j\omega\left(\dfrac{R}{L}\right)}{(j\omega)^2 + j\omega\left(\dfrac{R}{L}\right) + \dfrac{1}{LC}}$$

$$= \frac{(62.9 \times 10^3)j\omega}{-\omega^2 + (62.9 \times 10^3)j\omega + 39.6 \times 10^6}$$

which is the transfer function for a band-pass filter. At resonance, $\omega^2 = 1/LC$, and hence

$$\frac{\mathbf{V}_R}{\mathbf{V}_S} = 1$$

Now consider the gain $\mathbf{V}_L/\mathbf{V}_S$:

$$\frac{\mathbf{V}_L}{\mathbf{V}_S} = \frac{j\omega L}{j\omega L + R + 1/(j\omega C)} = \frac{-\omega^2}{(j\omega)^2 + j\omega\left(\dfrac{R}{L}\right) + \dfrac{1}{LC}}$$

$$= \frac{-\omega^2}{-\omega^2 + (62.9 \times 10^3)j\omega + 39.6 \times 10^6}$$

which is a second-order high-pass filter transfer function. Again, at resonance,

$$\frac{\mathbf{V}_L}{\mathbf{V}_S} = \frac{j\omega L}{R} = jQ = j0.1$$

Similarly, the gain $\mathbf{V}_C/\mathbf{V}_S$ is

$$\frac{\mathbf{V}_C}{\mathbf{V}_S} = \frac{1/(j\omega C)}{j\omega L + R + 1/(j\omega C)} = \frac{\dfrac{1}{LC}}{(j\omega)^2 + j\omega\left(\dfrac{R}{L}\right) + \dfrac{1}{LC}}$$

$$= \frac{39.6 \times 10^6}{-\omega^2 + (62.9 \times 10^3)j\omega + 39.6 \times 10^6}$$

which is a second-order low-pass filter transfer function. At resonant frequency,

$$\frac{\mathbf{V}_C}{\mathbf{V}_S} = \frac{1}{j\omega CR} = -jQ = -j0.1$$

Thus, one circuit produces three different filters depending on where the output is taken. This can be seen in the Bode plot for each of the three voltages in Fig. 11.42, where $\mathbf{V}_S$ is set to $1\underline{/0°}$ V.

We know that Kirchhoff's voltage law must be satisfied at all times. Note from the Bode plot that the $\mathbf{V}_R + \mathbf{V}_C + \mathbf{V}_L$ also equals $\mathbf{V}_S$ at all frequencies! Finally, let us demonstrate KVL by adding $\mathbf{V}_R$, $\mathbf{V}_L$, and $\mathbf{V}_C$.

$$\mathbf{V}_L + \mathbf{V}_R + \mathbf{V}_C = \frac{\left((j\omega)^2 + j\omega\left(\dfrac{R}{L}\right) + \dfrac{1}{\sqrt{LC}}\right)\mathbf{V}_S}{(j\omega)^2 + j\omega\left(\dfrac{R}{L}\right) + \dfrac{1}{\sqrt{LC}}} = \mathbf{V}_S$$

Thus, even though $\mathbf{V}_S$ is distributed between the resistor, capacitor, and inductor based on frequency, the sum of the three voltages completely reconstructs $\mathbf{V}_S$.

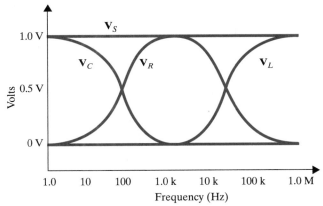

Figure 11.42
Bode plots for network in Fig. 11.41.

LEARNING Example 11.19

A telephone transmission system suffers from 60-Hz interference caused by nearby power utility lines. Let us use the network in Fig. 11.43 to design a simple notch filter to eliminate the 60-Hz interference.

SOLUTION The resistor R_{eq} represents the equivalent resistance of the telephone system to the right of the LC combination. The LC parallel combination has an equivalent impedance of

$$\mathbf{Z} = (j\omega L)//(1/j\omega C) = \frac{(L/C)}{j\omega L + 1/(j\omega C)}$$

Now the voltage transfer function is

$$\frac{\mathbf{V}_o}{\mathbf{V}_{in}} = \frac{R_{eq}}{R_{eq} + \mathbf{Z}} = \frac{R_{eq}}{R_{eq} + \dfrac{(L/C)}{j\omega L + (1/j\omega C)}}$$

(continued)

which can be written

$$\frac{\mathbf{V}_o}{\mathbf{V}_{in}} = \frac{(j\omega)^2 + \dfrac{1}{LC}}{(j\omega)^2 + \left(\dfrac{j\omega}{R_{eq}C}\right) + \dfrac{1}{LC}}$$

Figure 11.43 *Circuit used in Example 11.19.*

Note that at resonance, the numerator and thus $\mathbf{V}_o$ go to zero. We want resonance to occur at 60 Hz. Thus,

$$\omega_0 = \frac{1}{\sqrt{LC}} = 2\pi(60) = 120\pi$$

If we select $C = 100\ \mu\text{F}$, then the required value for L is 70.3 mH—both are reasonable values. To demonstrate the effectiveness of the filter, let the input voltage consist of a 60-Hz sinusoid and a 1000-Hz sinusoid of the form

$$v_{in}(t) = 1\ \sin\big[(2\pi)60t\big] + 0.2\ \sin\big[(2\pi)1000t\big]$$

The input and output waveforms are both shown in Fig. 11.44. Note that the output voltage, as desired, contains none of the 60-Hz interference.

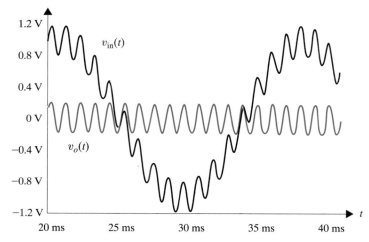

Figure 11.44
Transient analysis of the network in Fig. 11.43.

LEARNING EXTENSIONS

E11.16 Given the filter network shown in Fig. E11.16, sketch the magnitude characteristic of the Bode plot for $\mathbf{G}_v(j\omega)$.

Figure E11.16

ANSWER

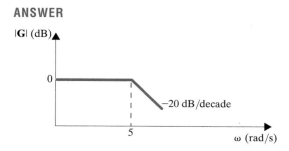

E11.17 Given the filter network in Fig. E11.17, sketch the magnitude characteristic of the Bode plot for $\mathbf{G}_v(j\omega)$.

ANSWER

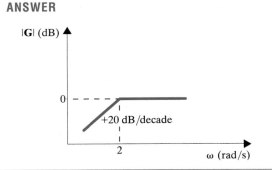

Figure E11.17

E11.18 A band-pass filter network is shown in Fig. E11.18. Sketch the magnitude characteristic of the Bode plot for $\mathbf{G}_v(j\omega)$.

ANSWER

Figure E11.18

ACTIVE FILTERS In the preceding section we saw that the four major classes of filters (*low pass, high pass, band pass,* and *band rejection*) are realizable with simple, passive element circuits. However, passive filters do have some serious drawbacks. One obvious problem is the inability to generate a network with a gain > 1 since a purely passive network cannot add energy to a signal. Another serious drawback of passive filters is the need in many topologies for inductive elements. Inductors are generally expensive and are not usually available in precise values. In addition, inductors usually come in odd shapes (toroids, bobbins, E-cores, etc.) and are not easily handled by existing automated printed circuit board assembly machines. By applying operational amplifiers in linear feedback circuits, one can generate all of the primary filter types using only resistors, capacitors, and the op-amp integrated circuits themselves.

The equivalent circuits for the operational amplifiers derived in Chapter 3 are valid in the sinusoidal steady-state case also, when we replace the attendant resistors with impedances. The equivalent circuits for the basic inverting and noninverting op-amp circuits are shown in Figs. 11.45a and b, respectively. Particular filter characteristics are obtained by judiciously selecting the impedances $\mathbf{Z}_1$ and $\mathbf{Z}_2$. However, due to the nature of the op-amp's internal circuitry, it has limitations and is not always the best amplifier choice. Two examples are high-frequency active filters and low-voltage ($< 3\,\text{V}$) circuitry. Given the evolution of the wireless market (cell phones, pagers, etc.) these applications will only grow in prominence. There is, however, an op-amp variant called the *operational transconductance amplifier*, or OTA, that performs excellently in these scenarios, allowing, for example, very advanced filters to be implemented on a single chip. In this text we will restrict our treatment of the OTA to this single application.

Advantages of the OTA over the op-amp can be deduced from the diagrams in Fig. 11.46. In the three-stage op-amp model, the input stage provides the large input resistance, converts

(a)

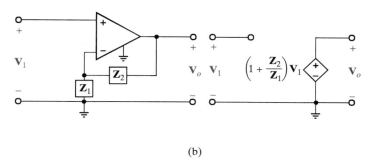

(b)

Figure 11.45
Equivalent circuits for the (a) inverting and (b) noninverting operational
amplifier circuits.

(a)

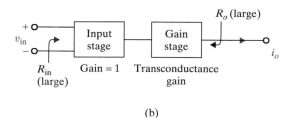

(b)

Figure 11.46
Block diagrams depicting the physical construction of the (a) op-amp
and (b) the OTA.

the differential input voltage $v_{in}(t)$ to a single-ended (referenced to ground) voltage and produces some voltage gain. The gain stage provides the bulk of the op-amp's voltage gain. Finally, the output stage, has little if any voltage gain, but produces a low output resistance. This three-stage model accurately depicts the physical design of most op-amps.

Now consider the two-stage OTA model. As in the op-amp, the input stage provides a large input resistance, but its voltage gain is minimal. The gain stage is very similar to that of the op-amp in that the value of R_o is large. Unlike the op-amp, the output signal is a current rather than a voltage, yielding an overall transconductance gain in A/V or siemens. With no output stage, the OTA is more compact and consumes less power than the op-amp and has an overall output resistance of R_o—a large value. Having all of the OTA's gain in a single stage further simplifies the internal design, resulting in a simple, fast, compact amplifier that can be efficiently replicated many times on a single silicon chip. The schematic symbol for the OTA and a simpler model are shown in Fig. 11.47a and b, respectively.

(a) (b)

Figure 11.47
The OTA schematic symbol
(a) and (b) simple model.

To compare the performance of the op-amp and OTA, consider the circuits in Fig. 11.48. For the op-amp, the overall voltage gain is

$$A = \frac{v_o}{v_{in}} = \left[\frac{R_{in}}{R_S + R_{in}}\right] A_V \left[\frac{R_L}{R_L + R_o}\right]$$

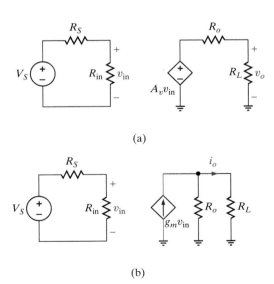

(a)

(b)

Figure 11.48
Simple circuits that demonstrate the relative strengths
of the (a) op-amp and (b) OTA.

Ideally, $R_{in} \to \infty$, $R_o \to 0$, and the output voltage is independent of external components R_S and R_L. The overall gain of the OTA is

$$G_m = \frac{i_o}{v_{in}} = \left[\frac{R_{in}}{R_S + R_{in}} \right] g_m \left[\frac{R_o}{R_L + R_o} \right]$$

For an ideal OTA, both R_{in} and $R_o \to \infty$, yielding an output current that is independent of R_S and R_L. Similarities and differences between ideal OTAs and ideal op-amps are listed in Table 11.2.

Table 11.2 A comparison of ideal Op-amps and OTA features

Amplifier Type	Ideal R_{in}	Ideal R_o	Ideal Gain	Input Currents	Input Voltage
Op-amp	∞	0	∞	0	0
OTA	∞	∞	g_m	0	nonzero

As with op-amps, OTAs can be used to create mathematical circuits. We will focus on three OTA circuits used extensively in active filters: the integrator, the simulated resistor, and the summer. To simplify our analyses, we assume that the OTA is ideal with infinite input and output resistances. The integrator in Fig. 11.49, which forms the heart of our OTA active filters, can be analyzed as follows.

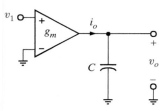

Figure 11.49
The OTA integrator.

$$i_o = g_m v_1 \qquad v_o = \frac{1}{C} \int i_o \, dt \qquad v_o = \frac{g_m}{C} \int v_1 \, dt$$

Or, in the frequency domain,

$$\mathbf{I}_o = g_m \mathbf{V}_1 \qquad \mathbf{V}_o = \frac{\mathbf{I}_o}{j\omega C} \qquad \mathbf{V}_o = \frac{g_m}{j\omega C} \mathbf{V}_1$$

An interesting aspect of IC fabrication is that resistors (especially large valued ones) are physically very large compared to other devices such as transistors. Additionally, producing accurate values is quite difficult. This has motivated designers to use OTAs to simulate resistors. One such circuit is the grounded resistor, shown in Fig. 11.50. Applying the ideal OTA conditions in Table 11.2,

Figure 11.50
The OTA simulated resistor.

$$i_o = g_m(0 - v_{in}) = -g_m v_{in} \qquad i_{in} = -i_o \qquad R_{eq} = \frac{v_{in}}{i_{in}} = \frac{1}{g_m} \qquad \textbf{11.55}$$

A simple summer circuit is shown in Fig. 11.51a, where OTA 3 is a simulated resistor. Based on Eq. (11.55), we produce the equivalent circuit in Fig. 11.51b. The analysis is straightforward.

$$i_{o1} = g_{m1} v_1 \qquad i_{o2} = g_{m2} v_2 \qquad i_o = i_{o1} + i_{o2} \qquad v_o = \frac{i_o}{g_{m3}} = \frac{g_{m1}}{g_{m3}} v_1 + \frac{g_{m2}}{g_{m3}} v_2 \qquad \textbf{11.56}$$

(a)

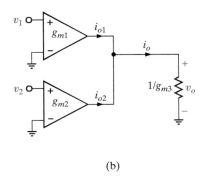

(b)

Figure 11.51
An OTA voltage summer.

At this point, we introduce our last important feature of the OTA—programmability. The transconductance, g_m, is linearly controlled by a current called the amplifier bias current, or I_{ABC}, as seen in Fig. 11.52. Unfortunately, the I_{ABC} input is not part of the schematic symbol. The sensitivity of g_m to I_{ABC} is typically 20 S/A but the range of g_m and its maximum value depend on the OTA design. Typical values are 10 mS for the maximum g_m and 3 to 7 powers of ten, or decades, for the transconductance range. For example, if the maximum g_m were 10 mS and the range were 4 decades, then the minimum g_m is 1 μS and the usable range of I_{ABC} is 0.05 μA to 0.5 mA.

Figure 11.52
A modified OTA schematic symbol showing the input bias current.

LEARNING Example 11.20

An ideal OTA has a $g_m - I_{ABC}$ sensitivity of 20, a maximum g_m of 4 mS, and a g_m range of 4 decades. Using the circuit in Fig. 11.50, produce an equivalent resistance of 25 kΩ, giving both g_m and I_{ABC}.

SOLUTION From Eq. (11.55), the equivalent resistance is $R_{eq} = 1/g_m = 25$ kΩ, yielding $g_m = 40$ μS. Since $g_m = 20I_{ABC}$, the required amplifier bias current is $I_{ABC} = 2$ μA.

LEARNING Example 11.21

The circuit in Fig. 11.53 is a floating simulated resistor. For an ideal OTA, find an expression for $R_{eq} = v_1/i_1$. Using the OTA described in Example 11.20, produce an 80-kΩ resistance. Repeat for a 10-MΩ resistor.

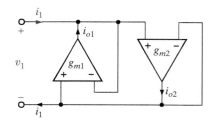

Figure 11.53
The floating simulated resistor.

SOLUTION For OTA 1, we have $i_{o1} = g_{m1}(-v_1)$ and $i_1 = -i_{o1}$. Thus, $R_{eq} = v_1/i_1 = 1/g_{m1}$. We must also consider the return current which is contributed by OTA 2, where $i_{o2} = g_{m2}(v_1)$ and $i_{o2} = i_1$. Now $R_{eq} = v_1/i_1 = 1/g_{m2}$. For proper operation, we must ensure that $g_{m1} = g_{m2}$.

For $R_{eq} = 1/g_m = 80$ kΩ, we have $g_{m1} = g_{m2} = g_m = 12.5$ μS. Since $g_m = 20I_{ABC}$, the required bias current for both OTAs is $I_{ABC} = 0.625$ μA. Changing to $R_{eq} = 1/g_m = 10$ MΩ, the transconductance becomes $g_m = 0.1$ μS. However, the minimum g_m for these OTAs is specified at 0.4 μS. We must find either suitable OTAs or a better circuit.

LEARNING Example 11.22

Using the summer in Fig. 11.51 and the OTAs specified in Example 11.20, produce the following function:

$$v_o = 10v_1 + 2v_2$$

Repeat for the function

$$v_o = 10v_1 - 2v_2$$

SOLUTION Comparing Eq. (11.56) to the desired function, we see that $g_{m1}/g_{m3} = 10$ and $g_{m2}/g_{m3} = 2$. With only two equations and three unknowns, we must choose one g_m value. Arbitrarily selecting $g_{m3} = 0.1$ mS yields $g_{m1} = 1$ mS and $g_{m2} = 0.2$ mS. The corresponding bias currents are $I_{ABC1} = 50$ μA, $I_{ABC2} = 10$ μA and $I_{ABC3} = 5$ μA.

For the second case, we simply invert the sign of v_2 as shown in Fig. 11.54. This is yet another advantage of OTA versus op-amps. Again choosing $g_{m3} = 0.1$ mS yields the same bias current as the first case.

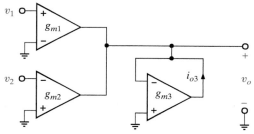

Figure 11.54 A slight modification of the adder in Fig. 11.51 yields this subtracting circuit.

Using the subcircuits in Fig. 11.49 and 11.50, one can create active filters called OTA-C filters, which contain only OTAs and capacitors. The lack of resistors makes OTA-C filters ideal for single-chip, or monolithic, implementations. As an introduction, consider the circuit in Fig. 11.55. For ideal OTAs, the transfer function can be determined as follows.

$$\mathbf{I}_{o1} = g_{m1}\mathbf{V}_{i1} \qquad \mathbf{I}_{o2} = -g_{m2}\mathbf{V} \qquad \mathbf{I}_C = \mathbf{V}_o(j\omega C) = \mathbf{I}_{o1} + \mathbf{I}_{o2}$$

$$\mathbf{V}_o = \frac{\mathbf{I}_{o1} + \mathbf{I}_{o2}}{j\omega C} = \frac{g_{m1}}{j\omega C}\mathbf{V}_{i1} - \frac{g_{m2}}{j\omega C}\mathbf{V}_o$$

Solving the transfer function yields the low-pass function

$$\frac{\mathbf{V}_o}{\mathbf{V}_{i1}} = \frac{g_{m1}/g_{m2}}{\dfrac{j\omega C}{g_{m2}} + 1}$$

11.57

Figure 11.55
A simple first-order low-pass OTA-C filter.

From Eq. (11.57), the circuit is a first-order low-pass filter with the asymptotic Bode plot shown in Fig. 11.56. Both the corner frequency, $f_C = g_{m2}/(2\pi C)$, and dc gain, $A_{dc} = g_{m1}/g_{m2}$, are programmable.

In monolithic OTA-C filters, the capacitors and OTAs are fabricated on a single chip. Typical OTA capacitor values range from about 1 pF up to 50 pF.

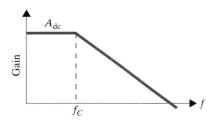

Figure 11.56
Asymptotic Bode plot for a first-order low-pass filter.

LEARNING Example 11.23

The low-pass filter in Fig. 11.55 is implemented using a 25-pF capacitor and OTAs with a $g_m - I_{ABC}$ sensitivity of 20, a maximum g_m of 1 mS and 3 decades of range. Find the required bias currents for the filter transfer function

$$\frac{\mathbf{V}_o}{\mathbf{V}_{in}} = \frac{4}{\dfrac{j\omega}{2\pi(10^5)} + 1} \qquad \textbf{11.58}$$

SOLUTION   Comparing Eq. (11.57) to the desired function, $g_{m2}/C = (2\pi)10^5$. For $C = 25$ pF, $g_{m2} = 15.7\ \mu S$. Since $g_m = 20I_{ABC}$, the bias current for OTA 2 is $I_{ABC2} = 0.785\ \mu A$. Finally $g_{m1}/g_{m2} = 4$ yields $I_{ABC1} = 3.14\ \mu A$.

Of the dozens of OTA filter topologies, a very popular one is the two-integrator biquad filter. The term biquad is short for biquadratic, which, in filter terminology, means the filter gain is a ratio of two quadratic functions such as

$$\frac{\mathbf{V}_o}{\mathbf{V}_{in}} = \frac{A(j\omega)^2 + B(j\omega) + C}{(j\omega)^2 + \dfrac{\omega_0}{Q}(j\omega) + \omega_0^2} \qquad \textbf{11.59}$$

By selecting appropriate values for A, B, and C, low-pass, band-pass, and high-pass functions can be created, as listed in Table 11.3. Figure 11.57 shows the most popular two-integrator biquad used in practice—the Tow–Thomas filter. Assuming ideal OTAs, we can derive the filter's transfer function. For OTA 1, an integrator,

$$\frac{\mathbf{V}_{o1}}{\mathbf{V}_{i1} - \mathbf{V}_{o2}} = \frac{g_{m1}}{j\omega C_1}$$

The output current of OTA 2 is

$$\mathbf{I}_{o2} = g_{m2}\left[\mathbf{V}_{o1} - \mathbf{V}_{i2}\right]$$

Applying KCL at the second output node, we find

$$\mathbf{I}_{o3} + \mathbf{I}_{o2} = \left(j\omega C_2\right)\mathbf{V}_{o2}$$

where

$$\mathbf{I}_{o3} = \left[\mathbf{V}_{i3} - \mathbf{V}_{o2}\right]g_{m3}$$

Solving for $\mathbf{V}_{o1}$ and $\mathbf{V}_{o2}$ yields

$$\mathbf{V}_{o1} = \frac{\left[\dfrac{j\omega C_2}{g_{m2}} + \dfrac{g_{m3}}{g_{m2}}\right]\mathbf{V}_{i1} + \mathbf{V}_{i2} - \left[\dfrac{g_{m3}}{g_{m2}}\right]\mathbf{V}_{i3}}{\left[\dfrac{C_1 C_2}{g_{m1}g_{m2}}\right](j\omega)^2 + \left[\dfrac{g_{m3}C_1}{g_{m2}g_{m1}}\right](j\omega) + 1}$$

Table 11.3 Various Tow–Thomas biquad filter possibilities

Filter Type	A	B	C
Low-pass	0	0	nonzero
Band-pass	0	nonzero	0
High-pass	nonzero	0	0

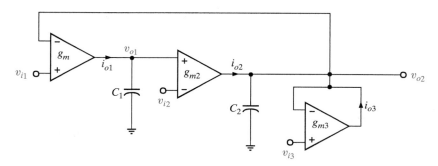

Figure 11.57
The Tow–Thomas OTA-C biquad filter.

and

$$
\mathbf{V}_{o2} = \frac{\mathbf{V}_{i1} - \left[\dfrac{j\omega C_1}{g_{m1}}\right]\mathbf{V}_{i2} + \left[\dfrac{j\omega C_1 g_{m3}}{g_{m1}g_{m2}}\right]\mathbf{V}_{i3}}{\left[\dfrac{C_1 C_2}{g_{m1}g_{m2}}\right](j\omega)^2 + \left[\dfrac{g_{m3}C_1}{g_{m2}g_{m1}}\right](j\omega) + 1}
\qquad \textbf{11.60}
$$

Note that this single circuit can implement both low-pass and band-pass filters depending on where the input is applied! Table 11.4 lists the possibilities. Comparing Eqs. (11.59) and (11.60), design equations for ω_0, Q, and the bandwidth can be written as

$$
\omega_0 = \sqrt{\frac{g_{m1}g_{m2}}{C_1 C_2}} \qquad \frac{\omega_0}{Q} = \mathrm{BW} = \frac{g_{m3}}{C_2} \qquad Q = \sqrt{\frac{g_{m1}g_{m2}}{g_{m3}^2}}\sqrt{\frac{C_2}{C_1}}
\qquad \textbf{11.61}
$$

Consider a Tow–Thomas band-pass filter. From Eq. (11.61), if $g_{m1} = g_{m2} = g_m$ and $C_1 = C_2 = C$, the following relationships are easily derived.

$$
\omega_0 = \frac{g_m}{C} = \frac{k}{C}I_{ABC} \qquad \frac{\omega_0}{Q} = \mathrm{BW} = \frac{g_{m3}}{C} = \frac{k}{C}I_{ABC3} \qquad Q = \frac{g_m}{g_{m3}} = \frac{I_{ABC}}{I_{ABC3}}
\qquad \textbf{11.62}
$$

where k is the $g_m - I_{ABC}$ sensitivity. Based on Eq. (11.62) we have efficient control over the filter characteristics. In particular, tuning I_{ABC} with I_{ABC3} fixed scales both the center frequency and Q directly without affecting bandwidth. Tuning I_{ABC3} only changes the bandwidth but not the center frequency. Finally, tuning all three bias currents scales both the center frequency and bandwidth proportionally, producing a constant Q factor.

Table 11.4 Low-pass and band-pass combinations for the Tow–Thomas biquad filter in Fig. 11.57

Filter Type	Input	Output	Sign
Low-pass	v_{i2}	v_{o1}	positive
	v_{i3}	v_{o1}	negative
	v_{i1}	v_{o2}	positive
Band-pass	v_{i2}	v_{o2}	negative
	v_{i3}	v_{o2}	positive

LEARNING Example 11.24

Using the OTAs specified in Example 11.20 and 50-pF capacitors, design a Tow–Thomas band-pass filter with a center frequency of 500 kHz, a bandwidth of 75 kHz, and a center frequency gain, A_C, of -5.

SOLUTION From Eq. (11.61), the filter characteristics for the v_{i2}–v_{o2} input-output pair are

$$
\mathrm{BW} = \frac{g_{m3}}{C} \qquad \omega_0^2 = \frac{g_{m1}g_{m2}}{C^2} \qquad A_C = -\frac{g_{m2}}{g_{m3}}
$$

(continued)

For the specifications given,

$$g_{m3} = (BW)(C) = (2\pi)(7.5 \times 10^4)(50 \times 10^{-12}) = 23.56 \ \mu S$$

$$g_{m2} = -A_C g_{m3} = 117.8 \ \mu S$$

$$g_{m1} = \frac{\omega_0^2 C^2}{g_{m2}} = \frac{(2\pi)^2(5 \times 10^5)^2(50 \times 10^{-12})^2}{117.8 \times 10^{-6}} = 209.5 \ \mu S$$

The required bias currents are $I_{ABC1} = 10.47 \ \mu A$, $I_{ABC} = 5.89 \mu A$, and $I_{ABC3} = 1.18 \ \mu A$. From the filter's Bode plot in Fig. 11.58, we see that the gain, bandwidth, and center frequency specifications are met.

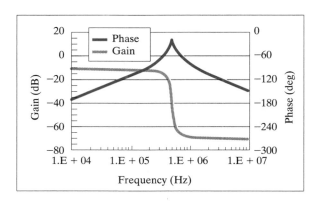

Figure 11.58
Bode plot for the Tow–Thomas band-pass filter of Example 11.24.

Learning by Application

LEARNING Example 11.25

The ac-dc converter in Fig. 11.59a is designed for use with a hand-held calculator. We will ignore the output load for now; it is addressed in Problem 11.70. Ideally, the circuit should convert a 120-V rms sinusoidal voltage to a 9-V dc output. In actuality the output is

$$v_o(t) = 9 + 0.5\sin 377t$$

Let us use a low-pass filter to reduce the 60-Hz component of $v_o(t)$.

SOLUTION The Thévenin equivalent circuit for the converter is shown in Fig. 11.59b. By placing a capacitor across the output terminals, as shown in Fig. 11.59c, we create a low-pass filter at the output. The transfer function of the filtered converter is

$$\frac{\mathbf{V}_{OF}}{\mathbf{V}_{Th}} = \frac{1}{1 + sR_{Th}C}$$

which has a pole at a frequency of $f = 1/2\pi R_{Th}C$. To obtain significant attenuation at 60 Hz, we choose to place the pole at 6 Hz, yielding the equation

$$\frac{1}{2\pi R_{Th}C} = 6$$

or

$$C = 53.05 \ \mu F$$

A transient simulation of the converter is used to verify performance.

Figure 11.59d shows the output without filtering $v_o(t)$, and with filtering, $v_{OF}(t)$. The filter has successfully reduced the unwanted 60-Hz component by a factor of roughly six.

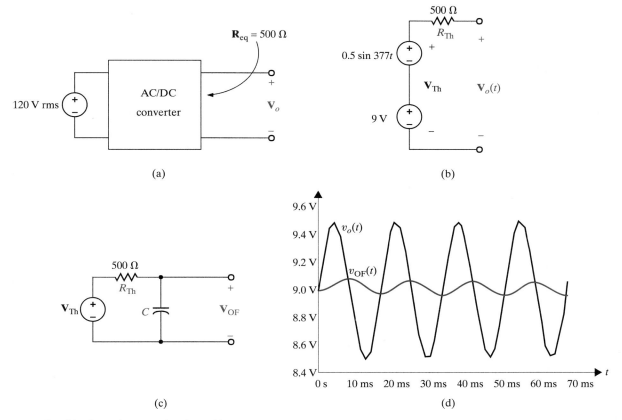

Figure 11.59 Circuits and output plots for ac/dc converter.

LEARNING Example 11.26

The antenna of an FM radio picks up stations across the entire FM frequency range—approximately 87.5 MHz to 108 MHz. The radio's circuitry must have the capability to first reject all of the stations except the one that the listener wants to hear and then to boost the minute antenna signal. A tuned amplifier incorporating parallel resonance can perform both tasks simultaneously.

The network in Fig. 11.60a is a circuit model for a single-stage tuned transistor amplifier where the resistor, capacitor, and inductor are discrete elements. Let us find the transfer function $V_o(s)/V_A(s)$, where $V_A(s)$ is the antenna voltage and the value of C for maximum gain at 91.1 MHz. Finally, we will simulate the results.

SOLUTION Since $V(s) = V_A(s)$, the transfer function is

$$\frac{V_o(s)}{V_A(s)} = -\frac{4}{1000}\left[R//sL//\frac{1}{sC}\right]$$

$$\frac{V_o(s)}{V_A(s)} = -\frac{4}{1000}\left[\frac{s/C}{s^2 + \dfrac{s}{RC} + \dfrac{1}{LC}}\right]$$

The parallel resonance network is actually a band-pass filter. Maximum gain occurs at the center frequency, f_0. This

(continued)

condition corresponds to a minimum value in the denominator. Isolating the denominator polynomial, $D(s)$, and letting $s = j\omega$, we have

$$D(j\omega) = \frac{R}{LC} - \omega^2 + \frac{j\omega}{C}$$

which has a minimum value when the real part goes to zero, or

$$\frac{1}{LC} - \omega_0^2 = 0$$

yielding a center frequency of

$$\omega_0 = \frac{1}{\sqrt{LC}}$$

Thus, for a center frequency of 91.1 MHz, we have

$$2\pi(91.1 \times 10^6) = \frac{1}{\sqrt{LC}}$$

and the required capacitor value is

$$C = 3.05 \text{ pF}$$

The Bode plot for the tuned amplifier, as shown in Fig. 11.60b, confirms the design, since the center frequency is 91.1 MHz, as specified.

(a)

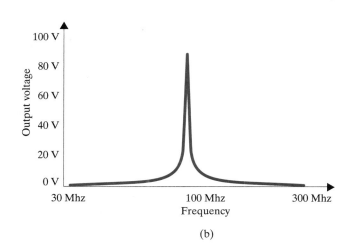

(b)

Figure 11.60
Circuit and Bode plot for parallel resonance tuned amplifier.

Learning by Design

Throughout this chapter we have presented a number of design examples. In this section we consider some additional ones that also have practical ramifications.

LEARNING Example 11.27

Compact disks (CDs) have become a very popular medium for recording and playing music. CDs store information in a digital manner; that is, the music is sampled at a very high rate, and the samples are recorded on the disc. The trick is to sample so quickly that the reproduction sounds continuous. The industry standard sampling rate is 44.1 kHz—one sample every 22.7 μs.

One interesting aspect regarding the analog-to-digital conversion that takes place inside the unit recording a CD is called the Nyquist criterion. This criterion states that in the analog conversion, any signal components at frequencies above half the sampling rate (22.05 kHz in this case) cannot be faithfully reproduced. Therefore, recording technicians filter these frequencies out before any sampling occurs, yielding higher fidelity to the listener.

Let us design a series of low-pass filters to perform this task.

SOLUTION   Suppose, for example, that our specification for the filter is unity gain at dc and 20 dB of attenuation at 22.05 kHz. Let us consider first the simple RC filter in Fig. 11.61.

Figure 11.61
Single-pole low-pass filter.

The transfer function is easily found to be

$$\mathbf{G}_{v1}(s) = \frac{\mathbf{V}_{o1}}{\mathbf{V}_{in}} = \frac{1}{1 + sRC}$$

Since a single-pole transfer function attenuates at 20 dB/decade, we should place the pole frequency one decade before the −20 dB point of 22.05 kHz.

Thus,

$$f_p = \frac{1}{2\pi RC} = 2.205 \text{ kHz}$$

If we arbitrarily choose $C = 1$ nF, the resulting value for R is 72.18 kΩ, which is reasonable. A Bode plot of the magnitude of $\mathbf{G}_{v1}(s)$ is shown in Fig. 11.62. All specifications are met but at

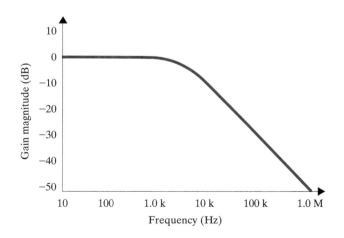

Figure 11.62
Bode plot for single-pole filter.

(continued)

the cost of severe attenuation in the audible frequency range. This is undesirable.

An improved filter is shown in Fig. 11.63. It is a two-stage low-pass filter with identical filter stages separated by a unity-gain buffer.

The presence of the op-amp permits us to consider the stages independently. Thus, the transfer function becomes

$$\mathbf{G}_{v2}(s) = \frac{\mathbf{V}_{o2}}{\mathbf{V}_{in}} = \frac{1}{[1 + sRC]^2}$$

To find the required pole frequencies, let us employ the equation for $\mathbf{G}_{v2}(s)$ at 22.05 kHz, since we know that the gain must be 0.1 (attenuated 20 dB) at that frequency. Using the substitution $s = j\omega$, we can express the magnitude of $\mathbf{G}_{v2}(s)$ as

$$|\mathbf{G}_{v2}| = \left\{ \frac{1}{1 + (22{,}050/f_p)^2} \right\} = 0.1$$

and the pole frequency is found to be 7.35 kHz. The corresponding resistor value is 21.65 kΩ. Bode plots for $\mathbf{G}_{v1}$ and $\mathbf{G}_{v2}$ are shown in Fig. 11.64. Note that the two-stage filter has a wider bandwidth, which improves the fidelity of the recording.

Let us try one more improvement—expanding the two-stage filter to a four-stage filter. Again, the gain magnitude is 0.1 at 22.05 kHz and can be written

$$|\mathbf{G}_{v3}| = \left\{ \frac{1}{[1 + (22{,}050/f_p)^2]^2} \right\} = 0.1$$

The resulting pole frequencies are at 15 kHz, and the required resistor value is 10.61 kΩ. Figure 11.65 shows all three Bode plots. Obviously, the four-stage filter, having the widest bandwidth, is the best option (discounting any extra cost associated with the additional active and passive circuit elements).

Figure 11.63
Two-stage buffered filter.

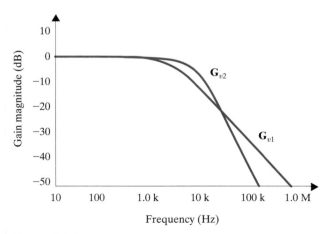

Figure 11.64
Bode plot for single- and two-stage filters.

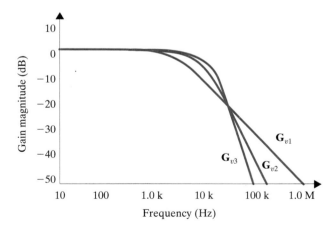

Figure 11.65
Bode plots for single-, two-, and four-stage filters.

LEARNING Example 11.28

The circuit in Fig. 11.66a is called a notch filter. From a sketch of its Bode plot in Fig. 11.66b, we see that at the notch frequency, f_n, the transfer function gain is zero, while at frequencies above and below f_n the gain is unity. Let us design a notch filter to remove an annoying 60-Hz hum from the output voltage of a cassette tape player and generate its Bode plot.

SOLUTION   Figure 11.66c shows a block diagram for the filter implementation. The tape output contains both the desired music and the undesired hum. After filtering, the voltage $\mathbf{V}_{amp}$ will have no 60-Hz component as well as some attenuation at frequencies around 60 Hz. An equivalent circuit for the block diagram including a Thévenin equivalent for the tape deck and an equivalent resistance for the power amp is shown in Fig. 11.66d. Applying voltage division, the transfer function is found to be

$$\frac{\mathbf{V}_{amp}}{\mathbf{V}_{tape}} = \frac{R_{amp}}{R_{amp} + R_{tape} + \left(sL//\dfrac{1}{Cs}\right)}$$

After some manipulation, the transfer function can be written as

$$\frac{\mathbf{V}_{amp}}{\mathbf{V}_{tape}} = \frac{R_{amp}}{R_{amp} + R_{tape}}\left[\frac{s^2LC + 1}{s^2LC + s\left(\dfrac{L}{R_{tape} + R_{amp}}\right) + 1}\right]$$

We see that the transfer function contains two zeros and two poles. Letting $s = j\omega$, the zero frequencies, ω_z, are found to be at

$$\omega_z = \pm\frac{1}{\sqrt{LC}}$$

Obviously, we would like the zero frequencies to be at 60 Hz. If we arbitrarily choose $C = 10\ \mu\text{F}$, then $L = 0.704$ mH.

The Bode plot, shown in Fig. 11.66e, confirms that there is indeed zero transmission at 60 Hz.

(a)

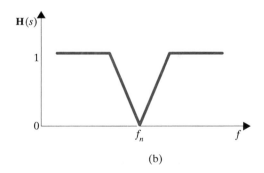

(b)

Figure 11.66
Circuits and Bode plots for 60-Hz notch filter.

(continued)

(c)

(d)

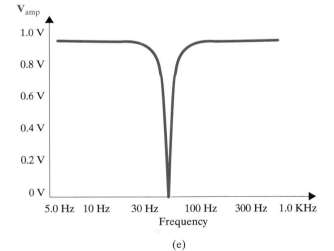

(e)

Figure 11.66 *Continued*

LEARNING Check

Summary

▶ There are four types of network or transfer functions:

1. $\mathbf{Z}(j\omega)$: the ratio of the output voltage to the input current

2. $\mathbf{Y}(j\omega)$: the ratio of the output current to the input voltage

3. $\mathbf{G}_v(j\omega)$: the ratio of the output voltage to the input voltage

4. $\mathbf{G}_1(j\omega)$: the ratio of the output current to the input current

▶ Driving-point functions are impedances or admittances defined at a single pair of terminals, such as the input impedance of a network.

▶ When the network function is expressed in the form

$$\mathbf{H}(s) = \frac{N(s)}{D(s)}$$

the roots of $N(s)$ cause $\mathbf{H}(s)$ to become zero and are called zeros of the function, and the roots of $D(s)$ cause $\mathbf{H}(s)$ to become infinite and are called poles of the function.

▶ Bode plots are semilog plots of the magnitude and phase of a transfer function as a function of frequency. Straight-line approximations can be used to sketch quickly the magnitude characteristic. The error between the actual characteristic and the straight-line approximation can be calculated when necessary.

▶ The resonant frequency, given by the expression

$$\omega_0 = \frac{1}{\sqrt{LC}}$$

is the frequency at which the impedance of a series RLC circuit or the admittance of a parallel RLC circuit is purely real.

▶ The quality factor is a measure of the sharpness of the resonant peak. The higher the Q, the sharper the peak.
  For series RLC circuits, $Q = (1/R)\sqrt{L/C}$. For parallel RLC circuits, $Q = R\sqrt{C/L}$.

▶ The half-power, cutoff, or break frequencies are the frequencies at which the magnitude characteristic of the Bode plot is $1/\sqrt{2}$ of its maximum value.

▶ The parameter values for passive circuit elements can be both magnitude and frequency scaled.

▶ The four common types of filters are low pass, high pass, band pass, and band rejection.

▶ The bandwidth of a band-pass or band-rejection filter is the difference in frequency between the half-power points; that is,

$$BW = \omega_{HI} - \omega_{LO}$$

For a series RLC circuit, $BW = R/L$. For a parallel RLC circuit, $BW = 1/RC$.

Problems

11.1 Determine the driving point impedance at the input terminals of the network shown in Fig. P11.1 as a function of s.

Figure P11.1

11.2 Determine the voltage transfer function $V_o(s)/V_i(s)$ as a function of s for the network shown in Fig. P11.1.

11.3 Determine the voltage transfer function $V_o(s)/V_i(s)$ as a function of s for the network shown in Fig.P11.3.

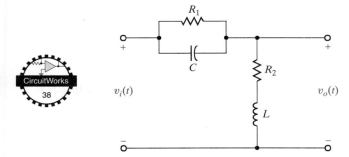

Figure P11.3

11.4 Find the transfer impedance $V_o(s)/I_S(s)$ for the network shown in Fig. P11.4.

Figure P11.4

11.5 Find the driving point impedance at the input terminals of the circuit in Fig. P11.5 as a function of s.

Figure P11.5

11.6 Draw the Bode plot for the network function

$$\mathbf{H}(j\omega) = \frac{j\omega 5 + 1}{j\omega 20 + 1}$$

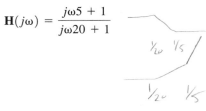

11.7 Draw the Bode plot for the network function

$$\mathbf{H}(j\omega) = \frac{j\omega 2 + 1}{j\omega 10 + 1}$$

11.8 Draw the Bode plot for the network function

$$\mathbf{H}(j\omega) = \frac{10(10j\omega + 1)}{(100j\omega + 1)(j\omega + 1)}$$

11.9 Draw the Bode plot for the network function

$$\mathbf{H}(j\omega) = \frac{j\omega}{(j\omega + 1)(0.1j\omega + 1)}$$

11.10 Draw the Bode plot for the network function

$$\mathbf{H}(j\omega) = \frac{10j\omega + 1}{j\omega(0.1j\omega + 1)}$$

11.11 Sketch the magnitude characteristic of the Bode plot for the transfer function

$$\mathbf{H}(j\omega) = \frac{10}{j\omega(0.1j\omega + 1)}$$

11.12 Sketch the magnitude characteristic of the Bode plot for the transfer function

$$\mathbf{H}(j\omega) = \frac{20(j\omega + 1)}{j\omega(0.1j\omega + 1)(0.01j\omega + 1)}$$

11.13 Sketch the magnitude characteristic of the Bode plot for the transfer function

$$\mathbf{H}(j\omega) = \frac{100(j\omega)}{(j\omega + 1)(j\omega + 10)(j\omega + 50)}$$

11.14 Draw the Bode plot for the network function

$$\mathbf{H}(j\omega) = \frac{16}{(j\omega)^2(j\omega 2 + 1)}$$

11.15 Sketch the magnitude characteristic of the Bode plot for the transfer function

$$\mathbf{H}(j\omega) = \frac{640(j\omega + 1)(0.01j\omega + 1)}{(j\omega)^2(j\omega + 10)}$$

11.16 Sketch the magnitude characteristic of the Bode plot for the transfer function

$$\mathbf{H}(j\omega) = \frac{10^5(5j\omega + 1)^2}{(j\omega)^2(j\omega + 10)(j\omega + 100)^2}$$

11.17 Sketch the magnitude characteristic of the Bode plot for the transfer function

$$\mathbf{G}(j\omega) = \frac{10j\omega}{(j\omega + 1)(j\omega + 10)^2}$$

11.18 Sketch the magnitude characteristic of the Bode plot for the transfer function

$$\mathbf{H}(j\omega) = \frac{10^2(j\omega)^2}{(j\omega + 1)(j\omega + 10)^2(j\omega + 50)}$$

11.19 Sketch the magnitude characteristic of the Bode plot for the transfer function

$$\mathbf{G}(j\omega) = \frac{-\omega^2 10^4}{(j\omega + 1)^2(j\omega + 10)(j\omega + 100)^2}$$

11.20 Sketch the magnitude characteristic of the Bode plot for the transfer function

$$\mathbf{G}(j\omega) = \frac{64(j\omega + 1)^2}{-j\omega^3(0.1j\omega + 1)}$$

11.21 Sketch the magnitude characteristic of the Bode plot for the transfer function

$$\mathbf{G}(j\omega) = \frac{-\omega^2}{(j\omega + 1)^3}$$

11.22 Draw the Bode plot for the network function

$$\mathbf{H}(j\omega) = \frac{72(j\omega + 2)}{j\omega[(j\omega)^2 + 2.4j\omega + 144]}$$

11.23 Sketch the magnitude characteristic of the Bode plot for the transfer function

$$G(j\omega) = \frac{10^4(j\omega + 1)(-\omega^2 + 6j\omega + 225)}{j\omega(j\omega + 50)^2(j\omega + 450)}$$

11.24 Sketch the magnitude characteristic of the Bode plot for the transfer function

$$H(j\omega) = \frac{+81(j\omega + 0.1)}{(j\omega)(-\omega^2 + 3.6j\omega + 81)}$$

11.25 Sketch the magnitude characteristic of the Bode plot for the transfer function

$$H(j\omega) = \frac{+6.4(j\omega)}{(j\omega + 1)(-\omega^2 + 8j\omega + 64)}$$

11.26 Determine $H(j\omega)$ if the amplitude characteristic for $H(j\omega)$ is shown in Fig. P11.26.

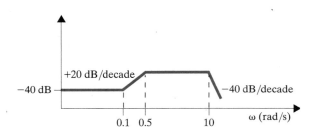

Figure P11.26

11.27 Find $H(j\omega)$ if its magnitude characteristic is shown in Fig. P11.27.

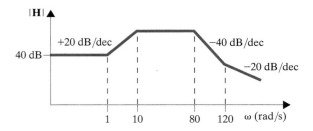

Figure P11.27

11.28 Find $H(j\omega)$ if its magnitude characteristic is shown in Fig. P11.28.

Figure P11.28

11.29 Find $H(j\omega)$ if its amplitude characteristic is shown in Fig. P11.29.

Figure P11.29

11.30 Find $H(j\omega)$ if its magnitude characteristic is shown in Fig. P11.30.

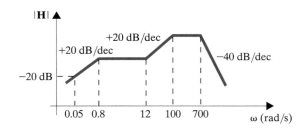

Figure P11.30

11.31 Find $\mathbf{H}(j\omega)$ if its amplitude characteristic is shown in Fig. P11.31.

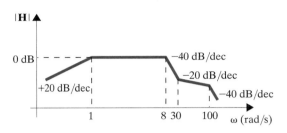

Figure P11.31

11.33 Find $\mathbf{G}(j\omega)$ if the amplitude characteristic for this function is shown in Fig. P11.33.

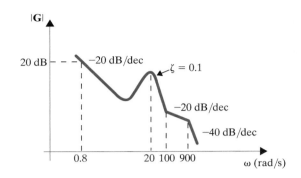

Figure P11.33

11.32 Given the magnitude characteristic in Fig. P11.32, find $\mathbf{G}(j\omega)$.

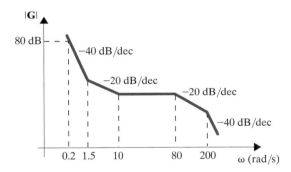

Figure P11.32

SECTION 11.3

11.34 A series RLC circuit resonates at 2000 rad/s. If $C = 20\ \mu\text{F}$ and it is known that the impedance at resonance is 2.4 Ω, compute the value of L, the Q of the circuit, and the bandwidth.

11.35 A series resonant circuit has a Q of 120 and a resonant frequency of 60,000 rad/s. Determine the half-power frequencies and the bandwidth of the circuit.

11.36 Given the series RLC circuit in Fig. P11.36, if $R = 10\ \Omega$, find the values of L and C such that the network will have a resonant frequency of 100 kHz and a bandwidth of 1 kHz.

Figure P11.36

11.37 Given the network in Fig. P11.37, find ω_0, Q, ω_{max}, and $|\mathbf{V}_o|_{max}$.

Figure P11.37

11.38 Repeat Problem 11.37 if the value of R is changed to 0.1 Ω.

11.39 A series RLC circuit is driven by a signal generator. The resonant frequency of the network is known to be 1600 rad/s, and at that frequency the impedance seen by the signal generator is 5 Ω. If $C = 20$ μF, find L, Q, and the bandwidth.

11.40 A variable-frequency voltage source drives the network in Fig. P11.40. Determine the resonant frequency, Q, BW, and the average power dissipated by the network at resonance.

Figure P11.40

11.41 In the network in Fig. P11.41, the inductor value is 10 mH, and the circuit is driven by a variable-frequency source. If the magnitude of the current at resonance is 12 A, $\omega_0 = 1000$ rad/sec, and $L = 10$ mH, find C, Q, and the bandwidth of the circuit.

Figure P11.41

11.42 Given the network in Fig. P11.42, find $|\mathbf{V}_o|_{max}$.

Figure P11.42

11.43 A parallel RLC resonant circuit with a resonant frequency of 20,000 rad/s has an admittance at resonance of 1 mS. If the capacitance of the network is 5 μF, find the values of R and L.

11.44 A parallel RLC resonant circuit has a resistance of 200 Ω. If it is known that the bandwidth is 80 rad/s and the lower half-power frequency is 800 rad/s, find the values of the parameters L and C.

11.45 A parallel RLC circuit, which is driven by a variable-frequency 2-A current source, has the following values: $R = 1$ kΩ, $L = 100$ mH, and $C = 10$ μF. Find the bandwidth of the network, the half-power frequencies, and the voltage across the network at the half-power frequencies.

11.46 A parallel RLC circuit, which is driven by a variable-frequency 10-A source, has the following parameters: $R = 500$ Ω, $L = 0.5$ mH, and $C = 20$ μF. Find the resonant frequency, the Q, the average power dissipated at the resonant frequency, the BW, and the average power dissipated at the half-power frequencies.

11.47 Consider the network in Fig. P11.47. If $R = 2$ kΩ, $L = 20$ mH, $C = 50$ μF, and $R_S = \infty$, determine the resonant frequency ω_0, the Q of the network, and the bandwidth of the network. What impact does an R_S of 10 kΩ have on the quantities determined?

Figure P11.47

11.48 The source in the network in Fig. P11.48 is
$i_S(t) = \cos 1000t + \cos 1500t$ A. $R = 200\ \Omega$ and
$C = 500\ \mu$F. If $\omega_0 = 1000$ rad/sec, find L, Q, and the
BW. Compute the output voltage $v_o(t)$ and discuss the
magnitude of the output voltage at the two input
frequencies.

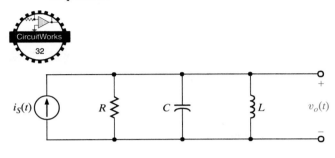

Figure P11.48

11.49 Determine the parameters of a parallel resonant circuit
that has the following properties: $\omega_0 = 2$ Mrad/s,
BW $= 20$ krad/s, and an impedance at resonance of
$2000\ \Omega$.

11.50 Determine the value of C in the network shown in
Fig. P11.50 for the circuit to be in resonance.

Figure P11.50

11.51 Determine the equation for the nonzero resonant
frequency of the impedance shown in Fig. P11.51.

Figure P11.51

SECTION 11.4

11.52 Determine the new parameters of the network shown in
Fig. P11.52 if $\mathbf{Z}_{\text{new}} = 10^4 \mathbf{Z}_{\text{old}}$.

Figure P11.52

11.53 Determine the new parameters of the network in
Problem 11.52 if $\omega_{\text{new}} = 10^4 \omega_{\text{old}}$.

SECTION 11.5

11.54 Given the network in Fig. P11.54, sketch the magnitude characteristic of the transfer function

$$\mathbf{G}_v(j\omega) = \frac{\mathbf{V}_o}{\mathbf{V}_1}(j\omega)$$

Identify the type of filter.

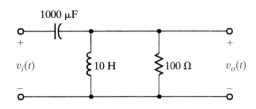

Figure P11.54

11.55 Given the network in Fig. P11.55, sketch the magnitude characteristic of the transfer function

$$\mathbf{G}_v(j\omega) = \frac{\mathbf{V}_o}{\mathbf{V}_1}(j\omega)$$

Identify the type of filter.

Figure P11.55

11.56 Determine what type of filter the network shown in Fig. P11.56 represents by determining the voltage transfer function.

Figure P11.56

11.57 Determine what type of filter the network shown in Fig. P11.57 represents by determining the voltage transfer function.

Figure P11.57

11.58 Given the lattice network shown in Fig. P11.58, determine what type of filter this network represents by determining the voltage transfer function.

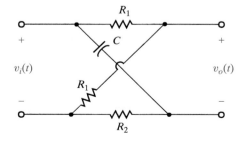

Figure P11.58

11.59 Given the network in Fig. P11.59, and employing the voltage follower analyzed in Chapter 3, determine the voltage transfer function and its magnitude characteristic. What type of filter does the network represent?

Figure P11.59

and determine what type of filter the network represents.

Figure P11.61

11.60 Determine the voltage transfer function and its magnitude characteristic for the network shown in Fig. P11.60 and identify the filter properties.

Figure P11.60

11.61 Given the network in Fig. P11.61, find the transfer function

$$\frac{\mathbf{V}_o}{\mathbf{V}_1}(j\omega)$$

11.62 Repeat Problem 11.54 for the network shown in Fig. P11.62.

Figure P11.62

In all OTA problems, the specifications are: $g_m - I_{ABC}$ sensitivity = 20, maximum g_m = 1 mS with range of 4 decades.

11.63 For the circuit in Fig. 11.50, find the g_m and I_{ABC} values required for a simulated resistance of 10 kΩ.

11.64 For the circuit in Fig. 11.53, find the g_m and I_{ABC} values required for a simulated resistance of 10 kΩ. Repeat for 750 kΩ.

11.65 Use the summing circuit in Fig. 11.51 to design a circuit that realizes the following function.

$$v_o = 7v_1 + 3v_2$$

11.66 Prove that the circuit in Fig. P11.66 is a simulated inductor. Find the inductance in terms of C, g_{m1} and g_{m2}.

Figure P11.66

11.67 In the Tow–Thomas biquad in Fig. 11.57, $C_1 = 20$ pF, $C_2 = 10$ pF, $g_{m1} = 10$ μS, $g_{m2} = 80$ μS, and $g_{m3} = 10$ μS. Find the low-pass filter transfer function for the $v_{i1} - v_{o2}$ input-output pair. Plot the corresponding Bode plot.

11.68 Find the transfer function of the OTA filter in Fig. P11.68. Express ω_0 and Q in terms of the capacitances and transconductances. What kind of filter is it?

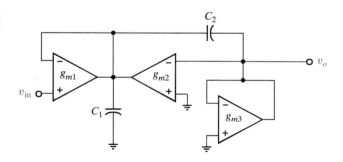

Figure P11.68

11.69 Find the transfer function of the OTA filter in Fig. P11.69. Express ω_0 and Q in terms of the capacitances and transconductances. What kind of filter is it?

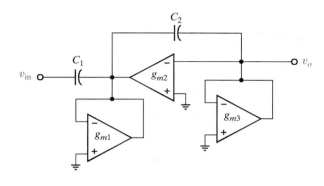

Figure P11.69

11.70 Refer to the ac/dc converter low-pass filter application of Example 11.25. If we put the converter to use powering a calculator, the load current can be modeled by a resistor as shown in Fig. P11.70. The load resistor will affect both the magnitude of the dc component of V_{OF} and the pole frequency. Plot both the pole frequency and the ratio of the 60-Hz component of the output voltage to the dc component of V_{OF} versus R_L for $100 \ \Omega \le R_L \le 100$ kΩ. Comment on the advisable limitations on R_L if **(a)** the dc component of V_{OF} is to remain within 20% of its 9-V ideal value; **(b)** the 60-Hz component of V_{OF} remains less than 15% of the dc component.

Figure P11.70

11.71 Referring to Example 11.28, design a notch filter for the tape deck for use in Europe, where power utilities generate at 50 Hz.

Typical Problems Found on the FE Exam

11FE-1 Determine the resonant frequency of the circuit in Fig. 11PFE-1, and find the voltage $\mathbf{V}_o$ at resonance.

Figure 11PFE-1

11FE-2 Given the series circuit in Fig. 11PFE-2, determine the resonant frequency, and find the value of R so that the BW of the network about the resonant frequency is 200 r/s.

Figure 11PFE-2

11FE-3 Given the low-pass filter circuit shown in Fig. 11PFE-3, find the frequency in Hz at which the output is down 3 dB from the dc, or very low frequency, output.

Figure 11PFE-3

11FE-4 Given the band-pass filter shown in Fig. 11PFE-4, find the components L and R necessary to provide a resonant frequency of 1000 r/s and a BW of 100 r/s.

Figure 11PFE-4

11FE-5 Given the low-pass filter shown in Fig. 11PFE-5, find the half-power frequency and the gain of this circuit, if the source frequency is 8 Hz.

Figure 11PFE-5

12

The Laplace Transform

LEARNING Goals

We will now introduce the Laplace transform. This is an extremely important technique in that for a given set of initial conditions, it will yield the total response of the circuit consisting of both the natural and forced responses in one operation.

Our use of the Laplace transform to solve circuit problems is analogous to that of using phasors in sinusoidal steady-state analysis. Using the Laplace transform, we transform the circuit problem from the time domain to the complex frequency domain, solve the problem using algebra in the complex frequency domain, and then convert the solution in the complex frequency domain back to the time domain. Therefore, as we shall see, the Laplace transform is an integral transform that converts a set of linear simultaneous integrodifferential equations to a set of simultaneous algebraic equations.

Our approach is to define the Laplace transform, derive some of the transform pairs, consider some of the important properties of the transform, illustrate the inverse transform operation, introduce the convolution integral, and finally apply the transform to circuit analysis.

12.1 Definition

The Laplace transform of a function $f(t)$ is defined by the equation

$$\mathcal{L}[f(t)] = \mathbf{F}(s) = \int_0^\infty f(t)e^{-st}\, dt \qquad \textbf{12.1}$$

where s is the complex frequency

$$s = \sigma + j\omega \qquad \textbf{12.2}$$

and the function $f(t)$ is assumed to possess the property that

$$f(t) = 0 \qquad \text{for } t < 0$$

Note that the Laplace transform is unilateral $(0 \le t < \infty)$, in contrast to the Fourier transform (see Chapter 14), which is bilateral $(-\infty < t < \infty)$. In our analysis of circuits using the Laplace transform, we will focus our attention on the time interval $t \ge 0$. It is important to note that it is the initial conditions that account for the operation of the circuit prior to $t = 0$, and therefore our analyses will describe the circuit operation for $t \ge 0$.

For a function $f(t)$ to possess a Laplace transform, it must satisfy the condition

$$\int_0^\infty e^{-\sigma t}\left|f(t)\right|\, dt < \infty \qquad \textbf{12.3}$$

for some real value of σ. Because of the convergence factor $e^{-\sigma t}$, a number of important functions have Laplace transforms, even though Fourier transforms for these functions do not exist. All of the inputs we will apply to circuits possess Laplace transforms. Functions that do not have Laplace transforms $\left(\text{e.g., } e^{t^2}\right)$ are of no interest to us in circuit analysis.

The inverse Laplace transform, which is analogous to the inverse Fourier transform, is defined by the relationship

$$\mathcal{L}^{-1}[\mathbf{F}(s)] = f(t) = \frac{1}{2\pi j}\int_{\sigma_1 - j\infty}^{\sigma_1 + j\infty} \mathbf{F}(s)e^{st}\, ds \qquad \textbf{12.4}$$

where σ_1 is real and $\sigma_1 > \sigma$ in Eq. (12.3). The evaluation of this integral is based on complex variable theory, and therefore we will circumvent its use by developing and using a set of Laplace transform pairs.

12.2 Two Important Singularity Functions

There are two singularity functions that are very important in circuit analysis: (1) the unit step function, $u(t)$, discussed in Chapter 6, and (2) the unit impulse or delta function, $\delta(t)$. They are called *singularity functions* because they are either not finite or they do not possess finite derivatives everywhere. They are mathematical models for signals that we employ in circuit analysis.

The *unit step function* $u(t)$ shown in Fig. 12.1a was defined in Section 6.2 as

$$u(t) = \begin{cases} 0 & t < 0 \\ 1 & t > 0 \end{cases}$$

(a)   (b)

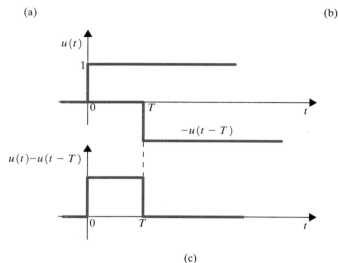

Figure 12.1
*Representations of the
unit step function.*

(c)

Recall that the physical analogy of this function, as illustrated earlier, corresponds to closing a switch at $t = 0$ and connecting a voltage source of 1 V or a current source of 1 A to a given circuit. The following example illustrates the calculation of the Laplace transform for unit step functions.

LEARNING Example 12.1

Let us determine the Laplace transform for the waveforms in Fig. 12.1.

SOLUTION   The Laplace transform for the unit step function in Fig. 12.1a is

$$\mathbf{F}(s) = \int_0^\infty u(t)e^{-st}\, dt$$

$$= \int_0^\infty 1e^{-st}\, dt$$

$$= -\frac{1}{s}e^{-st}\bigg|_0^\infty$$

$$= \frac{1}{s} \qquad \sigma > 0$$

Therefore,

$$\mathcal{L}[u(t)] = \mathbf{F}(s) = \frac{1}{s}$$

The Laplace transform of the time-shifted unit step function shown in Fig. 12.1b is

$$\mathbf{F}(s) = \int_0^\infty u(t - a)e^{-st}\, dt$$

Note that

$$u(t - a) = \begin{cases} 1 & a < t < \infty \\ 0 & t < a \end{cases}$$

Therefore,

$$\mathbf{F}(s) = \int_a^\infty e^{-st}\, dt$$

$$= \frac{e^{-as}}{s} \qquad \sigma > 0$$

Finally, the Laplace transform of the pulse shown in Fig. 12.1c is

$$\mathbf{F}(s) = \int_0^\infty \big[u(t) - u(t - T)\big]e^{-st}\, dt$$

$$= \frac{1 - e^{-Ts}}{s} \qquad \sigma > 0$$

Figure 12.2
Representations of the unit impulse.

The unit impulse function can be represented in the limit by the rectangular pulse shown in Fig. 12.2a as $a \rightarrow 0$. The function is defined by the following:

$$\delta(t - t_0) = 0 \qquad t \neq t_0$$

$$\int_{t_0 - \varepsilon}^{t_0 + \varepsilon} \delta(t - t_0)\, dt = 1 \qquad \varepsilon > 0$$

The unit impulse is zero except at $t = t_0$, where it is undefined, but it has unit area (sometimes referred to as *strength*). We represent the unit impulse function on a plot as shown in Fig. 12.2b.

An important property of the unit impulse function is what is often called the *sampling property*, which is exhibited by the following integral:

$$\int_{t_1}^{t_2} f(t)\delta(t - t_0)\, dt = \begin{cases} f(t_0) & t_1 < t_0 < t_2 \\ 0 & t_0 < t_1, t_0 > t_2 \end{cases}$$

for a finite t_0 and any $f(t)$ continuous at t_0. Note that the unit impulse function simply samples the value of $f(t)$ at $t = t_0$.

LEARNING by Doing

D 12.1 Evaluate the integral

$$\int_0^{2\pi} \cos t \; \delta(t - \pi)\, dt$$

ANSWER -1

D 12.2 Evaluate the integral

$$\int_0^{2\pi} e^{-t} \cos t \; \delta(t - 10\pi)\, dt$$

ANSWER 0

LEARNING Example 12.2

Let us determine the Laplace transform of an impulse function.

SOLUTION The Laplace transform of the impulse function is

$$\mathbf{F}(s) = \int_0^\infty \delta(t - t_0)e^{-st}\, dt$$

Using the sampling property of the delta function, we obtain

$$\mathcal{L}[\delta(t - t_0)] = e^{-t_0 s}$$

In the limit as $t_0 \rightarrow 0$, $e^{-t_0 s} \rightarrow 1$, and therefore

$$\mathcal{L}[\delta(t)] = \mathbf{F}(s) = 1$$

12.3 Transform Pairs

We will now illustrate the development of a number of basic transform pairs that are very useful in circuit analysis.

LEARNING Example 12.3

Let us find the Laplace transform of $f(t) = t$.

SOLUTION The Laplace transform of the function $f(t) = t$ is

$$\mathbf{F}(s) = \int_0^\infty t e^{-st} \, dt$$

Integrating the function by parts, we let

$$u = t \quad \text{and} \quad dv = e^{-st} \, dt$$

Then

$$du = dt \quad \text{and} \quad v = \int e^{-st} \, dt = -\frac{1}{s} e^{-st}$$

Therefore,

$$\mathbf{F}(s) = \frac{-t}{s} e^{-st} \Big|_0^\infty + \int_0^\infty \frac{e^{-st}}{s} \, dt$$

$$= \frac{1}{s^2} \qquad \sigma > 0$$

LEARNING Hint

$$t \leftrightarrow \frac{1}{s^2}$$

LEARNING Example 12.4

Let us determine the Laplace transform of the cosine function.

SOLUTION The Laplace transform for the cosine function is

$$\mathbf{F}(s) = \int_0^\infty \cos \omega t \, e^{-st} \, dt$$

$$= \int_0^\infty \frac{e^{+j\omega t} + e^{-j\omega t}}{2} e^{-st} \, dt$$

$$= \int_0^\infty \frac{e^{-(s-j\omega)t} + e^{-(s+j\omega)t}}{2} \, dt$$

$$= \frac{1}{2} \left(\frac{1}{s - j\omega} + \frac{1}{s + j\omega} \right) \qquad \sigma > 0$$

$$= \frac{s}{s^2 + \omega^2}$$

LEARNING Hint

$$\cos \omega t \leftrightarrow \frac{s}{s^2 + \omega^2}$$

A short table of useful Laplace transform pairs is shown in Table 12.1.

Once the transform pairs are known, we can easily move back and forth between the time domain and the complex frequency domain without having to use Eqs. (12.1) and (12.4).

Table 12.1 Short table of Laplace transform pairs

f(t)	F(s)
$\delta(t)$	1
$u(t)$	$\dfrac{1}{s}$
e^{-at}	$\dfrac{1}{s + a}$
t	$\dfrac{1}{s^2}$
$\dfrac{t^n}{n!}$	$\dfrac{1}{s^{n+1}}$
te^{-at}	$\dfrac{1}{(s + a)^2}$
$\dfrac{t^n e^{-at}}{n!}$	$\dfrac{1}{(s + a)^{n+1}}$

(continued on next page)

Table 12.1 *(continued)*

f(t)	F(s)
$\sin bt$	$\dfrac{b}{s^2 + b^2}$
$\cos bt$	$\dfrac{s}{s^2 + b^2}$
$e^{-at}\sin bt$	$\dfrac{b}{(s + a)^2 + b^2}$
$e^{-at}\cos bt$	$\dfrac{s + a}{(s + a)^2 + b^2}$

LEARNING EXTENSIONS

E12.1   If $f(t) = e^{-at}$, show that $\mathbf{F}(s) = 1/(s + a)$.

E12.2   If $f(t) = \sin \omega t$, show that $\mathbf{F}(s) = \omega/(s^2 + \omega^2)$.

12.4 Properties of the Transform

There are a number of useful theorems that describe important properties of the Laplace transform. We will first demonstrate a couple of these theorems, provide a concise listing of a number of them and, finally, illustrate their usefulness via several examples.

The *time-scaling theorem* states that

$$\mathcal{L}[f(at)] = \frac{1}{a}\mathbf{F}\left(\frac{s}{a}\right) \qquad a > 0 \qquad\qquad \textbf{12.5}$$

The Laplace transform of $f(at)$ *is*

$$\mathcal{L}[f(at)] = \int_0^\infty f(at)e^{-st}\,dt$$

Now let $\lambda = at$ *and* $d\lambda = a\,dt$. *Then*

$$\mathcal{L}[f(at)] = \int_0^\infty f(\lambda)e^{-(\lambda/a)s}\,\frac{d\lambda}{a}$$

$$= \frac{1}{a}\int_0^\infty f(\lambda)e^{-(s/a)\lambda}\,d\lambda$$

$$= \frac{1}{a}\mathbf{F}\left(\frac{s}{a}\right) \qquad a > 0$$

The *time-shifting theorem* states that

$$\mathcal{L}[f(t - t_0)u(t - t_0)] = e^{-t_0 s}\mathbf{F}(s) \qquad t_0 \geq 0 \qquad\qquad \textbf{12.6}$$

This theorem is illustrated as follows:

$$\mathcal{L}[f(t - t_0)u(t - t_0)] = \int_0^\infty f(t - t_0)u(t - t_0)e^{-st} \, dt$$

$$= \int_{t_0}^\infty f(t - t_0)e^{-st} \, dt$$

If we now let $\lambda = t - t_0$ *and* $d\lambda = dt$, *then*

$$\mathcal{L}[f(t - t_0)u(t - t_0)] = \int_0^\infty f(\lambda)e^{-s(\lambda + t_0)} \, d\lambda$$

$$= e^{-t_0 s} \int_0^\infty f(\lambda)e^{-s\lambda} \, d\lambda$$

$$= e^{-t_0 s}\mathbf{F}(s) \quad t_0 \geq 0$$

The *frequency-shifting* or *modulation theorem* states that

$$\mathcal{L}[e^{-at}f(t)] = \mathbf{F}(s + a) \tag{12.7}$$

By definition,

$$\mathcal{L}[e^{-at}f(t)] = \int_0^\infty e^{-at}f(t)e^{-st} \, dt$$

$$= \int_0^\infty f(t)e^{-(s+a)t} \, dt$$

$$= \mathbf{F}(s + a)$$

The three theorems we have demonstrated, together with a number of other important properties, are listed in a concise manner in Table 12.2. Let us now provide several simple examples that illustrate the manner in which these properties can be used.

Table 12.2 Some useful properties of the Laplace transform

Property Number	f(t)	F(s)
1.	$Af(t)$	$A\mathbf{F}(s)$
2.	$f_1(t) \pm f_2(t)$	$\mathbf{F}_1(s) \pm \mathbf{F}_2(s)$
3.	$f(at)$	$\dfrac{1}{a}\mathbf{F}\left(\dfrac{s}{a}\right), a > 0$
4.	$f(t - t_0)u(t - t_0), t_0 \geq 0$	$e^{-t_0 s}\mathbf{F}(s)$
5.	$f(t)u(t - t_0)$	$e^{-t_0 s}\mathcal{L}[f(t + t_0)]$
6.	$e^{-at}f(t)$	$\mathbf{F}(s + a)$
7.	$\dfrac{d^n f(t)}{dt^n}$	$s^n\mathbf{F}(s) - s^{n-1}f(0) - s^{n-2}f^1(0) \cdots s^0 f^{n-1}(0)$
8.	$tf(t)$	$-\dfrac{d\mathbf{F}(s)}{ds}$
9.	$\dfrac{f(t)}{t}$	$\displaystyle\int_s^\infty \mathbf{F}(\lambda) \, d\lambda$
10.	$\displaystyle\int_0^t f(\lambda) \, d\lambda$	$\dfrac{1}{s}\mathbf{F}(s)$
11.	$\displaystyle\int_0^t f_1(\lambda)f_2(t - \lambda) \, d\lambda$	$\mathbf{F}_1(s)\mathbf{F}_2(s)$

LEARNING Example 12.5

Use the Laplace transform of $\cos \omega t$ to find the Laplace transform of $e^{-at} \cos \omega t$.

SOLUTION   Since the Laplace transform of $\cos \omega t$ is known to be

$$\mathcal{L}[\cos \omega t] = \frac{s}{s^2 + \omega^2}$$

then using property number 6

$$\mathcal{L}[e^{-at} \cos \omega t] = \frac{s + a}{(s + a)^2 + \omega^2}$$

LEARNING Example 12.6

Let us demonstrate property number 9.

SOLUTION   If $f(t) = te^{-at}$, then

$$F(\lambda) = \frac{1}{(\lambda + a)^2}$$

Therefore,

$$\int_s^\infty F(\lambda) \, d\lambda = \int_s^\infty \frac{1}{(\lambda + a)^2} \, d\lambda = \frac{-1}{\lambda + a} \Big|_s^\infty = \frac{1}{s + a}$$

Hence,

$$f_1(t) = \frac{f(t)}{t} = \frac{te^{-at}}{t} = e^{-at} \quad \text{and} \quad F_1(s) = \frac{1}{s + a}$$

LEARNING Example 12.7

Let us employ the Laplace transform to solve the equation

$$\frac{dy(t)}{dt} + 2y(t) + \int_0^t y(\lambda) e^{-2(t-\lambda)} \, d\lambda = 10u(t) \qquad y(0) = 0$$

SOLUTION   Applying property numbers 7 and 11, we obtain

$$sY(s) + 2Y(s) + \frac{Y(s)}{s + 2} = \frac{10}{s}$$

$$Y(s)\left(s + 2 + \frac{1}{s + 2}\right) = \frac{10}{s}$$

$$Y(s) = \frac{10(s + 2)}{s(s^2 + 4s + 5)}$$

This is the solution of the linear constant-coefficient integrodifferential equation in the s-domain. However, we want the solution $y(t)$ in the time domain. $y(t)$ is obtained by performing the inverse transform, which is the topic of the next section, and the solution $y(t)$ is derived in Example 12.9.

LEARNING EXTENSIONS

E12.3   Find $F(s)$ if $f(t) = \frac{1}{2}(t - 4e^{-2t})$.

ANSWER
$$F(s) = \frac{1}{2s^2} - \frac{2}{s + 2}.$$

E12.4   If $f(t) = te^{-(t-1)}u(t - 1) - e^{-(t-1)}u(t - 1)$, determine $F(s)$ using the time-shifting theorem.

ANSWER
$$F(s) = \frac{e^{-s}}{(s + 1)^2}.$$

E12.5   Find $F(s)$ if $f(t) = e^{-4t}(t - e^{-t})$. Use property number 2.

ANSWER
$$F(s) = \frac{1}{(s + 4)^2} - \frac{1}{s + 5}.$$

12.5 Performing the Inverse Transform

As we begin our discussion of this topic, let us outline the procedure we will use in applying the Laplace transform to circuit analysis. First, we will transform the problem from the time domain to the complex frequency domain (that is, s-domain). Next we will solve the circuit equations algebraically in the complex frequency domain. Finally, we will transform the solution from the s-domain back to the time domain. It is this latter operation that we discuss now.

The algebraic solution of the circuit equations in the complex frequency domain results in a rational function of s of the form

$$\mathbf{F}(s) = \frac{\mathbf{P}(s)}{\mathbf{Q}(s)} = \frac{a_m s^m + a_{m-1} s^{m-1} + \cdots + a_1 s + a_0}{b_n s^n + b_{n-1} s^{n-1} + \cdots + b_1 s + b_0} \qquad \textbf{12.8}$$

The roots of the polynomial $\mathbf{P}(s)$ $\big(\text{i.e., } -z_1, -z_2 \cdots -z_m\big)$ are called the *zeros* of the function $\mathbf{F}(s)$ because at these values of s, $\mathbf{F}(s) = 0$. Similarly, the roots of the polynomial $\mathbf{Q}(s)$ $\big(\text{i.e., } -p_1, -p_2 \cdots -p_n\big)$ are called *poles* of $\mathbf{F}(s)$, since at these values of s, $\mathbf{F}(s)$ becomes infinite. If $\mathbf{F}(s)$ is a proper rational function of s, then $n > m$. However, if this is not the case, we simply divide $\mathbf{P}(s)$ by $\mathbf{Q}(s)$ to obtain a quotient and a remainder; that is,

$$\frac{\mathbf{P}(s)}{\mathbf{Q}(s)} = C_{m-n} s^{m-n} + \cdots + C_2 s^2 + C_1 s + C_0 + \frac{\mathbf{P}_1(s)}{\mathbf{Q}(s)} \qquad \textbf{12.9}$$

Now $\mathbf{P}_1(s)/\mathbf{Q}(s)$ is a proper rational function of s. Let us examine the possible forms of the roots of $\mathbf{Q}(s)$.

1. If the roots are simple, $\mathbf{P}_1(s)/\mathbf{Q}(s)$ can be expressed in partial fraction form as

$$\frac{\mathbf{P}_1(s)}{\mathbf{Q}(s)} = \frac{K_1}{s + p_1} + \frac{K_2}{s + p_2} + \cdots + \frac{K_n}{s + p_n} \qquad \textbf{12.10}$$

2. If $\mathbf{Q}(s)$ has simple complex roots, they will appear in complex-conjugate pairs and the partial fraction expansion of $\mathbf{P}_1(s)/\mathbf{Q}(s)$ for each pair of complex-conjugate roots will be of the form

$$\frac{\mathbf{P}_1(s)}{\mathbf{Q}_1(s)(s + \alpha - j\beta)(s + \alpha + j\beta)} = \frac{K_1}{s + \alpha - j\beta} + \frac{K_1^*}{s + \alpha + j\beta} + \cdots \quad \textbf{12.11}$$

where $\mathbf{Q}(s) = \mathbf{Q}_1(s)(s + a - j\beta)(s + \alpha + j\beta)$ and K_1^* in the complex conjugate of K_1.

3. If $\mathbf{Q}(s)$ has a root of multiplicity r, the partial fraction expansion for each such root will be of the form

$$\frac{\mathbf{P}_1(s)}{\mathbf{Q}_1(s)(s + p_1)^r} = \frac{K_{11}}{(s + p_1)} + \frac{K_{12}}{(s + p_1)^2} + \cdots + \frac{K_{1r}}{(s + p_1)^r} + \cdots \qquad \textbf{12.12}$$

The importance of these partial fraction expansions stems from the fact that once the function $\mathbf{F}(s)$ is expressed in this form, the individual inverse Laplace transforms can be obtained from known and tabulated transform pairs. The sum of these inverse Laplace transforms then yields the desired time function, $f(t) = \mathcal{L}^{-1}\big[\mathbf{F}(s)\big]$.

SIMPLE POLES Let us assume that all the poles of $\mathbf{F}(s)$ are simple, so that the partial fraction expansion of $\mathbf{F}(s)$ is of the form

$$\mathbf{F}(s) = \frac{\mathbf{P}(s)}{\mathbf{Q}(s)} = \frac{K_1}{s + p_1} + \frac{K_2}{s + p_2} + \cdots + \frac{K_n}{s + p_n} \qquad \textbf{12.13}$$

Then the constant K_i can be computed by multiplying both sides of this equation by $(s + p_i)$ and evaluating the equation at $s = -p_i$; that is,

$$\left.\frac{(s + p_i)\mathbf{P}(s)}{\mathbf{Q}(s)}\right|_{s=-p_i} = 0 + \cdots + 0 + K_i + 0 + \cdots + 0 \qquad i = 1,2,\ldots,n \quad \textbf{12.14}$$

Once all of the K_i terms are known, the time function $f(t) = \mathcal{L}^{-1}[\mathbf{F}(s)]$ can be obtained using the Laplace transform pair.

$$\mathcal{L}^{-1}\left[\frac{1}{s + a}\right] = e^{-at} \qquad\qquad \textbf{12.15}$$

LEARNING Example 12.8

Given that

$$\mathbf{F}(s) = \frac{12(s + 1)(s + 3)}{s(s + 2)(s + 4)(s + 5)}$$

let us find the function $f(t) = \mathcal{L}^{-1}[\mathbf{F}(s)]$.

SOLUTION   Expressing $\mathbf{F}(s)$ in a partial fraction expansion, we obtain

$$\frac{12(s + 1)(s + 3)}{s(s + 2)(s + 4)(s + 5)} = \frac{K_0}{s} + \frac{K_1}{s + 2} + \frac{K_2}{s + 4} + \frac{K_3}{s + 5}$$

To determine K_0, we multiply both sides of the equation by s to obtain the equation

$$\frac{12(s + 1)(s + 3)}{(s + 2)(s + 4)(s + 5)} = K_0 + \frac{K_1 s}{s + 2} + \frac{K_2 s}{s + 4} + \frac{K_3 s}{s + 5}$$

Evaluating the equation at $s = 0$ yields

$$\frac{(12)(1)(3)}{(2)(4)(5)} = K_0 + 0 + 0 + 0$$

or

$$K_0 = \frac{36}{40}$$

Similarly,

$$\left.(s + 2)\mathbf{F}(s)\right|_{s=-2} = \left.\frac{12(s + 1)(s + 3)}{s(s + 4)(s + 5)}\right|_{s=-2} = K_1$$

or

$$K_1 = 1$$

Using the same approach, we find that $K_2 = \frac{36}{8}$ and $K_3 = -\frac{32}{5}$. Hence $\mathbf{F}(s)$ can be written as

$$\mathbf{F}(s) = \frac{36/40}{s} + \frac{1}{s + 2} + \frac{36/8}{s + 4} - \frac{32/5}{s + 5}$$

Then $f(t) = \mathcal{L}^{-1}[\mathbf{F}(s)]$ is

$$f(t) = \left(\frac{36}{40} + 1e^{-2t} + \frac{36}{8}e^{-4t} - \frac{32}{5}e^{-5t}\right)u(t)$$

MATLAB can also be used to obtain the inverse Laplace transform. Two statements are required: the first states that the symbolic objects used in the workspace are s and t, and the second statement is the inverse Laplace (ilaplace) command. In this case, the two statements, together with the MATLAB results, are listed as follows.

```
>> syms s t
>> ilaplace
(12*(s+1)*(s+3)/(s*(s+2)*(s+4)*(s+5)
))`
ans =
9/10+exp (-2*t)+9/2*exp(-4*t)-
32/5*exp(-5*t)
```

$$f(t) = 0.9 + e^{-2t} + 4.5e^{-4t} - 6.4e^{-5t}$$

LEARNING EXTENSIONS

E12.6   Find $f(t)$ if $\mathbf{F}(s) = 10(s + 6)/(s + 1)(s + 3)$.

ANSWER
$f(t) = (25e^{-t} - 15e^{-3t})u(t)$.

E12.7   If $\mathbf{F}(s) = 12(s + 2)/s(s + 1)$, find $f(t)$.

ANSWER
$f(t) = (24 - 12e^{-t})u(t)$.

COMPLEX-CONJUGATE POLES Let us assume that $\mathbf{F}(s)$ has one pair of complex-conjugate poles. The partial fraction expansion of $\mathbf{F}(s)$ can then be written as

$$\mathbf{F}(s) = \frac{\mathbf{P}_1(s)}{\mathbf{Q}_1(s)(s + \alpha - j\beta)(s + \alpha + j\beta)} = \frac{K_1}{s + \alpha - j\beta} + \frac{K_1^*}{s + \alpha + j\beta} + \cdots \qquad 12.16$$

The constant K_1 can then be determined using the procedure employed for simple poles; that is,

$$(s + \alpha - j\beta)\mathbf{F}(s)\bigg|_{s=-\alpha+j\beta} = K_1 \qquad 12.17$$

In this case K_1 is in general a complex number that can be expressed as $|K_1|\underline{/\theta}$. Then $K_1^* = |K_1|\underline{/-\theta}$. Hence, the partial fraction expansion can be expressed in the form

$$\mathbf{F}(s) = \frac{|K_1|\underline{/\theta}}{s + \alpha - j\beta} + \frac{|K_1|\underline{/-\theta}}{s + \alpha - j\beta} + \cdots \qquad 12.18$$

$$= \frac{|K_1|e^{j\theta}}{s + \alpha - j\beta} + \frac{|K_1|e^{-j\theta}}{s + \alpha + j\beta} + \cdots$$

LEARNING Hint

Recall that

$$\cos x = \frac{e^{jx} + e^{-jx}}{2}$$

The corresponding time function is then of the form

$$f(t) = \mathcal{L}^{-1}[\mathbf{F}(s)] = |K_1|e^{j\theta}e^{-(\alpha-j\beta)t} + |K_1|e^{-j\theta}e^{-(\alpha+j\beta)t} + \cdots$$

$$= |K_1|e^{-\alpha t}[e^{j(\beta t+\theta)} + e^{-j(\beta t+\theta)}] + \cdots \qquad 12.19$$

$$= 2|K_1|e^{-\alpha t}\cos(\beta t + \theta) + \cdots$$

LEARNING Example 12.9

Let us determine the time function $y(t)$ for the function

$$\mathbf{Y}(s) = \frac{10(s + 2)}{s(s^2 + 4s + 5)}$$

SOLUTION Expressing the function in a partial fraction expansion, we obtain

$$\frac{10(s + 2)}{s(s + 2 - j1)(s + 2 + j1)}$$

$$= \frac{K_0}{s} + \frac{K_1}{s + 2 - j1} + \frac{K_1^*}{s + 2 - j1}$$

$$\frac{10(s + 2)}{s^2 + 4s + 5}\bigg|_{s=0} = K_0$$

$$4 = K_0$$

In a similar manner,

$$\frac{10(s + 2)}{s(s + 2 + j1)}\bigg|_{s=-2+j1} = K_1$$

$$2.236\,\underline{/-153.43°} = K_1$$

Therefore,

$$2.236\,\underline{/153.43°} = K_1^*$$

The partial fraction expansion of $\mathbf{Y}(s)$ is then

$$\mathbf{Y}(s) = \frac{4}{s} + \frac{2.236\,\underline{/-153.43°}}{s + 2 - j1} + \frac{2.236\,\underline{/153.43°}}{s + 2 + j1}$$

and therefore,

$$y(t) = [4 + 4.472e^{-2t}\cos(t - 153.43°)]u(t)$$

The MATLAB statements and resulting solution in this case are

```
>> syms s t
>>
ilaplace(10*(s+2)/(s*(s^2+4*s+5)))
ans =
4-4*exp(-2*t)*cos(t)+2*exp
(-2*t)*sin(t)
```

$$y(t) = 4 - 4e^{-2t}\cos(t) + 2e^{-2t}\sin(t)$$

Using Eq. (7.11) the reader can easily verify that this MATLAB result is the same as that obtained using the partial fraction expansion technique.

LEARNING EXTENSION

E12.8 Determine $f(t)$ if $\mathbf{F}(s) = s/(s^2 + 4s + 8)$.

ANSWER

$$f(t) = 1.41e^{-2t}\cos(2t + 45°)u(t).$$

MULTIPLE POLES Let us suppose that $\mathbf{F}(s)$ has a pole of multiplicity r. Then $\mathbf{F}(s)$ can be written in a partial fraction expansion of the form

$$\mathbf{F}(s) = \frac{\mathbf{P}_1(s)}{\mathbf{Q}_1(s)(s + p_1)^r}$$

$$= \frac{K_{11}}{s + p_1} + \frac{K_{12}}{(s + p_1)^2} + \cdots + \frac{K_{1r}}{(s + p_1)^r} + \cdots \qquad \textbf{12.20}$$

Employing the approach for a simple pole, we can evaluate K_{1r} as

$$(s + p_1)^r \mathbf{F}(s) \Big|_{s=-p_1} = K_{1r} \qquad \textbf{12.21}$$

To evaluate K_{1r-1} we multiply $\mathbf{F}(s)$ by $(s + p_1)^r$ as we did to determine K_{1r}; however, prior to evaluating the equation at $s = -p_1$, we take the derivative with respect to s. The proof that this will yield K_{1r-1} can be obtained by multiplying both sides of Eq. (12.20) by $(s + p_1)^r$ and then taking the derivative with respect to s. Now when we evaluate the equation at $s = -p_1$, the only term remaining on the right side of the equation is K_{1r-1}, and therefore,

$$\frac{d}{ds} \Big[(s + p_1)^r \mathbf{F}(s) \Big] \Big|_{s=-p_1} = K_{1r-1} \qquad \textbf{12.22}$$

K_{1r-2} can be computed in a similar fashion, and in that case the equation is

$$\frac{d^2}{ds^2} \Big[(s + p_1)^r \mathbf{F}(s) \Big] \Big|_{s=-p_1} = (2!)K_{1r-2} \qquad \textbf{12.23}$$

The general expression for this case is

$$K_{1j} = \frac{1}{(r - j)!} \frac{d^{r-j}}{ds^{r-j}} \Big[(s + p_1)^r \mathbf{F}(s) \Big] \Big|_{s=-p_1} \qquad \textbf{12.24}$$

Let us illustrate this procedure with an example.

LEARNING **Example 12.10**

Given the following function $\mathbf{F}(s)$, let us determine the corresponding time function $f(t) = \mathcal{L}^{-1}\big[\mathbf{F}(s)\big]$.

$$\mathbf{F}(s) = \frac{10(s + 3)}{(s + 1)^3(s + 2)}$$

SOLUTION Expressing $\mathbf{F}(s)$ as a partial fraction expansion, we obtain

$$\mathbf{F}(s) = \frac{10(s + 3)}{(s + 1)^3(s + 2)}$$

$$= \frac{K_{11}}{s + 1} + \frac{K_{12}}{(s + 1)^2} + \frac{K_{13}}{(s + 1)^3} + \frac{K_2}{s + 2}$$

Then

$$(s + 1)^3 \mathbf{F}(s) \Big|_{s=-1} = K_{13}$$

$$20 = K_{13}$$

K_{12} is now determined by the equation

$$\frac{d}{ds} \Big[(s + 1)^3 \mathbf{F}(s) \Big] \Big|_{s=-1} = K_{12}$$

$$\frac{-10}{(s + 2)^2} \Big|_{s=-1} = -10 = K_{12}$$

In a similar fashion K_{11} is computed from the equation

$$\frac{d^2}{ds^2} \Big[(s + 1)^3 \mathbf{F}(s) \Big] \Big|_{s=-1} = 2K_{11}$$

$$\frac{20}{(s + 2)^3} \Big|_{s=-1} = 20 = 2K_{11}$$

Therefore,

$$10 = K_{11}$$

In addition,

$$(s + 2)\mathbf{F}(s) \Big|_{s=-2} = K_2$$

$$-10 = K_2$$

(continued)

Hence, $\mathbf{F}(s)$ can be expressed as

$$\mathbf{F}(s) = \frac{10}{s + 1} - \frac{10}{(s + 1)^2} + \frac{20}{(s + 1)^3} - \frac{10}{s + 2}$$

Now we employ the transform pair

$$\mathcal{L}^{-1}\left[\frac{1}{(s + a)^{n+1}}\right] = \frac{t^n}{n!} e^{-at}$$

and hence,

$$f(t) = \left(10e^{-t} - 10te^{-t} + 10t^2e^{-t} - 10e^{-2t}\right)u(t)$$

Once again MATLAB can be used to obtain a solution as outlined next.

```
>> syms s t
>>
ilaplace(10*(s+3)/((s+1)^3*(s+2)))
ans =
10*t^2*exp(-t)-10*t*exp(-t)+10*exp
(-t)-10*exp(-2*t)
```

$$f(t) = 10t^2e^{-t} - 10te^{-t} + 10e^{-t} - 10e^{-2t}$$

LEARNING EXTENSIONS

E12.9 Determine $f(t)$ if $\mathbf{F}(s) = s/(s + 1)^2$.

ANSWER
$$f(t) = \left(e^{-t} - te^{-t}\right)u(t).$$

E12.10 If $\mathbf{F}(s) = (s + 2)/s^2(s + 1)$, find $f(t)$.

ANSWER
$$f(t) = \left(-1 + 2t + e^{-t}\right)u(t).$$

12.6 Convolution Integral

Convolution is a very important concept and has wide application in circuit and systems analysis. We first illustrate the connection that exists between the convolution integral and the Laplace transform. We then indicate the manner in which the convolution integral is applied in circuit analysis.

Property number 11 in Table 12.2 states the following.
If

$$f(t) = f_1(t) \otimes f_2(t) = \int_0^t f_1(t - \lambda)f_2(\lambda)\, d\lambda = \int_0^t f_1(\lambda)f_2(t - \lambda)\, d\lambda \qquad \textbf{12.25}$$

and

$$\mathcal{L}\left[f(t)\right] = \mathbf{F}(s), \mathcal{L}\left[f_1(t)\right] = \mathbf{F}_1(s) \quad \text{and} \quad \mathcal{L}\left[f_2(t)\right] = \mathbf{F}_2(s)$$

then

$$\mathbf{F}(s) = \mathbf{F}_1(s)\mathbf{F}_2(s) \qquad \textbf{12.26}$$

Our demonstration begins with the definition

$$\mathcal{L}\left[f(t)\right] = \int_0^\infty \left[\int_0^t f_1(t - \lambda)f_2(\lambda)\, d\lambda\right]e^{-st}\, dt$$

We now force the function into the proper format by introducing into the integral within the brackets the unit step function $u(t - \lambda)$. We can do this because

$$u(t - \lambda) = \begin{cases} 1 & \text{for } \lambda < t \\ 0 & \text{for } \lambda > t \end{cases} \qquad \textbf{12.27}$$

The first condition in Eq. (12.27) ensures that the insertion of the unit step function has no impact within the limits of integration. The second condition in Eq. (12.27) allows us to change the upper limit of integration from t to ∞. Therefore,

$$\mathcal{L}[f(t)] = \int_0^\infty \left[\int_0^\infty f_1(t - \lambda)u(t - \lambda)f_2(\lambda)d\lambda \right]e^{-st}\, dt$$

which can be written as

$$\mathcal{L}[f(t)] = \int_0^\infty f_2(\lambda)\left[\int_0^\infty f_1(t - \lambda)u(t - \lambda)e^{-st}\, dt \right]d\lambda$$

Note that the integral within the brackets is the time shifting theorem illustrated in Eq. (12.6). Hence, the equation can be written as

$$\mathcal{L}[f(t)] = \int_0^\infty f_2(\lambda)\mathbf{F}_1(s)e^{-s\lambda}\, d\lambda$$

$$= \mathbf{F}_1(s) \int_0^\infty f_2(\lambda)e^{-s\lambda}\, d\lambda$$

$$= \mathbf{F}_1(s)\mathbf{F}_2(s)$$

Note that convolution in the time domain corresponds to multiplication in the frequency domain.

Let us now illustrate the use of this property in the evaluation of an inverse Laplace transform.

LEARNING Example 12.11

The transfer function for a network is given by the expression

$$\mathbf{H}(s) = \frac{\mathbf{V}_o(s)}{\mathbf{V}_S(s)} = \frac{10}{s + 5}$$

The input is a unit step function $\mathbf{V}_S(s) = \dfrac{1}{s}$. Let us use convolution to determine the output voltage $v_o(t)$.

SOLUTION   Since $\mathbf{H}(s) = \dfrac{10}{(s + 5)}$, $h(t) = 10e^{-5t}$ and therefore

$$v_o(t) = \int_0^t 10u(\lambda)e^{-5(t-\lambda)}\, d\lambda$$

$$= 10e^{-5t} \int_0^t e^{5\lambda}\, d\lambda$$

$$= \frac{10e^{-5t}}{5}\left[e^{5t} - 1\right]$$

$$= 2\left[1 - e^{-5t}\right]u(t)$$

For comparison, let us determine $v_o(t)$ from $\mathbf{H}(s)$ and $\mathbf{V}_S(s)$ using the partial fraction expansion method. $\mathbf{V}_o(s)$ can be written as

$$\mathbf{V}_o(s) = \mathbf{H}(s)\mathbf{V}_S(s)$$

$$= \frac{10}{s(s + 5)} = \frac{K_0}{s} + \frac{K_1}{s + 5}$$

Evaluating the constants, we obtain $K_0 = 2$ and $K_1 = -2$. Therefore,

$$\mathbf{V}_o(s) = \frac{2}{s} - \frac{2}{s + 5}$$

and hence

$$v_o(t) = 2\left[1 - e^{-5t}\right]u(t)$$

Although we can employ convolution to derive an inverse Laplace transform, the example, although quite simple, illustrates that this is a very poor approach. If the function $\mathbf{F}(s)$ is very complicated, the mathematics can become unwieldy. Convolution is, however, a very powerful and useful tool. For example, if we know the impulse response of a network, we can use convolution to determine the network's response to an input that may be available only as an experimental curve obtained in the laboratory. Thus, convolution permits us to obtain the network response to inputs that cannot be written as analytical functions, but can be simulated on a digital computer. In addition, we can use convolution to model a circuit, which is completely unknown to us, and use this model to determine the circuit's response to some input signal.

LEARNING Example 12.12

To demonstrate the power of convolution, we will create a model for a "black-box" linear band-pass filter, shown as a block in Fig. 12.3. We have no details about the filter circuitry at all—no circuit diagram, no component list, no component values. As a result, our filter model must be based solely on measurements. Using our knowledge of convolution and the Laplace transform, let us discuss appropriate measurement techniques, the resulting model, and how to employ the model in subsequent simulations.

Figure 12.3
Conceptual diagram for a band-pass filter.

SOLUTION   Because the filter is linear, $v_o(t)$ can be written

$$v_o(t) = h(t) \otimes v_{in}(t) \qquad \textbf{12.28}$$

Thus, the function $h(t)$ will be our model for the filter. To determine $h(t)$, we must input some $v_{in}(t)$, measure the response, $v_o(t)$, and perform the appropriate mathematics. One obvious option for $v_{in}(t)$ is the impluse function, $\delta(t)$, then $\mathbf{V}_{in}(s)$ is 1, and the output is the desired model, $h(t)$.

$$v_o(t) = h(t)$$

Unfortunately, creating an adequate impulse, infinite amplitude, and zero width in the laboratory is nontrivial. It is much easier, and more common, to apply a step function such as $10\,u(t)$. Then $\mathbf{V}_{in}(s)$ is $10/s$, and the output can be expressed in the s-domain as

$$\mathbf{V}_o(s) = \mathbf{H}(s)\left[\frac{10}{s}\right]$$

or

$$\mathbf{H}(s) = \left[\frac{s}{10}\right]\mathbf{V}_o(s)$$

Since multiplication by s is equivalent to the time derivative, we have for $h(t)$

$$h(t) = \left[\frac{1}{10}\right]\frac{dv_o(t)}{dt} \qquad \textbf{12.29}$$

Thus, $h(t)$ can be obtained from the derivative of the filter response to a step input!

In the laboratory, the input $10\,u(t)$ was applied to the filter and the output voltage was measured using a digital oscilloscope. Data points for time and $v_o(t)$ were acquired every 50 μs over the interval 0 to 50 ms, that is, 1000 data samples. The digital oscilloscope formats the data as a text file, which can be transferred to a personal computer where the data can be processed. (In other

Table 12.3 The first five data samples of the step response and the evaluation of $h(t)$

n	Time (s)	Step Response (V)	h(t)
0	0.00E+00	0.00E+00	3.02E+02
1	5.00E−05	1.51E−01	8.98E+02
2	1.00E−04	6.00E−01	9.72E+02
3	1.50E−04	1.09E+00	9.56E+02
4	2.00E−04	1.56E+00	9.38E+02

words, we can find our derivative in Eq. (12.29), $dv_o(t)/dt$.) The results are shown in Table 12.3. The second and third columns in the table show the elapsed time and the output voltage for the first few data samples. To produce $h(t)$, the derivative was approximated in software using the simple algorithm,

$$\frac{dv_o(t)}{dt} \approx \frac{\Delta V_o}{\Delta t} = \frac{V_o\big[(n+1)T_S\big] - V_o\big[nT_S\big]}{T_S}$$

where T_S is the sample time, 50 μs, and n is the sample number. Results for $h(t)$ are shown in the fourth column of the table. At this point, $h(t)$ exists as a table of data points and the filter is now modeled.

To test our model, $h(t)$, we let the function $v_{in}(t)$ contain a combination of dc and sinusoid components such as

$$v_{in}(t) = \qquad\qquad \textbf{12.30}$$
$$\begin{cases} 1\sin\big[(2\pi)100t\big] + 1\sin\big[(2\pi)1234t\big] + 4 & 0 \le t < 25 \text{ ms} \\ 0 & t \ge 25 \text{ ms} \end{cases}$$

How will the filter perform? What will the output voltage look like? To find out, we must convolve $h(t)$ and $v_{in}(t)$. A data file for $v_{in}(t)$ can be created by simply evaluating the function in Eq. (12.30) every 50 μs. We will perform the convolution using MATLAB's c o n v function and simple file input/output functions. First, let us agree that text files for $h(t)$ and $v_{in}(t)$ have been saved as h t . t x t and v i n . t x t in MATLAB's Work subdirectory. A set of suitable commands is

```
h=textread('ht.txt','%f');
v=textread('vin.txt','%f');
out=conv(h,v);
dlmwrite('vout.txt',out,'\t')
```

The first two commands read files h t . t x t and v i n . t x t as MATLAB variables h and v, respectively. Entry ' % f ' specifies that each file's data are read as floating point numbers. The third command puts the convolution results into the MATLAB variable `out`. Finally, the last command puts the convolution results into a tab-delimited text file named `vout.txt`, in the Work subdirectory. The entry `'\t'` sets the delimiter to TAB.

Plots of the resulting $v_o(t)$ and $v_{in}(t)$ are shown in Fig. 12.4. An examination of the output waveform indicates that the 100-Hz component of $v_{in}(t)$ is amplified whereas the dc and 1234-Hz components are attenuated; that is, $v_o(t)$ has an amplitude of approximately 3 V and an average value of near zero. Indeed, the circuit performs as a band-pass filter. Remember that these waveforms are not measured; they are simulation results obtained from our model, $h(t)$.

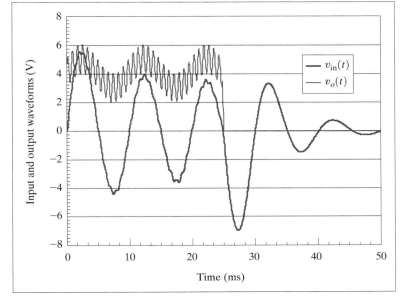

Figure 12.4
Plots of input and output waveforms reveal the nature of the band-pass filter—particularly, attenuation of dc and higher-frequency components.

12.7 Initial-Value and Final-Value Theorems

Suppose that we wish to determine the initial or final value of a circuit response in the time domain from the Laplace transform of the function in the s-domain without performing the inverse transform. If we determine the function $f(t) = \mathcal{L}^{-1}[\mathbf{F}(s)]$, we can find the initial value by evaluating $f(t)$ as $t \to 0$ and the final value by evaluating $f(t)$ as $t \to \infty$. It would be very convenient, however, if we could simply determine the initial and final values from $\mathbf{F}(s)$ without having to perform the inverse transform. The initial- and final-value theorems allow us to do just that.

The *initial-value theorem* states that

$$\lim_{t \to 0} f(t) = \lim_{s \to \infty} s\mathbf{F}(s) \qquad\qquad \textbf{12.31}$$

provided that $f(t)$ and its first derivative are transformable.

The proof of this theorem employs the Laplace transform of the function $df(t)/dt$.

$$\int_0^\infty \frac{df(t)}{dt} e^{-st} \, dt = s\mathbf{F}(s) - f(0)$$

Taking the limit of both sides as $s \to \infty$, we find that

$$\lim_{s \to \infty} \int_0^\infty \frac{df(t)}{dt} e^{-st} \, dt = \lim_{s \to \infty} \left[s\mathbf{F}(s) - f(0) \right]$$

and since

$$\int_0^\infty \frac{df(t)}{dt} \lim_{s \to \infty} e^{-st} \, dt = 0$$

then

$$f(0) = \lim_{s \to \infty} sF(s)$$

which is, of course,

$$\lim_{t \to 0} f(t) = \lim_{s \to \infty} sF(s)$$

The *final-value theorem* states that

$$\lim_{t \to \infty} f(t) = \lim_{s \to 0} sF(s) \qquad \text{12.32}$$

provided that $f(t)$ and its first derivative are transformable and that f(∞) exists. This latter requirement means that the poles of $F(s)$ must have negative real parts with the exception that there can be a simple pole at $s = 0$.

The proof of this theorem also involves the Laplace transform of the function $df(t)/dt$.

$$\int_0^\infty \frac{df(t)}{dt} e^{-st} \, dt = sF(s) - f(0)$$

Taking the limit of both sides as $s \to 0$ gives us

$$\lim_{s \to 0} \int_0^\infty \frac{df(t)}{dt} e^{-st} \, dt = \lim_{s \to 0} \left[sF(s) - f(0) \right]$$

Therefore,

$$\int_0^\infty \frac{df(t)}{dt} \, dt = \lim_{s \to 0} \left[sF(s) - f(0) \right]$$

and

$$f(\infty) - f(0) = \lim_{s \to 0} sF(s) - f(0)$$

and hence,

$$f(\infty) = \lim_{t \to \infty} f(t) = \lim_{s \to 0} sF(s)$$

LEARNING Example 12.13

Let us determine the initial and final values for the function

$$F(s) = \frac{10(s + 1)}{s(s^2 + 2s + 2)}$$

and corresponding time function

$$f(t) = 5 + 5\sqrt{2} \, e^{-t} \cos(t - 135°)$$

SOLUTION Applying the initial-value theorem, we have

$$f(0) = \lim_{s \to \infty} sF(s)$$

$$= \lim_{s \to \infty} \frac{10(s + 1)}{s^2 + 2s + 2}$$

$$= 0$$

The poles of $F(s)$ are $s = 0$ and $s = -1 \pm j1$, so the final-value theorem is applicable. Thus,

$$f(\infty) = \lim_{s \to 0} sF(s)$$

$$= \lim_{s \to 0} \frac{10(s + 1)}{s^2 + 2s + 2}$$

$$= 5$$

Note that these values could be obtained directly from the time function $f(t)$.

E12.11   Find the initial and final values of the function $f(t)$ if $\mathbf{F}(s) = \mathcal{L}[f(t)]$ is given by the expression

$$\mathbf{F}(s) = \frac{(s+1)^2}{s(s+2)(s^2+2s+2)}$$

ANSWER
$f(0) = 0$ and $f(\infty) = \frac{1}{4}$.

Learning by Application

As a prelude to Chapter 13 in which we will employ the power and versatility of the Laplace transform in a wide variety of circuit analysis problems, we will now demonstrate how the techniques outlined in this chapter can be used in the solution of a circuit problem via the differential equation that describes the network.

LEARNING Example 12.14

Consider the network shown in Fig. 12.5a. Assume that the network is in steady state prior to $t = 0$. Let us find the current $i(t)$ for $t > 0$.

SOLUTION   In steady state prior to $t = 0$, the network is as shown in Fig. 12.5b, since the inductor acts like a short circuit to dc and the capacitor acts like an open circuit to dc. From Fig. 12.5b we note that $i(0) = 4\,\text{A}$ and $v_C(0) = 4\,\text{V}$. For $t > 0$, the KVL equation for the network is

$$12u(t) = 2i(t) + 1\frac{di(t)}{dt} + \frac{1}{0.1}\int_0^t i(x)\,dx + v_C(0)$$

Using the results of Example 12.1 and properties 7 and 10, the transformed expression becomes

$$\frac{12}{s} = 2\mathbf{I}(s) + s\mathbf{I}(s) - i(0) + \frac{10}{s}\mathbf{I}(s) + \frac{v_C(0)}{s}$$

Using the initial conditions, the equation becomes

$$\frac{12}{s} = \mathbf{I}(s)\left(2 + s + \frac{10}{s}\right) - 4 + \frac{4}{s}$$

or

$$\mathbf{I}(s) = \frac{4(s+2)}{s^2+2s+10} = \frac{4(s+2)}{(s+1-j3)(s+1+j3)}$$

and then

$$K_1 = \left.\frac{4(s+2)}{s+1+j3}\right|_{s=-1+j3}$$

$$= 2.11\,\underline{/-18.4^\circ}$$

Therefore,

$$i(t) = 2(2.11)e^{-t}\cos(3t - 18.4^\circ)\,\text{A}$$

Note that this expression satisfies the initial condition $i(0) = 4\,\text{A}$.

Figure 12.5
Circuits used
in Example
12.14.

(a)   (b)

LEARNING EXTENSION

E12.12 Assuming the network in Fig. E12.12 is in steady state prior to $t = 0$, find $i(t)$ for $t > 0$. **ANSWER**
$$i(t) = 3 - e^{-2t} \text{ A.}$$

Figure E12.12

LEARNING Check

Summary

▶ In applying the Laplace transform, we convert an integro-differential equation in the time domain to an algebraic equation, which includes initial conditions, in the s-domain. We solve for the unknowns in the s-domain and convert the results back to the time domain.

▶ The Laplace transform is defined by the expression

$$\mathcal{L}[f(t)] = \mathbf{F}(s) = \int_0^\infty f(t)e^{-st}\,dt$$

▶ Laplace transform pairs, as listed in Table 12.1, can be used to convert back and forth between the time and frequency domains.

▶ The Laplace transform properties, as listed in Table 12.2, are useful in performing the Laplace transform and its inverse.

▶ The partial fraction expansion of a function in the s-domain permits the use of the transform pairs in Table 12.1 and the properties in Table 12.2 to convert the function to the time domain.

▶ The convolution of two functions in the time domain corresponds to a simple multiplication of the two functions in the s-domain.

▶ The initial and final values of a time-domain function can be obtained from its Laplace transform in the frequency domain.

Problems

SECTION 12.2

12.1 Find the Laplace transform of the function
$$f(t) = te^{-at}\delta(t - 1)$$

12.2 Find the Laplace transform of the function
$$f(t) = te^{-a(t-1)}\delta(t - 1).$$

SECTION 12.3

12.3 If $f(t) = e^{-at}\cos\omega t$, show that
$$\mathbf{F}(s) = (s + a)/[(s + a)^2 + \omega^2]$$

12.4 Find $\mathbf{F}(s)$ if $f(t) = e^{-at}\sin(\omega t)u(t - 1)$.

12.5 If $f(t) = t\cos(\omega t)u(t - 1)$, find $\mathbf{F}(s)$.

12.6 Find $\mathbf{F}(s)$ if $f(t) = te^{-at}u(t - 4)$.

SECTION 12.4

12.7 Use the time-shifting theorem to determine $\mathcal{L}[f(t)]$, where $f(t) = [e^{-(t-2)} - e^{-2(t-2)}]u(t - 2)$.

12.8 Use the time-shifting theorem to determine $\mathcal{L}[f(t)]$, where $f(t) = [t - 1 + e^{-(t-1)}]u(t - 1)$.

12.9 Use property number 5 to find $\mathcal{L}[f(t)]$ if
$$f(t) = e^{-at}u(t - 1).$$

12.10 Use property number 6 to find $\mathcal{L}[f(t)]$ if
$$f(t) = te^{-at}u(t - 1).$$

SECTION 12.5

12.11 Given the following functions $\mathbf{F}(s)$, find $f(t)$.

(a) $\mathbf{F}(s) = \dfrac{4}{(s+1)(s+2)}$

(b) $\mathbf{F}(s) = \dfrac{10s}{(s+1)(s+4)}$

12.12 Given the following functions $\mathbf{F}(s)$, find $f(t)$.

(a) $\mathbf{F}(s) = \dfrac{s+10}{(s+4)(s+6)}$

(b) $\mathbf{F}(s) = \dfrac{24}{(s+2)(s+8)}$

12.13 Given the following functions $\mathbf{F}(s)$, find $f(t)$.

(a) $\mathbf{F}(s) = \dfrac{s+1}{s(s+2)(s+3)}$

(b) $\mathbf{F}(s) = \dfrac{s^2+s+1}{s(s+1)(s+2)}$

12.14 Given the following functions $\mathbf{F}(s)$, find $f(t)$.

(a) $\mathbf{F}(s) = \dfrac{s^2+5s+4}{(s+2)(s+4)(s+6)}$

(b) $\mathbf{F}(s) = \dfrac{(s+3)(s+6)}{s(s^2+8s+12)}$

12.15 Given the following functions $\mathbf{F}(s)$, find $f(t)$.

(a) $\mathbf{F}(s) = \dfrac{s^2+7s+12}{(s+2)(s+4)(s+6)}$

(b) $\mathbf{F}(s) = \dfrac{(s+3)(s+6)}{s(s^2+10s+24)}$

12.16 Use MATLAB to solve Problem 12.15.

12.17 Given the following functions $\mathbf{F}(s)$, find $f(t)$.

(a) $\mathbf{F}(s) = \dfrac{10}{s^2+2s+2}$

(b) $\mathbf{F}(s) = \dfrac{10(s+2)}{s^2+4s+5}$

12.18 Given the following functions $\mathbf{F}(s)$, find the inverse Laplace transform of each function.

(a) $\mathbf{F}(s) = \dfrac{10(s+1)}{s^2+2s+2}$

(b) $\mathbf{F}(s) = \dfrac{s+1}{s(s^2+4s+5)}$

12.19 Given the following functions $\mathbf{F}(s)$, find $f(t)$.

(a) $\mathbf{F}(s) = \dfrac{s(s+6)}{(s+3)(s^2+6s+18)}$

(b) $\mathbf{F}(s) = \dfrac{(s+4)(s+8)}{s(s^2+8s+32)}$

12.20 Given the following functions $\mathbf{F}(s)$, find $f(t)$.

(a) $\mathbf{F}(s) = \dfrac{6s+12}{(s^2+4s+5)(s^2+4s+8)}$

(b) $\mathbf{F}(s) = \dfrac{s(s+2)}{s^2+2s+2}$

12.21 Use MATLAB to solve Problem 12.19.

12.22 Given the following functions $\mathbf{F}(s)$, find the inverse Laplace transform of each function.

(a) $\mathbf{F}(s) = \dfrac{s+1}{s^2(s+2)}$

(b) $\mathbf{F}(s) = \dfrac{s+3}{(s+1)^2(s+4)}$

12.23 Given the following functions $\mathbf{F}(s)$, find $f(t)$.

(a) $\mathbf{F}(s) = \dfrac{s+8}{s^2(s+6)}$

(b) $\mathbf{F}(s) = \dfrac{1}{s^2(s+1)^2}$

12.24 Given the following functions $\mathbf{F}(s)$, find $f(t)$.

(a) $\mathbf{F}(s) = \dfrac{s+4}{(s+2)^2}$

(b) $\mathbf{F}(s) = \dfrac{s+6}{s(s+1)^2}$

12.25 Given the following functions $\mathbf{F}(s)$, find $f(t)$.

(a) $\mathbf{F}(s) = \dfrac{s^2}{(s+1)^2(s+2)}$

(b) $\mathbf{F}(s) = \dfrac{s^2+9s+20}{s(s+4)^3(s+5)}$

12.26 Find $f(t)$ if $\mathbf{F}(s)$ is given by the expression

$$\mathbf{F}(s) = \frac{s(s+1)}{(s+2)^3(s+3)}$$

12.27 Find $f(t)$ if $\mathbf{F}(s)$ is given by the expression

$$\mathbf{F}(s) = \frac{12(s+2)}{s^2(s+1)(s^2+4s+8)}$$

12.28 Use MATLAB to solve Problem 12.25.

12.29 Find the inverse Laplace transform of the following functions.

(a) $\mathbf{F}(s) = \dfrac{e^{-s}}{s+1}$

(b) $\mathbf{F}(s) = \dfrac{1-e^{-2s}}{s}$

(c) $\mathbf{F}(s) = \dfrac{1-e^{-s}}{s+2}$

12.30 Find the inverse Laplace transform of the following functions.

(a) $F(s) = \dfrac{(s + 1)e^{-s}}{s(s + 2)}$

(b) $F(s) = \dfrac{10e^{-2s}}{(s + 1)(s + 3)}$

12.31 Find the inverse Laplace transform $f(t)$ if $F(s)$ is

$$F(s) = \frac{se^{-s}}{(s + 1)(s + 2)}$$

12.32 Find $f(t)$ if $F(s)$ is given by the following functions:

(a) $F(s) = \dfrac{(s^2 + 2s + 3)e^{-2s}}{s(s + 1)(s + 2)}$

(b) $F(s) = \dfrac{(s + 2)e^{-4s}}{s^2(s + 1)}$

12.33 Find $f(t)$ if $F(s)$ is given by the following functions:

(a) $F(s) = \dfrac{2(s + 1)e^{-s}}{(s + 2)(s + 3)}$

(b) $F(s) = \dfrac{10(s + 2)e^{-2s}}{(s + 1)(s + 4)}$

12.34 Solve the following differential equations using Laplace transforms.

(a) $\dfrac{dx(t)}{dt} + 3x(t) = e^{-2t}, \quad x(0) = 1$

(b) $\dfrac{dx(t)}{dt} + 3x(t) = 2u(t), \quad x(0) = 2$

12.35 Solve the following differential equations using Laplace transforms.

(a) $\dfrac{d^2y(t)}{dt^2} + \dfrac{2dy(t)}{dt} + y(t) = e^{-2t},$

$$y(0) = y'(0) = 0$$

(b) $\dfrac{d^2y(t)}{dt^2} + \dfrac{4dy(t)}{dt} + 4y(t) = u(t), \quad y(0) = 0,$

$$y'(0) = 1$$

12.36 Use Laplace transforms to find $y(t)$ if

$$\frac{dy(t)}{dt} + 5y(t) + 4\int_0^t y(\lambda)d\lambda = u(t), \quad y(0) = 0, \quad t > 0$$

12.37 Solve the following integrodifferential equation using Laplace transforms.

$$\frac{dy(t)}{dt} + 2y(t) + \int_0^t y(\lambda)d\lambda = 1 - e^{-2t}, \quad y(0) = 0, \quad t > 0$$

12.38 Determine $y(t)$ in the following equation if all initial conditions are zero.

$$\frac{d^3y(t)}{dt^3} + \frac{4d^2y(t)}{dt^2} + \frac{3dy(t)}{dt} = 10e^{-2t}$$

SECTION 12.6

12.39 Find $f(t)$ using convolution if $F(s)$ is

$$F(s) = \frac{1}{(s + 1)(s + 2)}$$

12.40 Use convolution to find $f(t)$ if

$$F(s) = \frac{1}{(s + 1)(s + 2)^2}$$

SECTION 12.7

12.41 Determine the initial and final values of $f(t)$ if $F(s)$ is given by the expressions

(a) $F(s) = \dfrac{2(s + 2)}{s(s + 1)}$

(b) $F(s) = \dfrac{2(s^2 + 2s + 6)}{(s + 1)(s + 2)(s + 3)}$

(c) $F(s) = \dfrac{2s^2}{(s + 1)(s^2 + 2s + 2)}$

12.42 Find the initial and final values of the time function $f(t)$ if $F(s)$ is given as

(a) $F(s) = \dfrac{10(s + 2)}{(s + 1)(s + 3)}$

(b) $F(s) = \dfrac{s^2 + 2s + 4}{(s + 6)(s^3 + 4s^2 + 8s + 10)}$

(c) $F(s) = \dfrac{2s}{s^2 + 2s + 2}$

12.43 Find the final values of the time function $f(t)$ given that

(a) $F(s) = \dfrac{10(s + 1)}{(s + 2)(s + 3)}$

(b) $F(s) = \dfrac{10}{s^2 + 4s + 4}$

SECTION 12.8

12.44 In the network in Fig. P12.44, the switch opens at $t = 0$. Use Laplace transforms to find $i(t)$ for $t > 0$.

Figure P12.44

12.45 The switch in the circuit in Fig. P12.45 opens at $t = 0$. Find $i(t)$ for $t > 0$ using Laplace transforms.

Figure P12.45

12.46 In the circuit in Fig. P12.46, the switch moves from position 1 to position 2 at $t = 0$. Use Laplace transforms to find $v(t)$ for $t > 0$.

Figure P12.46

12.47 In the network in Fig. P12.47, the switch closes at $t = 0$. Use Laplace transforms to find $v_C(t)$ for $t > 0$.

Figure P12.47

12.48 In the network in Fig. P12.48, the switch opens at $t = 0$. Use Laplace transforms to find $i_L(t)$ for $t > 0$.

Figure P12.48

12.49 In the network in Fig. P12.49, the switch opens at $t = 0$. Use Laplace transforms to find $v_o(t)$ for $t > 0$.

Figure P12.49

Typical Problems Found on the FE Exam

12FE-1 The output function of a network is expressed using Laplace transforms, in the following form.

$$\mathbf{V}_o(s) = \frac{12}{s(s^2 + 3s + 2)}$$

Find the output $v_o(t)$ as a function of time.

12FE-2 The Laplace transform function representing the output voltage of a network is expressed as

$$\mathbf{V}_o(s) = \frac{120}{s(s + 10)(s + 20)}$$

Determine the time-domain function and the value of $v_o(t)$ at $t = 100$ ms.

12FE-3 The Laplace transform function for the output voltage of a network is expressed in the following form.

$$\mathbf{V}_o(s) = \frac{12(s + 2)}{s(s + 1)(s + 3)(s + 4)}$$

Determine the final value of this voltage, i.e, $v_o(t)$ as $t \to \infty$.

13

Application of the Laplace Transform to Circuit Analysis

LEARNING Goals

We demonstrated in Chapter 12 how the Laplace transform can be used to solve linear constant-coefficient differential equations. Since all of the linear networks with which we are dealing can be described by linear constant-coefficient differential equations, use of the Laplace transform for circuit analysis would appear to be a feasible approach. The terminal characteristics for each circuit element can be described in the s-domain by transforming the appropriate time domain equations. Kirchhoff's laws when applied to a circuit produce a set of integrodifferential equations in terms of the terminal characteristics of the network elements, which when transformed yield a set of algebraic equations in the s-domain. Therefore, a complex frequency domain analysis, in which the passive network elements are represented by their transform impedance or admittance and sources (independent or dependent) are represented in terms of their transform variables, can be performed. Such an analysis in the s-domain is algebraic, and all of the techniques derived in dc analysis will apply. Therefore, the analysis is similar to the dc analysis of resistive networks, and all of the network analysis techniques and network theorems that were applied in dc analysis are valid in the s-domain (e.g., node analysis, loop analysis, superposition, source transformation, Thévenin's theorem, Norton's theorem, and combinations of impedance or admittance). Therefore, our approach will be to transform each of the circuit elements, draw an s-domain equivalent circuit, and then, using the transformed network, solve the circuit equations algebraically in the s-domain. Finally, a table of transform pairs can be employed to obtain a complete (transient plus steady-state) solution. Our approach in this chapter is to employ the circuit element models and demonstrate, using a variety of examples, many of the concepts and techniques presented earlier.

13.1 Laplace Circuit Solutions

To introduce the utility of the Laplace transform in circuit analysis, let us consider the RL series circuit shown in Fig. 13.1. In particular, let us find the current, $i(t)$.

$i(t)$ $R = 100\ \Omega$

$v_S(t) = 1\ u(t)$ $L = 100\ \text{mH}$

Figure 13.1
RL series network.

Using Kirchhoff's voltage law, we can write the time domain differential equation,

$$v_S(t) = L\left(\frac{di(t)}{dt}\right) + Ri(t)$$

The complementary differential equation is

$$L\left(\frac{di(t)}{dt}\right) + Ri(t) = 0 \qquad\qquad \textbf{13.1}$$

and has the solution

$$i_C(t) = K_C e^{-\alpha t}$$

Substituting $i_C(t)$ into the complementary equation yields the relationship

$$R - \alpha L = 0$$

or

$$\alpha = \frac{R}{L} = 1000$$

The particular solution is of the same form as the forcing function, $v_S(t)$.

$$i_p(t) = K_p$$

Substituting $i_p(t)$ into the original differential equation yields the expression

$$1 = RK_p$$

or

$$K_p = 1/R = 1/100$$

The final solution is the sum of $i_p(t)$ and $i_C(t)$,

$$i(t) = K_p + K_C e^{-\alpha t} = \frac{1}{100} + K_C e^{-1000t}$$

To find K_C, we must use the value of the current at some particular instant of time. For $t < 0$, the unit step function is zero and so is the current. At $t = 0$, the unit step goes to one; however, the inductor forces the current to instantaneously remain at zero. Therefore, at $t = 0$, we can write

$$i(0) = 0 = K_p + K_C$$

or

$$K_C = -K_p = -\frac{1}{100}$$

Thus, the current is

$$i(t) = 10(1 - e^{-1000t})u(t) \text{ mA}$$

Let us now try a different approach to the same problem. Making use of Table 12.2, let us take the Laplace transform of both sides of Eq. (13.1).

$$\mathcal{L}[v_S(t)] = \mathbf{V}_S(s) = L[s\mathbf{I}(s)] + R\mathbf{I}(s)$$

Now the circuit is represented not by a time domain differential equation, but rather by an algebraic expression in the s-domain. Solving for $\mathbf{I}(s)$, we can write

$$\mathbf{I}(s) = \frac{\mathbf{V}_S(s)}{sL + R} = \frac{1}{s[sL + R]}$$

We find $i(t)$ using the inverse Laplace transform. First, let us express $\mathbf{I}(s)$ as a sum of partial products

$$\mathbf{I}(s) = \frac{1/L}{s\left[s + \dfrac{R}{L}\right]} = \frac{1}{sR} - \frac{1}{R\left[s + \dfrac{R}{L}\right]}$$

The inverse transform is simply

$$i(t) = \frac{1}{R}\left(1 - e^{-Rt/L}\right)$$

Given the circuit element values in Fig. 13.1, the current is

$$i(t) = 10(1 - e^{-1000t})u(t) \text{ mA}$$

which is exactly the same as that obtained using the differential equation approach. Note carefully that the solution using the Laplace transform approach yields the entire solution in one step.

We have shown that the Laplace transform can be used to transform a differential equation into an algebraic equation. Since the voltage–current relationships for resistors, capacitors, and inductors involve only constants, derivatives, and integrals, we can represent and solve any circuit in the s-domain.

LEARNING by Doing

D 13.1 Find $v(t)$ for $t > 0$ in the following network:

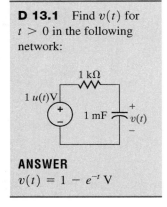

ANSWER
$v(t) = 1 - e^{-t} \text{ V}$

13.2 Circuit Element Models

The Laplace transform technique employed earlier implies that the terminal characteristics of circuit elements can be expressed as algebraic expressions in the s-domain. Let us examine these characteristics for the resistor, capacitor, and inductor.

The voltage–current relationship for a resistor in the time domain using the passive sign convention is

$$v(t) = Ri(t) \tag{13.2}$$

Using the Laplace transform, this relationship in the s-domain is

$$\mathbf{V}(s) = R\mathbf{I}(s) \tag{13.3}$$

Therefore, the time domain and complex frequency domain representations of this element are as shown in Fig. 13.2a.

The time domain relationships for a capacitor using the passive sign convention are

$$v(t) = \frac{1}{C}\int_0^t i(x)\,dx + v(0) \tag{13.4}$$

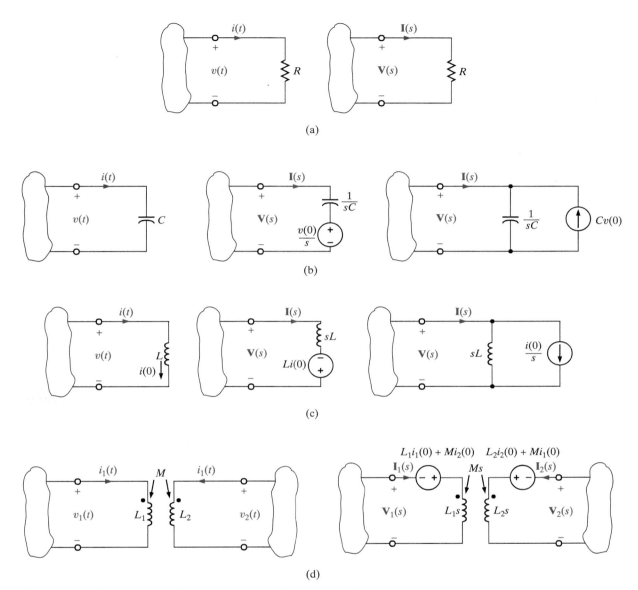

Figure 13.2   *Time domain and s-domain representations of circuit elements.*

$$i(t) = C\frac{dv(t)}{dt} \qquad\qquad \textbf{13.5}$$

The *s*-domain equations for the capacitor are then

$$\mathbf{V}(s) = \frac{\mathbf{I}(s)}{sC} + \frac{v(0)}{s} \qquad\qquad \textbf{13.6}$$

$$\mathbf{I}(s) = sC\mathbf{V}(s) - Cv(0) \qquad\qquad \textbf{13.7}$$

and hence the *s*-domain representation of this element is as shown in Fig. 13.2b.

For the inductor, the voltage–current relationships using the passive sign convention are

$$v(t) = L\frac{di(t)}{dt} \qquad \text{13.8}$$

$$i(t) = \frac{1}{L}\int_0^t v(x)\,dx + i(0) \qquad \text{13.9}$$

The relationships in the s-domain are then

$$\mathbf{V}(s) = sL\mathbf{I}(s) - Li(0) \qquad \text{13.10}$$

$$\mathbf{I}(s) = \frac{\mathbf{V}(s)}{sL} + \frac{i(0)}{s} \qquad \text{13.11}$$

The s-domain representation of this element is shown in Fig. 13.2c.

Using the passive sign convention, we find that the voltage–current relationships for the coupled inductors shown in Fig. 13.2d are

$$v_1(t) = L_1\frac{di_1(t)}{dt} + M\frac{di_2(t)}{dt} \qquad \text{13.12}$$

$$v_2(t) = L_2\frac{di_2(t)}{dt} + M\frac{di_1(t)}{dt}$$

The relationships in the s-domain are then

$$\mathbf{V}_1(s) = L_1 s\mathbf{I}_1(s) - L_1 i_1(0) + Ms\mathbf{I}_2(s) - Mi_2(0) \qquad \text{13.13}$$
$$\mathbf{V}_2(s) = L_2 s\mathbf{I}_2(s) - L_2 i_2(0) + Ms\mathbf{I}_1(s) - Mi_1(0)$$

Independent and dependent voltage and current sources can also be represented by their transforms; that is,

$$\mathbf{V}_1(s) = \mathcal{L}\big[v_1(t)\big]$$
$$\mathbf{I}_2(s) = \mathcal{L}\big[i_2(t)\big] \qquad \text{13.14}$$

and if $v_1(t) = Ai_2(t)$, which represents a current-controlled voltage source, then

$$\mathbf{V}_1(s) = A\mathbf{I}_2(s) \qquad \text{13.15}$$

Note carefully the direction of the current sources and the polarity of the voltage sources in the transformed network that result from the initial conditions. If the polarity of the initial voltage or direction of the initial current is reversed, the sources in the transformed circuit that results from the initial condition are also reversed.

LEARNING by Doing

D 13.2 The s-domain model for the network in (a) for $t > 0$ is shown in (b). Find the s-domain expression for the voltage across the inductor.

ANSWER $\mathbf{V}(s) = \dfrac{-1}{s+1}$.

(a)   (b)

13.3 Analysis Techniques

Now that we have the *s*-domain representation for the circuit elements, we are in a position to analyze networks using a transformed circuit.

LEARNING Example 13.1

Given the network in Fig. 13.3a, let us draw the *s*-domain equivalent circuit and find the output voltage in both the *s* and time domains.

SOLUTION   The *s*-domain network is shown in Fig. 13.3b. We can write the output voltage as

$$\mathbf{V}_o(s) = \left[R // \frac{1}{sC} \right] \mathbf{I}_S(s)$$

or

$$\mathbf{V}_o(s) = \left[\frac{1/C}{s + (1/RC)} \right] \mathbf{I}_S(s)$$

Given the element values, $\mathbf{V}_o(s)$ becomes

$$\mathbf{V}_o(s) = \left(\frac{40,000}{s + 4} \right) \left(\frac{0.003}{s + 1} \right) = \frac{120}{(s + 4)(s + 1)}$$

Expanding $\mathbf{V}_o(s)$ into partial fractions yields

$$\mathbf{V}_o(s) = \frac{120}{(s + 4)(s + 1)} = \frac{40}{s + 1} - \frac{40}{s + 4}$$

Performing the inverse Laplace transform yields the time domain representation

$$v_o(t) = 40\left[e^{-t} - e^{-4t} \right] u(t) \text{ V}$$

Figure 13.3
*Time domain
and s-domain
representations of an
RC parallel network.*

$i_S(t) = 3e^{-t}u(t) \text{ mA}$ $R = 10 \text{ k}\Omega$ $C = 25 \text{ μF}$ $v_o(t)$

(a)

$\mathbf{I}_S(s) = \dfrac{3}{s+1}$ $R = 10 \text{ k}$ $\dfrac{1}{sC} = \dfrac{40000}{s}$ $\mathbf{V}_o(s)$

(b)

Now that we have demonstrated the use of the Laplace transform in the solution of a simple circuit, let us consider the more general case. Note that in Fig. 13.2 we have shown two models for the capacitor and inductor when initial conditions are present. Let us now consider an example in which we will illustrate the use of these models in deriving both the node and loop equations for the circuit.

LEARNING Example 13.2

Given the circuits in Figs. 13.4a and b, we wish to write the mesh equations in the *s*-domain for the network in Fig. 13.4a and the node equations in the *s*-domain for the network in Fig. 13.4b.

SOLUTION   The transformed circuit for the network in Fig. 13.4a is shown in Fig. 13.4c. The mesh equations for this network are

$$\left(R_1 + \frac{1}{sC_1} + \frac{1}{sC_2} + sL_1 \right) \mathbf{I}_1(s) - \left(\frac{1}{sC_2} + sL_1 \right) \mathbf{I}_2(s)$$

$$= \mathbf{V}_A(s) - \frac{v_1(0)}{s} + \frac{v_2(0)}{s} - L_1 i_1(0)$$

$$-\left(\frac{1}{sC_2} + sL_1 \right) \mathbf{I}_1(s) + \left(\frac{1}{sC_2} + sL_1 + sL_2 + R_2 \right) \mathbf{I}_2(s)$$

$$= L_1 i_1(0) - \frac{v_2(0)}{s} - L_2 i_2(0) + \mathbf{V}_B(s)$$

The transformed circuit for the network in Fig. 13.4b is shown in Fig. 13.4d. The node equations for this network are

$$\left(G_1 + \frac{1}{sL_1} + sC_1 + \frac{1}{sL_2} \right) \mathbf{V}_1(s) - \left(\frac{1}{sL_2} + sC_1 \right) \mathbf{V}_2(s)$$

$$= \mathbf{I}_A(s) - \frac{i_1(0)}{s} + \frac{i_2(0)}{s} - C_1 v_1(0)$$

(continued)

$$-\left(\frac{1}{sL_2} + sC_1\right)\mathbf{V}_1(s) + \left(\frac{1}{sL_2} + sC_1 + G_2 + sC_2\right)\mathbf{V}_2(s)$$

$$= C_1 v_1(0) - \frac{i_2(0)}{s} - C_2 v_2(0) + \mathbf{I}_B(s)$$

LEARNING Hint

Note that the equations employ the same convention as that used in dc analysis.

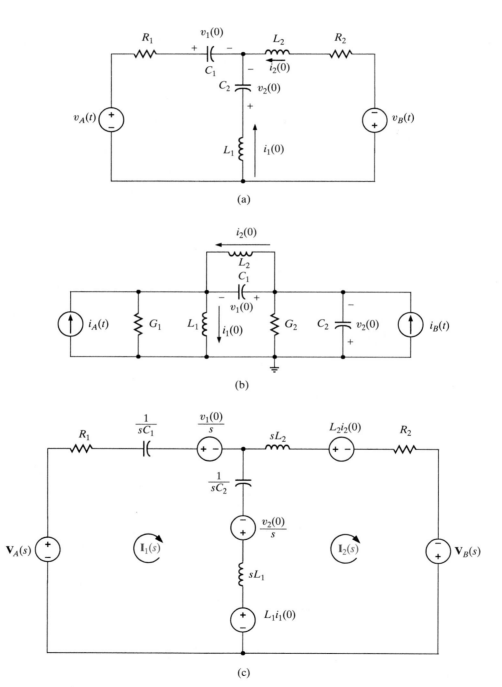

Figure 13.4 *Circuits used in Example 13.2.*

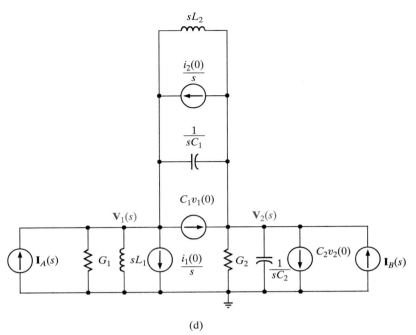

(d)

Figure 13.4 *(continued)*

Example 13.2 attempts to illustrate the manner in which to employ the two *s*-domain representations of the inductor and capacitor circuit elements when initial conditions are present. In the following examples we illustrate the use of a number of analysis techniques in obtaining the complete response of a transformed network. The circuits analyzed have been specifically chosen to demonstrate the application of the Laplace transform to circuits with a variety of passive and active elements.

LEARNING Example 13.3

Let us examine the network in Fig. 13.5a. We wish to determine the output voltage $v_o(t)$.

SOLUTION As a review of the analysis techniques presented earlier in this text, we will solve this problem using nodal analysis, mesh analysis, superposition, source exchange, Thévenin's theorem, and Norton's theorem.

The transformed network is shown in Fig. 13.5b. In our employment of nodal analysis, rather than writing KCL equations at the nodes labeled $\mathbf{V}_1(s)$ and $\mathbf{V}_o(s)$, we will use only the former node and use voltage division to find the latter.

KCL at the node labeled $\mathbf{V}_1(s)$ is

$$-\frac{4}{s} + \frac{\mathbf{V}_1(s) - \dfrac{12}{s}}{s} + \frac{\mathbf{V}_1(s)}{\dfrac{1}{s} + 2} = 0$$

Solving for $\mathbf{V}_1(s)$ we obtain

$$\mathbf{V}_1(s) = \frac{4(s + 3)(2s + 1)}{s(s^2 + 2s + 1)}$$

Now employing voltage division

$$\mathbf{V}_o(s) = \mathbf{V}_1(s)\left[\frac{2}{\dfrac{1}{s} + 2}\right] = \mathbf{V}_1(s)\left(\frac{2s}{2s + 1}\right)$$

$$= \frac{8(s + 3)}{(s + 1)^2}$$

In our mesh analysis we note that the current $\mathbf{I}_1(s)$ goes through the current source, and therefore KVL for the right-hand loop is

$$\frac{12}{s} - \left[\mathbf{I}_2(s) - \mathbf{I}_1(s)\right]s - \frac{\mathbf{I}_2(s)}{s} - 2\mathbf{I}_2(s) = 0$$

(continued)

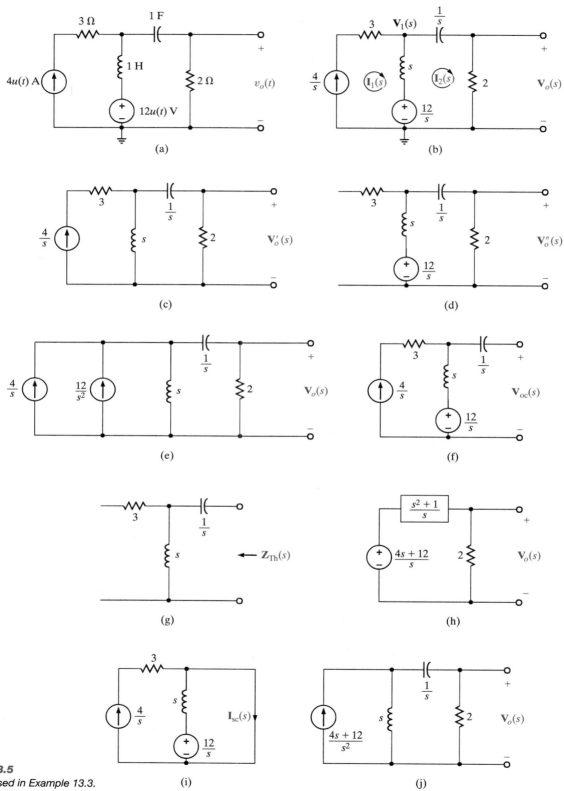

Figure 13.5
Circuits used in Example 13.3.

However, $\mathbf{I}_1(s) = 4/s$, and hence

$$\mathbf{I}_2(s) = \frac{4(s+3)}{(s+1)^2}$$

Therefore,

$$\mathbf{V}_o(s) = \frac{8(s+3)}{(s+1)^2}$$

It is interesting to note that the 3-Ω resistor never enters our equations. Furthermore, it will not enter our other analyses either. Why?

In using superposition, we first consider the current source acting alone as shown in Fig. 13.5c. Applying current division, we obtain

$$\mathbf{V}_o'(s) = \left[\frac{\dfrac{4}{s}(s)}{s + \dfrac{1}{s} + 2} \right](2)$$

$$= \frac{8s}{s^2 + 2s + 1}$$

With the voltage source acting alone, as shown in Fig. 13.5d, we obtain

$$\mathbf{V}_o''(s) = \left[\frac{\dfrac{12}{s}}{s + \dfrac{1}{s} + 2} \right](2)$$

$$= \frac{24}{s^2 + 2s + 1}$$

Hence,

$$\mathbf{V}_o(s) = \mathbf{V}_o'(s) + \mathbf{V}_o''(s)$$

$$= \frac{8(s+3)}{(s+1)^2}$$

In applying source exchange, we transform the voltage source and series inductor into a current source with the inductor in parallel as shown in Fig. 13.5e. Adding the current sources and applying current division yields

$$\mathbf{V}_o(s) = \left(\frac{12}{s^2} + \frac{4}{s} \right) \left[\frac{s}{s + \dfrac{1}{s} + 2} \right](2)$$

$$= \frac{\left(\dfrac{12}{s} + 4 \right)(2)}{s + \dfrac{1}{s} + 2}$$

$$\mathbf{V}_o(s) = \frac{8(s+3)}{(s+1)^2}$$

To apply Thévenin's theorem, we first find the open-circuit voltage shown in Fig. 13.5f. $\mathbf{V}_{oc}(s)$ is then

$$\mathbf{V}_{oc}(s) = \left(\frac{4}{s} \right)(s) + \frac{12}{s}$$

$$= \frac{4s + 12}{s}$$

The Thévenin equivalent impedance derived from Fig. 13.5g is

$$\mathbf{Z}_{Th}(s) = \frac{1}{s} + s$$

$$= \frac{s^2 + 1}{s}$$

Now, connecting the Thévenin equivalent circuit to the load produces the circuit shown in Fig. 13.5h. Then applying voltage division we obtain

$$\mathbf{V}_o(s) = \frac{4s + 12}{s} \left[\frac{2}{\dfrac{s^2 + 1}{s} + 2} \right]$$

$$= \frac{8(s+3)}{(s+1)^2}$$

In applying Norton's theorem, for simplicity we break the network to the right of the first mesh. In this case, the short-circuit current is obtained from the circuit in Fig. 13.5i, that is,

$$\mathbf{I}_{sc}(s) = \frac{\dfrac{12}{s}}{s} + \frac{4}{s}$$

$$= \frac{4s + 12}{s^2}$$

The Thévenin equivalent impedance in this application of Norton's theorem is $\mathbf{Z}_{Th}(s) = s$. Connecting the Norton equivalent circuit to the remainder of the original network yields the circuit in Fig. 13.5j. Then

$$\mathbf{V}_o(s) = \frac{4s + 12}{s^2} \left[\frac{s}{s + \dfrac{1}{s} + 2} \right](2)$$

$$= \frac{8(s+3)}{(s+1)^2}$$

Finally, $\mathbf{V}_o(s)$ can now be transformed to $v_o(t)$. $\mathbf{V}_o(s)$ can be written as

$$\mathbf{V}_o(s) = \frac{8(s+3)}{(s+1)^2} = \frac{K_{11}}{(s+1)^2} + \frac{K_{12}}{s+1}$$

(continued)

Evaluating the constants, we obtain

$$8(s + 3)|_{s=-1} = K_{11}$$

$$16 = K_{11}$$

and

$$\frac{d}{ds}[8(s + 3)]\Big|_{s=-1} = K_{12}$$

$$8 = K_{12}$$

Therefore,

$$v_o(t) = (16te^{-t} + 8e^{-t})u(t) \text{ V}$$

LEARNING Example 13.4

Consider the network shown in Fig. 13.6a. We wish to determine the output voltage $v_o(t)$.

SOLUTION As we begin to attack the problem we note two things. First, because the source $12u(t)$ is connected between $v_1(t)$ and $v_2(t)$, we have a supernode. Second, if $v_2(t)$ is known, $v_o(t)$ can be easily obtained by voltage division. Hence, we will use nodal analysis in conjunction with voltage division to obtain a solution. Then for purposes of comparison, we will find $v_o(t)$ using Thévenin's theorem.

The transformed network is shown in Fig. 13.6b. KCL for the supernode is

LEARNING Hint

Summing the currents leaving the supernode

$$\frac{V_1(s)}{2} + V_1(s)\frac{s}{2} - 2I(s) + \frac{V_2(s)}{s + 1} = 0$$

However,

$$I(s) = -\frac{V_1(s)}{2}$$

and

$$V_1(s) = V_2(s) - \frac{12}{s}$$

Substituting the last two equations into the first equation yields

$$\left[V_2(s) - \frac{12}{s}\right]\frac{s + 3}{2} + \frac{V_2(s)}{s + 1} = 0$$

or

$$V_2(s) = \frac{12(s + 1)(s + 3)}{s(s^2 + 4s + 5)}$$

Employing a voltage divider, we obtain

$$V_o(s) = V_2(s)\frac{1}{s + 1}$$

$$= \frac{12(s + 3)}{s(s^2 + 4s + 5)}$$

To apply Thévenin's theorem, we break the network to the right of the dependent current source as shown in Fig. 13.6c. KCL for the supernode is

(a)

(b)

(c)

(d)

(e)

Figure 13.6 *Circuits used in Example 13.4.*

$$\frac{\mathbf{V}_{oc}(s) - \dfrac{12}{s}}{2} + \frac{\mathbf{V}_{oc}(s) - \dfrac{12}{s}}{\dfrac{2}{s}} - 2\mathbf{I}'(s) = 0$$

where

$$\mathbf{I}'(s) = -\left(\frac{\mathbf{V}_{oc}(s) - \dfrac{12}{s}}{2}\right)$$

Solving these equations for $\mathbf{V}_{oc}(s)$ yields

$$\mathbf{V}_{oc}(s) = \frac{12}{s}$$

The short-circuit current is derived from the network in Fig. 13.6d as

$$\mathbf{I}_{sc}(s) = 2\mathbf{I}''(s) + \frac{\dfrac{12}{s}}{(2)\left(\dfrac{2}{s}\right)}{2 + \dfrac{2}{s}}$$

where

$$\mathbf{I}''(s) = \frac{\dfrac{12}{s}}{2}$$

Solving these equations for $\mathbf{I}_{sc}(s)$ yields

$$\mathbf{I}_{sc}(s) = \frac{6(s + 3)}{s}$$

The Thévenin equivalent impedance is then

$$\mathbf{Z}_{Th}(s) = \frac{\mathbf{V}_{oc}(s)}{\mathbf{I}_{sc}(s)}$$

$$= \frac{\dfrac{12}{s}}{\dfrac{6(s + 3)}{s}}$$

$$= \frac{2}{s + 3}$$

If we now connect the Thévenin equivalent circuit to the remainder of the original network, we obtain the circuit shown in Fig. 13.6e. Using voltage division,

$$\mathbf{V}_{oc}(s) = \frac{1}{\dfrac{2}{s + 3} + s + 1}\left(\frac{12}{s}\right)$$

$$= \frac{12(s + 3)}{s(s^2 + 4s + 5)}$$

or

$$\mathbf{V}_{oc}(s) = \frac{12(s + 3)}{s(s + 2 - j1)(s + 2 + j1)}$$

To obtain the inverse transform, the function is written as

$$\frac{12(s + 3)}{s(s + 2 - j1)(s + 2 + j1)}$$

$$= \frac{K_0}{s} + \frac{K_1}{s + 2 - j1} + \frac{K_1^*}{s + 2 + j1}$$

Evaluating the constants, we obtain

$$\frac{12(s + 3)}{s^2 + 4s + 5}\bigg|_{S=0} = K_0$$

$$\frac{36}{5} = K_0$$

and

$$\frac{12(s + 3)}{s(s + 2 + j1)}\bigg|_{S=-2+j1} = K_1$$

$$3.79\underline{/161.57°} = K_1$$

Therefore,

$$v_o(t) = \left[7.2 + 7.58e^{-2t}\cos(t + 161.57°)\right]u(t)\ \text{V}$$

LEARNING EXTENSIONS

E13.1 Find $i_o(t)$ in the network in Fig. E13.1 using node equations.

Figure E13.1

ANSWER
$i_o(t) = 6.53e^{-t/4}\cos\left[(\sqrt{15}/4)t - 156.72°\right]u(t)\ \text{A}.$

E13.2   Find $v_o(t)$ in the network in Fig. E13.2 using loop equations.

Figure E13.2

ANSWER

$$v_o(t) = \left(4 - 8.93e^{-3.73t} + 4.93e^{-0.27t}\right)u(t) \text{ V}.$$

We will now illustrate the use of the Laplace transform in the transient analysis of circuits. We will analyze networks such as those considered in Chapter 6. Our approach will first be to determine the initial conditions for the capacitors and inductors in the network, and then we will employ the element models that were specified at the beginning of this chapter together with the circuit analysis techniques to obtain a solution. The following example demonstrates the approach.

LEARNING Example 13.5

Let us determine the output voltage of the network shown in Fig. 13.7a for $t > 0$.

SOLUTION   At $t = 0$ the initial voltage across the capacitor is 1 V, and the initial current drawn through the inductor is 1 A. The circuit for $t > 0$ is shown in Fig. 13.7b with the initial conditions. The transformed network is shown in Fig. 13.7c.

The mesh equations for the transformed network are

$$(s + 1)\mathbf{I}_1(s) - s\mathbf{I}_2(s) = \frac{4}{s} + 1$$

$$-s\mathbf{I}_1(s) + \left(s + \frac{2}{s} + 1\right)\mathbf{I}_2(s) = \frac{-1}{s} - 1$$

which can be written in matrix form as

$$\begin{bmatrix} s + 1 & -s \\ -s & \dfrac{s^2 + s + 2}{s} \end{bmatrix} \begin{bmatrix} \mathbf{I}_1(s) \\ \mathbf{I}_2(s) \end{bmatrix} = \begin{bmatrix} \dfrac{s + 4}{s} \\ \dfrac{-(s + 1)}{s} \end{bmatrix}$$

Solving for the currents, we obtain

$$\begin{bmatrix} \mathbf{I}_1(s) \\ \mathbf{I}_2(s) \end{bmatrix} = \begin{bmatrix} s + 1 & -s \\ -s & \dfrac{s^2 + s + 2}{s} \end{bmatrix}^{-1} \begin{bmatrix} \dfrac{s + 4}{s} \\ \dfrac{-(s + 1)}{s} \end{bmatrix}$$

$$= \frac{s}{2s^2 + 3s + 2} \begin{bmatrix} \dfrac{s^2 + s + 2}{s} & s \\ s & s + 1 \end{bmatrix} \begin{bmatrix} \dfrac{s + 4}{s} \\ \dfrac{-(s + 1)}{s} \end{bmatrix}$$

$$= \begin{bmatrix} \dfrac{4s^2 + 6s + 8}{s(2s^2 + 3s + 2)} \\ \dfrac{2s - 1}{2s^2 + 3s + 2} \end{bmatrix}$$

The output voltage is then

$$\mathbf{V}_o(s) = \frac{2}{s}\mathbf{I}_2(s) + \frac{1}{s}$$

$$= \frac{2}{s}\left(\frac{2s - 1}{2s^2 + 3s + 2}\right) + \frac{1}{s}$$

$$= \frac{s + \frac{7}{2}}{s^2 + \frac{3}{2}s + 1}$$

This function can be written in a partial fraction expansion as

$$\frac{s + \frac{7}{2}}{s^2 + \frac{3}{2}s + 1} = \frac{K_1}{s + \frac{3}{4} - j(\sqrt{7}/4)} + \frac{K_1^*}{s + \frac{3}{4} + j(\sqrt{7}/4)}$$

Evaluating the constants, we obtain

$$\left. \frac{s + \frac{7}{2}}{s + \frac{3}{4} + j(\sqrt{7}/4)} \right|_{s = -(3/4) + j(\sqrt{7}/4)} = K_1$$

$$2.14 \underline{/-76.5^\circ} = K_1$$

Therefore,

$$v_o(t) = \left[4.29e^{-(3/4)t} \cos\left(\frac{\sqrt{7}}{4}t - 76.5^\circ\right) \right]u(t) \text{ V}$$

Figure 13.7 *Circuits employed in Example 13.5.*

E13.3 Solve Learning Extension E6.2 using Laplace transforms.

ANSWER $i_1(t) = 1e^{-9t}$ A.

E13.4 Solve Learning Extension E6.4 using Laplace transforms.

ANSWER
$v_o(t) = 6 - \frac{10}{3} e^{-2t}$ V.

13.4 Transfer Function

In Chapter 11 we introduced the concept of network or transfer function. It is essentially nothing more than the ratio of some output variable to some input variable. If both variables are voltages, the transfer function is a voltage gain. If both variables are currents, the transfer function is a current gain. If one variable is a voltage and the other is a current, the transfer function becomes a transfer admittance or impedance.

In deriving a transfer function, all initial conditions are set equal to zero. In addition, if the output is generated by more than one input source in a network, superposition can be employed in conjunction with the transfer function for each source.

To present this concept in a more formal manner, let us assume that the input/output relationship for a linear circuit is

$$b_n \frac{d^n y_o(t)}{dt^n} + b_{n-1} \frac{d^{n-1} y_o(t)}{dt^{n-1}} + \cdots + b_1 \frac{dy_o(t)}{dt} + b_0 y_o(t)$$

$$= a_m \frac{d^m x_i(t)}{dt^m} + a_{m-1} \frac{d^{m-1} x_i(t)}{dt^{m-1}} + \cdots + a_1 \frac{dx_i(t)}{dt} + a_0 x_i(t)$$

If all the initial conditions are zero, the transform of the equation is

$$(b_n s^n + b_{n-1} s^{n-1} + \cdots + b_1 s + b_0)\mathbf{Y}_o(s) = (a_m s^m + a_{m-1} s^{m-1} + \cdots + a_1 s + a_0)\mathbf{X}_i(s)$$

or

$$\frac{\mathbf{Y}_o(s)}{\mathbf{X}_i(s)} = \frac{a_m s^m + a_{m-1} s^{m-1} + \cdots + a_1 s + a_0}{b_n s^n + b_{n-1} s^{n-1} + \cdots + b_1 s + b_0}$$

This ratio of $\mathbf{Y}_o(s)$ to $\mathbf{X}_i(s)$ is called the *transfer* or *network function*, which we denote as $\mathbf{H}(s)$; that is,

$$\frac{\mathbf{Y}_o(s)}{\mathbf{X}_i(s)} = \mathbf{H}(s)$$

or

$$\mathbf{Y}_o(s) = \mathbf{H}(s)\mathbf{X}_i(s) \qquad\qquad \textbf{13.16}$$

This equation states that the output response $\mathbf{Y}_o(s)$ is equal to the network function multiplied by the input $\mathbf{X}_i(s)$. Note that if $x_i(t) = \delta(t)$ and therefore $\mathbf{X}_i(s) = 1$, the impulse response is equal to the inverse Laplace transform of the network function. This is an extremely important concept because it illustrates that if we know the impulse response of a network, we can find the response due to some other forcing function using Eq. (13.16).

At this point it is informative to review briefly the natural response of both first-order and second-order networks. We demonstrated in Chapter 6 that if only a single storage element is present, the natural response of a network to an initial condition is always of the form

$$x(t) = X_0 e^{-t/\tau}$$

where $x(t)$ can be either $v(t)$ or $i(t)$, X_0 is the initial value of $x(t)$, and τ is the time constant of the network. We also found that the natural response of a second-order network is controlled by the roots of the *characteristic equation*, which is of the form

$$s^2 + 2\zeta\omega_0 s + \omega_0^2 = 0$$

where ζ is the *damping ratio* and ω_0 is the *undamped natural frequency*. These two key factors, ζ and ω_0, control the response, and there are basically three cases of interest.

Case 1, $\zeta > 1$: Overdamped Network   The roots of the characteristic equation are $s_1, s_2 = -\zeta\omega_0 \pm \omega_0 \sqrt{\zeta^2 - 1}$ and, therefore, the network response is of the form

$$x(t) = K_1 e^{-(\zeta\omega_0 + \omega_0 \sqrt{\zeta^2-1})t} + K_2 e^{-(\zeta\omega_0 - \omega_0 \sqrt{\zeta^2-1})t}$$

Case 2, $\zeta > 1$: Underdamped Network   The roots of the characteristic equation are $s_1, s_2 = -\zeta\omega_0 \pm j\omega_0 \sqrt{1 - \zeta^2}$ and, therefore, the network response is of the form

$$x(t) = K e^{-\zeta\omega_0 t} \cos(\omega_0 \sqrt{1 - \zeta^2}\, t + \phi)$$

Case 3, $\zeta = 1$: Critically Damped Network   The roots of the characteristic equation are $s_1, s_2 = -\omega_0$ and, hence, the response is of the form

$$x(t) = K_1 t e^{-\omega_0 t} + K_2 e^{-\omega_0 t}$$

It is very important that the reader note that the characteristic equation is the denominator of the transfer function $\mathbf{H}(s)$, and the roots of this equation, which are the poles of the network, determine the form of the network's natural response.

A convenient method for displaying the network's poles and zeros in graphical form is the use of a pole-zero plot. A pole-zero plot of a function can be accomplished using what is commonly called the *complex* or *s-plane*. In the complex plane the abscissa is σ and the ordinate is $j\omega$. Zeros are represented by 0's and poles are represented by $\times$'s. Although we are concerned only with the finite poles and zeros specified by the network or response function, we should point out that a rational function must have the same number of poles and zeros. Therefore, if $n > m$, there are $n - m$ zeros at the point at infinity, and if $n < m$, there are $m - n$ poles at the point at infinity. A systems engineer can tell a lot about the operation of a network or system by simply examining its pole-zero plot.

In order to correlate the natural response of a network to an initial condition with the network's pole locations, we have illustrated in Fig. 13.8 the correspondence for all three cases: overdamped, underdamped, and critically damped. Note that if the network poles are real and unequal, the response is slow and, therefore, $x(t)$ takes a long time to reach zero. If the network poles are complex conjugates, the response is fast; however, it overshoots and is eventually damped out. The dividing line between the overdamped and underdamped cases is the critically damped case in which the roots are real and equal. In this case the transient response dies out as quickly as possible, with no overshoot.

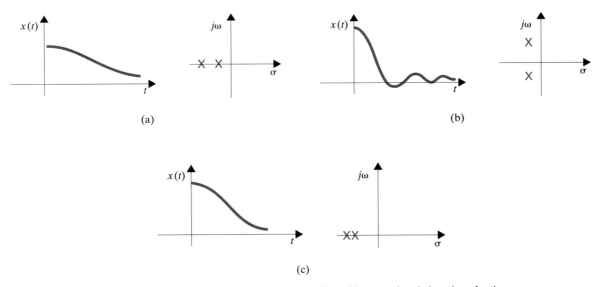

(a)

(b)

(c)

Figure 13.8   *Natural response of a second-order network together with network pole locations for the three cases: (a) overdamped, (b) underdamped, and (c) critically damped.*

LEARNING Example 13.6

If the impulse response of a network is $h(t) = e^{-t}$, let us determine the response $v_o(t)$ to an input $v_i(t) = 10e^{-2t}u(t)$ V.

SOLUTION   The transformed variables are

$$\mathbf{H}(s) = \frac{1}{s + 1}$$

$$\mathbf{V}_i(s) = \frac{10}{s + 2}$$

(continued)

Therefore,

$$\mathbf{V}_o(s) = \mathbf{H}(s)\mathbf{V}_i(s)$$

$$= \frac{10}{(s+1)(s+2)}$$

and hence,

$$v_o(t) = 10(e^{-t} - e^{-2t})u(t) \text{ V}$$

The importance of the transfer function stems from the fact that it provides the systems engineer with a great deal of knowledge about the system's operation, since its dynamic properties are governed by the system poles.

LEARNING Example 13.7

Let us derive the transfer function $\mathbf{V}_o(s)/\mathbf{V}_i(s)$ for the network in Fig. 13.9a.

SOLUTION Our output variable is the voltage across a variable capacitor and the input voltage is a unit step. The transformed network is shown in Fig. 13.9b. The mesh equations for the network are

$$2\mathbf{I}_1(s) - \mathbf{I}_2(s) = \mathbf{V}_i(s)$$

$$-\mathbf{I}_1(s) + \left(s + \frac{1}{sC} + 1\right)\mathbf{I}_2(s) = 0$$

and the output equation is

$$\mathbf{V}_o(s) = \frac{1}{sC}\mathbf{I}_2(s)$$

From these equations we find that the transfer function is

$$\frac{\mathbf{V}_o(s)}{\mathbf{V}_i(s)} = \frac{1/2C}{s^2 + \frac{1}{2}s + 1/C}$$

Since the transfer function is dependent on the value of the capacitor, let us examine the transfer function and the output response for three values of the capacitor.

(a) $C = 8 \text{ F}$

$$\frac{\mathbf{V}_o(s)}{\mathbf{V}_i(s)} = \frac{\frac{1}{16}}{\left(s^2 + \frac{1}{2}s + \frac{1}{8}\right)}$$

$$= \frac{\frac{1}{16}}{\left(s + \frac{1}{4} - j\frac{1}{4}\right)\left(s + \frac{1}{4} + j\frac{1}{4}\right)}$$

The output response is

$$\mathbf{V}_o(s) = \frac{\frac{1}{16}}{s\left(s + \frac{1}{4} - j\frac{1}{4}\right)\left(s + \frac{1}{4} + j\frac{1}{4}\right)}$$

As illustrated in Chapter 6, the poles of the transfer function, which are the roots of the characteristic equation, are complex conjugates, as shown in Fig. 13.9c; therefore, the output response will be *underdamped*. The output response as a function of time is

$$v_o(t) = \left[\frac{1}{2} + \frac{1}{\sqrt{2}}e^{-t/4}\cos\left(\frac{t}{4} + 135°\right)\right]u(t) \text{ V}$$

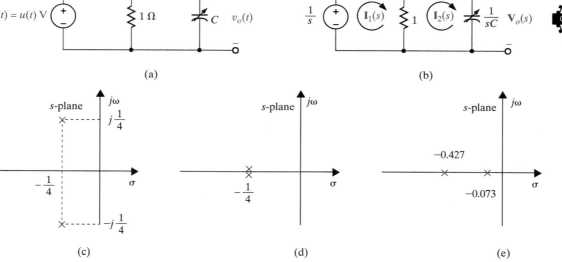

(a)   (b)   (c)   (d)   (e)

Figure 13.9   Networks and pole zero plots used in Example 13.7.

Note that for large values of time the transient oscillations, represented by the second term in the response, become negligible and the output settles out to a value of $\frac{1}{2}$ V. This can also be seen directly from the circuit since for large values of time the input looks like a dc source, the inductor acts like a short circuit, the capacitor acts like an open circuit, and the resistors form a voltage divider.

(b) C = 16 F

$$\frac{\mathbf{V}_o(s)}{\mathbf{V}_i(s)} = \frac{\frac{1}{32}}{s^2 + \frac{1}{2}s + \frac{1}{16}}$$

$$= \frac{\frac{1}{32}}{\left(s + \frac{1}{4}\right)^2}$$

The output response is

$$\mathbf{V}_o(s) = \frac{\frac{1}{32}}{s\left(s + \frac{1}{4}\right)^2}$$

Since the poles of the transfer function are real and equal as shown in Fig. 13.9d, the output response will be *critically damped*. $v_o(t) = \mathcal{L}^{-1}\big[\mathbf{V}_o(s)\big]$ is

$$v_o(t) = \left[\frac{1}{2} - \left(\frac{t}{8} + \frac{1}{2}\right)e^{-t/4}\right]u(t) \text{ V}$$

(c) C = 32 F

$$\frac{\mathbf{V}_o(s)}{\mathbf{V}_i(s)} = \frac{\frac{1}{64}}{s^2 + \frac{1}{2}s + \frac{1}{32}}$$

$$= \frac{\frac{1}{64}}{(s + 0.427)(s + 0.073)}$$

The output response is

$$\mathbf{V}_o(s) = \frac{\frac{1}{64}}{s(s + 0.427)(s + 0.073)}$$

The poles of the transfer function are real and unequal, as shown in Fig. 13.9e and, therefore, the output response will be *over-damped*. The response as a function of time is

$$v_o(t) = \left(0.5 + 0.103e^{-0.427t} - 0.603e^{-0.073t}\right)u(t) \text{ V}$$

Although the values selected for the network parameters are not very practical, remember that both magnitude and frequency scaling, as outlined in Chapter 11, can be applied here also.

LEARNING Example 13.8

For the network in Fig. 13.10a let us compute (a) the transfer function, (b) the type of damping exhibited by the network, and (c) the unit step response.

SOLUTION   Recall that the voltage across the op-amp input terminals is zero and therefore KCL at the node labeled $\mathbf{V}_1(s)$ in Fig. 13.10b yields the following equation:

$$\frac{\mathbf{V}_S(s) - \mathbf{V}_1(s)}{1} = s\mathbf{V}_1(s) + \frac{\mathbf{V}_1(s) - \mathbf{V}_o(s)}{1} + \frac{\mathbf{V}_1(s)}{1}$$

Since the current into the negative input terminal of the op-amp is zero, KCL requires that

$$s\mathbf{V}_o(s) = -\frac{\mathbf{V}_1(s)}{1}$$

Combining the two equations yields the transfer function

$$\frac{\mathbf{V}_o(s)}{\mathbf{V}_S(s)} = \frac{-1}{s^2 + 3s + 1}$$

which can be expressed in the form

$$\frac{\mathbf{V}_o(s)}{\mathbf{V}_S(s)} = \frac{-1}{(s + 2.62)(s + 0.38)}$$

Since the roots are real and unequal, the step response of the network will be overdamped. The step response is

$$\mathbf{V}_o(s) = \frac{-1}{s(s + 2.62)(s + 0.38)}$$

$$= \frac{-1}{s} + \frac{-0.17}{s + 2.62} + \frac{1.17}{s + 0.38}$$

Therefore,

$$v_o(t) = \left(-1 - 0.17e^{-2.62t} + 1.17e^{-0.38t}\right)u(t) \text{ V}$$

Figure 13.10
Circuits used in
Example 13.8.

(a)   (b)

LEARNING EXTENSIONS

E13.5 If the unit impulse response of a network is known to be $\frac{10}{9}(e^{-t} - e^{-10t})$, determine the unit step response.

ANSWER $x(t) = (1 - \frac{10}{9}e^{-t} + \frac{1}{9}e^{-10t})u(t)$.

E13.6 The transfer function for a network is

$$\mathbf{H}(s) = \frac{s + 10}{s^2 + 4s + 8}$$

Determine the pole-zero plot of $\mathbf{H}(s)$, the type of damping exhibited by the network, and the unit step response of the network.

ANSWER The network is underdamped;
$$x(t) = \left[\frac{10}{8} + 1.46e^{-2t}\cos(2t - 210.96°)\right]u(t).$$

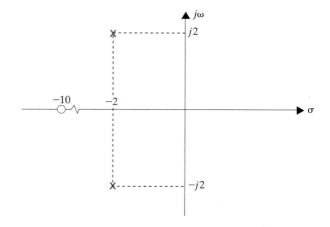

Recall from our previous discussion that if a second-order network is underdamped, the characteristic equation of the network is of the form

$$s^2 + 2\zeta\omega_0 s + \omega_0^2 = 0$$

and the roots of this equation, which are the network poles, are of the form

$$s_1, s_2 = -\zeta\omega_0 \pm j\omega_0 \sqrt{1 - \zeta^2}$$

The roots s_1 and s_2, when plotted in the s-plane, generally appear as shown in Fig. 13.11, where

$\zeta = $ damping ratio

$\omega_0 = $ undamped natural frequency

and as shown in Fig. 13.11,

$$\zeta = \cos\theta$$

It is important to note that the damping ratio and the undamped natural frequency are exactly the same quantities as those employed in Chapter 11 when determining a network's frequency response. We find that it is these same quantities that govern the network's transient response.

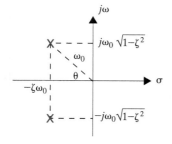

Figure 13.11
Pole locations for second-order underdamped network.

LEARNING Example 13.9

Let us examine the effect of pole position in the s-plane on the transient response of the second-order RLC series network shown in Fig. 13.12.

Figure 13.12
RLC series network.

SOLUTION The voltage gain transfer function is

$$\mathbf{G}_v(s) = \cfrac{\cfrac{1}{LC}}{s^2 + s\left(\cfrac{R}{L}\right) + \cfrac{1}{LC}} = \cfrac{\omega_0^2}{s^2 + 2\zeta\omega_0 s + \omega_0^2}$$

For this analysis we will let $\omega_0 = 2000$ rad/s for $\zeta = 0.25, 0.50, 0.75$, and 1.0. From the preceding equation we see that

$$LC = \frac{1}{\omega_0^2} = 2.5 \times 10^{-7}$$

and

$$R = 2\zeta\sqrt{\frac{L}{C}}$$

If we arbitrarily let $L = 10$ mH, then $C = 25$ μF. Also, for $\zeta = 0.25, 0.50, 0.75$, and 1.0, $R = 10\ \Omega, 20\ \Omega, 30\ \Omega$, and $40\ \Omega$, respectively. Over the range of ζ values, the network ranges from underdamped to critically damped. Since poles are complex for underdamped systems, the real and imaginary components and the magnitude of the poles of $\mathbf{G}_v(s)$ are given in Table 13.1 for the ζ values listed previously.

Table 13.1 Pole locations for $\zeta = 0.25$ to 1.0

Damping Ratio	Real	Imaginary	Magnitude
1.00	2000.0	0.0	2000.0
0.75	1500.0	1322.9	2000.0
0.50	1000.0	1732.1	2000.0
0.25	500.0	1936.5	2000.0

Fig. 13.13 shows the pole-zero diagrams for each value of ζ. Note first that all the poles lie on a circle; thus, the pole magnitudes are constant, consistent with Table 13.1. Second, as ζ decreases, the real part of the pole decreases while the imaginary part increases. In fact, when ζ goes to zero, the poles become imaginary.

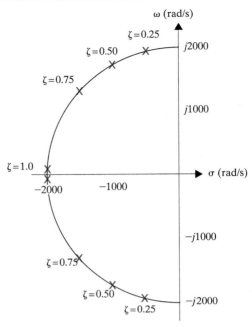

Figure 13.13 Pole-zero diagrams for $\zeta = 0.25$ to 1.0.

A PSPICE simulation of a unit step transient excitation for all four values of R is shown in Fig. 13.14. We see that as ζ decreases, the overshoot in the output voltage increases. Furthermore, when the network is critically damped ($\zeta = 1$) there is no overshoot at all. In most applications excessive overshoot is not desired. To correct this, the damping ratio, ζ, should be increased, which for this circuit would require an increase in the resistor value.

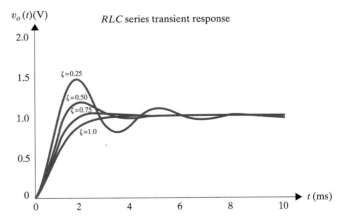

Figure 13.14 PSPICE *transient response output for* $\zeta = 0.25$ *to* 1.0.

LEARNING Example 13.10

Let us revisit the Tacoma Narrows Bridge disaster examined in Example 11.12. A photograph of the bridge as it collapsed is shown in Fig. 13.15.

In Chapter 11 we assumed that the bridge's demise was brought on by winds oscillating back and forth at a frequency near that of the bridge (0.2 Hz). We found that we could create an *RLC* circuit, shown in Fig. 11.29, that resonates at 0.2 Hz and has an output voltage consistent with the vertical deflection of the bridge. This kind of forced resonance never happened at Tacoma Narrows. The real culprit was not so much wind fluctuations, but the bridge itself. This is thoroughly explained in the paper "Resonance, Tacoma Narrows Bridge Failure, and Undergraduate Physics Textbooks," by K. Y. Billah and R. H. Scalan published in the *American Journal of Physics*, vol. 59, no. 2 p. 118–124, in which the authors determined that changes in wind speed affected the coefficients of the second-order differential equation that models the resonant behavior. In particular, the damping ratio, ζ, was dependent on the wind speed and is roughly given as

$$\zeta = 0.00460 - 0.00013U \qquad \textbf{13.17}$$

where U is the wind speed in mph. Note, as shown in Fig. 13.16, that ζ becomes negative at wind speeds in excess of 35 mph—a point we will demonstrate later. Furthermore, Billah and Scalan report that the bridge resonated in a twisting mode, which can be easily seen in Fig. 11.28 and is described by the differential equation

$$\frac{d^2\theta(t)}{dt^2} + 2\zeta\omega_0\frac{d\theta(t)}{dt} + \omega_0^2\theta(t) = 0$$

or

$$\ddot{\theta} + 2\zeta\omega_0\dot{\theta} + \omega_0^2\theta = 0 \qquad \textbf{13.18}$$

where $\theta(t)$ is the angle of twist in degrees and wind speed is implicit in ζ through Eq. (13.17). Billah and Scalan list the following data obtained either by direct observation at the bridge site or through scale model experiments afterward.

Wind speed at failure $\approx$ 42 mph

Twist at failure $\approx$ ±12°

Time to failure $\approx$ 45 minutes

Figure 13.15 *Tacoma Narrows Bridge as it collapsed on November 7, 1940. (Used with permission from Special Collection Division, University of Washington Libraries. Photo by Farguharson, negative number 12.)*

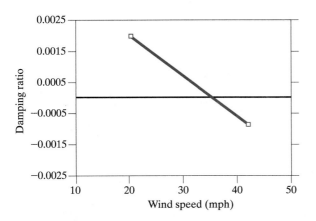

Figure 13.16 *Damping ratio versus wind speed for the second-order twisting model of the Tacoma Narrows Bridge.*

We will start the twisting oscillations using an initial condition on θ(0) and see whether the bridge oscillations decrease or increase over time. Let us now design a network that will simulate the true Tacoma Narrows disaster.

SOLUTION First, we solve for $\ddot{\theta}(t)$ in Eq. (13.18)

$$\ddot{\theta} = -2\zeta\omega_0\dot{\theta} - \omega_0^2\theta \qquad \textbf{13.19}$$

$$\ddot{\theta} = -2(2\pi)(0.2)(0.0046 - 0.00013U)\dot{\theta} - \left[2(2\pi)(0.2)\right]^2\theta$$

or

$$\ddot{\theta} = -(0.01156 - 0.00033U)\dot{\theta} - 1.579\theta$$

We now wish to model this equation to produce a voltage proportional to $\ddot{\theta}(t)$. We can accomplish this using the op-amp integrator circuit shown in Fig. 13.17.

The circuit's operation can perhaps be best understood by first assigning the voltage v_α to be proportional to $\ddot{\theta}(t)$, where 1 V represents 1 deg/s². Thus, the output of the first integrator, v_ω must be

$$v_\omega = -\frac{1}{R_\omega C_\omega}\int v_\alpha\, dt$$

or, since $R_\omega = 1\ \Omega$ and $C_\omega = 1\ F$,

$$v_\omega = -\int v_\alpha\, dt$$

So v_ω is proportional to $-\ddot{\theta}(t)$ and 1 V equals −1 deg/s. Similarly, the output of the second integrator must be

$$v_\theta = -\int v_\omega\, dt$$

where $v_\theta(t)$ is proportional to $\theta(t)$ and 1 V equals 1 degree. The outputs of the integrators are then fed back as inputs to the summing op-amp. Note that the dependent sources, E_ω and E_{wind}, re-create the coefficient on $\dot{\theta}(t)$ in Eq. (13.17); that is,

$$2\zeta\omega_0 = (2)(0.2)(2\pi)\zeta = 0.01156 - 0.00033U$$

To simulate various wind speeds, we need only change the gain factor of E_{wind}. Finally, we can solve the circuit for $v_\alpha(t)$.

$$v_\alpha(t) = -\left(\frac{R_f}{R_2}\right)(E_\omega - E_{wind}) - \left(\frac{R_f}{R_1}\right)v_\theta$$

which matches Eq. (13.19) if

$$\frac{R_f}{R_1} = \omega_0^2 = \left[2\pi(0.2)\right]^2 = 1.579$$

and

$$\frac{R_f}{R_2}\left[E_\omega - E_{wind}\right] = 2\zeta\omega_0$$

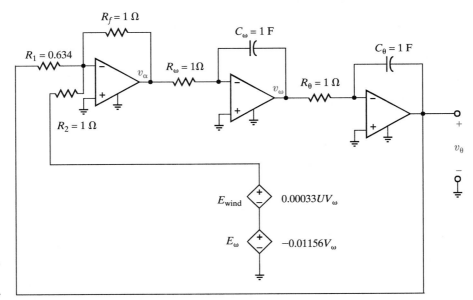

Figure 13.17 *Circuit diagram for Tacoma Narrows Bridge simulations.*

(continued)

or

$$\frac{R_f}{R_2} = 1$$

Thus, if $R_f = R_2 = 1$ and $R_1 = 0.634$, the circuit will simulate the bridge's twisting motion. We will start the twisting oscillations using an initial condition $\theta(0)$ and see whether the bridge oscillations decrease or increase over time.

The first simulation is for a wind speed of 20 mph and one degree of twist. The corresponding output voltage is shown in Fig. 13.18. Note that the bridge twists at a frequency of 0.2 Hz and the oscillations decrease exponentially, indicating a nondestructive situation.

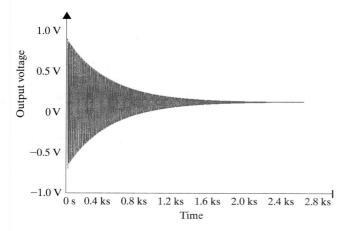

Figure 13.18   *Tacoma Narrows Bridge simulation at 20 mph wind speed and one degree twist initial condition.*

Fig. 13.19 shows the output for 35-mph winds and an initial twist of one degree. Notice that the oscillations neither increase nor decrease. This indicates that the damping ratio is zero.

Finally, the simulation at a wind speed of 42 mph and one degree initial twist is shown in Fig. 13.20. The twisting becomes worse and worse until after 45 minutes, the bridge is twisting

Figure 13.19   *Tacoma Narrows Bridge simulation at 35-mph winds and one degree of initial twist.*

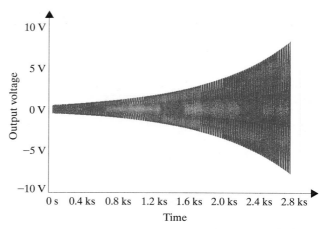

Figure 13.20   *Tacoma Narrows Bridge simulation at 42 mph wind speed and one degree of initial twist.*

± 12.5 degrees, which matches values reported by Billah and Scalan for collapse.

The dependency of the damping ratio on wind speed can also be demonstrated by investigating how the system poles change with the wind. The characteristic equation for the system is

$$s^2 + 2\zeta\omega_0 s + \omega_0^2 = 0$$

or

$$s^2 + (0.01156 - 0.00033U)s + 1.579 = 0$$

The roots of the characteristic equation yield the pole locations. Fig. 13.21 shows the system poles at wind speeds of 20, 35, and 42 mph. Note that at 20 mph, the stable situation is shown in Fig. 13.18, and the poles are in the left-half of the *s*-plane. At 35 mph ($\zeta = 0$) the poles are on the $j\omega$ axis and the system is oscillatory, as shown in Fig. 13.19. Finally, at 42 mph, we see that the poles are in the right half of the *s*-plane and, from Fig. 13.20 we know this is an unstable system. This relationship between pole location and transient response is true for all systems—right-half plane poles result in unstable systems.

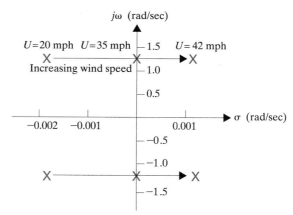

Figure 13.21   *Pole-zero plot for Tacoma Narrows Bridge second-order model at wind speeds of 20, 35, and 42 mph.*

13.5 Pole-Zero Plot/Bode Plot Connection

In Chapter 11 we introduced the Bode plot as an analysis tool for sinusoidal frequency-response studies. Let us now investigate the relationship between the s-plane pole-zero plot and the Bode plot. As an example, consider the transfer function of the RLC high-pass filter shown in Fig. 13.22.

$R = 2 \, \Omega \quad C = 0.2 \, F$

$v_{in}(t)$

$L = 1 \, H \quad v_o(t)$

Figure 13.22
RLC high-pass filter.

The transfer function is

$$\mathbf{G}_v(s) = \frac{sL}{sL + R + \dfrac{1}{sC}} = \frac{s^2}{s^2 + s\left(\dfrac{R}{L}\right) + \dfrac{1}{LC}}$$

Using the element values, the transfer function becomes

$$\mathbf{G}_v(s) = \frac{s^2}{s^2 + 2s + 5} = \frac{s^2}{(s + 1 + j2)(s + 1 - j2)}$$

We see that the transfer function has two zeros at the origin $(s = 0)$ and two complex-conjugate poles $s = -1 \pm j2$. The standard pole-zero plot for this function is shown in Fig. 13.23a. A three-dimensional s-plane plot of the magnitude of $\mathbf{G}_v(s)$ is shown in Fig. 13.23b. Note carefully that when $s = 0$, $\mathbf{G}_v(s) = 0$ and when $s = -1 \pm j2$, the function is infinite.

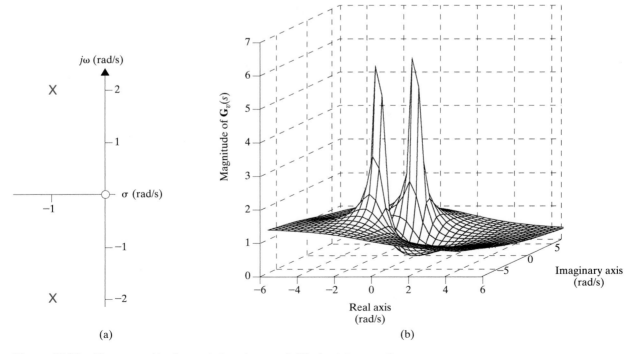

(a)

(b)

Figure 13.23 *Figures used to demonstrate pole-zero plot/Bode plot connection.*

(c)

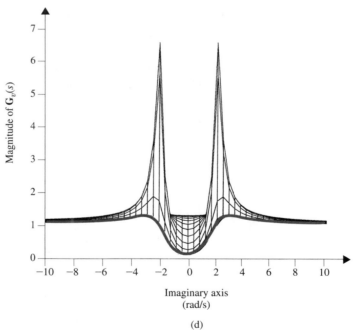

Figure 13.23
(continued)

(d)

Recall that the Bode plot of a transfer function's magnitude is in reality a plot of the magnitude of the gain versus frequency. The frequency domain, where $s = j\omega$, corresponds to the $j\omega$-axis in the s-plane obtained by setting σ, the real part of s, to zero. Thus, the frequency domain corresponds directly to that part of the s-domain where $\sigma = 0$, as illustrated in the three-dimensional plot in Fig. 13.23c.

Let us develop the Bode plot by first rotating Fig. 13.23c such that the real axis is perpendicular to the page as shown in Fig. 13.23d. Note that the transfer function maximum occurs

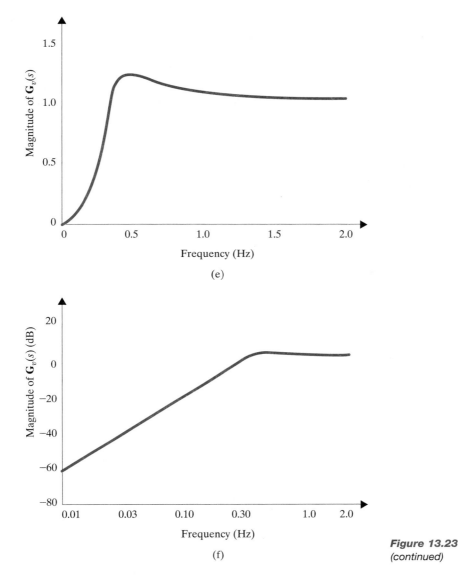

Figure 13.23
(continued)

(e)

(f)

at $\omega = \sqrt{5} = 2.24$ rad/s, which is the magnitude of the complex pole frequencies. In addition, the symmetry of the pole around the real axis becomes readily apparent. As a result of this symmetry, we can restrict our analysis to positive values of $j\omega$, with no loss of information. This plot for $\omega \geq 0$ is shown in Fig. 13.23e where frequency is plotted in Hz rather than rad/s. Finally, converting the transfer function magnitude to dB and using a log axis for frequency, we produce the Bode plot in Fig. 13.23f.

13.6 Steady-State Response

In Section 13.3 we have demonstrated, using a variety of examples, the power of the Laplace transform technique in determining the complete response of a network. This complete response is composed of transient terms, which disappear as $t \to \infty$, and steady-state terms, which are present at all times. Let us now examine a method by which to determine the steady-state response of a network directly. Recall from previous examples that the network response can be written as

$$\mathbf{Y}(s) = \mathbf{H}(s)\mathbf{X}(s) \qquad\qquad \textbf{13.20}$$

where $\mathbf{Y}(s)$ is the output or response, $\mathbf{X}(s)$ is the input or forcing function, and $\mathbf{H}(s)$ is the network function or transfer function defined in Section 11.1. The transient portion of the response $\mathbf{Y}(s)$ results from the poles of $\mathbf{H}(s)$, and the steady-state portion of the response results from the poles of the input or forcing function.

As a direct parallel to the sinusoidal response of a network as outlined in Section 7.2, we assume that the forcing function is of the form

$$x(t) = X_M e^{j\omega_0 t} \qquad\qquad \textbf{13.21}$$

which by Euler's identity can be written as

$$x(t) = X_M \cos \omega_0 t + j X_M \sin \omega_0 t \qquad\qquad \textbf{13.22}$$

The Laplace transform of Eq. (13.21) is

$$\mathbf{X}(s) = \frac{X_M}{s - j\omega_0} \qquad\qquad \textbf{13.23}$$

and therefore,

$$\mathbf{Y}(s) = \mathbf{H}(s)\left(\frac{X_M}{s - j\omega_0}\right) \qquad\qquad \textbf{13.24}$$

At this point we tacitly assume that $\mathbf{H}(s)$ does not have any poles of the form $(s - j\omega_k)$. If, however, this is the case, we simply encounter difficulty in defining the steady-state response.

Performing a partial fraction expansion of Eq. (13.24) yields

$$\mathbf{Y}(s) = \frac{X_M \mathbf{H}(j\omega_0)}{s - j\omega_0} + \text{terms that occur due to the poles of } \mathbf{H}(s) \qquad\qquad \textbf{13.25}$$

The first term to the right of the equal sign can be expressed as

$$\mathbf{Y}(s) = \frac{X_M |\mathbf{H}(j\omega_0)| e^{j\phi(j\omega_0)}}{s - j\omega_0} + \cdots \qquad\qquad \textbf{13.26}$$

since $\mathbf{H}(j\omega_0)$ is a complex quantity with a magnitude and phase that are a function of $j\omega_0$.

Performing the inverse transform of Eq. (13.26), we obtain

$$y(t) = X_M |\mathbf{H}(j\omega_0)| e^{j\omega_0 t} e^{j\phi(j\omega_0)} + \cdots \qquad\qquad \textbf{13.27}$$

$$= X_M |\mathbf{H}(j\omega_0)| e^{(j\omega_0 t + \phi(j\omega_0))} + \cdots$$

and hence the steady-state response is

LEARNING Hint

The transient terms disappear in steady state.

$$y_{ss}(t) = X_M |\mathbf{H}(j\omega_0)| e^{j(\omega_0 t + \phi(j\omega_0))} \qquad\qquad \textbf{13.28}$$

Since the actual forcing function is $X_M \cos \omega_0(t)$, which is the real part of $X_M e^{j\omega_0 t}$, the steady-state response is the real part of Eq. (13.28).

$$y_{ss}(t) = X_M |\mathbf{H}(j\omega_0)| \cos[\omega_0 t + \phi(j\omega_0)] \qquad\qquad \textbf{13.29}$$

In general, the forcing function may have a phase angle θ. In this case, θ is simply added to $\phi(j\omega_0)$ so that the resultant phase of the response is $\phi(j\omega_0) + \theta$.

LEARNING Example 13.11

For the circuit shown in Fig. 13.24a, we wish to determine the steady-state voltage $v_{oss}(t)$ for $t > 0$ if the initial conditions are zero.

SOLUTION   As illustrated earlier, this problem could be solved using a variety of techniques, such as node equations, mesh equations, source transformation, and Thévenin's theorem. We will employ node equations to obtain the solution. The transformed network using the impedance values for the parameters is shown in Fig. 13.24b. The node equations for this network are

$$\left(\frac{1}{2} + \frac{1}{s} + \frac{s}{2}\right)\mathbf{V}_1(s) - \left(\frac{s}{2}\right)\mathbf{V}_o(s) = \frac{1}{2}\mathbf{V}_i(s)$$

$$-\left(\frac{s}{2}\right)\mathbf{V}_1(s) + \left(\frac{s}{2} + 1\right)\mathbf{V}_o(s) = 0$$

Solving these equations for $\mathbf{V}_o(s)$, we obtain

$$\mathbf{V}_o(s) = \frac{s^2}{3s^2 + 4s + 4}\mathbf{V}_i(s)$$

Note that this equation is in the form of Eq. (13.20), where $\mathbf{H}(s)$ is

$$\mathbf{H}(s) = \frac{s^2}{3s^2 + 4s + 4}$$

Since the forcing function is $10\cos 2t\, u(t)$, then $V_M = 10$ and $\omega_0 = 2$. Hence,

$$\mathbf{H}(j2) = \frac{(j2)^2}{3(j2)^2 + 4(j2) + 4}$$

$$= 0.354\,\underline{/45°}$$

Therefore,

$$|\mathbf{H}(j2)| = 0.354$$

$$\phi(j2) = 45°$$

and, hence, the steady-state response is

$$v_{oss}(t) = V_M|\mathbf{H}(j2)|\cos\left[2t + \phi(j2)\right]$$

$$= 3.54\cos(2t + 45°)$$

The complete (transient plus steady-state) response can be obtained from the expression

$$\mathbf{V}_o(s) = \frac{s^2}{3s^2 + 4s + 4}\mathbf{V}_i(s)$$

$$= \frac{s^2}{3s^2 + 4s + 4}\left(\frac{10s}{s^2 + 4}\right)$$

$$= \frac{10s^3}{(s^2 + 4)(3s^2 + 4s + 4)}$$

Determining the inverse Laplace transform of this function using the techniques of Chapter 12, we obtain

$$v_o(t) = 3.54\cos(2t + 45°)$$

$$+ 1.44e^{-(2/3)t}\cos\left(\frac{2\sqrt{2}}{3}t - 55°\right)\text{V}$$

Note that as $t \to \infty$ the second term approaches zero, and thus the steady-state response is

$$v_{oss}(t) = 3.54\cos(2t + 45°)\text{ V}$$

which can easily be checked using a phasor analysis.

(a)   (b)

Figure 13.24   *Circuits used in Example 13.11.*

LEARNING EXTENSION

E13.7   Determine the steady-state voltage $v_{oss}(t)$ in the network in Fig. E13.7 for $t > 0$ if the initial conditions in the network are zero.

ANSWER
$v_{oss}(t) = 3.95\cos(2t - 99.46°)$ V.

Figure E13.7

Learning by Application

LEARNING Example 13.12

The Recording Industry Association of America (RIAA) uses standardized recording and playback filters to improve the quality of phonographic disk recordings. This process is demonstrated in Fig. 13.25. During a recording session, the voice or music signal is passed through the recording filter, which deemphasizes the bass content. This filtered signal is then recorded into the vinyl. On playback, the phonograph needle assembly senses the recorded message and reproduces the filtered signal, which proceeds to the playback filter. The purpose of the playback filter is to emphasize the bass content and reconstruct the original voice/music signal. Next, the reconstructed signal can be amplified and sent on to the speakers.

Let us examine the pole-zero diagrams for the record and playback filters.

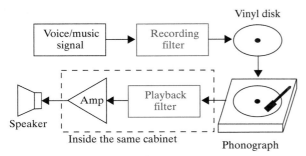

Figure 13.25 *Block diagram for phonograph disk recording and playback.*

SOLUTION The transfer function for the recording filter is

$$\mathbf{G}_{vR}(s) = \frac{K(1 + s\tau_{z1})(1 + s\tau_{z2})}{1 + s\tau_p}$$

where the time constants are $\tau_{z1} = 75$ µs, $\tau_{z2} = 3180$ µs, and $\tau_p = 318$ µs, and K is a constant chosen such that $\mathbf{G}_{vR}(s)$ has a magnitude of 1 at 1000 Hz. The resulting pole and zero frequencies in radians/second are

$$\omega_{z1} = 1/\tau_{z1} = 13.33 \text{ kr/s}$$
$$\omega_{z2} = 1/\tau_{z2} = 313.46 \text{ r/s}$$
$$\omega_p = 1/\tau_p = 3.14 \text{ kr/s}$$

Fig. 13.26a shows the pole-zero diagram for the recording filter.

The playback filter transfer function is the reciprocal of the record transfer function.

$$\mathbf{G}_{vp}(s) = \frac{1}{\mathbf{G}_{vR}(s)} = \frac{A_o(1 + s\tau_z)}{(1 + s\tau_{p1})(1 + s\tau_{p2})}$$

where the time constants are now $\tau_{p1} = 75$ µs, $\tau_{p2} = 3180$ µs, $\tau_z = 318$ µs, and A_o is $1/K$. Pole and zero frequencies, in radians/second, are

$$\omega_{p1} = 1/\tau_{z1} = 13.33 \text{ krad/s}$$
$$\omega_{p2} = 1/\tau_{z2} = 313.46 \text{ rad/s}$$
$$\omega_z = 1/\tau_p = 3.14 \text{ krad/s}$$

which yields the pole-zero diagram in Fig. 13.26b. The voice/music signal eventually passes through both filters before proceeding to the amplifier. In the s-domain, this is equivalent to multiplying $\mathbf{V}_s(s)$ by both $\mathbf{G}_{vR}(s)$, and $\mathbf{G}_{vp}(s)$. In the pole-zero diagram, we simply superimpose the pole-zero diagrams of the two filters, as shown in Fig. 13.26c. Note that at each pole frequency there is a zero and vice versa. The pole-zero pairs cancel one another, yielding a pole-zero diagram that contains no poles and no zeros. This effect can be seen mathematically by multiplying the two transfer functions, $\mathbf{G}_{vR}(s)\mathbf{G}_{vp}(s)$, which yields a product independent of s. Thus, the original voice/music signal is reconstructed and fidelity is preserved.

(a)

(b)

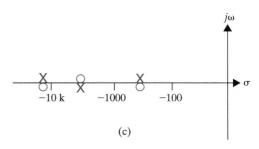

(c)

Figure 13.26 *Pole-zero diagrams for RIAA phonographic filters.*

Learning by Design

LEARNING Example 13.13

In a large computer network, two computers are transferring digital data on a single wire at a rate of 1000 bits/s. The voltage waveform, v_{data}, in Fig. 13.27 shows a possible sequence of bits alternating between "high" and "low" values. Also present in the environment is a source of 100 kHz (628 krad/s) noise, which is corrupting the data.

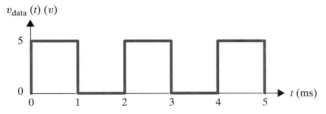

Figure 13.27   *1000 bit/s digital data waveform.*

It is necessary to filter out the high-frequency noise without destroying the data waveform. Let us place the second-order low-pass active filter of Fig. 13.28 in the data path so that the data and noise signals will pass through it.

Figure 13.28   *Second-order low-pass filter.*

SOLUTION   The filter's transfer function is found to be

$$
\mathbf{G}_v(s) = \frac{\mathbf{V}_o(s)}{\mathbf{V}_{\text{data}}(s)} = \frac{-\left(\dfrac{R_3}{R_1}\right)\left(\dfrac{1}{R_2 R_3 C_1 C_2}\right)}{s^2 + s\left(\dfrac{1}{R_1 C_1} + \dfrac{1}{R_2 C_1} + \dfrac{1}{R_3 C_1}\right) + \dfrac{1}{R_2 R_3 C_1 C_2}}
$$

To simplify our work, let $R_1 = R_2 = R_3 = R$. From our work in Chapter 11, we know that the characteristic equation of a second-order system can be expressed

$$
s^2 + 2s\zeta\omega_0 + \omega_0^2 = 0
$$

Comparing the two preceding equations, we find that

$$
\omega_0 = \frac{1}{R\sqrt{C_1 C_2}}
$$

$$
2\zeta\omega_0 = \frac{3}{RC_1}
$$

and therefore,

$$
\zeta = \frac{3}{2}\sqrt{\frac{C_2}{C_1}}
$$

The poles of the filter are at

$$
s_1, s_2 = -\zeta\omega_0 \pm \omega_0\sqrt{\zeta^2 - 1}
$$

To eliminate the 100-kHz noise, at least one pole should be well below 100 kHz, as shown in the Bode plot sketched in Fig. 13.29. By placing a pole well below 100 kHz, the gain of the filter will be quite small at 100 kHz, effectively filtering the noise.

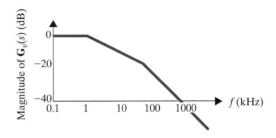

Figure 13.29   *Bode plot sketch for a second-order low-pass filter.*

If we arbitrarily choose an overdamped system with $\omega_0 = 25$ krad/s and $\zeta = 2$, the resulting filter is overdamped with poles at $s_1 = -6.7$ krad/s and $s_2 = -93.3$ krad/s. The pole-zero diagram for the filter is shown in Fig. 13.30.

If we let $R = 40$ kΩ, then we may write

$$
\omega_0 = 25{,}000 = \frac{1}{40{,}000\sqrt{C_1 C_2}}
$$

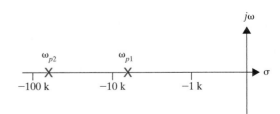

Figure 13.30   *Pole-zero diagram for low-pass filter.*

(continued)

or

$$C_1 C_2 = 10^{-18}$$

Also,

$$\zeta = 2 = \frac{3}{2}\sqrt{\frac{C_2}{C_1}}$$

which can be expressed as

$$\frac{C_2}{C_1} = \frac{16}{9}$$

Solving for C_1 and C_2 yields

$$C_1 = 0.75 \text{ nF}$$

$$C_2 = 1.33 \text{ nF}$$

The circuit used to simulate the filter is shown in Fig. 13.31. The sinusoidal source has a frequency of 100 kHz and is used to represent the noise source.

Figure 13.31 *Circuit for second-order filter.*

Plots for the input to the filter and the output voltage for 2 ms are shown in Fig. 13.32. Note that output indeed contains much less of the 100-kHz noise. Also, the fast rise and fall times of the data signal are slower in the output voltage. Despite this slower response, the output voltage is fast enough to keep pace with the 1000-bits/s transfer rate.

Let us now increase the data transfer rate from 1000 to 25,000 bits/s, as shown in Fig. 13.33. The total input and output signals are plotted in Fig. 13.34 for 200 μs. Now the output cannot keep pace with the input and the data information is lost. Let us investigate why this occurs. We know that the filter is second order with poles at s_1 and s_2. If we represent the data input as a 5-V step function, the output voltage is

$$\mathbf{V}_o(s) = \mathbf{G}_v(s)\left(\frac{5}{s}\right) = \frac{K}{(s + s_1)(s + s_2)}\left(\frac{5}{s}\right)$$

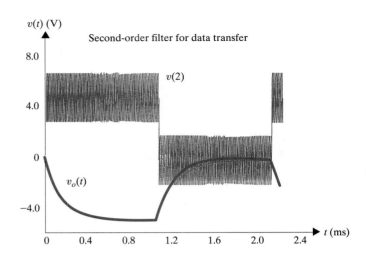

Figure 13.32 *Simulation outputs for node 2 and $v_o(t)$.*

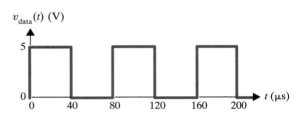

Figure 13.33 *25,000-bit/s digital data waveform.*

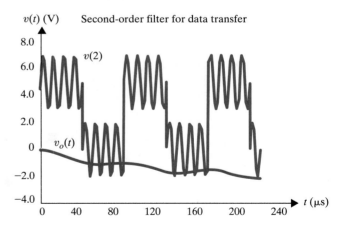

Figure 13.34 *Simulation output for node 2 and $v_o(t)$ with 25,000-bit/s data transfer rate.*

where K is a constant. Since the filter is overdamped, s_1 and s_2 are real and positive. A partial fraction expansion of $\mathbf{V}_o(s)$ is of the form

$$\mathbf{V}_o(s) = \frac{K_1}{s} + \frac{K_2}{(s + s_1)} + \frac{K_3}{(s + s_2)}$$

yielding the time domain expression

$$v_o(t) = \left[K_1 + K_2 e^{-s_1 t} + K_3 e^{-s_2 t}\right]u(t) \text{ V}$$

where K_1, K_2, and K_3 are real constants. The exponential time constants are the reciprocals of the pole frequencies.

$$\tau_1 = \frac{1}{s_1} = \frac{1}{6.7k} = 149 \text{ μs}$$

$$\tau_2 = \frac{1}{s_2} = \frac{1}{93.3k} = 10.7 \text{ μs}$$

Since exponentials reach steady state in roughly 5τ, the exponential associated with τ_2 affects the output for about 50 μs and the τ_1 exponential will reach steady state after about 750 μs. From Fig. 13.33 we see that at a 25,000-bits/s data transfer rate, each bit (a "high" or "low" voltage value) occupies a 40-μs time span. Therefore, the exponential associated with s_1, and thus $v_o(t)$, is still far from its steady-state condition when the next bit is transmitted. In short, s_1 is too small.

Let us remedy this situation by increasing the pole frequencies and changing to a critically damped system, $\zeta = 1$. If we select $\omega_0 = 125$ krad/s, the poles will be at $s_1 = s_2 = -125$ krad/s or 19.9 kHz—both below the 100-kHz noise we wish to filter out. Figure 13.35 shows the new pole positions moved to the left of their earlier positions, which we expect will result in a quicker response to the v_{data} pulse train.

Now the expressions for ω_0 and ζ are

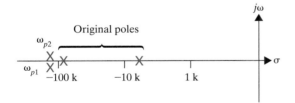

Figure 13.35 *Pole-zero diagram for both original and critically damped systems.*

$$\omega_0 = 125,000 = \frac{1}{40,000\sqrt{C_1 C_2}}$$

or

$$C_1 C_2 = 4 \times 10^{-20}$$

Also,

$$\zeta = 1 = \frac{3}{2}\sqrt{\frac{C_2}{C_1}}$$

which can be expressed

$$\frac{C_2}{C_1} = \frac{4}{9}$$

Solving for C_1 and C_2 yields

$$C_1 = 300 \text{ pF}$$
$$C_2 = 133.3 \text{ pF}$$

A simulation using these new capacitor values produces the input-output data shown in Fig. 13.36. Now the output voltage just reaches the "high" and "low" levels just before v_{data} makes its next transition and the 100-kHz noise is still much reduced.

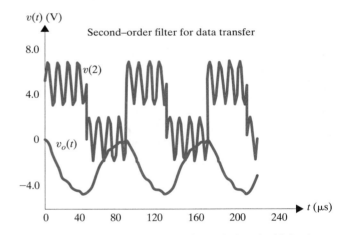

Figure 13.36 *Simulation outputs for node 2 and $v_o(t)$ for the critically damped system.*

LEARNING Check

Summary

▶ The use of *s*-domain models for circuit elements permits us to describe them with algebraic, rather than differential, equations.

▶ All the dc analysis techniques, including the network theorems, are applicable in the *s*-domain. Once the *s*-domain solution is obtained, the inverse transform is used to obtain a time domain solution.

▶ The dc properties of the storage elements, L and C, can be used to obtain initial and final conditions. The initial conditions are required as a part of the *s*-domain model, and final conditions are often useful in verifying a solution.

▶ The Laplace transform solution for the network response is composed of transient terms, which disappear as $t \to \infty$, and steady-state terms, which are present at all times.

▶ The roots of the network's characteristic equation (i.e., the poles) determine the type of network response. A plot of these roots in the left half of the s-plane provides an immediate indication of the network's behavior. The relationship between the pole-zero plot and the Bode plot provides further insight.

▶ The transfer (network) function for a network is expressed as

$$H(s) = \frac{Y(s)}{X(s)}$$

where $Y(s)$ is the network response and $X(s)$ is the input forcing function. If the transfer function is known, the output response is simply given by the product $H(s)X(s)$. If the input is an impulse function so that $X(s) = 1$, the impulse response is equal to the inverse Laplace transform of the network function.

▶ The network response can be expressed as

$$Y(s) = H(s)X(s)$$

▶ The transient portion of the response $Y(s)$ results from the poles of $H(s)$, and the steady-state portion of the response results from the poles of the forcing function $X(s)$.

Problems

SECTION 13.3

13.1 Find the input impedance $Z(s)$ of the network in Fig. P13.1.

Figure P13.1

13.2 Find the input impedance $Z(s)$ of the network in Fig. P13.2 **(a)** when the terminals $B\text{-}B'$ are open circuited and **(b)** when the terminals $B\text{-}B'$ are short circuited.

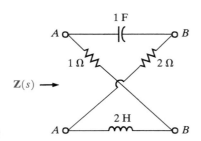

Figure P13.2

13.3 Use Laplace transforms to find $v(t)$ for $t > 0$ in the network shown in Fig. P13.3. Assume zero initial conditions.

Figure P13.3

13.4 Use Laplace transforms and nodal analysis to find $i_1(t)$ for $t > 0$ in the network shown in Fig. P13.4. Assume zero initial conditions.

Figure P13.4

13.5 For the network shown in Fig. P13.5, find $i_o(t)$, $t > 0$.

Figure P13.5

13.6 For the network shown in Fig. P13.6, find $v_o(t)$, $t > 0$.

Figure P13.6

13.7 For the network in Fig. P13.7, find $v(t)$ for $t > 0$.

Figure P13.7

13.8 Find $v_o(t)$, $t > 0$, in the network shown in Fig. P13.8 using node equations.

Figure P13.8

13.9 Find $v_o(t)$, $t > 0$, in the network shown in Fig. P13.9 using nodal analysis.

Figure P13.9

13.10 For the network shown in Fig. P13.10, find $v_o(t)$, $t > 0$, using mesh equations.

Figure P13.10

13.11 Given the network in Fig. P13.11, find $i_o(t)$, $t > 0$, using mesh equations.

Figure P13.11

13.12 Use loop equations to find $i_o(t)$, $t > 0$, in the network shown in Fig. P13.12.

Figure P13.12

13.13 Use mesh equations to find $v_o(t)$, $t > 0$, in the network in Fig. P13.13.

Figure P13.13

13.14 Given the network in Fig. P13.14, find $i_o(t)$, $t > 0$, using mesh equations.

Figure P13.14

13.15 For the network shown in Fig. P13.15, find $v_o(t)$, $t > 0$, using loop equations.

Figure P13.15

13.16 Use loop analysis to find $v_o(t)$ for $t > 0$ in the network in Fig. P13.16.

Figure P13.16

13.17 Use mesh analysis to find $v_o(t)$, $t > 0$, in the network in Fig. P13.17.

Figure P13.17

13.18 Use loop equations to find $v_o(t)$, $t > 0$, in the network shown in Fig. P13.18.

Figure P13.18

13.19 Use superposition to find $v_o(t)$, $t > 0$ in the network shown in Fig. P13.19.

Figure P13.19

13.20 Use source transformation to solve Problem 13.19.

13.21 Use Thévenin's theorem to solve Problem 13.19.

13.22 Use Thévenin's theorem to solve Problem 13.11.

13.23 Use Thévenin's theorem to find $v_o(t)$, $t > 0$, in the network in Fig. P13.23.

Figure P13.23

13.24 Use Thévenin's theorem to determine $i_o(t)$, $t > 0$ in the circuit shown in Fig. P13.24.

Figure P13.24

13.25 Use Thévenin's theorem to find $v_o(t)$, $t > 0$ in the network in Fig. P13.25.

Figure P13.25

13.26 Find $v_o(t)$, for $t > 0$ in the network in Fig. P13.26 using Thévenin's theorem.

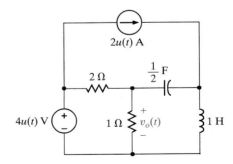

Figure P13.26

13.27 Use Thévenin's theorem to find $v_o(t), t > 0$, in the network shown in Fig. P13.27.

Figure P13.27

13.28 Find $v_o(t), t > 0$, in the network shown in Fig. P13.28 using Laplace transforms. Assume that the circuit has reached steady state at $t = 0-$.

Figure P13.28

13.29 Find $i_o(t), t > 0$, in the network shown in Fig. P13.29.

Figure P13.29

13.30 Find $i_o(t), t > 0$, in the network shown in Fig. P13.30.

Figure P13.30

13.31 Find $v_o(t), t > 0$, in the circuit shown in Fig. P13.31.

Figure P13.31

13.32 Find $i_o(t), t > 0$, in the network in Fig. P13.32.

Figure P13.32

13.33 Find $v_o(t), t > 0$, in the circuit in Fig. P13.33.

Figure P13.33

13.34 Find $v_o(t)$, $t > 0$, in the network in Fig. P13.34.

Figure P13.34

13.35 Find $v_o(t)$, for $t > 0$, in the network in Fig. P13.35.

Figure P13.35

13.36 Find $v_o(t)$, for $t > 0$, in the network in Fig. P13.36.

Figure P13.36

13.37 Find $v_o(t)$ for $t > 0$, in the network in Fig. P13.37.

Figure P13.37

13.38 Find $v_o(t)$ for $t > 0$ in the network in Fig. P13.38.

Figure P13.38

13.39 Find $i_o(t)$, for $t > 0$, in the network in Fig. P13.39.

Figure P13.39

13.40 Find $v_o(t)$, for $t > 0$, in the network in Fig. P13.40.

Figure P13.40

13.41 Find $v_o(t)$ for $t > 0$, in the network in Fig. P13.41.

Figure P13.41

13.42 Determine the output voltage $v_o(t)$ in the network in Fig. P13.42a if the input is given by the source in Fig. P13.42b.

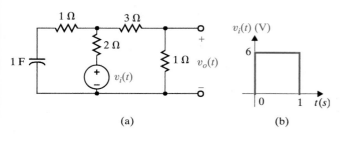

Figure P13.42

13.43 Find the output voltage $v_o(t)$, $t > 0$, in the network in Fig. P13.43a if the input is represented by the waveform shown in Fig. P13.43b.

(a)   (b)

Figure P13.43

13.44 Determine the output voltage, $v_o(t)$, in the circuit in Fig. P13.44a if the input is given by the source described in Fig. P13.44b.

(a)   (b)

Figure P13.44

SECTION 13.4

13.45 Find the transfer function $\mathbf{V}_o(s)/\mathbf{V}_i(s)$ for the network shown in Fig. P13.45.

Figure P13.45

13.46 Find the transfer function for the network in Fig. P13.46.

Figure P13.46

13.47 Find the transfer function for the network shown in Fig. P13.47.

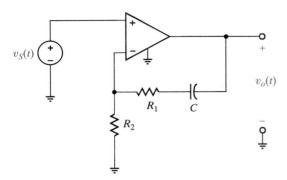

Figure P13.47

13.48 Find the transfer function for the network shown in Fig. P13.48.

Figure P13.48

13.49 Find the transfer function for the network in Fig. P13.49.

Figure P13.49

13.50 Determine the transfer function for the network shown in Fig. P13.50. If a step function is applied to the network, what type of damping will the network exhibit?

Figure P13.50

13.51 Find the transfer function for the network in Fig. P13.51. If a step function is applied to the network, will the response be overdamped, underdamped, or critically damped?

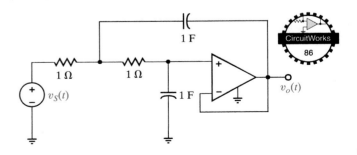

Figure P13.51

13.52 The voltage response of the network to a unit step input is

$$V_o(s) = \frac{2(s + 1)}{s(s^2 + 12s + 27)}$$

Is the response overdamped?

13.53 The transfer function of the network is given by the expression

$$G(s) = \frac{100s}{s^2 + 22s + 40}$$

Determine the damping ratio, the undamped natural frequency, and the type of response that will be exhibited by the network.

13.54 The current response of a network to a unit step input is

$$I_o(s) = \frac{10(s + 2)}{s^2(s^2 + 11s + 30)}$$

Is the response underdamped?

13.55 The voltage response of a network to a unit step input is

$$V_o(s) = \frac{10}{s(s^2 + 8s + 16)}$$

Is the response critically damped?

13.56 For the network in Fig. P13.56, choose the value of C for critical damping.

Figure P13.56

13.57 For the filter in Fig. P13.57, choose the values of C_1 and C_2 to place poles at $s = -2$ and $s = -5$ rad/s.

Figure P13.57

SECTION 13.6

13.58 Find the steady-state response $v_o(t)$ for the network in Fig. P13.58.

Figure P13.58

13.59 Find the steady-state response $v_o(t)$ for the circuit shown in Fig. P13.59.

Figure P13.59

13.60 Determine the steady-state response $v_o(t)$ for the network in Fig. P13.60.

Figure P13.60

13.61 Find the steady-state response $i_o(t)$ for the network shown in Fig. P13.61.

Figure P13.61

13.62 Determine the steady-state response $i_o(t)$ for the network in Fig. P13.62.

Figure P13.62

13.63 Find the steady-state response $v_o(t)$, for $t > 0$, in the network in Fig. P13.63.

Figure P13.63

13.64 Find the steady-state response $v_o(t)$, for $t > 0$, in the network in Fig. P13.64.

Figure P13.64

Typical Problems Found on the FE Exam

13FE-1 A single loop, second-order circuit is described by the following differential equation.

$$2\frac{dv^2(t)}{dt^2} + 4\frac{dv(t)}{dt} + 4v(t) = 12u(t) \qquad t > 0$$

Which is the correct form of the total (natural plus forced) response?

(a) $v(t) = K_1 + K_2 e^{-t}$

(b) $v(t) = K_1 \cos t + K_2 \sin t$

(c) $v(t) = K_1 + K_2 t e^{-t}$

(d) $v(t) = K_1 + K_2 e^{-t} \cos t + K_3 e^{-t} \sin t$

13FE-2 If all initial conditions are zero in the network in Fig. 13PFE-2, find the transfer function $\mathbf{V}_o(s)/\mathbf{V}_S(s)$, and determine the type of damping exhibited by the network.

Figure 13PFE-2

13FE-3 The initial conditions in the circuit in Fig. 13PFE-3 are zero. Find the transfer function $\mathbf{I}_o(s)/\mathbf{I}_s(s)$, and determine the type of damping exhibited by the circuit.

Figure 13PFE-3

Fourier Analysis Techniques | 14

In this chapter we examine two very important topics: the Fourier series and the Fourier transform. These two techniques vastly expand our circuit analysis capabilities because they provide a means of effectively dealing with nonsinusoidal periodic signals and aperiodic signals. Using the Fourier series, we show that we can determine the steady-state response of a network to a nonsinusoidal periodic input. The Fourier transform will allow us to analyze circuits with aperiodic inputs by transforming the problem to the frequency domain, solving it algebraically, and then transforming back to the time domain in a manner similar to that used with Laplace transforms.

LEARNING Goals

14.1 Fourier Series The exponential and trigonometric Fourier series are introduced. The role of symmetrical properties in determining trigonometric Fourier series is also examined. The power of PSPICE is also demonstrated in deriving the Fourier series of a given waveform. The usefulness of various aspects of the Fourier series is explained and its use in steady-state network response is demonstrated...*Page 562*

14.2 Fourier Transform The Fourier transform and its use in describing a periodic function is examined. Various properties of the transform are outlined, and the use of Parseval's theorem for determining the energy content of a signal is demonstrated...*Page 586*

Learning by Application...*Page 593*

Learning by Design...*Page 595*

Learning Check...*Page 600*

   Summary...*Page 600*

   Problems...*Page 601*

561

14.1 Fourier Series

A periodic function is one that satisfies the relationship

$$f(t) = f(t + nT_0), \qquad n = \pm 1, \pm 2, \pm 3, \ldots$$

for every value of t where T_0 is the period. As we have shown in previous chapters, the sinusoidal function is a very important periodic function. However, there are many other periodic functions that have wide applications. For example, laboratory signal generators produce the pulse-train and square-wave signals shown in Figs. 14.1a and b, respectively, which are used for testing circuits. The oscilloscope is another laboratory instrument and the sweep of its electron beam across the face of the cathode ray tube is controlled by a triangular signal of the form shown in Fig. 14.1c.

The techniques we will explore are based on the work of Jean Baptiste Joseph Fourier. Although our analyses will be confined to electric circuits, it is important to point out that the techniques are applicable to a wide range of engineering problems. In fact, it was Fourier's work in heat flow that led to the techniques that will be presented here.

In his work, Fourier demonstrated that a periodic function $f(t)$ could be expressed as a sum of sinusoidal functions. Therefore, given this fact and the fact that if a periodic function is expressed as a sum of linearly independent functions, each function in the sum must be periodic with the same period, and the function $f(t)$ can be expressed in the form

$$f(t) = a_0 + \sum_{n=1}^{\infty} D_n \cos(n\omega_0 t + \theta_n) \qquad \textbf{14.1}$$

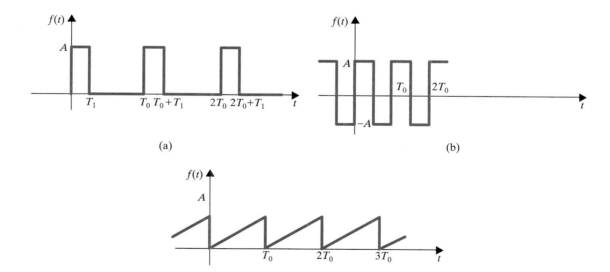

Figure 14.1
Some useful periodic signals.

where $\omega_0 = 2\pi/T_0$ and a_0 is the average value of the waveform. An examination of this expression illustrates that all sinusoidal waveforms that are periodic with period T_0 have been included. For example, for $n = 1$, one cycle covers T_0 seconds and $D_1 \cos(\omega_0 t + \theta_1)$ is called the *fundamental*. For $n = 2$, two cycles fall within T_0 seconds and the term $D_2 \cos(2\omega_0 t + \theta_2)$ is called the *second harmonic*. In general, for $n = k$, k cycles fall within T_0 seconds and $D_k \cos(k\omega_0 t + \theta_k)$ is the *kth harmonic term*.

Since the function $\cos(n\omega_0 t + \theta_k)$ can be written in exponential form using Euler's identity or as a sum of cosine and sine terms of the form $\cos n\omega_0 t$ and $\sin n\omega_0 t$ as demonstrated in Chapter 7, the series in Eq. (14.1) can be written as

$$f(t) = a_0 + \sum_{\substack{n=-\infty \\ n \neq 0}}^{\infty} \mathbf{c}_n e^{jn\omega_0 t} = \sum_{n=-\infty}^{\infty} \mathbf{c}_n e^{jn\omega_0 t} \qquad \textbf{14.2}$$

Using the real-part relationship employed as a transformation between the time domain and the frequency domain, we can express $f(t)$ as

$$f(t) = a_0 + \sum_{n=1}^{\infty} \mathrm{Re}\left[(D_n \underline{/\theta_n}) e^{jn\omega_0 t}\right] \qquad \textbf{14.3}$$

$$= a_0 + \sum_{n=1}^{\infty} \mathrm{Re}\left(2\mathbf{c}_n e^{jn\omega_0 t}\right) \qquad \textbf{14.4}$$

$$= a_0 + \sum_{n=1}^{\infty} \mathrm{Re}\left[(a_n - jb_n) e^{jn\omega_0 t}\right] \qquad \textbf{14.5}$$

$$= a_0 + \sum_{n=1}^{\infty} a_n \cos n\omega_0 t + b_n \sin n\omega_0 t \qquad \textbf{14.6}$$

These equations allow us to write the Fourier series in a number of equivalent forms. Note that the *phasor* for the nth harmonic is

$$D_n \underline{/\theta_n} = 2\mathbf{c}_n = a_n - jb_n \qquad \textbf{14.7}$$

The approach we will take will be to represent a nonsinusoidal periodic input by a sum of complex exponential functions, which because of Euler's identity is equivalent to a sum of sines and cosines. We will then use (1) the superposition property of linear systems and (2) our knowledge that the steady-state response of a time-invariant linear system to a sinusoidal input of frequency ω_0 is a sinusoidal function of the same frequency to determine the response of such a system.

To illustrate the manner in which a nonsinusoidal periodic signal can be represented by a Fourier series, consider the periodic function shown in Fig. 14.2a. In Figs. 14.2b–d we can see the impact of using a specific number of terms in the series to represent the original function. Note that the series more closely represents the original function as we employ more and more terms.

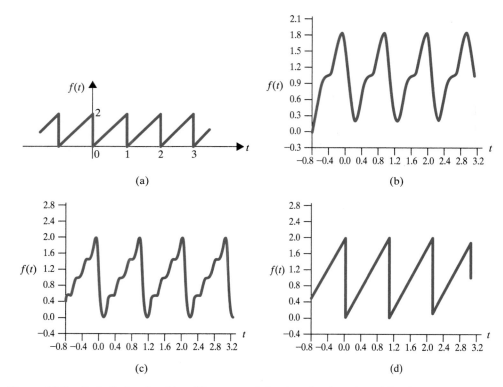

Figure 14.2 *Periodic function (a) and its representation by a fixed number of Fourier series terms, (b) 2 terms, (c) 4 terms, (d) 100 terms.*

EXPONENTIAL FOURIER SERIES *Any physically realizable periodic signal* may be represented over the interval $t_1 < t < t_1 + T_0$ by the *exponential Fourier series*

$$f(t) = \sum_{n=-\infty}^{\infty} \mathbf{c}_n e^{jn\omega_0 t} \qquad\qquad \textbf{14.8}$$

where the $\mathbf{c}_n$ are the complex (phasor) Fourier coefficients. These coefficients are derived as follows. Multiplying both sides of Eq. (14.8) by $e^{-jk\omega_0 t}$ and integrating over the interval t_1 to $t_1 + T_0$, we obtain

$$\int_{t_1}^{t_1+T_0} f(t) e^{-jk\omega_0 t}\, dt = \int_{t_1}^{t_1+T_0} \left(\sum_{n=-\infty}^{\infty} \mathbf{c}_n e^{jn\omega_0 t} \right) e^{-jk\omega_0 t}\, dt$$

$$= \mathbf{c}_k T_0$$

since

$$\int_{t_1}^{t_1+T_0} e^{j(n-k)\omega_0 t}\, dt = \begin{cases} 0 & \text{for } n \neq k \\ T_0 & \text{for } n = k \end{cases}$$

Therefore, the Fourier coefficients are defined by the equation

$$c_n = \frac{1}{T_0} \int_{t_1}^{t_1+T_0} f(t)e^{-jn\omega_0 t}\, dt \qquad\qquad \textbf{14.9}$$

The following example illustrates the manner in which we can represent a periodic signal by an exponential Fourier series.

LEARNING Example 14.1

We wish to determine the exponential Fourier series for the periodic voltage waveform shown in Fig. 14.3.

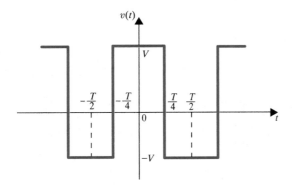

Figure 14.3 *Periodic voltage waveform.*

SOLUTION The Fourier coefficients are determined using Eq. (14.9) by integrating over one complete period of the waveform.

$$c_n = \frac{1}{T} \int_{-T/2}^{T/2} f(t)e^{-jn\omega_0 t}\, dt$$

$$= \frac{1}{T} \int_{-T/2}^{-T/4} -Ve^{-jn\omega_0 t}\, dt$$

$$+ \int_{-T/4}^{T/4} Ve^{-jn\omega_0 t}\, dt + \int_{T/4}^{T/2} -Ve^{-jn\omega_0 t}\, dt$$

$$= \frac{V}{jn\omega_0 T} \left[+e^{-jn\omega_0 t}\Big|_{-T/2}^{-T/4} - e^{-jn\omega_0 t}\Big|_{-T/4}^{T/4} + e^{-jn\omega_0 t}\Big|_{T/4}^{T/2} \right]$$

$$= \frac{V}{jn\omega_0 T} \left(2e^{jn\pi/2} - 2e^{-jn\pi/2} + e^{-jn\pi} - e^{+jn\pi} \right)$$

$$= \frac{V}{n\omega_0 T} \left[4\sin\frac{n\pi}{2} - 2\sin(n\pi) \right]$$

CircuitWorks
1

$$= 0 \qquad \text{for } n \text{ even}$$

$$= \frac{2V}{n\pi} \sin\frac{n\pi}{2} \qquad \text{for } n \text{ odd}$$

c_0 corresponds to the average value of the waveform. This term can be evaluated using the original equation for c_n. Therefore,

$$c_0 = \frac{1}{T} \int_{-\frac{T}{2}}^{\frac{T}{2}} v(t)\, dt$$

$$= \frac{1}{T} \left[\int_{-\frac{T}{2}}^{-\frac{T}{4}} -V\, dt + \int_{-\frac{T}{4}}^{\frac{T}{4}} V\, dt + \int_{\frac{T}{4}}^{\frac{T}{2}} -V\, dt \right]$$

$$= \frac{1}{T} \left[-\frac{VT}{4} + \frac{VT}{2} - \frac{VT}{4} \right] = 0$$

Therefore,

$$v(t) = \sum_{\substack{n=-\infty \\ n \neq 0 \\ n \text{ odd}}}^{\infty} \frac{2V}{n\pi} \sin\frac{n\pi}{2}\, e^{jn\omega_0 t}$$

This equation can be written as

$$v(t) = \sum_{\substack{n=1 \\ n \text{ odd}}}^{\infty} \frac{2V}{n\pi} \sin\frac{n\pi}{2}\, e^{jn\omega_0 t} + \sum_{\substack{n=-1 \\ n \text{ odd}}}^{-\infty} \frac{2V}{n\pi} \sin\frac{n\pi}{2}\, e^{jn\omega_0 t}$$

$$= \sum_{\substack{n=1 \\ n \text{ odd}}}^{\infty} \left(\frac{2V}{n\pi} \sin\frac{n\pi}{2} \right) e^{jn\omega_0 t} + \left(\frac{2V}{n\pi} \sin\frac{n\pi}{2} \right)^{*} e^{-jn\omega_0 t}$$

(continued)

Since a number plus its complex conjugate is equal to two times the real part of the number, $v(t)$ can be written as

$$v(t) = \sum_{\substack{n=1 \\ n \text{ odd}}}^{\infty} 2\,\mathrm{Re}\left(\frac{2V}{n\pi}\sin\frac{n\pi}{2}\,e^{jn\omega_0 t}\right)$$

or

$$v(t) = \sum_{\substack{n=1 \\ n \text{ odd}}}^{\infty} \frac{4V}{n\pi}\sin\frac{n\pi}{2}\cos n\omega_0 t$$

Note that this same result could have been obtained by integrating over the interval $-T/4$ to $3T/4$.

LEARNING EXTENSIONS

E14.1   Find the Fourier coefficients for the waveform in Fig. E14.1.

ANSWER $\mathbf{c}_n = \dfrac{1 - e^{-jn\pi}}{j2\pi n}$, $\mathbf{c}_0 = \dfrac{1}{2}$.

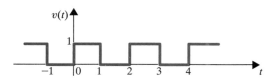

Figure E14.1

E14.2   Find the Fourier coefficients for the waveform in Fig. E14.2.

ANSWER

$$\mathbf{c}_n = \frac{2}{n\pi}\left(2\sin\frac{2\pi n}{3} - \sin\frac{n\pi}{3}\right),\ \mathbf{c}_0 = 2.$$

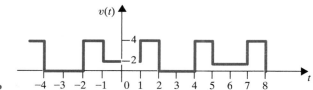

Figure E14.2

TRIGONOMETRIC FOURIER SERIES   Let us now examine another form of the Fourier series. Since

$$2\mathbf{c}_n = a_n - jb_n \qquad\qquad \textbf{14.10}$$

we will examine this quantity $2\mathbf{c}_n$ and separate it into its real and imaginary parts. Using Eq. (14.9), we find that

$$2\mathbf{c}_n = \frac{2}{T_0}\int_{t_1}^{t_1+T_0} f(t)e^{-jn\omega_0 t}\,dt \qquad\qquad \textbf{14.11}$$

Using Euler's identity, we can write this equation in the form

$$2\mathbf{c}_n = \frac{2}{T_0}\int_{t_1}^{t_1+T_0} f(t)\left(\cos n\omega_0 t - j\sin n\omega_0 t\right)dt$$

$$= \frac{2}{T_0} \int_{t_1}^{t_1+T_0} f(t)\cos n\omega_0 t \, dt - j\frac{2}{T_0} \int_{t_1}^{t_1+T_0} f(t) \sin n\omega_0 t \, dt$$

From Eq. (14.10) we note then that

$$a_n = \frac{2}{T_0} \int_{t_1}^{t_1+T_0} f(t) \cos n\omega_0 t \, dt \qquad \textbf{14.12}$$

$$b_n = \frac{2}{T_0} \int_{t_1}^{t_1+T_0} f(t) \sin n\omega_0 t \, dt \qquad \textbf{14.13}$$

These are the coefficients of the Fourier series described by Eq. (14.6), which we call the *trigonometric Fourier series*. These equations are derived directly in most textbooks using the orthogonality properties of the cosine and sine functions. Note that we can now evaluate $\mathbf{c}_n$, a_n, b_n, and since

$$2\mathbf{c}_n = D_n \underline{/\theta_n} \qquad \textbf{14.14}$$

we can derive the coefficients for the *cosine Fourier series* described by Eq. (14.1). This form of the Fourier series is particularly useful because it allows us to represent each harmonic of the function as a phasor.

From Eq. (14.9) we note that $\mathbf{c}_0$, which is written as a_0, is

$$a_0 = \frac{1}{T} \int_{t_1}^{t_1+T_0} f(t) \, dt \qquad \textbf{14.15}$$

This is the average value of the signal $f(t)$ and can often be evaluated directly from the waveform.

SYMMETRY AND THE TRIGONOMETRIC FOURIER SERIES
If a signal exhibits certain symmetrical properties, we can take advantage of these properties to simplify the calculations of the Fourier coefficients. There are three types of symmetry: (1) even-function symmetry, (2) odd-function symmetry, and (3) half-wave symmetry.

Even-Function Symmetry A function is said to be even if

$$f(t) = f(-t) \qquad \textbf{14.16}$$

An even function is symmetrical about the vertical axis and a notable example is the function $\cos n\omega_0 t$. Note that the waveform in Fig. 14.3 also exhibits even-function symmetry. Let us now determine the expressions for the Fourier coefficients if the function satisfies Eq. (14.16).

If we let $t_1 = -T_0/2$ in Eq. (14.15), we obtain

$$a_0 = \frac{1}{T_0} \int_{-T_0/2}^{T_0/2} f(t) \, dt$$

which can be written as

$$a_0 = \frac{1}{T_0} \int_{-T_0/2}^{0} f(t)\, dt + \frac{1}{T_0} \int_{0}^{T_0/2} f(t)\, dt$$

If we now change the variable on the first integral (i.e., let $t = -x$), then $f(-x) = f(x)$, $dt = -dx$, and the range of integration is from $x = T_0/2$ to 0. Therefore, the preceding equation becomes

$$a_0 = \frac{1}{T_0} \int_{T_0/2}^{0} f(x)(-dx) + \frac{1}{T_0} \int_{0}^{T_0/2} f(t)\, dt$$

$$= \frac{1}{T_0} \int_{0}^{T_0/2} f(x)\, dx + \frac{1}{T_0} \int_{0}^{T_0/2} f(t)\, dt \qquad \textbf{14.17}$$

$$= \frac{2}{T_0} \int_{0}^{T_0/2} f(t)\, dt$$

The other Fourier coefficients are derived in a similar manner. The a_n coefficient can be written

$$a_n = \frac{2}{T_0} \int_{-T_0/2}^{0} f(t) \cos n\omega_0 t\, dt + \frac{2}{T_0} \int_{0}^{T_0/2} f(t) \cos n\omega_0 t\, dt$$

Employing the change of variable that led to Eq. (14.17), we can express the preceding equation as

$$a_n = \frac{2}{T_0} \int_{T_0/2}^{0} f(x) \cos(-n\omega_0 x)(-dx) + \frac{2}{T_0} \int_{0}^{T_0/2} f(t) \cos n\omega_0 t\, dt$$

$$= \frac{2}{T_0} \int_{0}^{T_0/2} f(x) \cos n\omega_0 x\, dx + \frac{2}{T_0} \int_{0}^{T_0/2} f(t) \cos n\omega_0 t\, dt$$

$$a_n = \frac{4}{T_0} \int_{0}^{T_0/2} f(t) \cos n\omega_0 t\, dt \qquad \textbf{14.18}$$

Once again, following the preceding development, we can write the equation for the b_n coefficient as

$$b_n = \frac{2}{T_0} \int_{-T_0/2}^{0} f(t) \sin n\omega_0 t\, dt + \int_{0}^{T_0/2} f(t) \sin n\omega_0 t\, dt$$

The variable change employed previously yields

$$b_n = \frac{2}{T_0} \int_{T_0/2}^{0} f(x) \sin(-n\omega_0 x)(-dx) + \frac{2}{T_0} \int_{0}^{T_0/2} f(t) \sin n\omega_0 t\, dt$$

$$= \frac{-2}{T_0} \int_0^{T_0/2} f(x) \sin n\omega_0 x \, dx + \frac{2}{T_0} \int_0^{T_0/2} f(t) \sin n\omega_0 t \, dt$$

$$b_n = 0 \qquad\qquad\qquad \textbf{14.19}$$

The preceding analysis indicates that the Fourier series for an even periodic function consists only of a constant term and cosine terms. Therefore, if $f(t)$ is even, $b_n = 0$ and from Eqs. (14.10) and (14.14), c_n are real and θ_n are multiples of $180°$.

Odd-Function Symmetry   A function is said to be odd if

$$f(t) = -f(-t) \qquad\qquad\qquad \textbf{14.20}$$

An example of an odd function is $\sin n\omega_0 t$. Another example is the waveform in Fig. 14.4a. Following the mathematical development that led to Eqs. (14.17) to (14.19), we can show that for an odd function the Fourier coefficients are

$$a_0 = 0 \qquad\qquad\qquad \textbf{14.21}$$

$$a_n = 0 \qquad \text{for all } n > 0 \qquad\qquad \textbf{14.22}$$

$$b_n = \frac{4}{T_0} \int_0^{T_0/2} f(t) \sin n\omega_0 t \, dt \qquad\qquad \textbf{14.23}$$

(a)

(b)

(c)

Figure 14.4   *Three waveforms; (a) and (c) possess half-wave symmetry.*

Therefore, if $f(t)$ is odd, $a_n = 0$ and, from Eqs. (14.10) and (14.14), $\mathbf{c}_n$ are pure imaginary and θ_n are odd multiples of 90°.

Half-Wave Symmetry A function is said to possess *half-wave symmetry* if

$$f(t) = -f\left(t - \frac{T_0}{2}\right) \qquad\qquad \textbf{14.24}$$

Basically, this equation states that each half-cycle is an inverted version of the adjacent half-cycle; that is, if the waveform from $-T_0/2$ to 0 is inverted, it is identical to the waveform from 0 to $T_0/2$. The waveforms shown in Figs. 14.4a and c possess half-wave symmetry.

Once again we can derive the expressions for the Fourier coefficients in this case by repeating the mathematical development that led to the equations for even-function symmetry using the change of variable $t = x + T_0/2$ and Eq. (14.24). The results of this development are the following equations:

$$a_0 = 0 \qquad\qquad \textbf{14.25}$$

$$a_n = b_n = 0 \qquad\qquad \text{for } n \text{ even} \qquad\qquad \textbf{14.26}$$

$$a_n = \frac{4}{T_0} \int_0^{T_0/2} f(t) \cos n\omega_0 t \, dt \qquad \text{for } n \text{ odd} \qquad\qquad \textbf{14.27}$$

$$b_n = \frac{4}{T_0} \int_0^{T_0/2} f(t) \sin n\omega_0 t \, dt \qquad \text{for } n \text{ odd} \qquad\qquad \textbf{14.28}$$

The following equations are often useful in the evaluation of the trigonometric Fourier series coefficients:

$$\int \sin ax \, dx = -\frac{1}{a} \cos ax$$

$$\int \cos ax \, dx = \frac{1}{a} \sin ax$$

$$\int x \sin ax \, dx = \frac{1}{a^2} \sin ax - \frac{1}{a} x \cos ax \qquad\qquad \textbf{14.29}$$

$$\int x \cos ax \, dx = \frac{1}{a^2} \cos ax + \frac{1}{a} x \sin ax$$

LEARNING Example 14.2

We wish to find the trigonometric Fourier series for the periodic signal in Fig. 14.3.

SOLUTION The waveform exhibits even-function symmetry and therefore

$$a_0 = 0$$

$$b_n = 0 \quad \text{for all } n$$

The waveform exhibits half-wave symmetry and therefore

$$a_n = 0 \quad \text{for } n \text{ even}$$

Hence,

$$a_n = \frac{4}{T_0} \int_0^{T/2} f(t) \cos n\omega_0 t \, dt \quad \text{for } n \text{ odd}$$

$$= \frac{4}{T} \left(\int_0^{T/4} V \cos n\omega_0 t \, dt - \int_{T/4}^{T/2} V \cos n\omega_0 t \, dt \right)$$

$$= \frac{4V}{n\omega_0 T} \left(\sin n\omega_0 t \Big|_0^{T/4} - \sin n\omega_0 t \Big|_{T/4}^{T/2} \right)$$

$$= \frac{4V}{n\omega_0 T} \left(\sin \frac{n\pi}{2} - \sin n\pi + \sin \frac{n\pi}{2} \right)$$

$$= \frac{8V}{n2\pi} \sin \frac{n\pi}{2} \quad \text{for } n \text{ odd}$$

$$= \frac{4V}{n\pi} \sin \frac{n\pi}{2} \quad \text{for } n \text{ odd}$$

The reader should compare this result with that obtained in Example 14.1.

LEARNING Example 14.3

Let us determine the trigonometric Fourier series expansion for the waveform shown in Fig. 14.4a.

SOLUTION The function not only exhibits odd-function symmetry, but it possesses half-wave symmetry as well. Therefore, it is necessary to determine only the coefficients b_n for n odd. Note that

$$v(t) = \begin{cases} \dfrac{4Vt}{T_0} & 0 \le t \le T_0/4 \\[2mm] 2V - \dfrac{4Vt}{T_0} & T_0/4 < t \le T_0/2 \end{cases}$$

The b_n coefficients are then

$$b_n = \frac{4}{T_0} \int_0^{T_0/4} \frac{4Vt}{T_0} \sin n\omega_0 t \, dt$$

$$+ \frac{4}{T_0} \int_{T_0/4}^{T_0/2} \left(2V - \frac{4Vt}{T_0} \right) \sin n\omega_0 t \, dt$$

The evaluation of these integrals is tedious but straightforward and yields

$$b_n = \frac{8V}{n^2\pi^2} \sin \frac{n\pi}{2} \quad \text{for } n \text{ odd}$$

Hence, the Fourier series expansion is

$$v(t) = \sum_{\substack{n=1 \\ n \text{ odd}}}^{\infty} \frac{8V}{n^2\pi^2} \sin \frac{n\pi}{2} \sin n\omega_0 t$$

LEARNING Example 14.4

We wish to find the trigonometric Fourier series expansion of the waveform in Fig. 14.4b.

SOLUTION Note that this waveform has an average value of $\frac{3}{2}$.

Therefore, instead of determining the Fourier series expansion of $f(t)$, we will determine the Fourier series for $f(t) - 3/2$, which is the waveform shown in Fig. 14.4c. The latter waveform

(continued)

possesses half-wave symmetry. The function is also odd and therefore

$$b_n = \frac{4}{T_0} \int_0^{T_0/2} \tfrac{1}{2} \sin n\omega_0 t \, dt$$

$$= \frac{2}{T_0} \left(\frac{-1}{n\omega_0} \cos n\omega_0 t \bigg|_0^{T_0/2} \right)$$

$$= \frac{-2}{n\omega_0 T_0} (\cos n\pi - 1)$$

$$= \frac{2}{n\pi} \quad \text{for } n \text{ odd}$$

Therefore, the Fourier series expansion for $f(t) - \frac{3}{2}$ is

$$f(t) - \tfrac{3}{2} = \sum_{\substack{n=1 \\ n \text{ odd}}}^{\infty} \frac{2}{n\pi} \sin n\omega_0 t$$

or

$$f(t) = \tfrac{3}{2} + \sum_{\substack{n=1 \\ n \text{ odd}}}^{\infty} \frac{2}{n\pi} \sin n\omega_0 t$$

LEARNING EXTENSIONS

E14.3 Determine the type of symmetry exhibited by the waveform in Figs. E14.2 and E14.3.

ANSWER Figure E14.2, even symmetry; Fig. E14.3, half-wave symmetry.

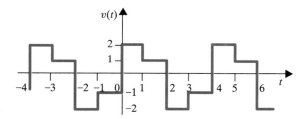

Figure E14.3

E14.4 Find the trigonometric Fourier series for the voltage waveform in Fig. E14.2.

ANSWER

$$v(t) = 2 + \sum_{n=1}^{\infty} \frac{4}{n\pi} \left(2 \sin \frac{2\pi n}{3} - \sin \frac{n\pi}{3} \right) \cos \frac{n\pi}{3} t \text{ V.}$$

E14.5 Find the trigonometric Fourier series for the voltage waveform in Fig. E14.3.

ANSWER $v(t) = \sum_{\substack{n=1 \\ n \text{ odd}}}^{\infty} \frac{2}{n\pi} \sin \frac{n\pi}{2} \cos \frac{n\pi}{2} t$

$$+ \frac{2}{n\pi} (2 - \cos n\pi) \sin \frac{n\pi}{2} t \text{ V.}$$

Fourier Series Via PSPICE Simulation

Most of the circuit simulators are capable of performing Fourier series calculations. The following example illustrates the manner in which PSPICE determines the Fourier components for a specified waveform.

LEARNING Example 14.5

Let us use PSPICE to determine the Fourier series of the waveform in Figure 14.5a.

SOLUTION Our procedure consists of 4 steps: (1) create the PSPICE schematic; (2) create the waveform of interest; (3) set up

(a)

(b)

(c)

(d)

Figure 14.5 *(a) The waveform for Example 14.5. (b) The PSPICE schematic, (c) the VPWL_FILE source attributes, (d) the PWL1.TXT file, and (e) the Transient setup requirements for the Fourier analysis simulation.*

(continued)

(e)

Figure 14.5 (continued)

the simulation particulars; and (4) view the results. The schematic in Fig. 14.5b satisfies step 1 where the piecewise linear source VPWL_FILE from the SOURCE library will provide the signal in Fig. 14.5a.

For step 2, we edit the source attributes by double-clicking on the VPWL_FILE symbol to open the dialog box in Fig. 14.5c. Attributes **DC** and **AC** are the voltage values for **DC** and **AC** Sweep analyses, respectively. Since the Fourier series is calculated from a transient analysis, both the **DC** and **AC** attributes can be set to zero. The **FILE** attribute specifies the filename for the piecewise linear waveform. Here, we have chosen the name PWL1.TXT. Attributes **TSF** and **VSF** are scale factors for the time and voltage axes, respectively, of the waveform in PWL1.TXT. For example, if **VSF=2**, then the voltage values in PWL1.TXT are multiplied by 2. Finally, **REPEAT_VALUE** is the number of cycles the waveform repeats. Next, we must enter (time, voltage) data pairs into PWL1.TXT to create the actual waveform. Double-click on the FILE=PWL1.TXT text and, when asked whether you would like to create the file, select **OK**. The NOTEPAD text editor will open as shown in Fig. 14.5d where the required data pairs for the waveform of interest have already

been entered. When finished, save and exit the file to return to PSPICE.

In step 3, we must specify a transient analysis with the appropriate Fourier series information. To do this, select SETUP from the **Analysis** menu, then choose TRANSIENT. A dialog box similar to that in Fig. 14.5e will appear. The box has already been edited to (1) enable the Fourier analysis, (2) set the fundamental frequency at 1 Hz (PSPICE calls this the Center Frequency), (3) request 20 harmonics, and (4) specify the voltage Vs for Fourier analysis.

To simulate the circuit, select SIMULATE from the **Analysis** menu. The Fourier results are in the output file, which can be accessed from the *Schematics* menu **Analysis/Examine Output**, or from the PROBE menu **View/Output File**. The results are shown in Table 14.1 for the trigonometric series

$$v_S(t) = a_0 + \sum_{n=1}^{\infty} b_n \sin(n\omega_0 t + \theta_n)$$

Note that the dc component, a_0, is essentially zero, as expected from Fig. 14.5a. The 5th, 10th, 15th, ... harmonics are also zero. Additionally, all phases are either zero (very small) or 180 degrees.

Table 14.1 Fourier analysis results for Example 14.5. FOURIER COMPONENTS OF TRANSIENT RESPONSE V(Vs)
DC COMPONENT = −1.353267E-08

Harmonic No.	Frequency (Hz)	Fourier Component	Normalized Component	Phase (deg)	Normalized Phase (deg)
1	1.000E+00	4.222E+00	1.000E+00	2.969E−07	0.000E+00
2	2.000E+00	1.283E+00	3.039E−01	1.800E+02	1.800E+02
3	3.000E+00	4.378E−01	1.037E−0	−1.800E+02	−1.800E+02
4	4.000E+00	3.838E−01	9.090E−02	4.620E−07	−7.254E−07
5	5.000E+00	1.079E−04	2.556E−05	1.712E−03	1.711E−03
6	6.000E+00	1.703E−01	4.034E−02	1.800E+02	1.800E+02
7	7.000E+00	8.012E−02	1.898E−02	−9.548E−06	−1.163E−05
8	8.000E+00	8.016E−02	1.899E−02	5.191E−06	2.816E−06
9	9.000E+00	5.144E−02	1.218E−02	−1.800E+02	−1.800E+02
10	1.000E+01	1.397E−04	3.310E−05	1.800E+02	1.800E+02
11	1.100E+01	3.440E−02	8.149E−03	−1.112E−04	−1.145E−04
12	1.200E+01	3.531E−02	8.364E−03	1.800E+02	1.800E+02
13	1.300E+01	2.343E−02	5.549E−03	1.800E+02	1.800E+02
14	1.400E+01	3.068E−02	7.267E−03	−3.545E−05	−3.960E−05
15	1.500E+01	3.379E−04	8.003E−05	−3.208E−03	−3.212E−03
16	1.600E+01	2.355E−02	5.579E−03	−1.800E+02	−1.800E+02
17	1.700E+01	1.309E−02	3.101E−03	2.905E−04	2.854E−04
18	1.800E+01	1.596E−02	3.781E−03	−5.322E−05	−5.856E−05
19	1.900E+01	1.085E−02	2.569E−03	−1.800E+02	−1.800E+02
20	2.000E+01	2.994E−04	7.092E−05	1.800E+02	1.800E+02

Total harmonic distortion = 3.378352E+01 percent

This procedure is very useful in validating calculations done by hand. As an aside, the accuracy of the Fourier analysis is affected by the TRANSIENT setup parameters in Fig. 14.5e. In general, accuracy increases as the number of cycles in the simulation decreases and as the number of data points increases. The **Final Time** determines the number of cycles, and increasing the **Step Ceiling** will increase the number of data points.

TIME-SHIFTING Let us now examine the effect of time-shifting a periodic waveform $f(t)$ defined by the equation

$$f(t) = \sum_{n=-\infty}^{\infty} c_n e^{jn\omega_0 t}$$

Note that

$$f(t - t_0) = \sum_{n=-\infty}^{\infty} c_n e^{jn\omega_0(t-t_0)}$$

$$f(t - t_0) = \sum_{n=-\infty}^{\infty} \left(c_n e^{-jn\omega_0 t_0}\right) e^{jn\omega_0 t} \qquad \textbf{14.30}$$

Since $e^{-jn\omega_0 t_0}$ corresponds to a phase shift, the Fourier coefficients of the time-shifted function are the Fourier coefficients of the original function with the angle shifted by an amount directly proportional to frequency. Therefore, time shift in the time domain corresponds to phase shift in the frequency domain.

LEARNING Example 14.6

Let us time delay the waveform in Fig. 14.3 by a quarter period and compute the Fourier series.

SOLUTION The waveform in Fig. 14.3 time delayed by $T_0/4$ is shown in Fig. 14.6. Since the time delay is $T_0/4$,

$$n\omega_0 t_d = n \frac{2\pi}{T_0} \frac{T_0}{4} = n \frac{\pi}{2} = n\,90°$$

Therefore, using Eq. (14.30) and the results of Example 14.1, the Fourier coefficients for the time-shifted waveform are

$$c_n = \frac{2V}{n\pi} \sin \frac{n\pi}{2} \underline{/-n\,90°} \qquad n \text{ odd}$$

and therefore,

$$v(t) = \sum_{\substack{n=1 \\ n \text{ odd}}}^{\infty} \frac{4V}{n\pi} \sin \frac{n\pi}{2} \cos(n\omega_0 t - n\,90°)$$

If we compute the Fourier coefficients for the time-shifted waveform in Fig. 14.6, we obtain

$$c_n = \frac{1}{T_0} \int_{-T_0/2}^{T_0/2} f(t) e^{-jn\omega_0 t}\, dt$$

$$= \frac{1}{T_0} \int_{-T_0/2}^{0} -V e^{-jn\omega_0 t}\, dt + \frac{1}{T_0} \int_{0}^{T_0/2} V e^{-jn\omega_0 t}\, dt$$

$$= \frac{2V}{jn\pi} \qquad \text{for } n \text{ odd}$$

Therefore,

$$c_n = \frac{2V}{n\pi} \underline{/-90°} \qquad n \text{ odd}$$

Since n is odd, we can show that this expression is equivalent to the one obtained earlier.

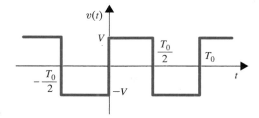

Figure 14.6 *Waveform in Fig. 14.3 time-shifted by $T_0/4$.*

In general, we can compute the phase shift in degrees using the expression

$$\text{phase shift(deg)} = \omega_0 t_d = (360°)\frac{t_d}{T_0} \qquad \textbf{14.31}$$

so that a time shift of one-quarter period corresponds to a 90° phase shift.

As another interesting facet of the time shift, consider a function $f_1(t)$ that is nonzero in the interval $0 \le t \le T_0/2$ and is zero in the interval $T_0/2 < t \le T_0$. For purposes of illustration, let us assume that $f_1(t)$ is the triangular waveform shown in Fig. 14.7a. $f_1(t - T_0/2)$ is then shown in Fig. 14.7b. Then the function $f(t)$ defined as

$$f(t) = f_1(t) - f_1\left(t - \frac{T_0}{2}\right) \qquad \textbf{14.32}$$

is shown in Fig. 14.7c. Note that $f(t)$ has half-wave symmetry. In addition, note that if

$$f_1(t) = \sum_{n=-\infty}^{\infty} \mathbf{c}_n e^{-jn\omega_0 t}$$

then

$$f(t) = f_1(t) - f_1\left(t - \frac{T_0}{2}\right) = \sum_{n=-\infty}^{\infty} \mathbf{c}_n\left(1 - e^{-jn\pi}\right)e^{jn\omega_0 t}$$

$$\qquad \qquad \textbf{14.33}$$

$$= \begin{cases} \displaystyle\sum_{n=-\infty}^{\infty} 2\mathbf{c}_n e^{jn\omega_0 t} & n \text{ odd} \\ 0 & n \text{ even} \end{cases}$$

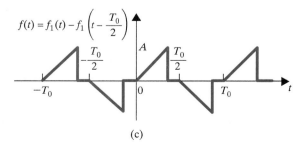

Figure 14.7   *Waveforms that illustrate the generation of half-wave symmetry.*

Therefore, we see that any function with half-wave symmetry can be expressed in the form of Eq. (14.32), where the Fourier series is defined by Eq. (14.33), and c_n is the Fourier coefficient for $f_1(t)$.

LEARNING EXTENSION

E14.6 If the waveform in Fig. E14.1 is time-delayed 1 s, we obtain the waveform in Fig. E14.6. Compute the exponential Fourier coefficients for the waveform in Fig. E14.6 and show that they differ from the coefficients for the waveform in Fig. E14.1 by an angle $n(180°)$.

ANSWER $c_0 = \dfrac{1}{2}$,

$$c_n = -\left(\frac{1 - e^{-jn\pi}}{j2\pi n}\right).$$

Figure E14.6

WAVEFORM GENERATION The magnitude of the harmonics in a Fourier series is independent of the time scale for a given wave shape. Therefore, the equations for a variety of waveforms can be given in tabular form without expressing a specific time scale. Table 14.2 is a set of commonly occurring periodic waves where the advantage of symmetry has been used to simplify the coefficients. These waveforms can be used to generate other waveforms. The level of a wave can be adjusted by changing the average value component; the time can be shifted by adjusting the angle of the harmonics; and two waveforms can be added to produce a third waveform; for example, the waveforms in Figs. 14.8a and b can be added to produce the waveform in Fig. 14.8c.

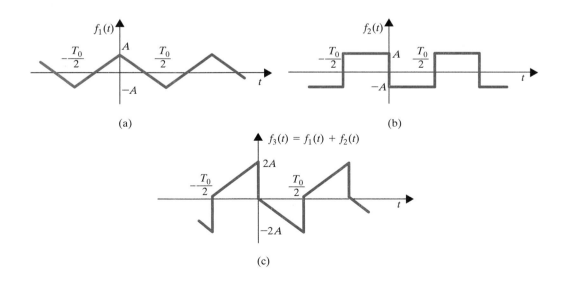

Figure 14.8
Example of waveform generation.

Table 14.2 Fourier series for some common waveforms

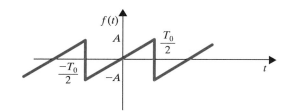

$$f(t) = \sum_{n=1}^{\infty} (-1)^{n+1} \frac{2A}{n\pi} \sin n\omega_0 t$$

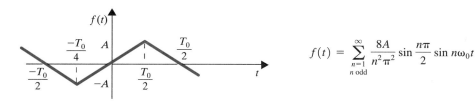

$$f(t) = \sum_{\substack{n=1 \\ n \text{ odd}}}^{\infty} \frac{8A}{n^2\pi^2} \sin \frac{n\pi}{2} \sin n\omega_0 t$$

$$f(t) = \sum_{n=-\infty}^{\infty} \frac{A}{n\pi} \sin \frac{n\pi\delta}{T_0} e^{jn\omega_0[t-(\delta/2)]}$$

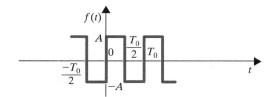

$$f(t) = \sum_{\substack{n=1 \\ n \text{ odd}}}^{\infty} \frac{4A}{n\pi} \sin n\omega_0 t$$

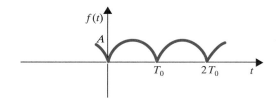

$$f(t) = \frac{2A}{\pi} + \sum_{n=1}^{\infty} \frac{4A}{\pi(1 - 4\pi^2)} \cos n\omega_0 t$$

$$f(t) = \frac{A}{\pi} + \frac{A}{2} \sin \omega_0 t + \sum_{\substack{n=2 \\ n \text{ even}}}^{\infty} \frac{2A}{\pi(1 - n^2)} \cos n\omega_0 t$$

(continued)

Table 14.2 (continued)

$$f(t) = \frac{A}{2} + \sum_{\substack{n=-\infty \\ n \neq 0 \\ n \text{ odd}}}^{\infty} \frac{-2A}{n^2 \pi^2} e^{jn\omega_0 t}$$

$$f(t) = \frac{A}{2} + \sum_{n=1}^{\infty} \frac{-A}{n\pi} \sin n\omega_0 t$$

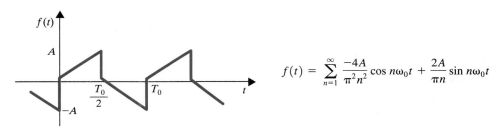

$$f(t) = \sum_{n=1}^{\infty} \frac{-4A}{\pi^2 n^2} \cos n\omega_0 t + \frac{2A}{\pi n} \sin n\omega_0 t$$

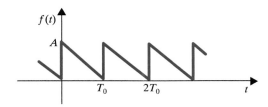

$$f(t) = \frac{A}{2} + \sum_{n=1}^{\infty} \frac{A}{\pi n} \sin n\omega_0 t$$

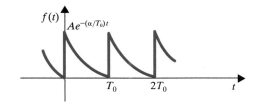

$$f(t) = \sum_{n=-\infty}^{\infty} \frac{A(1 - e^{-\alpha})}{\alpha + j2\pi n} e^{jn\omega_0 t}$$

E14.7   Two periodic waveforms are shown in Fig. E14.7. Compute the exponential Fourier series for each waveform, and then add the results to obtain the Fourier series for the waveform in Fig. E14.2.

ANSWER

$$v_1(t) = \frac{2}{3} + \sum_{\substack{n=-\infty \\ n \neq 0}}^{\infty} \frac{2}{n\pi} \sin \frac{n\pi}{3} e^{jn\omega_0 t},$$

$$v_2(t) = \frac{4}{3} + \sum_{n=-\infty}^{\infty} -\frac{4}{n\pi} \left(\sin \frac{n\pi}{3} - \sin \frac{2n\pi}{3} \right) e^{jn\omega_0 t}.$$

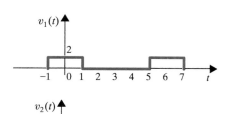

Figure E14.7

FREQUENCY SPECTRUM   The *frequency spectrum* of the function $f(t)$ expressed as a Fourier series consists of a plot of the amplitude of the harmonics versus frequency, which we call the *amplitude spectrum*, and a plot of the phase of the harmonics versus frequency, which we call the *phase spectrum*. Since the frequency components are discrete, the spectra are called *line spectra*. Such spectra illustrate the frequency content of the signal. Plots of the amplitude and phase spectra are based on Eqs. (14.1), (14.3), and (14.7) and represent the amplitude and phase of the signal at specific frequencies.

LEARNING Example 14.7

The Fourier series for the triangular-type waveform shown in Fig. 14.8c with $A = 5$ is given by the equation

$$v(t) = \sum_{\substack{n=1 \\ n \text{ odd}}}^{\infty} \left(\frac{20}{n\pi} \sin n\omega_0 t - \frac{40}{n^2\pi^2} \cos n\omega_0 t \right)$$

We wish to plot the first four terms of the amplitude and phase spectra for this signal.

SOLUTION   Since $D_n \underline{/\theta_n} = a_n - jb_n$, the first four terms for this signal are

$$D_1 \underline{/\theta_1} = -\frac{40}{\pi^2} - j\frac{20}{\pi} = 7.5 \underline{/-122°}$$

$$D_3 \underline{/\theta_3} = -\frac{40}{9\pi^2} - j\frac{20}{3\pi} = 2.2 \underline{/-102°}$$

$$D_5 \underline{/\theta_5} = -\frac{40}{25\pi^2} - j\frac{20}{5\pi} = 1.3 \underline{/-97°}$$

$$D_7 \underline{/\theta_7} = -\frac{40}{49\pi^2} - j\frac{20}{7\pi} = 0.91 \underline{/-95°}$$

Therefore, the plots of the amplitude and phase versus ω are as shown in Fig. 14.9.

(continued)

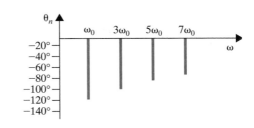

Figure 14.9 *Amplitude and phase spectra.*

LEARNING EXTENSION

E14.8 Determine the trigonometric Fourier series for the voltage waveform in Fig. E14.8 and plot the first four terms of the amplitude and phase spectra for this signal.

ANSWER $a_0 = \frac{1}{2}$,
$D_1 = -j(1/\pi)$,
$D_2 = -j(1/2\pi)$,
$D_3 = -j(1/3\pi)$,
$D_4 = -j(1/4\pi)$.

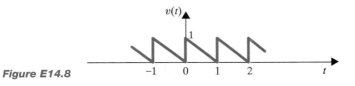

Figure E14.8

STEADY-STATE NETWORK RESPONSE If a periodic signal is applied to a network, the steady-state voltage or current response at some point in the circuit can be found in the following manner. First, we represent the periodic forcing function by a Fourier series. If the input forcing function for a network is a voltage, the input can be expressed in the form

$$v(t) = v_0 + v_1(t) + v_2(t) + \cdots$$

and therefore represented in the time domain as shown in Fig. 14.10. Each source has its own amplitude and frequency. Next we determine the response due to each component of the input Fourier series; that is, we use phasor analysis in the frequency domain to determine the network response due to each source. The network response due to each source in the frequency domain is then transformed to the time domain. Finally, we add the time domain solutions due to each source using the Principle of Superposition to obtain the Fourier series for the total *steady-state* network response.

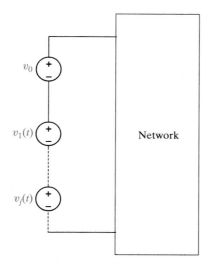

Figure 14.10 Network with a periodic voltage forcing function.

LEARNING Example 14.8

We wish to determine the steady-state voltage $v_o(t)$ in Fig. 14.11 if the input voltage $v(t)$ is given by the expression

$$v(t) = \sum_{\substack{n=1 \\ n\ \text{odd}}}^{\infty} \left(\frac{20}{n\pi} \sin 2nt - \frac{40}{n^2\pi^2} \cos 2nt \right) \text{V}$$

SOLUTION Note that this source has no constant term, and therefore its dc value is zero. The amplitude and phase for the first four terms of this signal are given in Example 14.7, and therefore the signal $v(t)$ can be written as

$$v(t) = 7.5 \cos(2t - 122°) + 2.2 \cos(6t - 102°)$$
$$+ 1.3 \cos(10t - 97°) + 0.91 \cos(14t - 95°) + \cdots$$

From the network we find that

$$\mathbf{I} = \frac{\mathbf{V}}{2 + \dfrac{2/j\omega}{2 + 1/j\omega}} = \frac{\mathbf{V}(1 + 2j\omega)}{4 + 4j\omega}$$

$$\mathbf{I}_1 = \frac{\mathbf{I}(1/j\omega)}{2 + 1/j\omega} = \frac{\mathbf{I}}{1 + 2j\omega}$$

$$\mathbf{V}_o = (1)\mathbf{I}_1 = 1 \cdot \frac{\mathbf{V}(1 + 2j\omega)}{4 + 4j\omega} \frac{1}{1 + 2j\omega} = \frac{\mathbf{V}}{4 + 4j\omega}$$

Therefore, since $\omega_0 = 2$,

$$\mathbf{V}_o(n) = \frac{\mathbf{V}(n)}{4 + j8n}$$

Figure 14.11
RC circuit employed in Example 14.8.

(continued)

The individual components of the output due to the components of the input source are then

$$\mathbf{V}_o(\omega_0) = \frac{7.5 \,\underline{/-122°}}{4 + j8} = 0.84 \,\underline{/-185.4°} \text{ V}$$

$$\mathbf{V}_o(3\omega_0) = \frac{2.2 \,\underline{/-102°}}{4 + j24} = 0.09 \,\underline{/-182.5°} \text{ V}$$

$$\mathbf{V}_o(5\omega_0) = \frac{1.3 \,\underline{/-97°}}{4 + j40} = 0.03 \,\underline{/-181.3°} \text{ V}$$

$$\mathbf{V}_o(7\omega_0) = \frac{0.91 \,\underline{/-95°}}{4 + j56} = 0.016 \,\underline{/-181°} \text{ V}$$

Hence, the steady-state output voltage $v_o(t)$ can be written as

$$v_o(t) = 0.84 \cos(2t - 185.4°) + 0.09 \cos(6t - 182.5°)$$
$$+ 0.03 \cos(10t - 181.3°) + 0.016 \cos(14t - 181°) + \cdots$$

LEARNING EXTENSION

E14.9   Determine the expression for the steady-state current $i(t)$ in Fig. E14.9 if the input voltage $v_S(t)$ is given by the expression

$$v_S(t) = \frac{20}{\pi} + \sum_{n=1}^{\infty} \frac{-40}{\pi(4n^2 - 1)} \cos 2nt \text{ V}$$

ANSWER $i(t) = 2.12$
$$+ \sum_{n=1}^{\infty} \frac{-40}{\pi(4n^2 - 1)} \frac{1}{A_n} \cos(2nt - \theta_n) \text{ A}.$$

CircuitWorks
16

Figure E14.9

AVERAGE POWER   We have shown that when a linear network is forced with a nonsinusoidal periodic signal, voltages and currents throughout the network are of the form

$$v(t) = V_{dc} + \sum_{n=1}^{\infty} V_n \cos\left(n\omega_0 t - \theta_{v_n}\right)$$

and

$$i(t) = I_{dc} + \sum_{n=1}^{\infty} I_n \cos\left(n\omega_0 t - \theta_{i_n}\right)$$

If we employ the passive sign convention and assume that the voltage across an element and the current through it are given by the preceding equations, then from Eq. (9.6),

$$P = \frac{1}{T} \int_{t_0}^{t_0+T} p(t) \, dt$$

$$= \frac{1}{T} \int_{t_0}^{t_0+T} v(t)i(t) \, dt$$

14.34

Note that the integrand involves the product of two infinite series. However, the determination of the average power is actually easier than it appears. First, note that the product $V_{dc}I_{dc}$ when integrated over a period and divided by the period is simply $V_{dc}I_{dc}$. Second, the product of V_{dc} and any harmonic of the current or I_{dc} and any harmonic of the voltage when integrated over a period yields zero. Third, the product of any two *different* harmonics of the voltage and the current when integrated over a period yields zero. Finally, nonzero terms result only from the products of voltage and current at the *same* frequency. Hence, using the mathematical development that follows Eq. (9.6), we find that

$$P = V_{dc}I_{dc} + \sum_{n=1}^{\infty} \frac{V_n I_n}{2} \cos(\theta_{v_n} - \theta_{i_n}) \qquad \text{14.35}$$

LEARNING Example 14.9

In the network in Fig. 14.12, $v(t) = 42 + 16 \cos(377t + 30°) + 12 \cos(754t - 20°)$ V. We wish to compute the current $i(t)$ and determine the average power absorbed by the network.

SOLUTION The capacitor acts as an open circuit to dc, and therefore $I_{dc} = 0$. At $\omega = 377$ rad/s.

$$\frac{1}{j\omega C} = \frac{1}{j(377)(100)(10)^{-6}} = -j26.53 \ \Omega$$

$$j\omega L = j(377)(20)10^{-3} = j7.54 \ \Omega$$

Hence,

$$\mathbf{I}_{377} = \frac{16 \ /30°}{16 + j7.54 - j26.53} = 0.64 \ /79.88° \ \text{A}$$

At $\omega = 754$ rad/s,

$$\frac{1}{j\omega C} = \frac{1}{j(754)(100)(10)^{-6}} = -j13.26 \ \Omega$$

$$j\omega L = j(754)(20)10^{-3} = j15.08 \ \Omega$$

Hence,

$$\mathbf{I}_{754} = \frac{12 \ /20°}{16 + j15.08 - j13.26} = 0.75 \ /-26.49° \ \text{A}$$

Therefore, the current $i(t)$ is

$$i(t) = 0.64 \cos(377t + 79.88°)$$
$$+ 0.75 \cos(754t + 26.49°) \ \text{A}$$

and the average power absorbed by the network is

$$P = (42)(0) + \frac{(16)(0.64)}{2} \cos(30° - 79.88°)$$
$$+ \frac{(12)(0.75)}{2} \cos(-20° + 26.49°)$$
$$= 7.77 \ \text{W}$$

Figure 14.12 Network used in Example 14.9.

E14.10 At the input terminals of a network, the voltage $v(t)$ and the current $i(t)$ are given by the following expression:

ANSWER $P_{ave} = 16.91$ W.

$$v(t) = 64 + 36\cos(377t + 60°) - 24\cos(754t + 102°) \text{ V}$$

$$i(t) = 1.8\cos(377t + 45°) + 1.2\cos(754t + 100°) \text{ A}$$

Find the average power absorbed by the network.

14.2 Fourier Transform

The preceding sections of this chapter have illustrated that the exponential Fourier series can be used to represent a periodic signal for all time. We will now consider a technique for representing an aperiodic signal for all values of time.

Suppose that an aperiodic signal $f(t)$ is as shown in Fig. 14.13a. We now construct a new signal $f_p(t)$ that is identical to $f(t)$ in the interval $-T/2$ to $T/2$ but is *periodic* with period T, as shown in Fig. 14.13b. Since $f_p(t)$ is periodic, it can be represented in the interval $-\infty$ to ∞ by an exponential Fourier series;

$$f_p(t) = \sum_{n=-\infty}^{\infty} \mathbf{c}_n e^{jn\omega_0 t} \qquad \textbf{14.36}$$

where

$$\mathbf{c}_n = \frac{1}{T} \int_{-T/2}^{T/2} f_p(t) e^{-jn\omega_0 t} \, dt \qquad \textbf{14.37}$$

and

$$\omega_0 = \frac{2\pi}{T} \qquad \textbf{14.38}$$

At this point we note that if we take the limit of the function $f_p(t)$ as $T \rightarrow \infty$, the periodic signal in Fig. 14.13b approaches the aperiodic signal in Fig. 14.13a; that is, the repetitious signals centered at $-T$ and $+T$ in Fig. 14.13b are moved to infinity.

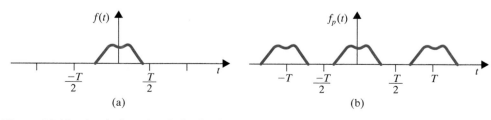

Figure 14.13 *Aperiodic and periodic signals.*

The line spectrum for the periodic signal exists at harmonic frequencies $(n\omega_0)$ and the incremental spacing between the harmonics is

$$\Delta\omega = (n+1)\omega_0 - n\omega_0 = \omega_0 = \frac{2\pi}{T} \qquad \textbf{14.39}$$

As $T \to \infty$ the lines in the frequency spectrum for $f_p(t)$ come closer and closer together, $\Delta\omega$ approaches the differential $d\omega$, and $n\omega_0$ can take on any value of ω. Under these conditions the line spectrum becomes a continuous spectrum. Since as $T \to \infty$, $\mathbf{c}_n \to 0$ in Eq. (14.37), we will examine the product $\mathbf{c}_n T$, where

$$\mathbf{c}_n T = \int_{-T/2}^{T/2} f_p(t)e^{-jn\omega_0 t}\, dt$$

In the limit as $T \to \infty$,

$$\lim_{T\to\infty}(\mathbf{c}_n T) = \lim_{T\to\infty}\int_{-T/2}^{T/2} f_p(t)e^{-jn\omega_0 t}\, dt$$

which in view of the previous discussion can be written as

$$\lim_{T\to\infty}(\mathbf{c}_n T) = \int_{-\infty}^{\infty} f(t)e^{-j\omega t}\, dt$$

This integral is the Fourier transform of $f(t)$, which we will denote as $\mathbf{F}(\omega)$, and hence

$$\mathbf{F}(\omega) = \int_{-\infty}^{\infty} f(t)e^{-j\omega t}\, dt \qquad \textbf{14.40}$$

Similarly, $f_p(t)$ can be expressed as

$$f_p(t) = \sum_{n=-\infty}^{\infty} \mathbf{c}_n e^{jn\omega_0 t}$$

$$= \sum_{n=-\infty}^{\infty} (\mathbf{c}_n T)e^{jn\omega_0 t}\,\frac{1}{T}$$

$$= \sum_{n=-\infty}^{\infty} (\mathbf{c}_n T)e^{jn\omega_0 t}\,\frac{\Delta\omega}{2\pi}$$

which in the limit as $T \to \infty$ becomes

$$f(t) = \frac{1}{2\pi}\int_{-\infty}^{\infty} \mathbf{F}(\omega)e^{j\omega t}\, d\omega \qquad \textbf{14.41}$$

Equations (14.40) and (14.41) constitute what is called the *Fourier transform pair*. Since $\mathbf{F}(\omega)$ is the Fourier transform of $f(t)$ and $f(t)$ is the inverse Fourier transform of $\mathbf{F}(\omega)$, they are normally expressed in the form

$$\mathbf{F}(\omega) = \mathcal{F}[f(t)] = \int_{-\infty}^{\infty} f(t)e^{-j\omega t}\, dt \qquad \textbf{14.42}$$

$$f(t) = \mathcal{F}^{-1}[\mathbf{F}(\omega)] = \frac{1}{2\pi} \int_{-\infty}^{\infty} \mathbf{F}(\omega)e^{j\omega t}\, d\omega \qquad \textbf{14.43}$$

SOME IMPORTANT TRANSFORM PAIRS There are a number of important Fourier transform pairs, and in the following material we derive a number of them and then list some of the more common ones in tabular form.

LEARNING Example 14.10

We wish to derive the Fourier transform for the voltage pulse shown in Fig. 14.14a.

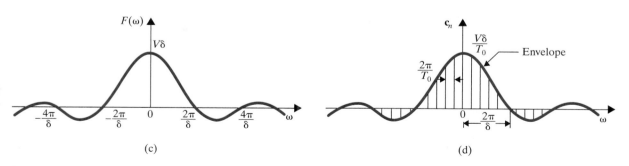

Figure 14.14 *Pulses and their spectra.*

SOLUTION Using Eq. (14.42), the Fourier transform is

$$\mathbf{F}(\omega) = \int_{-\delta/2}^{\delta/2} V e^{-j\omega t} \, dt$$

$$= \frac{V}{-j\omega} e^{-j\omega t} \Big|_{-\delta/2}^{\delta/2}$$

$$= V \frac{e^{-j\omega\delta/2} - e^{+j\omega\delta/2}}{-j\omega}$$

$$= V\delta \frac{\sin(\omega\delta/2)}{\omega\delta/2}$$

Therefore, the Fourier transform for the function

$$f(t) = \begin{cases} 0 & -\infty < t \le -\dfrac{\delta}{2} \\[2mm] V & -\dfrac{\delta}{2} < t \le \dfrac{\delta}{2} \\[2mm] 0 & \dfrac{\delta}{2} < t < \infty \end{cases}$$

is

$$\mathbf{F}(\omega) = V\delta \frac{\sin(\omega\delta/2)}{\omega\delta/2}$$

A plot of this function is shown in Fig. 14.14c. Let us explore this example even further. Consider now the pulse train shown in Fig. 14.14b. Using the techniques that have been demonstrated earlier, we can show that the Fourier coefficients for this waveform are

$$\mathbf{c}_n = \frac{V\delta}{T_0} \frac{\sin(n\omega_0\delta/2)}{n\omega_0\delta/2}$$

The line spectrum for $T_0 = 5\delta$ is shown in Fig. 14.14d.

The equations and figures in this example indicate that as $T_0 \to \infty$ and the periodic function becomes aperiodic, the lines in the discrete spectrum become denser and the amplitude gets smaller, and the amplitude spectrum changes from a line spectrum to a continuous spectrum. Note that the envelope for the discrete spectrum has the same shape as the continuous spectrum. Since the Fourier series represents the amplitude and phase of the signal at specific frequencies, the Fourier transform also specifies the frequency content of a signal.

LEARNING Example 14.11

Find the Fourier transform for the unit impulse function $\delta(t)$.

SOLUTION The Fourier transform of the unit impulse function $\delta(t - a)$ is

$$\mathbf{F}(\omega) = \int_{-\infty}^{\infty} \delta(t - a) e^{-j\omega t} \, dt$$

Using the sampling property of the unit impulse, we find that

$$\mathbf{F}(\omega) = e^{-j\omega a}$$

and if $a = 0$, then

$$\mathbf{F}(\omega) = 1$$

Note then that the $\mathbf{F}(\omega)$ for $f(t) = \delta(t)$ is *constant for all frequencies*. This is an important property, as we shall see later.

LEARNING Example 14.12

We wish to determine the Fourier transform of the function $f(t) = e^{j\omega_0 t}$.

SOLUTION In this case note that if $\mathbf{F}(\omega) = 2\pi\delta(\omega - \omega_0)$, then

$$f(t) = \frac{1}{2\pi} \int_{-\infty}^{\infty} 2\pi\delta(\omega - \omega_0) e^{j\omega t} \, d\omega$$

$$= e^{j\omega_0 t}$$

Therefore, $f(t) = e^{j\omega_0 t}$ and $\mathbf{F}(\omega) = 2\pi\delta(\omega - \omega_0)$ represent a Fourier transform pair.

E14.11 If $f(t) = \sin \omega_0 t$, find $\mathbf{F}(\omega)$.

ANSWER $\mathbf{F}(\omega) = \pi j[\delta(\omega + \omega_0) - \delta(\omega - \omega_0)]$.

A number of useful Fourier transform pairs are shown in Table 14.3.

Table 14.3 Fourier transform pairs

$f(t)$	$\mathbf{F}(\omega)$
$\delta(t - a)$	$e^{-j\omega a}$
A	$2\pi A\delta(\omega)$
$e^{j\omega_0 t}$	$2\pi\delta(\omega - \omega_0)$
$\cos \omega_0 t$	$\pi\delta(\omega - \omega_0) + \pi\delta(\omega + \omega_0)$
$\sin \omega_0 t$	$j\pi\delta(\omega + \omega_0) - j\pi\delta(\omega - \omega_0)$
$e^{-at}u(t), a > 0$	$\dfrac{1}{a + j\omega}$
$e^{-\alpha|t|}, a > 0$	$\dfrac{2a}{a^2 + \omega^2}$
$e^{-at}\cos \omega_0 t u(t), a > 0$	$\dfrac{j\omega + a}{(j\omega + a)^2 + \omega_0^2}$
$e^{-at}\sin \omega_0 t u(t), a > 0$	$\dfrac{\omega_0}{(j\omega + a)^2 + \omega_0^2}$

SOME PROPERTIES OF THE FOURIER TRANSFORM The Fourier transform defined by the equation

$$\mathbf{F}(\omega) = \int_{-\infty}^{\infty} f(t)e^{-j\omega t}\, dt$$

has a number of important properties. Table 14.4 provides a short list of a number of these properties.

The proofs of these properties are generally straightforward; however, as an example we will demonstrate the time convolution property.

If $\mathcal{F}[f_1(t)] = \mathbf{F}_1(\omega)$ and $\mathcal{F}[f_2(t)] = \mathbf{F}_2(\omega)$, then

$$\mathcal{F}\left[\int_{-\infty}^{\infty} f_1(x)f_2(t - x)\, dx\right] = \int_{t=-\infty}^{\infty}\int_{x=-\infty}^{\infty} f_1(x)f_2(t - x)\, dx\, e^{-j\omega t}\, dt$$

$$= \int_{x=-\infty}^{\infty} f_1(x)\int_{t=-\infty}^{\infty} f_2(t - x)e^{-j\omega t}\, dt\, dx$$

Table 14.4 Properties of the Fourier transform

f(t)	F(ω)	Property
$Af(t)$	$A\mathbf{F}(\omega)$	Linearity
$f_1(t) \pm f_2(t)$	$\mathbf{F}_1(\omega) \pm \mathbf{F}_2(\omega)$	
$f(at)$	$\dfrac{1}{a}\mathbf{F}\!\left(\dfrac{\omega}{a}\right), a > 0$	Time-scaling
$f(t - t_0)$	$e^{-j\omega t_0}\mathbf{F}(\omega)$	Time-shifting
$e^{j\omega t_0} f(t)$	$\mathbf{F}(\omega - \omega_0)$	Modulation
$\dfrac{d^n f(t)}{dt^n}$	$(j\omega)^n \mathbf{F}(\omega)$	
$t^n f(t)$	$(j)^n \dfrac{d^n \mathbf{F}(\omega)}{d\omega^n}$	Differentiation
$\displaystyle\int_{-\infty}^{\infty} f_1(x) f_2(t - x)\, dx$	$\mathbf{F}_1(\omega)\mathbf{F}_2(\omega)$	Convolution
$f_1(t) f_2(t)$	$\dfrac{1}{2\pi}\displaystyle\int_{-\infty}^{\infty} \mathbf{F}_1(x)\mathbf{F}_2(\omega - x)\, dx$	

If we now let $u = t - x$, then

$$\mathcal{F}\left[\int_{-\infty}^{\infty} f_1(x) f_2(t - x)\, dx\right] = \int_{x=-\infty}^{\infty} f_1(x) \int_{u=-\infty}^{\infty} f_2(u) e^{-j\omega(u+x)}\, du\, dx$$

$$= \int_{x=-\infty}^{\infty} f_1(x) e^{-j\omega x} \int_{u=-\infty}^{\infty} f_2(u) e^{-j\omega u}\, du\, dx \qquad \textbf{14.44}$$

$$= \mathbf{F}_1(\omega)\mathbf{F}_2(\omega)$$

We should note very carefully the time convolution property of the Fourier transform. With reference to Fig. 14.15, this property states that if $\mathbf{V}_i(\omega) = \mathcal{F}[v_i(t)]$, $\mathbf{H}(\omega) = \mathcal{F}[h(t)]$, and $\mathbf{V}_o(\omega) = \mathcal{F}[v_o(t)]$, then

$$\mathbf{V}_o(\omega) = \mathbf{H}(\omega)\mathbf{V}_i(\omega) \qquad \textbf{14.45}$$

where $\mathbf{V}_i(\omega)$ represents the input signal, $\mathbf{H}(\omega)$ is the network transfer function, and $\mathbf{V}_o(\omega)$ represents the output signal. Equation (14.45) tacitly assumes that the initial conditions of the network are zero.

Figure 14.15   *Representation of the time convolution property.*

E14.12 Determine the output $v_o(t)$ in Fig. E14.12 if the signal $v_i(t) = e^{-t}u(t)$, the network impulse response $h(t) = e^{-2t}u(t)$, and all initial conditions are zero.

ANSWER

$v_o(t) = (e^{-t} - e^{-2t})u(t)$ V.

Figure E14.12

PARSEVAL'S THEOREM   A mathematical statement of Parseval's theorem is

$$\int_{-\infty}^{\infty} f^2(t)\, dt = \frac{1}{2\pi} \int_{-\infty}^{\infty} |\mathbf{F}(\omega)|^2\, d\omega \qquad 14.46$$

This relationship can be easily derived as follows:

$$\int_{-\infty}^{\infty} f^2(t)\, dt = \int_{-\infty}^{\infty} f(t) \frac{1}{2\pi} \int_{-\infty}^{\infty} \mathbf{F}(\omega) e^{j\omega t}\, d\omega\, dt$$

$$= \frac{1}{2\pi} \int_{-\infty}^{\infty} \mathbf{F}(\omega) \int_{-\infty}^{\infty} f(t) e^{-j(-\omega)t}\, dt\, d\omega$$

$$= \int_{-\infty}^{\infty} \frac{1}{2\pi} \mathbf{F}(\omega)\mathbf{F}(-\omega)\, d\omega$$

$$= \int_{-\infty}^{\infty} \frac{1}{2\pi} \mathbf{F}(\omega)\mathbf{F}^*(\omega)\, d\omega$$

$$= \int_{-\infty}^{\infty} \frac{1}{2\pi} |\mathbf{F}(\omega)|^2\, d\omega$$

The importance of Parseval's theorem can be seen if we imagine that $f(t)$ represents the current in a 1-Ω resistor. Since $f^2(t)$ is power and the integral of power over time is energy, Eq. (14.46) shows that we can compute this 1-Ω energy or normalized energy in either the time domain or the frequency domain.

Learning by Application

LEARNING Example 14.13

Consider the network shown in Fig. 14.16a. This network represents a simple low-pass filter as shown in Chapter 11. We wish to illustrate the impact of this network on the input signal by examining the frequency characteristics of the output signal and the relationship between the 1-Ω or normalized energy at the input and output of the network.

SOLUTION The network transfer function is

$$\mathbf{H}(\omega) = \frac{1/RC}{1/RC + j\omega} = \frac{5}{5 + j\omega} = \frac{1}{1 + 0.2j\omega}$$

The Fourier transform of the input signal is

$$\mathbf{V}_i(\omega) = \frac{20}{20 + j\omega} = \frac{1}{1 + 0.05j\omega}$$

Then, using Eq. (14.45), the Fourier transform of the output is

$$\mathbf{V}_o(\omega) = \frac{1}{(1 + 0.2j\omega)(1 + 0.05j\omega)}$$

Using the techniques of Chapter 11, we note that the straight-line log-magnitude plot (frequency characteristic) for these functions is shown in Figs. 14.16b–d. Note that the low-pass filter passes the low frequencies of the input signal but attenuates the high frequencies.

The normalized energy at the filter input is

$$W_i = \int_0^\infty \left(20e^{-20t}\right)^2 dt$$

$$= \frac{400}{-40} e^{-40t} \Big|_0^\infty$$

$$= 10 \text{ J}$$

The normalized energy at the filter output can be computed using Parseval's theorem. Since

$$\mathbf{V}_o(\omega) = \frac{100}{(5 + j\omega)(20 + j\omega)}$$

and

$$|\mathbf{V}_o(\omega)|^2 = \frac{10^4}{(\omega^2 + 25)(\omega^2 + 400)}$$

(a)

(b)

(c)

(d)

Figure 14.16 Low-pass filter, its frequency characteristic, and its output spectra.

(continued)

$\left|\mathbf{V}_o(\omega)\right|^2$ is an even function, and therefore

$$W_o = 2\left(\frac{1}{2\pi}\right)\int_0^\infty \frac{10^4\,d\omega}{(\omega^2 + 25)(\omega^2 + 400)}$$

However, we can use the fact that

$$\frac{10^4}{(\omega^2 + 25)(\omega^2 + 400)} = \frac{10^4/375}{\omega^2 + 25} - \frac{10^4/375}{\omega^2 + 400}$$

Then

$$W_o = \frac{1}{\pi}\left(\int_0^\infty \frac{10^4/375}{\omega^2 + 25}\,d\omega - \int_0^\infty \frac{10^4/375}{\omega^2 + 400}\,d\omega\right)$$

$$= \frac{10^4}{375}\left(\frac{1}{\pi}\right)\left[\frac{1}{5}\left(\frac{\pi}{2}\right) - \frac{1}{20}\left(\frac{\pi}{2}\right)\right]$$

$$= 2.0 \text{ J}$$

Example 14.13 illustrates the effect that $\mathbf{H}(\omega)$ has on the frequency spectrum of the input signal. In general, $\mathbf{H}(\omega)$ can be selected to shape that spectrum in some prescribed manner. As an illustration of this effect, consider the *ideal* frequency spectrums shown in Fig. 14.17. Figure 14.17a shows an ideal input magnitude spectrum $\left|\mathbf{V}_i(\omega)\right|$. $\left|\mathbf{H}(\omega)\right|$ and the output magnitude spectrum $\left|\mathbf{V}_o(\omega)\right|$, which are related by Eq. (14.45), are shown in Figs. 14.17b–e for an *ideal* low-pass, high-pass, band-pass, and band-elimination filter, respectively.

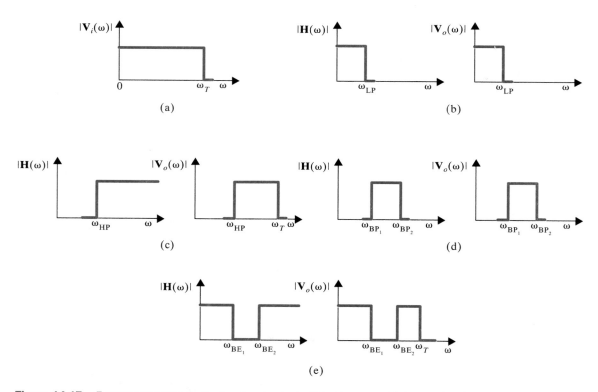

Figure 14.17 Frequency spectra for the input and output of ideal low-pass, high-pass, band-pass, and band-elimination filters.

E14.13 Compute the total 1-Ω energy content of the signal $v_i(t) = e^{-2t}u(t)$ **V** using both the time domain and frequency domain approaches. **ANSWER** $W_T = 0.25$ J.

E14.14 Compute the 1-Ω energy content of the signal $v_i(t) = e^{-2t}u(t)$ **V** in the frequency from 0 to 1 rad/s. **ANSWER** $W = 0.07$ J.

We note that by using Parseval's theorem we can compute the total energy content of a signal using either a time domain or frequency domain approach. However, the frequency domain approach is more flexible in that it permits us to determine the energy content of a signal within some specified frequency band.

Learning by Design

LEARNING Example 14.14

Two nearby AM stations are broadcasting at carrier frequencies

$$f_1 = 900 \text{ kHz}$$

and

$$f_2 = 960 \text{ kHz}$$

respectively. To simplify the analysis, we will assume that the information signals, $s_1(t)$ and $s_2(t)$, are identical. It follows that the Fourier transforms $\mathbf{S}_1(\omega)$ and $\mathbf{S}_2(\omega)$ are also identical, and a sketch of what they might look like is shown in Fig. 14.18.

The broadcast waveforms are

$$v_1(t) = \left[1 + s_1(t)\right]\cos(\omega_1 t)$$

and

$$v_2(t) = \left[1 + s_2(t)\right]\cos(\omega_2 t)$$

An antenna in the vicinity will "pick up" both broadcasts. Assuming that $v_1(t)$ and $v_2(t)$ are of equal strength at the antenna, the voltage received is

$$v_r(t) = K\left[v_1(t) + v_2(t)\right]$$

where K is a constant much less than 1. (Typical antenna voltages are in the μV-to-mV range). A sketch of the Fourier transform of $v_r(t)$ is shown in Fig. 14.19.

Before passing $v_r(t)$ on to amplifying and decoding circuitry, we must first employ a tuner to select a particular station. Let us design an *RLC* band-pass filter that contains a variable capacitor to serve as our tuner. Such a circuit is shown in Fig. 14.20.

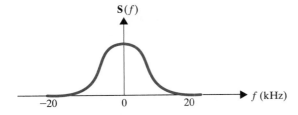

Figure 14.18 *Sketch of an arbitrary AM Fourier transform.*

Figure 14.19 *Fourier transform of the antenna waveform, $v_r(t)$.*

(continued)

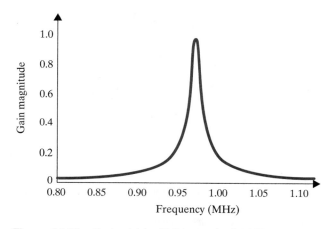

Figure 14.20 *RLC band-pass filter-tuner circuit.*

SOLUTION The transfer function is easily found to be

$$\mathbf{G}_v(s) = \frac{\mathbf{V}_o(s)}{\mathbf{V}_r(s)} = \frac{s\left(\dfrac{R}{L}\right)}{s^2 + s\left(\dfrac{R}{L}\right) + \dfrac{1}{LC}}$$

As shown in Chapter 11, the center frequency and bandwidth can be expressed in hertz as

$$f_o = \frac{1}{2\pi \sqrt{LC}}$$

and

$$\mathrm{BW} = \frac{1}{2\pi}\frac{R}{L}$$

Since the two carrier frequencies are separated by only 60 kHz, the filter bandwidth should be less than 60 kHz. Let us arbitrarily choose a bandwidth of 10 kHz and $R = 10\ \Omega$. Based on this selection, our design involves determining the resulting L and C values. From the expression for bandwidth,

$$L = \frac{1}{2\pi}\frac{R}{\mathrm{BW}}$$

or

$$L = 159.2\ \mu\mathrm{H}$$

Placing the center frequency at 900 kHz, we find C to be

$$C = \frac{1}{L\left[2\pi f_o\right]^2}$$

or

$$C = 196.4\ \mathrm{pF}$$

To tune to 960 kHz we need only change C to 172.6 pF and the bandwidth is unchanged. A Bode plot of the magnitude of $\mathbf{G}_v(s)$ tuned to 960 kHz is shown in Fig. 14.21. Note that the broadcast at 900 kHz, although attenuated, is not completely eliminated. If this is a problem, we can either narrow the bandwidth through R and/or L or design a more complex tuner filter.

Figure 14.21 *Bode plot for RLC tuner circuit of Fig. 14.20.*

LEARNING Example 14.15

The signal expressed in Eq. (14.47) describes a 10-kHz signal swamped in noise that has two frequency components—1 kHz and 100 kHz. From the equation we see that the signal amplitude is only 1/10 that of the noise components. Let us use the circuit in Fig. 14.22 to design a band-pass filter such that the signal amplitude is 100 times that of the noise components. Assume that the op-amps are ideal.

$$v_S(t) = 0.1 \sin\left[(2\pi)10^3 t\right]$$
$$+ 0.01 \sin\left[(2\pi)10^4 t\right] + 0.1 \sin\left[(2\pi)10^5 t\right] \quad \textbf{14.47}$$

SOLUTION Note that the band-pass filter in Fig. 14.22 consists of two identical cascaded stages. We need only determine the gain of a single stage, $\mathbf{A}_1(j\omega)$, since the total gain is

$$\mathbf{A}(j\omega) = \mathbf{A}_1(j\omega)\mathbf{A}_1(j\omega) = \left[\mathbf{A}_1(j\omega)\right]^2$$

Applying KCL at the first op-amp's inverting input, we have

$$\mathbf{V}_A j\omega C = \frac{-\mathbf{V}_o}{R_2}$$

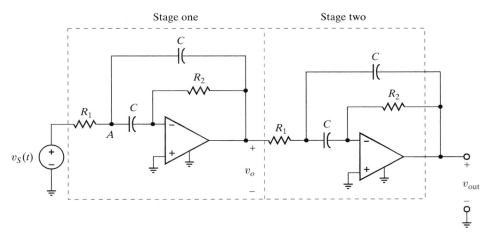

Figure 14.22   *A two-stage, fourth-order band-pass filter.*

or

$$\mathbf{V_A} = \frac{-\mathbf{V}_o}{j\omega C R_2} \qquad \textbf{14.48}$$

Using KCL at node A yields

$$\frac{\mathbf{V}_S - \mathbf{V_A}}{R_1} = (\mathbf{V_A} - \mathbf{V}_o)j\omega C + \mathbf{V_A}j\omega C$$

Multiplying both sides by R_1 and collecting terms produces

$$\mathbf{V}_S = \mathbf{V_A}\big[2j\omega C R_1 + 1\big] - \mathbf{V}_o j\omega C R_1$$

Substituting from Eq. (14.48) and solving for the gain, $\mathbf{A}_1(j\omega) = \mathbf{V}_o/\mathbf{V}_S$,

$$\mathbf{A}_1(j\omega) = \frac{\mathbf{V}_o}{\mathbf{V}_S} = \frac{-j\omega C R_2}{-\omega^2 C^2 R_1 R_2 + j\omega 2 C R_1 + 1}$$

Finally, we rearrange the gain expression in the form

$$\frac{\mathbf{V}_o}{\mathbf{V}_S} = \frac{-\left[\dfrac{1}{C R_1}\right] j\omega}{-\omega^2 + \left[\dfrac{2}{C R_2}\right] j\omega + \dfrac{1}{C^2 R_1 R_2}} \qquad \textbf{14.49}$$

which matches the general form of a band-pass filter given by the expression

$$\frac{\mathbf{V}_o}{\mathbf{V}_S} = \frac{A_o\left[\dfrac{\omega_0}{Q}\right] j\omega}{-\omega^2 + \left[\dfrac{\omega_0}{Q}\right] j\omega + \omega_0^2} \qquad \textbf{14.50}$$

Comparing Eqs. (14.49) and (14.50), we find

$$\omega_0 = \frac{1}{C\sqrt{R_1 R_2}} \qquad \frac{\omega_0}{Q} = \frac{2}{C R_2}$$

$$\textbf{14.51}$$

$$Q = \frac{1}{2}\sqrt{\frac{R_2}{R_1}} \qquad A_o = \frac{-R_2}{2R_1}$$

There are two requirements for the filter performance. First, given the signal and noise amplitudes at the filter input, producing the desired ratio of signal to noise components at the output requires the ratio of the center frequency gain, A_o, to the gains at 1 kHz and 100 kHz to be 1000/1. Since the band-pass gain is symmetric about the center frequency on a log axis, the gains at 1 kHz and 100 kHz will be the same. Thus, we will focus on the gain at 1 kHz only. From Eq. (14.50) the ratio of the single-stage gain at ω_0 and $\omega_0/10$ is

(continued)

$$\frac{A(\omega_0)}{A(\omega_0/10)} = \frac{-A_o}{\dfrac{-A_o \omega_0^2}{10Q}}$$
$$\left| \omega_0^2 - \dfrac{\omega_0^2}{100} + j\dfrac{\omega_0^2}{10Q} \right|$$

$$= \frac{\left| \omega_0^2 - \dfrac{\omega_0^2}{100} + j\dfrac{\omega_0^2}{10Q} \right|}{\dfrac{\omega_0^2}{10Q}} = \sqrt{1000}$$

For simplicity sake, we will assume that $\omega_0^2 \gg \omega_0^2/100$. Solving for Q yields

$$\sqrt{100Q^2 + 1} = \sqrt{1000}$$

or, employing Eq. (14.51),

$$Q \approx \sqrt{10} = \frac{1}{2}\sqrt{\frac{R_2}{R_1}}$$

Thus, the gain requirement forces $R_2 = 40R_1$. Arbitrarily choosing $R_1 = 1$ kΩ fixes R_2 at 40 kΩ. The second requirement is that $\omega_0/2\pi$ must equal 10 kHz. From Eq. (14.51) and our resistor values, we have

$$\omega_0 = (2\pi)10^4 = \frac{1}{C\sqrt{R_1 R_2}} = \frac{1}{10^3 \sqrt{40}C}$$

which yields $C = 2.5$ nF.

The resulting Bode plot is shown in Fig. 14.23a where the center frequency is at 10 kHz, the gain at 10 kHz is roughly 400, and the gains at 1 kHz and 100 kHz are 0.4—a ratio of 1000/1. Output voltage results from a PSPICE transient analysis are shown in Fig. 14.23b for 10 cycles of the 10 kHz signal. Appropriately, the waveform is a 4-V sinewave at 10 kHz with little visible distortion. To view the Fourier components of the output, simply click on the Fast Fourier Transform hot button 🔳 at the top of the PROBE window. From the resulting plot, shown in Fig. 14.23c, we confirm that the signal amplitude is 1000 times larger than the noise components. It should be mentioned that the FFT in Fig. 14.23c is the result of a 5-ms transient simulation. That is 50 cycles at 10 kHz. In general, the more cycles in the transient analysis, the better is the frequency resolution of the FFT. Op-amps from the ANALOG library with gains of 10^6 and supply rails of ± 15 V were used in the simulation.

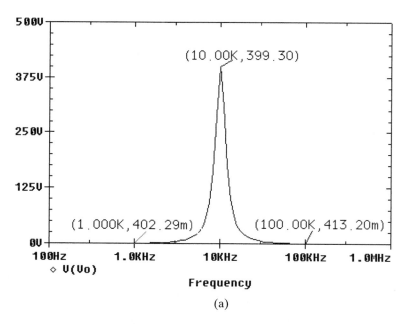

(a)

Figure 14.23 (a) Results from a PSPICE AC Sweep analysis showing the amplification of the signal over the noise. (b) PSPICE transient simulation results and (c) the corresponding FFT.

(b)

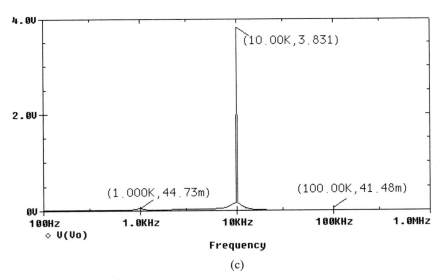

(c)

Figure 14.23   (continued)

LEARNING Check

Summary

A periodic function, its representation using a Fourier series, and some of the useful properties of a Fourier series are outlined here.

- **A periodic function**

$$f(t) = f(t + nT_0), \quad n = 1, 2, 3, \ldots \text{ and } T_0 \text{ is the period}$$

- **Exponential Fourier series of a periodic function**

$$f(t) = \sum_{n=-\infty}^{\infty} c_n e^{jn\omega_0 t}, \quad c_n = \frac{1}{T_0} \int_{t_1}^{t_1 + T_0} f(t) e^{-jn\omega_0 t} \, dt$$

- **Trigonometric Fourier series of a periodic function**

$$f(t) = a_0 + \sum_{n=1}^{\infty} (a_n \cos n\omega_0 t + b_n \sin n\omega_0 t)$$

$$a_n = \frac{2}{T_0} \int_{t_1}^{t_1 + T_0} f(t) \cos n\omega_0 t \, dt,$$

$$b_n = \frac{2}{T_0} \int_{t_1}^{t_1 + T_0} f(t) \sin n\omega_0 t \, dt,$$

and

$$a_0 = \frac{1}{T_0} \int_{t_1}^{t_1 + T_0} f(t) \, dt$$

- **Even symmetry of a periodic function**

$$f(t) = f(-t)$$

$$a_n = \frac{4}{T_0} \int_0^{T_0/2} f(t) \cos n\omega_0 t \, dt,$$

$$b_n = 0,$$

and

$$a_0 = \frac{2}{T_0} \int_0^{T_0/2} f(t) \, dt$$

- **Odd symmetry of a periodic function**

$$f(t) = -f(-t)$$

$$a_n = 0, \quad b_n = \frac{4}{T_0} \int_0^{T_0/2} f(t) \sin n\omega_0 t \, dt, \quad \text{and} \quad a_0 = 0$$

- **Half-wave symmetry of a periodic function**

$$f(t) = -f(t - T_0/2)$$

$$a_n = b_n = 0, \quad \text{for } n \text{ even}$$

$$a_n = \frac{4}{T_0} \int_0^{T_0/2} f(t) \cos n\omega_0 t \, dt \qquad \text{for } n \text{ odd}$$

$$b_n = \frac{4}{T_0} \int_0^{T_0/2} f(t) \sin n\omega_0 t \, dt \qquad \text{for } n \text{ odd and } a_0 = 0$$

- **Time-shifting of a periodic function**

$$f(t - t_0) = \sum_{n=-\infty}^{\infty} (c_n e^{-jn\omega_0 t_0}) e^{jn\omega_0 t}$$

- **Frequency spectrum of a periodic function**
  A Fourier series contains discrete frequency components, called line spectra.

- **Steady-state response of a periodic function input** The periodic function input is expressed as a Fourier series, and phasor analysis is used to determine the response of each component of the series. Each component is transformed to the time domain, and superposition is used to determine the total output.

The Fourier transform, its features and properties, as well as its use in circuit analysis are outlined here.

▶ **Fourier transform for an aperiodic function**

$$\mathbf{F}(\omega) = \int_{-\infty}^{\infty} f(t)e^{-j\omega t} \quad \text{and} \quad f(t) = \frac{1}{2\pi} \int_{-\infty}^{\infty} \mathbf{F}(\omega)e^{j\omega t} d\omega$$

▶ **Fourier transform pairs and properties**

The Fourier transform pairs in Table 14.3 and the properties in Table 14.4 can be used together to transform time domain functions to the frequency domain and vice versa.

▶ **Parseval's theorem for determining the energy content of a signal**

$$\int_{-\infty}^{\infty} f^2(t) \, dt = \frac{1}{2\pi} \int_{-\infty}^{\infty} |\mathbf{F}(\omega)|^2 d\omega$$

▶ **Network response to an aperiodic input** An aperiodic input $x(t)$ can be transformed to the frequency domain as $\mathbf{X}(\omega)$. Then using the network transfer function $\mathbf{H}(\omega)$, the output can be computed as $\mathbf{Y}(\omega) = \mathbf{H}(\omega)\mathbf{X}(\omega)$. $y(t)$ can be obtained transforming $\mathbf{Y}(\omega)$ to the time domain.

Problems

SECTION 14.1

14.1 Find the exponential Fourier series for the signal shown in Fig. P14.1.

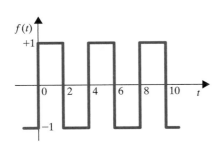

Figure P14.1

14.2 Find the exponential Fourier series for the periodic pulse train shown in Fig. P14.2.

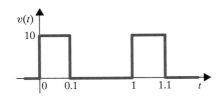

Figure P14.2

14.3 Find the exponential Fourier series for the periodic signal shown in Fig. P14.3.

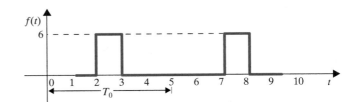

Figure P14.3

14.4 Compute the exponential Fourier series for the waveform that is the sum of the two waveforms in Fig. P14.4 by computing the exponential Fourier series of the two waveforms and adding them.

(a)

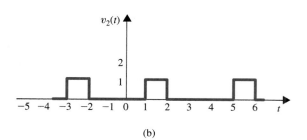

(b)

Figure P14.4

14.5 Find the exponential Fourier series for the signal shown in Fig. P14.5.

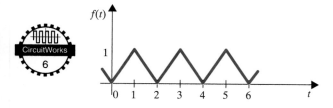

Figure P14.5

14.6 Given the waveform in Fig. P14.6, determine the type of symmetry that exists if the origin is selected at (a) l_1 and (b) l_2.

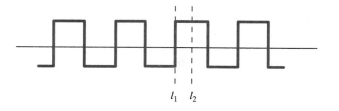

Figure P14.6

14.7 What type of symmetry is exhibited by the two waveforms in Fig. P14.7?

(a)

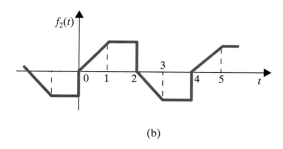

(b)

Figure P14.7

14.8 Derive the trigonometric Fourier series for the waveform shown in Fig. P14.8.

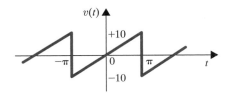

Figure P14.8

14.9 Find the trigonometric Fourier series coefficients for the waveform in Fig. P14.9.

Figure P14.9

14.10 Find the trigonometric Fourier series coefficients for the waveform in Fig. P14.10.

Figure P14.10

14.11 Find the trigonometric Fourier series coefficients for the waveform in Fig. P14.11.

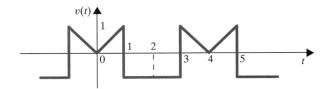

Figure P14.11

14.12 Find the trigonometric Fourier series for the waveform shown in Fig. P14.12.

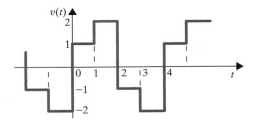

Figure P14.12

14.13 Find the trigonometric Fourier series for the waveform shown in Fig. P14.13.

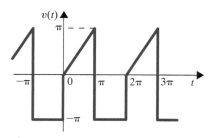

Figure P14.13

14.14 Find the trigonometric Fourier series coefficients for the waveform in Fig. P14.14.

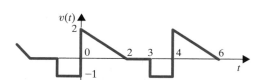

Figure P14.14

14.15 Find the trigonometric Fourier series coefficients for the waveform in Fig. P14.15.

Figure P14.15

14.16 Find the trigonometric Fourier series coefficients for the waveform in Fig. P14.16.

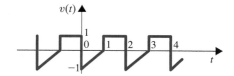

Figure P14.16

14.17 Derive the trigonometric Fourier series for the function shown in Fig. P14.17.

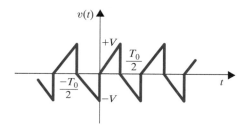

Figure P14.17

14.18 Derive the trigonometric Fourier series for the function $v(t) = A|\sin t|$ as shown in Fig. P14.18.

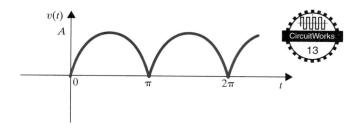

Figure P14.18

14.19 Derive the trigonometric Fourier series for the waveform shown in Fig. P14.19.

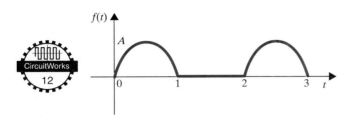

Figure P14.19

14.20 Use PSPICE to determine the Fourier series of the waveform in Fig. P14.20 in the form

$$v_S(t) = a_0 + \sum_{n=1}^{\infty} b_n \sin(n\omega_t + \theta_n)$$

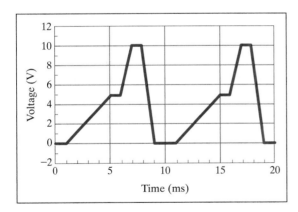

Figure P14.20

14.21 Use PSPICE to determine the Fourier series of the waveform in Fig. P14.21 in the form

$$i_S(t) = a_0 + \sum_{n=1}^{\infty} b_n \sin(n\omega_0 t + \theta_n)$$

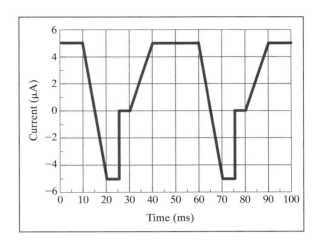

Figure P14.21

14.22 The discrete line spectrum for a periodic function $f(t)$ is shown in Fig. P14.22. Determine the expression for $f(t)$.

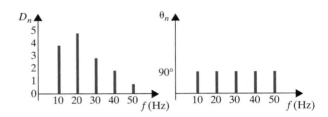

Figure P14.22

14.23 The amplitude and phase spectra for a periodic function $v(t)$ that has only a small number of terms is shown in Fig. P14.23. Determine the expression for $v(t)$ if $T_0 = 0.1$ s.

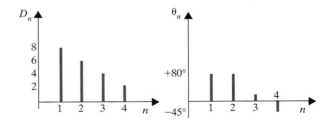

Figure P14.23

14.24 Plot the first four terms of the amplitude and phase spectra for the signal

$$f(t) = \sum_{\substack{n=1 \\ n \text{ odd}}}^{\infty} \frac{-2}{n\pi} \sin \frac{n\pi}{2} \cos n\omega_0 t + \frac{6}{n\pi} \sin n\omega_0 t$$

14.25 Determine the steady-state response of the current $i_o(t)$ in the circuit shown in Fig. P14.25 if the input voltage is described by the waveform shown in Problem 14.8.

Figure P14.25

14.26 If the input voltage in Problem 14.25 is

$$v_S(t) = 1 - \frac{2}{\pi} \sum_{n=1}^{\infty} \frac{1}{n} \sin 0.2\pi n t \text{ V}$$

find the expression for the steady-state current $i_o(t)$.

14.27 Determine the first three terms of the steady-state voltage $v_o(t)$ in Fig. P14.27 if the input voltage is a periodic signal of the form

$$v(t) = \frac{1}{2} + \sum_{n=1}^{\infty} \frac{1}{n\pi} (\cos n\pi - 1) \sin nt \text{ V}$$

Figure P14.27

14.28 The current $i_S(t)$ shown in Fig. P14.28a is applied to the circuit shown in Fig. P14.28b. Determine the expression for the steady-state current $i_o(t)$ using the first four harmonics.

(a)

(b)

Figure P14.28

14.29 Find the average power absorbed by the network in Fig. P14.29 if $v(t) = 60 + 36 \cos(377t + 45°) + 24 \cos(754t - 60°)$ V.

Figure P14.29

14.30 Find the average power absorbed by the 12-Ω resistor in the network in Fig. P14.29 if
$$v(t) = 50 + 25 \cos(377t - 45°)$$
$$+ 12.5 \cos(754t + 45°) \text{ V}.$$

SECTION 14.2

14.31 Determine the Fourier transform of the waveform shown in Fig. P14.31.

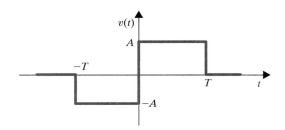

Figure P14.31

14.32 Derive the Fourier transform for the following functions:
(a) $f(t) = e^{-2t} \cos 4t u(t)$
(b) $f(t) = e^{-2t} \sin 4t u(t)$

14.33 Show that
$$\mathcal{F}[f_1(t)f_2(t)] = \frac{1}{2\pi} \int_{-\infty}^{\infty} \mathbf{F}_1(x)\mathbf{F}_2(\omega - x)dx$$

14.34 Find the Fourier transform of the function $f(t) = e^{-a|t|}$.

14.35 Find the Fourier transform of the function $f(t) = 12e^{-2|t|} \cos 4t$.

14.36 Determine the output signal $v_o(t)$ of a network with input signal $v_i(t) = 3e^{-t}u(t)$ and network impulse response $h(t) = e^{-2t}u(t)$. Assume that all initial conditions are zero.

14.37 The input signal to a network is $v_i(t) = e^{-3t}u(t)$. The transfer function of the network is $\mathbf{H}(j\omega) = 1/(j\omega + 4)$. Find the output of the network $v_o(t)$ if the initial conditions are zero.

14.38 The input signal for the network in Fig. P14.38 is $v_i(t) = 10e^{-5t}u(t)$ V. Determine the total 1-Ω energy content of the output $v_o(t)$.

Figure P14.38

14.39 Use the Fourier transform to find $i(t)$ in the network in Fig. P14.39 if $v_i(t) = 2e^{-t}u(t)$.

Figure P14.39

14.40 Determine the voltage $v_o(t)$ in the circuit shown in Fig. P14.40, using the Fourier transform if $i_i(t) = 2e^{-4t}u(t)$ A.

Figure P14.40

14.41 Compute the 1-Ω energy content of the signal $v_o(t)$ in Fig. P14.38 in the frequency range from $\omega = 2$ to $\omega = 4$ rad/s.

14.42 Determine the 1-Ω energy content of the signal $v_o(t)$ in Fig. P14.38 in the frequency range from 0 to 1 rad/s.

Typical Problems Found on the FE Exam

14FE-1 Given the waveform in Fig. 14PFE-1, determine which of the trigonometric Fourier coefficients have zero value, which have nonzero value, and why.

14FE-2 Given the waveform in Fig. 14PFE-2, describe the type of symmetry and its impact on the trigonometric coefficients in the Fourier series—that is, a_0, a_n, and b_n.

Figure 14PFE-1

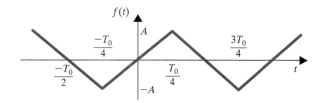

Figure 14PFE-2

Two-Port Networks | 15

We study two-ports and the parameters that describe them for a number of reasons. For example, most circuits or systems have at least two ports. We may put an input signal into one port and obtain an output signal from another. The parameters of the two-port completely describe its behavior in terms of the voltage and current at each port. Thus, knowing the parameters of a two-port network permits us to describe its operation when it is connected into a larger network. Two-port networks are also important in modeling electronic devices and system components. For example, in electronics, two-port networks are employed to model such things as transistors and op-amps. Other examples of electrical components modeled by two-ports are transformers and transmission lines.

In general, we describe the two-port as a network consisting of R, L, and C elements, transformers, op-amps, dependent sources, but no independent sources. The network has an input port and an output port, and as is the case with an op-amp, one terminal may be common to both ports.

Four popular types of two-port parameters are examined: admittance, impedance, hybrid, and transmission. We demonstrate the usefulness of each set of parameters, show how they are related to one another, and finally illustrate how two-port networks can be interconnected in parallel, series, or cascade.

15.1 Admittance Parameters

We say that the linear network in Fig. 15.1a has a single *port*—that is, a single pair of terminals. The pair of terminals *A-B* that constitutes this port could represent a single element (e.g., *R*, *L*, or *C*), or it could be some interconnection of these elements. The linear network in Fig. 15.1b is called a two-port. As a general rule the terminals *A-B* represent the input port and the terminals *C-D* represent the output port.

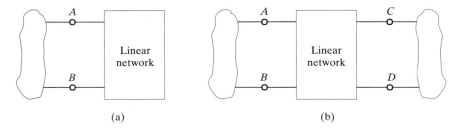

Figure 15.1
(a) Single-port network;
(b) two-port network.

(a)   (b)

In the two-port network shown in Fig. 15.2, it is customary to label the voltages and currents as shown; that is, the upper terminals are positive with respect to the lower terminals, the currents are into the two-port at the upper terminals, and, because KCL must be satisfied at each port, the current is out of the two-port at the lower terminals. Since the network is linear and contains no independent sources, the principle of superposition can be applied to determine the current $\mathbf{I}_1$, which can be written as the sum of two components, one due to $\mathbf{V}_1$ and one due to $\mathbf{V}_2$. Using this principle, we can write

$$\mathbf{I}_1 = \mathbf{y}_{11}\mathbf{V}_1 + \mathbf{y}_{12}\mathbf{V}_2$$

where $\mathbf{y}_{11}$ and $\mathbf{y}_{12}$ are essentially constants of proportionality with units of siemens. In a similar manner $\mathbf{I}_2$ can be written as

$$\mathbf{I}_2 = \mathbf{y}_{21}\mathbf{V}_1 + \mathbf{y}_{22}\mathbf{V}_2$$

Therefore, the two equations that describe the two-port network are

$$\mathbf{I}_1 = \mathbf{y}_{11}\mathbf{V}_1 + \mathbf{y}_{12}\mathbf{V}_2$$
$$\mathbf{I}_2 = \mathbf{y}_{21}\mathbf{V}_1 + \mathbf{y}_{22}\mathbf{V}_2$$

15.1

or in matrix form,

$$\begin{bmatrix} \mathbf{I}_1 \\ \mathbf{I}_2 \end{bmatrix} = \begin{bmatrix} \mathbf{y}_{11} & \mathbf{y}_{12} \\ \mathbf{y}_{21} & \mathbf{y}_{22} \end{bmatrix} \begin{bmatrix} \mathbf{V}_1 \\ \mathbf{V}_2 \end{bmatrix}$$

Note that subscript 1 refers to the input port and subscript 2 refers to the output port, and the equations describe what we will call the *Y parameters* for a network. If these parameters $\mathbf{y}_{11}, \mathbf{y}_{12}, \mathbf{y}_{21}$, and $\mathbf{y}_{22}$ are known, the input/output operation of the two-port is completely defined.

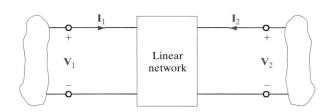

Figure 15.2
Generalized two-port network.

From Eq. (15.1) we can determine the Y parameters in the following manner. Note from the equations that $\mathbf{y}_{11}$ is equal to $\mathbf{I}_1$ divided by $\mathbf{V}_1$ with the output short-circuited (i.e., $\mathbf{V}_2 = 0$).

$$\mathbf{y}_{11} = \left.\frac{\mathbf{I}_1}{\mathbf{V}_1}\right|_{\mathbf{V}_2=0} \qquad\qquad \textbf{15.2}$$

Since $\mathbf{y}_{11}$ is an admittance at the input measured in siemens with the output short-circuited, it is called the *short-circuit input admittance*. The equations indicate that the other Y parameters can be determined in a similar manner:

$$\mathbf{y}_{12} = \left.\frac{\mathbf{I}_1}{\mathbf{V}_2}\right|_{\mathbf{V}_1=0}$$

$$\mathbf{y}_{21} = \left.\frac{\mathbf{I}_2}{\mathbf{V}_1}\right|_{\mathbf{V}_2=0} \qquad\qquad \textbf{15.3}$$

$$\mathbf{y}_{22} = \left.\frac{\mathbf{I}_2}{\mathbf{V}_2}\right|_{\mathbf{V}_1=0}$$

$\mathbf{y}_{12}$ and $\mathbf{y}_{21}$ are called the *short-circuit transfer admittances*, and $\mathbf{y}_{22}$ is called the *short-circuit output admittance*. As a group the Y parameters are referred to as the *short-circuit admittance parameters*. Note that by applying the preceding definitions, these parameters could be determined experimentally for a two-port network whose actual configuration is unknown.

LEARNING Example 15.1

We wish to determine the Y parameters for the two-port network shown in Fig. 15.3a. Once these parameters are known, we will determine the current in a 4-Ω load, which is connected to the output port when a 2-A current source is applied at the input port.

SOLUTION   From Fig. 15.3b, we note that

$$\mathbf{I}_1 = \mathbf{V}_1\left(\tfrac{1}{1} + \tfrac{1}{2}\right)$$

Therefore,

$$\mathbf{y}_{11} = \tfrac{3}{2}\,\mathrm{S}$$

Figure 15.3   *Networks employed in Example 15.1.*

(continued)

As shown in Fig. 15.3c,

$$\mathbf{I}_1 = -\frac{\mathbf{V}_2}{2}$$

and hence,

$$\mathbf{y}_{12} = -\tfrac{1}{2}\,\mathrm{S}$$

Also, $\mathbf{y}_{21}$ is computed from Fig. 15.3b using the equation

$$\mathbf{I}_2 = -\frac{\mathbf{V}_1}{2}$$

and therefore,

$$\mathbf{y}_{21} = -\tfrac{1}{2}\,\mathrm{S}$$

Finally, $\mathbf{y}_{22}$ can be derived from Fig. 15.3c using

$$\mathbf{I}_2 = \mathbf{V}_2(\tfrac{1}{3} + \tfrac{1}{2})$$

and

$$\mathbf{y}_{22} = \tfrac{5}{6}\,\mathrm{S}$$

Therefore, the equations that describe the two-port itself are

$$\mathbf{I}_1 = \tfrac{3}{2}\mathbf{V}_1 - \tfrac{1}{2}\mathbf{V}_2$$

$$\mathbf{I}_2 = -\tfrac{1}{2}\mathbf{V}_1 + \tfrac{5}{6}\mathbf{V}_2$$

These equations can now be employed to determine the operation of the two-port for some given set of terminal conditions. The terminal conditions we will examine are shown in Fig. 15.3d. From this figure we note that

$$\mathbf{I}_1 = 2\,\mathrm{A} \qquad \text{and} \qquad \mathbf{V}_2 = -4\mathbf{I}_2$$

Combining these with the preceding two-port equations yields

$$2 = \tfrac{3}{2}\mathbf{V}_1 - \tfrac{1}{2}\mathbf{V}_2$$

$$0 = -\tfrac{1}{2}\mathbf{V}_1 + \tfrac{13}{12}\mathbf{V}_2$$

or in matrix form

$$\begin{bmatrix} \tfrac{3}{2} & -\tfrac{1}{2} \\ -\tfrac{1}{2} & \tfrac{13}{12} \end{bmatrix} \begin{bmatrix} \mathbf{V}_1 \\ \mathbf{V}_2 \end{bmatrix} = \begin{bmatrix} 2 \\ 0 \end{bmatrix}$$

Note carefully that these equations are simply the nodal equations for the network in Fig. 15.3d. Solving the equations, we obtain $\mathbf{V}_2 = \tfrac{8}{11}\,\mathrm{V}$ and therefore $\mathbf{I}_2 = -\tfrac{2}{11}\,\mathrm{A}$.

LEARNING EXTENSIONS

E15.1 Find the Y parameters for the two-port network shown in Fig. E15.1.

Figure E15.1

ANSWER $\mathbf{y}_{11} = \tfrac{1}{14}\,\mathrm{S}$, $\mathbf{y}_{12} = \mathbf{y}_{21} = -\tfrac{1}{21}\,\mathrm{S}$, $\mathbf{y}_{22} = \tfrac{1}{7}\,\mathrm{S}$.

E15.2 If a 10-A source is connected to the input of the two-port network in Fig. E15.1, find the current in a 5-Ω resistor connected to the output port.

ANSWER $\mathbf{I}_2 = -4.29\,\mathrm{A}$.

15.2 Impedance Parameters

Once again if we assume that the two-port network is a linear network that contains no independent sources, then by means of superposition we can write the input and output voltages as the sum of two components, one due to $\mathbf{I}_1$ and one due to $\mathbf{I}_2$:

$$\mathbf{V}_1 = \mathbf{z}_{11}\mathbf{I}_1 + \mathbf{z}_{12}\mathbf{I}_2$$

$$\mathbf{V}_2 = \mathbf{z}_{21}\mathbf{I}_1 + \mathbf{z}_{22}\mathbf{I}_2$$

15.4

These equations, which describe the two-port network, can also be written in matrix form as

$$\begin{bmatrix} \mathbf{V}_1 \\ \mathbf{V}_2 \end{bmatrix} = \begin{bmatrix} \mathbf{z}_{11} & \mathbf{z}_{12} \\ \mathbf{z}_{21} & \mathbf{z}_{22} \end{bmatrix} \begin{bmatrix} \mathbf{I}_1 \\ \mathbf{I}_2 \end{bmatrix}$$

15.5

Like the Y parameters, these *Z parameters* can be derived as follows:

$$\mathbf{z}_{11} = \left.\frac{\mathbf{V}_1}{\mathbf{I}_1}\right|_{\mathbf{I}_2=0}$$

$$\mathbf{z}_{12} = \left.\frac{\mathbf{V}_1}{\mathbf{I}_2}\right|_{\mathbf{I}_1=0}$$

$$\mathbf{z}_{21} = \left.\frac{\mathbf{V}_2}{\mathbf{I}_1}\right|_{\mathbf{I}_2=0}$$

$$\mathbf{z}_{22} = \left.\frac{\mathbf{V}_2}{\mathbf{I}_2}\right|_{\mathbf{I}_1=0}$$

15.6

In the preceding equations, setting $\mathbf{I}_1$ or $\mathbf{I}_2 = 0$ is equivalent to open-circuiting the input or output port. Therefore, the Z parameters are called the *open-circuit impedance parameters*. $\mathbf{z}_{11}$ is called the *open-circuit input impedance*, $\mathbf{z}_{22}$ is called the *open-circuit output impedance*, and $\mathbf{z}_{12}$ and $\mathbf{z}_{21}$ are termed *open-circuit transfer impedances*.

LEARNING Example 15.2

We wish to find the Z parameters for the network in Fig. 15.4a. Once the parameters are known, we will use them to find the current in a 4-Ω resistor that is connected to the output terminals when a $12\,\underline{/0°}$-V source with an internal impedance of $1 + j0\ \Omega$ is connected to the input.

SOLUTION   From Fig. 15.4a we note that

$$\mathbf{z}_{11} = 2 - j4\ \Omega$$
$$\mathbf{z}_{12} = -j4\ \Omega$$
$$\mathbf{z}_{21} = -j4\ \Omega$$
$$\mathbf{z}_{22} = -j4 + j2 = -j2\ \Omega$$

The equations for the two-port network are, therefore,

$$\mathbf{V}_1 = (2 - j4)\mathbf{I}_1 - j4\mathbf{I}_2$$
$$\mathbf{V}_2 = -j4\mathbf{I}_1 - j2\mathbf{I}_2$$

The terminal conditions for the network shown in Fig. 15.4b are

$$\mathbf{V}_1 = 12\,\underline{/0°} - (1)\mathbf{I}_1$$
$$\mathbf{V}_2 = -4\mathbf{I}_2$$

Combining these with the two-port equations yields

$$12\,\underline{/0°} = (3 - j4)\mathbf{I}_1 - j4\mathbf{I}_2$$
$$0 = -j4\mathbf{I}_1 + (4 - j2)\mathbf{I}_2$$

It is interesting to note that these equations are the mesh equations for the network. If we solve the equations for $\mathbf{I}_2$, we obtain $\mathbf{I}_2 = 1.61\,\underline{/137.73°}$ A, which is the current in the 4-Ω load.

Figure 15.4
*Circuits
employed in
Example 15.2.*

(a)

(b)

LEARNING EXTENSION

E15.3   Find the Z parameters for the network in Fig. E15.3. Then compute the current in a 4-Ω load if a $24\,\underline{/0°}$-V source is connected at the input port.

ANSWER
$\mathbf{I}_2 = -0.73\,\underline{/0°}$ A.

Figure E15.3

15.3 Hybrid Parameters

Under the assumptions used to develop the Y and Z parameters, we can obtain what are commonly called the *hybrid parameters*. In the pair of equations that define these parameters, $\mathbf{V}_1$ and $\mathbf{I}_2$ are the independent variables. Therefore, the two-port equations in terms of the hybrid parameters are

$$\mathbf{V}_1 = \mathbf{h}_{11}\mathbf{I}_1 + \mathbf{h}_{12}\mathbf{V}_2$$

$$\mathbf{I}_2 = \mathbf{h}_{21}\mathbf{I}_1 + \mathbf{h}_{22}\mathbf{V}_2$$

15.7

or in matrix form,

$$\begin{bmatrix} \mathbf{V}_1 \\ \mathbf{I}_2 \end{bmatrix} = \begin{bmatrix} \mathbf{h}_{11} & \mathbf{h}_{12} \\ \mathbf{h}_{21} & \mathbf{h}_{22} \end{bmatrix} \begin{bmatrix} \mathbf{I}_1 \\ \mathbf{V}_2 \end{bmatrix}$$

15.8

These parameters are especially important in transistor circuit analysis. The parameters are determined via the following equations:

$$\mathbf{h}_{11} = \left.\frac{\mathbf{V}_1}{\mathbf{I}_1}\right|_{\mathbf{V}_2=0}$$

$$\mathbf{h}_{12} = \left.\frac{\mathbf{V}_1}{\mathbf{V}_2}\right|_{\mathbf{I}_1=0}$$

15.9

$$\mathbf{h}_{21} = \left.\frac{\mathbf{I}_2}{\mathbf{I}_1}\right|_{\mathbf{V}_2=0}$$

$$\mathbf{h}_{22} = \left.\frac{\mathbf{I}_2}{\mathbf{V}_2}\right|_{\mathbf{I}_1=0}$$

The parameters $\mathbf{h}_{11}$, $\mathbf{h}_{12}$, $\mathbf{h}_{21}$, and $\mathbf{h}_{22}$ represent the *short-circuit input impedance*, the *open-circuit reverse voltage gain*, the *short-circuit forward current gain*, and the *open-circuit output admittance*, respectively. Because of this mix of parameters, they are called *hybrid parameters*. In transistor circuit analysis the parameters $\mathbf{h}_{11}$, $\mathbf{h}_{12}$, $\mathbf{h}_{21}$, and $\mathbf{h}_{22}$ are normally labeled $\mathbf{h}_i$, $\mathbf{h}_r$, $\mathbf{h}_f$, and $\mathbf{h}_o$.

LEARNING Example 15.3

An equivalent circuit for the op-amp in Fig. 15.5a is shown in Fig. 15.5b. We will determine the hybrid parameters for this network.

SOLUTION Parameter $\mathbf{h}_{11}$ is derived from Fig. 15.5c. With the output shorted, $\mathbf{h}_{11}$ is a function of only R_i, R_1, and R_2 and

$$\mathbf{h}_{11} = R_i + \frac{R_1 R_2}{R_1 + R_2}$$

Figure 15.5d is used to derive $\mathbf{h}_{12}$. Since $\mathbf{I}_1 = 0$, $\mathbf{V}_i = 0$ and the relationship between $\mathbf{V}_1$ and $\mathbf{V}_2$ is a simple voltage divider.

$$\mathbf{V}_1 = \frac{V_2 R_1}{R_1 + R_2}$$

Therefore,

$$\mathbf{h}_{12} = \frac{R_1}{R_1 + R_2}$$

KVL and KCL can be applied to Fig. 15.5c to determine $\mathbf{h}_{21}$. The two equations that relate $\mathbf{I}_2$ to $\mathbf{I}_1$ are

$$\mathbf{V}_i = \mathbf{I}_1 R_i$$

$$\mathbf{I}_2 = \frac{-A\mathbf{V}_i}{R_o} - \frac{\mathbf{I}_1 R_1}{R_1 + R_2}$$

Therefore,

$$\mathbf{h}_{21} = -\left(\frac{AR_i}{R_o} + \frac{R_1}{R_1 + R_2}\right)$$

(a)

(b)

(c)

(d)

Figure 15.5 *Circuits employed in Example 15.3.*

Finally, the relationship between I_2 and V_2 in Fig. 15.5d is

$$\frac{V_2}{I_2} = \frac{R_o(R_1 + R_2)}{R_o + R_1 + R_2}$$

and therefore,

$$h_{22} = \frac{R_o + R_1 + R_2}{R_o(R_1 + R_2)}$$

The network equations are, therefore,

$$V_1 = \left(R_i + \frac{R_1 R_2}{R_1 + R_2}\right)I_1 + \frac{R_1}{R_1 + R_2} V_2$$

$$I_2 = -\left(\frac{AR_i}{R_o} + \frac{R_1}{R_1 + R_2}\right)I_1 + \frac{R_o + R_1 + R_2}{R_o(R_1 + R_2)}$$

LEARNING EXTENSIONS

E15.4 Find the hybrid parameters for the network shown in Fig. E15.3.

ANSWER $h_{11} = 14\ \Omega$, $h_{12} = \frac{2}{3}$, $h_{21} = -\frac{2}{3}$, $h_{22} = \frac{1}{9}\ S$.

E15.5 If a 4-Ω load is connected to the output port of the network examined in Extension 15.4, determine the input impedance of the two-port with the load connected.

ANSWER $Z_i = 15.23\ \Omega$.

15.4 Transmission Parameters

The final parameters we will discuss are called the *transmission parameters*. They are defined by the equations

$$V_1 = AV_2 - BI_2$$

$$I_1 = CV_2 - DI_2$$

15.10

or in matrix form,

$$\begin{bmatrix} \mathbf{V}_1 \\ \mathbf{I}_1 \end{bmatrix} = \begin{bmatrix} \mathbf{A} & \mathbf{B} \\ \mathbf{C} & \mathbf{D} \end{bmatrix} \begin{bmatrix} \mathbf{V}_2 \\ -\mathbf{I}_2 \end{bmatrix}$$

15.11

These parameters are very useful in the analysis of circuits connected in cascade, as we will demonstrate later. The parameters are determined via the following equations:

$$\mathbf{A} = \left. \frac{\mathbf{V}_1}{\mathbf{V}_2} \right|_{\mathbf{I}_2=0}$$

$$\mathbf{B} = \left. \frac{\mathbf{V}_1}{-\mathbf{I}_2} \right|_{\mathbf{V}_2=0}$$

15.12

$$\mathbf{C} = \left. \frac{\mathbf{I}_1}{\mathbf{V}_2} \right|_{\mathbf{I}_2=0}$$

$$\mathbf{D} = \left. \frac{\mathbf{I}_1}{-\mathbf{I}_2} \right|_{\mathbf{V}_2=0}$$

A, **B**, **C**, and **D** represent the *open-circuit voltage ratio*, the *negative short-circuit transfer impedance*, the *open-circuit transfer admittance*, and the *negative short-circuit current ratio*, respectively. For obvious reasons the transmission parameters are commonly referred to as the *ABCD parameters*.

LEARNING Example 15.4

We will now determine the transmission parameters for the network in Fig. 15.6.

Figure 15.6
Circuit used in
Example 15.4.

SOLUTION Let us consider the relationship between the variables under the conditions stated in the parameters in Eq. (15.12). For example, with $\mathbf{I}_2 = 0$, $\mathbf{V}_2$ can be written as

$$\mathbf{V}_2 = \frac{\mathbf{V}_1}{1 + 1/j\omega} \left(\frac{1}{j\omega} \right)$$

or

$$\mathbf{A} = \left. \frac{\mathbf{V}_1}{\mathbf{V}_2} \right|_{\mathbf{I}_2=0} = 1 + j\omega$$

Likewise, with $\mathbf{V}_2 = 0$, the relationship between $\mathbf{I}_2$ and $\mathbf{V}_1$ is

$$-\mathbf{I}_2 = \frac{\mathbf{V}_1}{1 + \dfrac{1/j\omega}{1 + 1/j\omega}} \left(\frac{1/j\omega}{1 + 1/j\omega} \right)$$

or

$$\mathbf{B} = \frac{\mathbf{V}_1}{-\mathbf{I}_2} = 2 + j\omega$$

In a similar manner, we can show that $\mathbf{C} = j\omega$ and $\mathbf{D} = 1 + j\omega$.

LEARNING EXTENSION

E15.6 Compute the transmission parameters for the two-port in Fig. E15.1.

ANSWER $\mathbf{A} = 3$,
$\mathbf{B} = 21\ \Omega$, $\mathbf{C} = \frac{1}{6}\,\text{S}$, $\mathbf{D} = \frac{3}{2}$.

15.5 Parameter Conversions

If all the two-port parameters for a network exist, it is possible to relate one set of parameters to another since the parameters interrelate the variables $\mathbf{V}_1$, $\mathbf{I}_1$, $\mathbf{V}_2$, and $\mathbf{I}_2$.

Table 15.1 lists all the conversion formulas that relate one set of two-port parameters to another. Note that Δ_Z, Δ_Y, Δ_H, and Δ_T refer to the determinants of the matrices for the Z, Y, hybrid, and ABCD parameters, respectively. Therefore, given one set of parameters for a network, we can use Table 15.1 to find others.

Table 15.1 Two-port parameter conversion formulas

$$\begin{bmatrix} \mathbf{z}_{11} & \mathbf{z}_{12} \\ \mathbf{z}_{21} & \mathbf{z}_{22} \end{bmatrix} \quad \begin{bmatrix} \dfrac{\mathbf{y}_{22}}{\Delta_Y} & \dfrac{-\mathbf{y}_{12}}{\Delta_Y} \\ \dfrac{-\mathbf{y}_{21}}{\Delta_Y} & \dfrac{\mathbf{y}_{11}}{\Delta_Y} \end{bmatrix} \quad \begin{bmatrix} \dfrac{\mathbf{A}}{\mathbf{C}} & \dfrac{\Delta_T}{\mathbf{C}} \\ \dfrac{1}{\mathbf{C}} & \dfrac{\mathbf{D}}{\mathbf{C}} \end{bmatrix} \quad \begin{bmatrix} \dfrac{\Delta_H}{\mathbf{h}_{22}} & \dfrac{\mathbf{h}_{12}}{\mathbf{h}_{22}} \\ \dfrac{-\mathbf{h}_{21}}{\mathbf{h}_{22}} & \dfrac{1}{\mathbf{h}_{22}} \end{bmatrix}$$

$$\begin{bmatrix} \dfrac{\mathbf{z}_{22}}{\Delta_Z} & \dfrac{-\mathbf{z}_{12}}{\Delta_Z} \\ \dfrac{-\mathbf{z}_{21}}{\Delta_Z} & \dfrac{\mathbf{z}_{11}}{\Delta_Z} \end{bmatrix} \quad \begin{bmatrix} \mathbf{y}_{11} & \mathbf{y}_{12} \\ \mathbf{y}_{21} & \mathbf{y}_{22} \end{bmatrix} \quad \begin{bmatrix} \dfrac{\mathbf{D}}{\mathbf{B}} & \dfrac{-\Delta_T}{\mathbf{B}} \\ \dfrac{1}{\mathbf{B}} & \dfrac{\mathbf{A}}{\mathbf{B}} \end{bmatrix} \quad \begin{bmatrix} \dfrac{1}{\mathbf{h}_{11}} & \dfrac{-\mathbf{h}_{12}}{\mathbf{h}_{11}} \\ \dfrac{\mathbf{h}_{21}}{\mathbf{h}_{11}} & \dfrac{\Delta_H}{\mathbf{h}_{11}} \end{bmatrix}$$

$$\begin{bmatrix} \dfrac{\mathbf{z}_{11}}{\mathbf{z}_{21}} & \dfrac{\Delta_Z}{\mathbf{z}_{21}} \\ \dfrac{1}{\mathbf{z}_{21}} & \dfrac{\mathbf{z}_{22}}{\mathbf{z}_{21}} \end{bmatrix} \quad \begin{bmatrix} \dfrac{-\mathbf{y}_{22}}{\mathbf{y}_{21}} & \dfrac{-1}{\mathbf{y}_{21}} \\ \dfrac{-\Delta_Y}{\mathbf{y}_{21}} & \dfrac{-\mathbf{y}_{11}}{\mathbf{y}_{21}} \end{bmatrix} \quad \begin{bmatrix} \mathbf{A} & \mathbf{B} \\ \mathbf{C} & \mathbf{D} \end{bmatrix} \quad \begin{bmatrix} \dfrac{-\Delta_H}{\mathbf{h}_{21}} & \dfrac{-\mathbf{h}_{11}}{\mathbf{h}_{21}} \\ \dfrac{-\mathbf{h}_{22}}{\mathbf{h}_{21}} & \dfrac{-1}{\mathbf{h}_{21}} \end{bmatrix}$$

$$\begin{bmatrix} \dfrac{\Delta_Z}{\mathbf{z}_{22}} & \dfrac{\mathbf{z}_{12}}{\mathbf{z}_{22}} \\ \dfrac{-\mathbf{z}_{21}}{\mathbf{z}_{22}} & \dfrac{1}{\mathbf{z}_{22}} \end{bmatrix} \quad \begin{bmatrix} \dfrac{1}{\mathbf{y}_{11}} & \dfrac{-\mathbf{y}_{12}}{\mathbf{y}_{11}} \\ \dfrac{\mathbf{y}_{21}}{\mathbf{y}_{11}} & \dfrac{\Delta_Y}{\mathbf{y}_{11}} \end{bmatrix} \quad \begin{bmatrix} \dfrac{\mathbf{B}}{\mathbf{D}} & \dfrac{\Delta_T}{\mathbf{D}} \\ \dfrac{1}{\mathbf{D}} & \dfrac{\mathbf{C}}{\mathbf{D}} \end{bmatrix} \quad \begin{bmatrix} \mathbf{h}_{11} & \mathbf{h}_{12} \\ \mathbf{h}_{21} & \mathbf{h}_{22} \end{bmatrix}$$

LEARNING EXTENSION

E15.7 Determine the Y parameters for a two-port if the Z parameters are

$$\mathbf{Z} = \begin{bmatrix} 18 & 6 \\ 6 & 9 \end{bmatrix}$$

ANSWER $\mathbf{y}_{11} = \frac{1}{14}\,\text{S}$, $\mathbf{y}_{12} = \mathbf{y}_{21} = -\frac{1}{21}\,\text{S}$, $\mathbf{y}_{22} = \frac{1}{7}\,\text{S}$.

15.6 Interconnection of Two-Ports

Interconnected two-port circuits are important because when designing complex systems it is generally much easier to design a number of simpler subsystems that can then be interconnected to form the complete system. If each subsystem is treated as a two-port network, the interconnection techniques described in this section provide some insight into the manner in which a total system may be analyzed and/or designed. Thus, we will now illustrate techniques for treating a network as a combination of subnetworks. We will, therefore, analyze a two-port network as an interconnection of simpler two-ports. Although two-ports can be interconnected in a variety of ways, we will treat only three types of connections: parallel, series, and cascade.

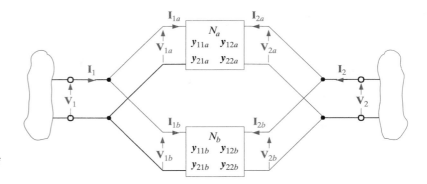

Figure 15.7
Parallel interconnection of two-ports.

For the two-port interconnections to be valid, they must satisfy certain specific requirements that are outlined in the book *Network Analysis and Synthesis* by L. Weinberg, McGraw-Hill, 1962. The following examples will serve to illustrate the interconnection techniques.

In the parallel interconnection case, a two-port N is composed of two-ports N_a and N_b, connected as shown in Fig. 15.7. *Provided that the terminal characteristics of the two networks N_a and N_b are not altered by the interconnection illustrated in the figure*, then the Y parameters for the total network are

$$\begin{bmatrix} \mathbf{y}_{11} & \mathbf{y}_{12} \\ \mathbf{y}_{21} & \mathbf{y}_{22} \end{bmatrix} = \begin{bmatrix} \mathbf{y}_{11a} + \mathbf{y}_{11b} & \mathbf{y}_{12a} + \mathbf{y}_{12b} \\ \mathbf{y}_{21a} + \mathbf{y}_{21b} & \mathbf{y}_{22a} + \mathbf{y}_{22b} \end{bmatrix}$$ **15.13**

and hence to determine the Y parameters for the total network, we simply add the Y parameters of the two networks N_a and N_b.

Likewise, if the two-port N is composed of the series connection of N_a and N_b, as shown in Fig. 15.8, then *once again, as long as the terminal characteristics of the two networks N_a and N_b are not altered by the series interconnection*, the Z parameters for the total network are

$$\begin{bmatrix} \mathbf{z}_{11} & \mathbf{z}_{12} \\ \mathbf{z}_{21} & \mathbf{z}_{22} \end{bmatrix} = \begin{bmatrix} \mathbf{z}_{11a} + \mathbf{z}_{11b} & \mathbf{z}_{12a} + \mathbf{z}_{12b} \\ \mathbf{z}_{21a} + \mathbf{z}_{21b} & \mathbf{z}_{22a} + \mathbf{z}_{22b} \end{bmatrix}$$ **15.14**

Therefore, the Z parameters for the total network are equal to the sum of the Z parameters for the networks N_a and N_b.

Figure 15.8
Series interconnection of two-ports.

Finally, if a two-port N is composed of a cascade interconnection of N_a and N_b, as shown in Fig. 15.9, the equations for the total network are

$$\begin{bmatrix} \mathbf{V}_1 \\ \mathbf{I}_1 \end{bmatrix} = \begin{bmatrix} \mathbf{A}_a & \mathbf{B}_a \\ \mathbf{C}_a & \mathbf{D}_a \end{bmatrix} \begin{bmatrix} \mathbf{A}_b & \mathbf{B}_b \\ \mathbf{C}_b & \mathbf{D}_b \end{bmatrix} \begin{bmatrix} \mathbf{V}_2 \\ -\mathbf{I}_2 \end{bmatrix}$$ **15.15**

Figure 15.9
Cascade interconnection of networks.

Hence, the transmission parameters for the total network are derived by matrix multiplication as indicated previously. The order of the matrix multiplication is important and is performed in the order in which the networks are interconnected.

The cascade interconnection is very useful. Many large systems can be conveniently modeled as the cascade interconnection of a number of stages. For example, the very weak signal picked up by a radio antenna is passed through a number of successive stages of amplification—each of which can be modeled as a two-port subnetwork. In addition, in contrast to the other interconnection schemes, no restrictions are placed on the parameters of N_a and N_b in obtaining the parameters of the two-port resulting from their interconnection.

LEARNING Example 15.5

We wish to determine the Y parameters for the network shown in Fig. 15.10a by considering it to be a parallel combination of two networks as shown in Fig. 15.10b. The capacitive network will be referred to as N_a, and the resistive network will be referred to as N_b.

SOLUTION The Y parameters for N_a are

$$\mathbf{y}_{11a} = j\tfrac{1}{2}\,\mathrm{S} \qquad \mathbf{y}_{12a} = -j\tfrac{1}{2}\,\mathrm{S}$$

$$\mathbf{y}_{21a} = -j\tfrac{1}{2}\,\mathrm{S} \qquad \mathbf{y}_{22a} = j\tfrac{1}{2}\,\mathrm{S}$$

and the Y parameters for N_b are

$$\mathbf{y}_{11b} = \tfrac{3}{5}\,\mathrm{S} \qquad \mathbf{y}_{12b} = -\tfrac{1}{5}\,\mathrm{S}$$

$$\mathbf{y}_{21b} = -\tfrac{1}{5}\,\mathrm{S} \qquad \mathbf{y}_{22b} = \tfrac{2}{5}\,\mathrm{S}$$

Hence, the Y parameters for the network in Fig. 15.10 are

$$\mathbf{y}_{11} = \tfrac{3}{5} + j\tfrac{1}{2}\,\mathrm{S} \qquad \mathbf{y}_{12} = -\left(\tfrac{1}{5} + j\tfrac{1}{2}\right)\mathrm{S}$$

$$\mathbf{y}_{21} = -\left(\tfrac{1}{5} + j\tfrac{1}{2}\right)\mathrm{S} \qquad \mathbf{y}_{22} = \tfrac{2}{5} + j\tfrac{1}{2}\,\mathrm{S}$$

To gain an appreciation for the simplicity of this approach, you need only try to find the Y parameters for the network in Fig. 15.10a directly.

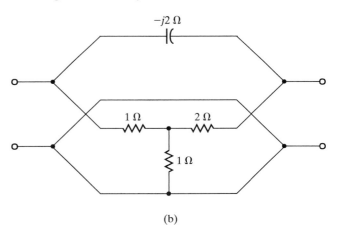

(a)   (b)

Figure 15.10   Network composed of the parallel combination of two subnetworks.

LEARNING Example 15.6

Let us determine the Z parameters for the network shown in Fig. 15.10a. The circuit is redrawn in Fig. 15.11, illustrating a series interconnection. The upper network will be referred to as N_a, and the lower network as N_b.

SOLUTION The Z parameters for N_a, are

$$\mathbf{z}_{11a} = \frac{2 - 2j}{3 - 2j}\,\Omega \qquad \mathbf{z}_{12a} = \frac{2}{3 - 2j}\,\Omega$$

Figure 15.11
Network in Fig. 15.10a redrawn as a series interconnection of two networks.

(continued)

$$\mathbf{z}_{21a} = \frac{2}{3 - 2j} \, \Omega \qquad \mathbf{z}_{22a} = \frac{2 - 4j}{3 - 2j} \, \Omega$$

and the Z parameters for N_b are

$$\mathbf{z}_{11b} = \mathbf{z}_{12b} = \mathbf{z}_{21b} = \mathbf{z}_{22b} = 1 \, \Omega$$

Hence the Z parameters for the total network are

$$\mathbf{z}_{11} = \frac{5 - 4j}{3 - 2j} \, \Omega \qquad \mathbf{z}_{12} = \frac{5 - 2j}{3 - 2j} \, \Omega$$

$$\mathbf{z}_{21} = \frac{5 - 2j}{3 - 2j} \, \Omega \qquad \mathbf{z}_{22} = \frac{5 - 6j}{3 - 2j} \, \Omega$$

We could easily check these results against those obtained in Example 15.5 by applying the conversion formulas in Table 15.1.

LEARNING Example 15.7

Let us derive the two-port parameters of the network in Fig. 15.12 by considering it to be a cascade connection of two networks as shown in Fig. 15.6.

Figure 15.12   Circuit used in Example 15.7.

SOLUTION   The ABCD parameters for the identical T networks were calculated in Example 15.4 to be

$$\mathbf{A} = 1 + j\omega \qquad \mathbf{B} = 2 + j\omega$$

$$\mathbf{C} = j\omega \qquad \mathbf{D} = 1 + j\omega$$

Therefore, the transmission parameters for the total network are

$$\begin{bmatrix} \mathbf{A} & \mathbf{B} \\ \mathbf{C} & \mathbf{D} \end{bmatrix} = \begin{bmatrix} 1 + j\omega & 2 + j\omega \\ j\omega & 1 + j\omega \end{bmatrix} \begin{bmatrix} 1 + j\omega & 2 + j\omega \\ j\omega & 1 + j\omega \end{bmatrix}$$

Performing the matrix multiplication, we obtain

$$\begin{bmatrix} \mathbf{A} & \mathbf{B} \\ \mathbf{C} & \mathbf{D} \end{bmatrix} = \begin{bmatrix} 1 + 4j\omega - 2\omega^2 & 4 + 6j\omega - 2\omega^2 \\ 2j\omega - 2\omega^2 & 1 + 4j\omega - 2\omega^2 \end{bmatrix}$$

Learning by Application

LEARNING Example 15.8

Figure 15.13 is a per phase model used in the analysis of three-phase high-voltage transmission lines. As a general rule in these systems, the voltage and current at the receiving end are known and it is the conditions at the sending end that we wish to find. The transmission parameters perfectly fit this scenario. Thus, we will find the transmission parameters for a reasonable transmission line model, and, then, given the receiving end voltages, power, and power factor, we will find the receiving end current, sending end voltage and current, and the transmission efficiency. Finally, we will plot the efficiency versus the power factor.

SOLUTION   Given a 150-mile-long transmission line, reasonable values for the π circuit elements of the transmission line model are $C = 1.326 \, \mu\text{F}$, $R = 9.0 \, \Omega$, and $L = 264.18 \, \text{mH}$. The transmission parameters can be easily found using the circuits in Fig. 15.14. At 60 Hz, the transmission parameters are

$$\mathbf{A} = 0.9590 \, \underline{/0.27^\circ} \qquad \mathbf{C} = 975.10 \, \underline{/90.13^\circ} \, \mu\text{S}$$

$$\mathbf{B} = 100.00 \, \underline{/84.84^\circ} \, \Omega \qquad \mathbf{D} = 0.9590 \, \underline{/0.27^\circ}$$

Figure 15.13
A π circuit model for
power transmission lines.

$$\left.\frac{\mathbf{V}_2}{\mathbf{V}_1}\right|_{\mathbf{I}_2=0} = \frac{\mathbf{Z}_C}{\mathbf{Z}_C + \mathbf{Z}_L + R}$$

$$\mathbf{A} = \left.\frac{\mathbf{V}_1}{\mathbf{V}_2}\right|_{\mathbf{I}_2=0} = \frac{\mathbf{Z}_C + \mathbf{Z}_L + R}{\mathbf{Z}_C} = 0.9590\,\underline{/0.27°}$$

$$\left.\frac{-\mathbf{I}_2}{\mathbf{V}_1}\right|_{\mathbf{V}_2=0} = \frac{1}{\mathbf{Z}_L + R}$$

$$\mathbf{B} = \left.\frac{\mathbf{V}_1}{-\mathbf{I}_2}\right|_{\mathbf{V}_2=0} = \mathbf{Z}_L + R = 100.00\,\underline{/84.84°}\ \Omega$$

$$\left.\frac{\mathbf{V}_2}{\mathbf{I}_1}\right|_{\mathbf{I}_2=0} = \frac{\mathbf{Z}_C^2}{2\mathbf{Z}_C + \mathbf{Z}_L + R}$$

$$\mathbf{C} = \left.\frac{\mathbf{V}_2}{\mathbf{I}_1}\right|_{\mathbf{I}_2=0} = \frac{2\mathbf{Z}_C + \mathbf{Z}_L + R}{\mathbf{Z}_C^2} = 975.10\,\underline{/90.13°}\ \mu S$$

Figure 15.14
Equivalent circuits used to determine the transmission parameters.

$$\left.\frac{-\mathbf{I}_2}{\mathbf{I}_1}\right|_{\mathbf{V}_2=0} = \frac{\mathbf{Z}_C}{\mathbf{Z}_C + \mathbf{Z}_L + R}$$

$$\mathbf{D} = \left.\frac{\mathbf{I}_1}{-\mathbf{I}_2}\right|_{\mathbf{V}_2=0} = \frac{\mathbf{Z}_C + \mathbf{Z}_L + R}{\mathbf{Z}_C} = 0.950\,\underline{/0.27°}$$

To use the transmission parameters, we must know the receiving end current, $\mathbf{I}_2$. Using standard three-phase circuit analysis outlined in Chapter 10, we find the line current to be

$$\mathbf{I}_2 = -\frac{600\,\underline{/\cos^{-1}(\text{pf})}}{\sqrt{3}(300)(\text{pf})} = -1.215\,\underline{/-18.19°}\ \text{kA}$$

where the line-to-neutral (i.e., phase) voltage at the receiving end, $\mathbf{V}_2$, is assumed to have zero phase. Now, we can use the transmission parameters to determine the sending end voltage and power. Since the line-to-neutral voltage at the receiving end is $300/\sqrt{3} = 173.21$ kV, the results are

$$\begin{aligned}\mathbf{V}_1 &= \mathbf{AV}_2 - \mathbf{BI}_2 = (0.9590\,\underline{/0.27°})(173.21\,\underline{/0°})\\&\quad + (100.00\,\underline{/84.84°})(1.215\,\underline{/-18.19°}) = 241.92\,\underline{/27.67°}\ \text{kV}\end{aligned}$$

$$\begin{aligned}\mathbf{I}_1 &= \mathbf{CV}_2 - \mathbf{DI}_2 = (975.10 \times 10^{-6}\,\underline{/90.13°})(173.21\,\underline{/0°})\\&\quad + (0.9590\,\underline{/0.27°})(1.215\,\underline{/-18.19°}) = 1.12\,\underline{/-9.71°}\ \text{kA}\end{aligned}$$

At the sending end, the power factor and power are

$$\text{pf} = \cos(27.67 - (-9.71)) = \cos(37.38) = 0.80 \text{ lagging}$$

$$P_1 = 3V_1I_1(\text{pf}) = (3)(241.92)(1.12)(0.80) = 650.28 \text{ MW}$$

Finally, the transmission efficiency is

$$\eta = \frac{P_2}{P_1} = \frac{600}{650.28} = 92.3\%$$

This entire analysis can be easily programmed into an EXCEL spreadsheet or MATLAB. A plot of the transmission efficiency versus power factor at the receiving end is shown in Fig. 15.15. We see that as the power factor decreases, the transmission efficiency drops, which increases the cost of production for the power utility. This is precisely why utilities encourage industrial customers to operate as close to unity power factor as possible.

Figure 15.15 *The results of an EXCEL simulation showing the effect of the receiving-end power factor on the transmission efficiency. Because the EXCEL simulation used more significant digits, slight differences exist between the values in the plot and those in the text.*

LEARNING Example 15.9

We have available the noninverting op-amp circuit shown in Fig. 15.16 with the following parameters: $A = 20,000$, $R_i = 1\ M\Omega$, $R_o = 500\ \Omega$, $R_1 = 1\ k\Omega$, and $R_2 = 49\ k\Omega$. To determine the possible applications for this network configuration, we will determine the effect of the load R_L on the gain and the gain error (a comparison of the actual gain with the ideal gain).

Figure 15.16
The classic noninverting gain configuration with load.

SOLUTION In Example 15.3, the hybrid parameters for the noninverting op-amp were found to be

$$\mathbf{h}_{11} = R_i + \frac{R_1 R_2}{R_1 + R_2} \qquad \mathbf{h}_{12} = \frac{R_1}{R_1 + R_2}$$

$$\mathbf{h}_{21} = -\left[\frac{AR_i}{R_o} + \frac{R_1}{R_1 + R_2}\right] \qquad \mathbf{h}_{22} = \frac{R_o + R_1 + R_2}{R_o(R_1 + R_2)}$$

If we solve the hybrid parameter two-port Eqs. (15.7) for $\mathbf{V}_2$, we obtain

$$\mathbf{V}_2 = \frac{-\mathbf{h}_{21}\mathbf{V}_1 + \mathbf{h}_{11}\mathbf{I}_2}{\mathbf{h}_{11}\mathbf{h}_{22} - \mathbf{h}_{12}\mathbf{h}_{21}}$$

Since the op-amp is connected to a load R_L, then

$$\mathbf{I}_2 = -\frac{\mathbf{V}_2}{R_L}$$

Combining these expressions, we obtain the equation for the gain,

$$\frac{\mathbf{V}_2}{\mathbf{V}_1} = \frac{-\mathbf{h}_{21}}{\mathbf{h}_{11}\mathbf{h}_{22} - \mathbf{h}_{12}\mathbf{h}_{21} + \dfrac{\mathbf{h}_{11}}{R_L}}$$

Using the parameter values, the equation becomes

$$\frac{\mathbf{V}_2}{\mathbf{V}_1} = \frac{4 \times 10^7}{8.02 \times 10^5 + \dfrac{10^6}{R_L}} = \frac{49.88}{1 + \dfrac{1.247}{R_L}} \qquad \textbf{15.16}$$

If the term involving R_L remains small compared to unity, then the gain will be largely independent of R_L.

It is convenient to view the gain of the amplifier with respect to its ideal value of

$$A_{ideal} = \frac{\mathbf{V}_2}{\mathbf{V}_1}\bigg|_{ideal\ op\text{-}amp} = 1 + \frac{R_2}{R_1} = 1 + \frac{49}{1} = 50$$

From Eq. (15.16), if R_L is infinite, the gain is only 49.88. This deviation from the ideal performance is caused by nonideal values for the op-amp gain, input resistance, and output resistance. We define the gain error as

$$\text{Gain error} = \frac{A_{actual} - A_{ideal}}{A_{ideal}} = \frac{0.998}{1 + \dfrac{1.247}{R_L}} - 1 \qquad \textbf{15.17}$$

A plot of the gain and the gain error versus R_L is shown in Fig. 15.17. Note that as the load resistance decreases, the gain drops and the error increases—consistent with Eq. (15.17). In addition, as R_L increases, the gain asymptotically approaches the ideal value, never quite reaching it.

To identify specific uses for this amplifier, recognize that at a gain of 50, a 0.1-V input will produce a 5-V output. Three possible applications are as follows:

1. Low-budget audio preamplifier—Amplifies low voltages from tape heads and phonograph needle cartridges to levels suitable for power amplification to drive speakers.
2. Sensor amplifier—In many sensors—for example, temperature dependent resistors—changes in the electrical characteristic (resistance) can be much less than changes in the environmental parameter (temperature). The resulting output voltage changes are also small and usually require amplification.
3. Current sensing—Monitoring large currents can be done inexpensively by using low-value sense resistors and a voltmeter. By Ohm's law, the resulting voltage is IR_{sense} where I is the current of interest. A voltmeter can be used to measure the voltage and, knowing the value of R_{sense}, the current can be determined. The power lost in the sense resistor is $I^2 R_{sense}$. Thus, low power loss implies low sense voltage values, which most inexpensive voltmeters cannot accurately measure. Our simple amplifier can boost the sense voltage to more reasonable levels.

Figure 15.17 *The gain and gain error of the noninverting gain configuration described in Example 15.9.*

Learning by Design

LEARNING Example 15.10

For a particular application, we need an amplifier with a gain of 10,000 when connected to a 1-kΩ load. We have available to us some noninverting op-amps that could be used for this application.

SOLUTION The noninverting op-amp, together with some available components that, based on the results of the previous example (i.e., the ideal gain formula), should yield a gain of 10,000, are shown in Fig. 15.18. Using the hybrid parameter equations for the amplifier, as outlined in the previous example, yields

$$\mathbf{h}_{11} = 1.001 \text{ M}\Omega \qquad \mathbf{h}_{12} = 1.000 \times 10^{-4}$$

$$\mathbf{h}_{21} = -4.000 \times 10^7 \qquad \mathbf{h}_{22} = 2.000 \text{ mS}$$

Op-amp Specifications	Components
$A = 20,000$	$R_1 = 1 \text{ k}\Omega$
$R_i = 1 \text{ M}\Omega$	$R_2 = 9.999 \text{ M}\Omega$
$R_o = 500 \text{ }\Omega$	

Figure 15.18 *A single-stage amplifier that should have a gain of 10,000.*

Using Table 15.1, we can convert to the transmission parameters.

$$\mathbf{A} = \frac{-\Delta_H}{\mathbf{h}_{21}} = \frac{\mathbf{h}_{12}\mathbf{h}_{21} - \mathbf{h}_{11}\mathbf{h}_{22}}{\mathbf{h}_{21}} \qquad \mathbf{B} = \frac{-\mathbf{h}_{11}}{\mathbf{h}_{21}}$$

$$\mathbf{C} = \frac{-\mathbf{h}_{22}}{\mathbf{h}_{21}} \qquad \mathbf{D} = \frac{-1}{\mathbf{h}_{21}}$$

Based on the hybrid parameter values above, we find

$$\mathbf{A} = 1.501 \times 10^{-4} \qquad \mathbf{B} = 2.502 \times 10^{-2}$$

$$\mathbf{C} = 5.000 \times 10^{-11} \qquad \mathbf{D} = 2.500 \times 10^{-8}$$

The circuit is now modeled by the two-port equations

$$\begin{bmatrix} \mathbf{V}_1 \\ \mathbf{I}_1 \end{bmatrix} = \begin{bmatrix} \mathbf{A} & \mathbf{B} \\ \mathbf{C} & \mathbf{D} \end{bmatrix} \begin{bmatrix} \mathbf{V}_2 \\ -\mathbf{I}_2 \end{bmatrix}$$

Since $\mathbf{V}_2 = -\mathbf{I}_2 R_L$, we can write the equation for $\mathbf{V}_1$ as

$$\mathbf{V}_1 = \mathbf{A}\mathbf{V}_2 + \frac{\mathbf{B}}{R_L}\mathbf{V}_2$$

and the gain as

$$\frac{\mathbf{V}_2}{\mathbf{V}_1} = \frac{1}{\mathbf{A} + \dfrac{\mathbf{B}}{R_L}} = \frac{6667}{1 + \dfrac{166.7}{R_L}}$$

Although the ideal model predicts a gain of 10,000, the actual gain for infinite R_L is only 6667—a rather large discrepancy! A careful analysis of the parameters indicates two problems: (1) the gain of the op-amp is on the same order as the circuit gain, and (2) R_2 is actually larger than R_i. Recall that the ideal op-amp assumptions are that both A and R_i should approach infinity, or, in essence, A should be much larger than the overall gain and R_i should be the largest resistor in the circuit—neither condition exists. We will address these issues by cascading two op-amps. We will design each stage for a gain of 100 by selecting R_2 to be 99 kΩ, thus alleviating the two issues above. Since the stages are cascaded, the ideal overall gain should be $100 \times 100 = 10,000$.

The transmission parameters with the new values of R_2 are

$$\mathbf{A} = 1.005 \times 10^{-2} \qquad \mathbf{B} = 2.502 \times 10^{-2}$$

$$\mathbf{C} = 5.025 \times 10^{-11} \qquad \mathbf{D} = 2.500 \times 10^{-8}$$

Since the two stages are cascaded, the transmission parameter equations that describe the overall circuit are

$$\begin{bmatrix} \mathbf{V}_1 \\ \mathbf{I}_1 \end{bmatrix} = \begin{bmatrix} \mathbf{A}_a & \mathbf{B}_a \\ \mathbf{C}_a & \mathbf{D}_a \end{bmatrix} \begin{bmatrix} \mathbf{A}_b & \mathbf{B}_b \\ \mathbf{C}_b & \mathbf{D}_b \end{bmatrix} \begin{bmatrix} \mathbf{V}_2 \\ -\mathbf{I}_2 \end{bmatrix}$$

or

$$\begin{bmatrix} \mathbf{V}_1 \\ \mathbf{I}_1 \end{bmatrix} = \begin{bmatrix} \mathbf{A}_a\mathbf{A}_b - \mathbf{B}_a\mathbf{C}_b & \mathbf{A}_a\mathbf{B}_b - \mathbf{B}_a\mathbf{D}_b \\ \mathbf{C}_a\mathbf{A}_b - \mathbf{D}_a\mathbf{C}_b & \mathbf{C}_a\mathbf{B}_b - \mathbf{D}_a\mathbf{D}_b \end{bmatrix} \begin{bmatrix} \mathbf{V}_2 \\ -\mathbf{I}_2 \end{bmatrix}$$

where the subscripts a and b indicate the first and second op-amp stages. Since the stages are identical, we can just use $\mathbf{A}$, $\mathbf{B}$, $\mathbf{C}$, and $\mathbf{D}$. Still, $\mathbf{V}_2 = -\mathbf{I}_2 R_L$, and the gain is

$$\frac{\mathbf{V}_2}{\mathbf{V}_1} = \frac{1}{\mathbf{A}^2 - \mathbf{B}\mathbf{C} + \dfrac{\mathbf{A}\mathbf{B} - \mathbf{B}\mathbf{D}}{R_L}}$$

(continued)

Using our transmission parameter values, when R_L is infinite, the gain is 9900.75, an error of less than 1%. This is a significant improvement over the single-stage amplifier. Figure 15.19 shows the single- and two-stage gains versus load resistance, R_L. As R_L decreases, the superiority of the two-stage amp is even more marked.

Figure 15.19
The voltage gain of the single- and two-stage op-amp circuits versus load resistance.

LEARNING Check

Summary

▶ Four of the most common parameters used to describe a two-port network are the admittance, impedance, hybrid, and transmission parameters.

▶ If all the two-port parameters for a network exist, a set of conversion formulas can be used to relate one set of two-port parameters to another.

▶ When interconnecting two-ports, the Y parameters are added for a parallel connection, the Z parameters are added for a series connection, and the transmission parameters in matrix form are multiplied together for a cascade connection.

Problems

SECTIONS 15.1 and 15.2

15.1 Given the two networks in Fig. P15.1, find the Y parameters for the circuit in (a) and the Z parameters for the circuit in (b).

(a)

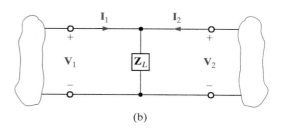

(b)

Figure P15.1

15.2 Find the Y parameters for the two-port network shown in Fig. P15.2.

Figure P15.2

15.3 Find the Y parameters for the two-port network shown in Fig. P15.3.

Figure P15.3

15.4 If a 12-A source is connected at the input port of the network shown in Fig. P15.2, find the current in a 4-Ω load resistor.

15.5 Find the Y parameters for the two-port network in Fig. P15.5.

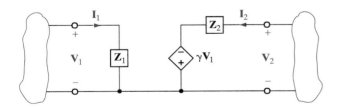

Figure P15.5

15.6 Determine the Y parameters for the network shown in Fig. P15.6.

Figure P15.6

15.7 Determine the admittance parameters for the network shown in Fig. P15.7.

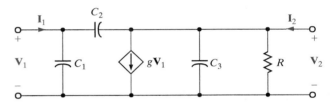

Figure P15.7

15.8 Find the Z parameters for the two-port network in Fig. P15.8.

Figure P15.8

15.9 Find the Z parameters for the network in Fig. P15.5.

15.10 Determine the Z parameters for the two-port network shown in Fig. P15.10.

Figure P15.10

15.11 Find the Z parameters for the two-port shown in Fig. P15.11 and determine the voltage gain of the entire circuit with a 4-kΩ load attached to the output.

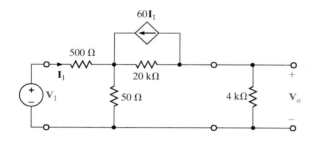

Figure P15.11

15.12 Find the Z parameters for the two-port network shown in Fig. P15.12.

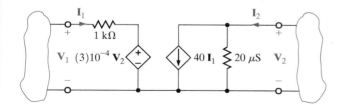

Figure P15.12

15.13 Find the voltage gain of the two-port network in Fig. P15.12 if a 12-kΩ load is connected to the output port.

15.14 Find the input impedance of the network in Fig. P15.12.

15.15 Find the Z parameters of the two-port network in Fig. P15.15.

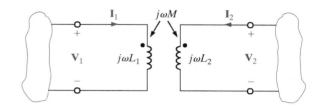

Figure P15.15

15.16 Determine the Z parameters for the two-port network in Fig. P15.16.

Figure P15.16

SECTION 15.3

15.17 Compute the hybrid parameters for the network in Fig. E15.1.

15.18 Consider the network in Fig. P15.18. The two-port network is a hybrid model for a basic transistor. Determine the voltage gain of the entire network, V_2/V_S, if a source V_S with internal resistance R_1 is applied at the input to the two-port network and a load R_L is connected at the output port.

Figure P15.18

15.19 Find the hybrid parameters for the network in Problem 15.2.

15.20 Determine the hybrid parameters for the network shown in Fig. P15.20.

Figure P15.20

SECTION 15.4

15.21 Find the ABCD parameters for the networks in Fig. P15.1.

15.22 Find the transmission parameters for the network in Fig. P15.22.

Figure P15.22

15.23 Find the transmission parameters for the network shown in Fig. P15.3.

15.24 Find the ABCD parameters for the circuit in Fig. P15.5.

15.25 Determine the transmission parameters for the network in Fig. P15.25.

Figure P15.25

15.26 Find the transmission parameters for the circuit in Fig. P15.26.

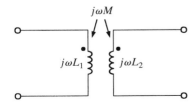

Figure P15.26

15.27 Given the network in Fig. P15.27, find the transmission parameters for the two-port network and then find $\mathbf{I}_o$ using the terminal conditions.

Figure P15.27

SECTION 15.5

15.28 Following are the hybrid parameters for a network.

$$\begin{bmatrix} \mathbf{h}_{11} & \mathbf{h}_{12} \\ \mathbf{h}_{21} & \mathbf{h}_{22} \end{bmatrix} = \begin{bmatrix} \frac{11}{5} & \frac{2}{5} \\ -\frac{2}{5} & \frac{1}{5} \end{bmatrix}$$

Determine the Y parameters for the network.

15.29 If the Y parameters for a network are known to be

$$\begin{bmatrix} \mathbf{y}_{11} & \mathbf{y}_{12} \\ \mathbf{y}_{21} & \mathbf{y}_{22} \end{bmatrix} = \begin{bmatrix} \frac{5}{11} & -\frac{2}{11} \\ -\frac{2}{11} & \frac{3}{11} \end{bmatrix}$$

find the Z parameters.

15.30 Find the Z parameters in terms of the ABCD parameters.

15.31 Find the hybrid parameters in terms of the impedance parameters.

SECTION 15.6

15.32 Find the Y parameters of the network in Fig. E15.1 by considering the network to be a parallel interconnection of two two-port networks as shown in Fig. P15.32.

Figure P15.32

15.33 Find the Y parameters of the two-port network in Fig. P15.33. Find the input admittance of the network when the capacitor is connected to the output port.

Figure P15.33

15.34 Find the Y parameters for the network in Fig. P15.34.

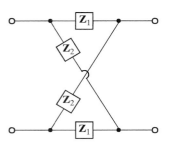

Figure P15.34

15.35 Determine the Y parameters for the network shown in Fig. P15.35.

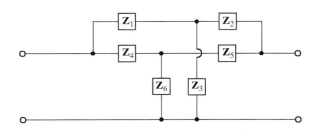

Figure P15.35

15.36 Find the Z parameters of the network in Fig. E15.3 by considering the circuit to be a series interconnection of two two-port networks as shown in Fig. P15.36.

Figure P15.36

15.37 Find the transmission parameters of the network in Fig. E15.3 by considering the circuit to be a cascade interconnection of three two-port networks as shown in Fig. P15.37.

Figure P15.37

15.38 Find the ABCD parameters for the circuit in Fig. P15.38.

Figure P15.38

15.39 Find the transmission parameters for the two-port network in Fig. P15.39.

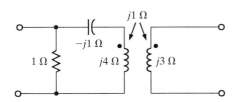

Figure P15.39

15.40 Find the transmission parameters of the two-port network in Fig. P15.40 and then use the terminal conditions to compute $\mathbf{I}_o$.

Figure P15.40

Typical Problems Found on the FE Exam

15FE-1 A two-port network is known to have the following parameters:

$$y_{11} = \tfrac{1}{14}\,\text{S} \qquad y_{12} = y_{21} = -\tfrac{1}{21}\,\text{S} \qquad y_{22} = \tfrac{1}{7}\,\text{S}$$

If a 2-A current source is connected to the input terminals as shown in Fig. 15PFE.1, find the voltage across this current source.

Figure 15PFE-1

15FE-2 Find the Thévenin equivalent circuit at the output terminals of the network in Fig. 15PFE.1.

Basic Semiconductor Electronic Circuits

16

In this chapter we briefly introduce two electronic elements that are widely used in circuits of all types. These elements are the diode and the transistor. Our discussion of the diode focuses on the construction of the device and four techniques that are used for modeling this element. The models are then used in circuit examples.

Two popular types of transistors are presented: the MOSFET (Metal Oxide Semiconductor Field Effect Transistor) and the BJT (Bipolar Junction Transistor). We first discuss the construction of the devices and their characteristics. Then we focus our attention on the MOSFET and consider its usefulness in both switching and amplifier applications.

16.1 Introduction

In this chapter we provide an introduction to some basic semiconductor electronic devices. We first illustrate how the standard passive elements are constructed, or fabricated, in semiconductor "chips." We then introduce two new semiconductor devices: the diode and the transistor. The use of these two components in conjunction with standard passive elements provides us with almost limitless capabilities in the application and design of circuits.

Figure 16.1 illustrates the layout of the passive elements on a silicon chip. Figure 16.1a shows a resistor, and a capacitor and a spiral inductor are shown in Figs. 16.1b and c, respectively. The pads shown in some of the figures are external connection points. Inspecting Fig. 16.1, we see that each device is fabricated on the silicon surface—a technique known as *planar processing*. Planar processing and the lack of magnetic core materials are inherent problems in fabricating large-valued inductors on chip. Thus, inductors are rarely used in semiconductor electronics and designers will typically use a combination of other passive and active devices to obtain the desired effect. Although this use of multiple components as a substitute for an inductor may seem extravagant, it is not. It is much easier to fabricate these other devices than it is to fabricate an inductor and, thus, it is common practice to do so in most cases. One notable exception is in hand-held wireless products such as cell phones, where transmission frequencies approach and exceed 1 GHz. At such high frequencies, effective L-C filters can be constructed on-chip using inductors in the nH range.

In the following sections, we present the semiconductor diode—a nonlinear circuit element. Our approach will be to describe the operation of this device using four distinct models and illustrate the unique effect the diode has when used in circuits. Finally, we introduce the semiconductor transistor, another nonlinear device. In particular, we will discuss one specific device, the MOSFET—metal oxide semiconductor field effect transistor. Once again, we will model this device using equivalent linear circuits and describe its use in several applications. Throughout this chapter you will find the many concepts and theorems, presented earlier, employed with these new semiconductor electronic devices.

Silicon is the bottom plate!

Serpentine resistor structure

Contact points

Top capacitor plate

Contact points

Spiral inductor

Contact points

(a)   (b)   (c)

Figure 16.1 *Examples of passive-device construction in integrated circuit processing. (a) The resistor snakes back and forth in a serpentine pattern. This produces a larger resistor value in a compact area. (b) Capacitors require two conductive plates separated by an insulator. Here the bottom plate is the silicon itself and the insulator, hidden by the top plate, is silicon dioxide. (c) A planar inductor is simple to fabricate but yields a low inductance value.*

16.2 Diodes

STRUCTURE   Silicon, an element in column IV of the periodic table (shown partially in Fig. 16.2), is a poor conductor of current. If we dope silicon with elements from column III, such as boron, we get what is known as a p-type material. Doping with an element from column V, such as phosphorus, yields what is called an n-type material. P-type doping generates positive charge carriers, whereas n-type doping generates negative charge carriers. A carrier is a charged particle that can move about, creating current. A p-n junction is formed if the doping changes from n-type to p-type over a short distance. This p-n junction forms what is called a diode. The diode's symbol and general structure are shown in Fig. 16.3a, and a photograph of an actual diode is shown in Fig. 16.3b. In concert with the old vacuum tube technology, the p-side of the junction is called the anode and the n-side is called the cathode.

Column III	Column IV	Column V	Column VI
5	6	7	8
Boron	Carbon	Nitrogen	Oxygen
13	14	15	16
Aluminum	Silicon	Phosphorus	Sulfur
31	32	33	34
Gallium	Germanium	Arsenic	Selenium

Figure 16.2   *A portion of the periodic table emphasizing the location of silicon and its typical dopants, boron and phosphorus.*

(a)   (b)

Figure 16.3
(a) Diode symbol and
(b) structure.

MODELING TECHNIQUES   Diodes can be modeled using two distinct approaches: a nonlinear model based on solid-state physics, or a linear approximation to the voltage–current performance of the device. There are three popular linear models for the diode: the ideal model, the constant voltage model, and the piecewise linear model. In the ideal model, shown in Fig. 16.4a, the device is treated as a simple switch. When acting like a closed switch, as shown in Fig. 16.4b, the diode is said to be forward-biased with V_D zero and I_D positive. Alternatively, when reverse-biased, the condition in Fig. 16.4c, I_D is zero, V_D is negative, and the "switch"

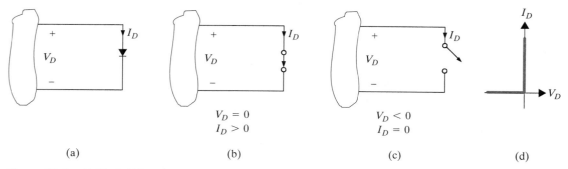

Figure 16.4 *(a) Diode I-V polarities definitions. (b) Ideal diode I-V conditions and equivalent circuits in forward bias and (c) reverse bias. (d) The I-V curve for the ideal diode.*

is open. Thus, the ideal diode passes current from anode to cathode and blocks current in the opposite direction. The *I-V* characteristic for the ideal diode is shown in Fig. 16.4d. The discontinuity in the characteristic clearly indicates that this element is a nonlinear device, and thus care must be exercised in our analysis of the circuits containing these elements.

The voltage–current characteristic, or *I-V* curve, for the actual diode is shown in Fig. 16.5 and is described by a mathematical equation called the diode equation

$$I_D = I_S\left[e^{qV_D/nkT} - 1\right]$$ **16.1**

where I_D and V_D are the diode current and voltage defined in Fig. 16.4a, I_S is the reverse saturation current, q is the charge on an electron, k is Boltzmann's constant, T is temperature in degrees Kelvin, and n is the ideality factor. Typical values for I_S and n are around 10^{-14} A and 1, respectively. However, in diodes designed for large currents, approximately 10^{-10} A and 2 are more representative. At room temperature, q/kT is approximately 39. Therefore, the diode equation can be expressed as

$$I_D = I_S\left[e^{39V_D} - 1\right]$$

In the forward-bias condition, the exponential term has a positive exponent and this term dominates the function. Thus, the *I-V* characteristic is exponential in nature. However, some positive voltage is required before any reasonable amount of current flow is obtained. In the reverse-bias condition, the negative value of V_D makes the exponential term negative and I_D asymptotically approaches the value of $-I_S$. The diode equation in Eq. (16.1) forms the core of the diode model used in simulators such as PSPICE.

The constant voltage model, a modification of the ideal model, is the most popular model for pen-and-paper analysis. Here, the diode is modeled as a dc voltage source in series with an

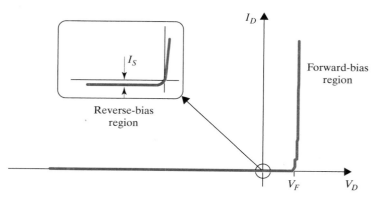

Figure 16.5
Current-voltage characteristics for an actual diode.

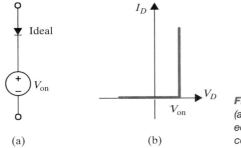

Figure 16.6
(a) The constant voltage model equivalent circuit and (b) the corresponding I-V curve.

ideal diode, as shown in Fig. 16.6a. The resulting *I-V* characteristic is shown in Fig. 16.6b. When forward-biased, the diode voltage is V_{on}, typically chosen at 0.6 or 0.7 V. Note that V_D values below V_{on} produce no current. Therefore, the diode is reverse-biased for $V_D < V_{on}$.

Neither the ideal nor constant voltage models treat the finite slope of the *I-V* curve in forward-bias. The piecewise linear model does so by adding a series resistance, r_d, to the constant voltage model, as shown in Fig. 16.7a. Figure 16.7b shows the resulting *I-V* characteristic where the slope is $1/r_d$. The diode is reverse-biased with no current until V_D reaches V_F. At that point, the diode is forward-biased and current increases linearly with V_D.

A comparison of these four models is shown in Fig. 16.8. We see that the accuracy of the models with respect to the diode equation improves as we progress from the ideal to the constant voltage and finally to the piecewise linear model. Unfortunately, this accuracy is bought at the price of model complexity. Tradeoffs such as these are common in engineering and should not be viewed pessimistically. Instead, consider the options a blessing, investigate the trade-offs, and identify the strengths and weaknesses of each option. Then use your knowledge to choose the best model for the analysis at hand.

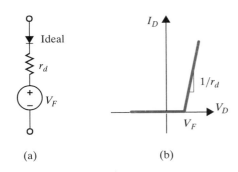

Figure 16.7
(a) The piecewise linear model equivalent circuit and (b) the corresponding I-V curve.

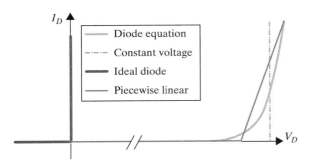

Figure 16.8
A comparison of I-V curves for our diode models.

LEARNING Example 16.1

Using the ideal diode model, find V_1 for the circuits in Fig. 16.9.

Figure 16.9 Circuits used to study the ideal diode model.

SOLUTION In Fig. 16.10a, the voltage source produces a clockwise current, which forward-biases the diode. Therefore, by inspection, $V_1 = 12$ V.

In Fig. 16.10b, the current source should produce a clockwise current and I should be positive. However, the diode is reverse-biased, conducting no current, and all the source current flows through the resistor. Thus, the voltage is $V_1 = (0.1)(100) = 10$ V.

In Fig. 16.10c, the voltage source produces clockwise currents, which will forward-bias the diode. Under this condition, the 3-kΩ resistor is shorted by the diode and, hence, $V_1 = 4$ V.

In Fig. 16.10d, the voltage source will produce counterclockwise currents. Since the diode cannot conduct in that direction, it is reverse-biased, and V_1 can be found by voltage division as

$$V = -4\left[\frac{1000}{1000 + 3000}\right] = -1 \text{ V}$$

Figure 16.10 Equivalent circuits for those in Fig. 16.9 using the ideal diode model.

LEARNING Example 16.2

Diodes are commonly used to protect sensitive circuitry from excessive over/under-voltage conditions. The circuit in Fig. 16.11a is typical where $v_S(t)$ and R_S make up the Thévenin equivalent of the driving network. Using all four diode models, let us deter-

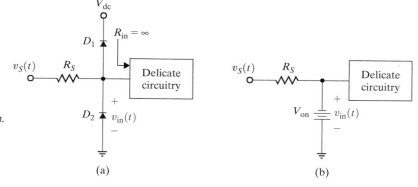

Figure 16.11
(a) A diode circuit that provides voltage protection.
(b) Modeling the under-voltage condition with the diode constant voltage model.

mine the maximum and minimum voltages at the input to the protected circuit.

SOLUTION   For simplicity, we will assume that the protected circuit has infinite input resistance; that is, if the diodes are reverse-biased, then $v_{in}(t) = v_S(t)$. Diodes D_1 and D_2 provide over/under-voltage protection as follows. Suppose $v_S(t)$ drops below ground to the point that D_2 becomes forward-biased. Figure 16.11b shows the situation for the constant voltage model. Now $v_{in}(t)$ is "pinned" at $-V_{on}$ and the circuit is protected. Should $v_S(t)$ exceed V_{dc}, then D_1 will be forward-biased and, in a similar manner, $v_{in}(t)$ pins at $V_{dc} + V_{on}$.

If we use the ideal diode model, $v_{in}(t)$ pins exactly at ground and V_{dc}. For the piecewise linear model, the circuit in Fig. 16.12 models the under-voltage scenario. By superposition, v_{in} can be written as

$$v_{inu} = \frac{v_S r_d}{r_d + R_S} - \frac{V_F R_S}{r_d + R_S}$$

A similar analysis in over-voltage yields

$$v_{ino} = \frac{v_S r_d}{r_d + R_S} + \frac{(V_{dc} + V_F)R_S}{r_d + R_S}$$

where v_{inu} and v_{ino} denote the under- and over-voltage conditions, respectively. Plots for v_{in} versus v_S for $V_{dc} = 3$ V are shown in Fig. 16.13 for the ideal, constant voltage $(V_{on} = 0.7$ V$)$ and the piecewise linear model $(V_F = 0.6$ V, $r_d = 10\ \Omega$, and $R_S = 40\ \Omega)$.

Let us now analyze the under-voltage scenario using the diode equation for diodes with $I_S = 10^{-14}$ A and $n = 1$. The nodal equation at v_{in} is

$$i_D = \frac{(v_{in} - v_S)}{R_S}$$

and the diode equation is

$$i_D = I_S\left[e^{39v_D} - 1\right]$$

From the circuit diagram, $v_D = -v_{in}$. Substituting, eliminating I_D and solving for v_{in} (the variable of interest) yields

$$v_{in} = -\frac{1}{39}\ln\left[\frac{v_{in} - v_S}{R_S I_S} + 1\right] \qquad \textbf{16.2}$$

This equation cannot be solved directly and is obviously the motivation for the linear diode models we have discussed. We can, however, iterate to find v_{in}; that is, we guess. Rather than a hit-and-miss approach, we can use Eq. (16.2) to make the next guess automatically. Thus, the approach is

1. Make a guess for v_{in}.
2. Using the right-hand side of Eq. (16.2), calculate the next guess for v_{in}.
3. Put the new v_{in} into the right-hand side of Eq. (16.2) and calculate another guess.
4. Repeat (3) until two successive guesses are within a predetermined tolerance.

Figure 16.12   *The under-voltage scenario using the piecewise linear model.*

Figure 16.13   *Input voltage using each of the three diode models showing protection against over- and under-voltage inputs.*

(continued)

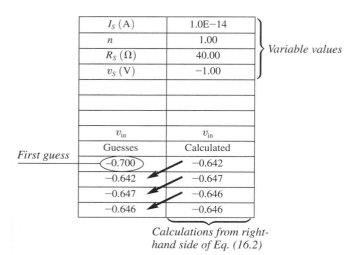

I_S (A)	1.0E−14
n	1.00
R_S (Ω)	40.00
v_S (V)	−1.00

} *Variable values*

v_{in}	v_{in}
Guesses	Calculated
−0.700	−0.642
−0.642	−0.647
−0.647	−0.646
−0.646	−0.646

First guess

Calculations from right-hand side of Eq. (16.2)

Figure 16.14 *The iterative EXCEL spreadsheet for Example 16.2.*

The diode voltage, if forward-biased, should be around 0.6 to 0.8 V. Thus, an original guess of $v_{in} = -0.7$ and a tolerance of 0.001 V are reasonable. Although a calculator could be used to perform the calculations in steps 1–4, an automated process is preferred. We will use Microsoft's EXCEL spreadsheet program. Figure 16.14 shows the results for $v_S = -1$ V. The first column of the iteration table contains the guesses and the second column the calculations from the right-hand side of Eq. (16.2). Variables I_S, n, and v_{in} have been included as well. Note that using $kT/q = 1/39$ implies that the circuit operates at room temperature. From Fig. 16.14, we see that after only three iterations, $v_{in} = -0.646$ V, which corresponds to a diode current of 874 μA. To produce a plot similar to Fig. 16.13, iterations must be performed for a range of v_S values.

16.3 Transistors

THE STRUCTURE OF MOSFETS AND BJTS We live in the information age— a world of marvels such as personal wireless communications (cell phones), "instant" access to data, both vital and trivial (the Internet), and powerful personal computing. It is not unreasonable to state that the transistor is the single technological advancement that made this revolution possible. Transistors are the core components of electronic circuits from simple amplifiers to the most advanced microprocessors. In the former application, the device amplifies an input signal, producing a scaled replica; in the latter, the transistor acts as a voltage-controlled switch, making complex yet compact logic circuits possible.

The transistor is a three-terminal nonlinear device where current flow between two terminals is controlled by a voltage or current at the third. This phenomenon is sometimes called transistor action. We will consider the two most popular transistor types: the MOSFET (metal oxide semiconductor field-effect transistor) and the BJT (bipolar junction transistor). A cross-section of the MOSFET, shown in Fig. 16.15a, reveals the inner workings of the device, which consist of two diodes and a MOS (metal oxide semiconductor) sandwich. Figure 16.15b shows a photograph of three MOSFETs. Figure 16.15c shows the schematic symbol for both the four- and three-terminal MOSFETs. If many MOSFETs share the same silicon die, as shown in Fig. 16.15b, then the four-terminal symbol is used. When MOSFETs are packaged individually, manufacturers usually connect the bulk and source internally and the three-terminal symbol is appropriate. The oxide is an insulator, allowing zero current into the gate terminal. Transistor action occurs when the voltage at the gate terminal regulates the current from drain to source with no current escaping through the bulk terminal. To understand how this occurs, consider the situation in Fig. 16.16a, where the source, drain, gate, and bulk are grounded. Since all terminal voltages are zero, all currents are zero as well, and from source to drain the MOSFET appears to be an "open" switch. For obvious reasons, these terminal conditions are called cutoff. Now increase V_{GS} as shown in Fig. 16.16b. Eventually, the negatively charged electrons in the n-type source and drain are attracted into the area underneath the gate, forming a thin conduction path from source to drain. The gate-to-source voltage at which the channel forms is called the threshold voltage, V_T, and is typically between 0.25 V (low-voltage digital applications like microprocessors) and 5 V

(a)

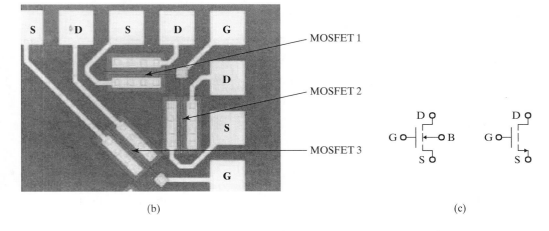

(b)   (c)

Figure 16.15
*(a) A cross-sectional
diagram of an n-channel
MOSFET showing the four
terminals. (b) A photograph
of three MOSFETs and (c)
schematic symbols for the
four-terminal and three-
terminal device.*

(high-voltage/high-current applications like power supplies). Since electrons form the conducting "channel," this device is called an n-channel MOSFET. As V_{GS} increases further, more electrons are added as illustrated in Fig. 16.16c, which makes the path less resistive. If V_{DS} is increased slightly, as shown in Fig. 16.17a, electrons in the channel will be attracted to the more positive voltage at the drain, creating a current. The source terminal is the source of the electrons, and the drain terminal drains them out. Thus, a positive current, I_D, flows drain to source and the device acts like a "closed" switch.

Recall that current is the movement of charge and may be increased by increasing either the amount of moving charge or the velocity of the charge, or both. In the MOSFET, we have direct control over these two means. If V_{DS} increases, the electric field from source to drain also increases, the electrons move faster, and the drain to source current, I_D, increases. The relationship between V_{DS} and I_D is nearly linear. Alternatively, should we increase V_{GS}, more electrons will be attracted to the thin layer between the source and drain, and more charge yields more current.

(a)   (b)   (c)

Figure 16.16
*(a) The MOSFET at cutoff,
(b) the transition from cutoff
to "on," and (c) "on."*

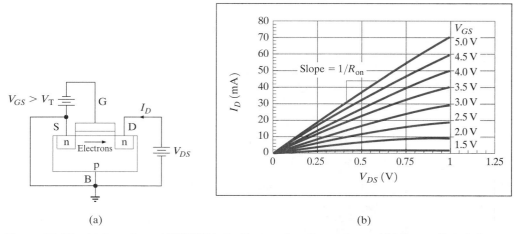

Figure 16.17 *(a) An n-channel MOSFET in the linear region of operation and (b) the resulting device terminal characteristics.*

Interestingly, I_D is nearly linearly dependent on V_{GS} as well. The effects of V_{DS} and V_{GS} on I_D can be seen in the plot in Fig. 16.17b. We now define a new parameter, called the on-resistance, as

$$R_{on} = \frac{V_{DS}}{I_D}$$

On-resistance is exactly that—the resistance between source and drain when the FET is conducting (the switch is closed). This resistance should be small. Since the current is linearly related to V_{DS}, this region of operation is called the linear region. Note that in the linear region, V_{DS} is small. However, if V_{DS} increases, I_D will deviate from the linear curves in Fig. 16.17b and will eventually reach a maximum as shown in Fig. 16.18. This particular plot of I_D versus V_{DS} for various V_{GS}, called the output curve of the transistor, completely describes the MOSFET's operation. Since the current has reached a maximum value, these terminal conditions are called the saturation region of operation. The principle application for the saturation region is amplification, which will be described in more detail in the section Modeling Amplifier Applications. Note that similar arguments can be made for the p-channel MOSFET where all voltages and currents change signs.

A photograph of a bipolar transistor is shown in Fig. 16.19a. The corresponding output curve and its schematic symbol are shown in Figs. 16.19b and c, indicating an important difference between the BJT and MOSFET. The BJT is a current-controlled device in which the

Figure 16.18
MOSFET output curves for $V_T = 1.0$ V indicating cutoff (A), linear (B), and saturation (C).

(a)   (b)   (c)

Figure 16.19   *(a) A photograph of a bipolar transistor. (b) The output curves and (c) schematic symbol for an npn bipolar junction transistor.*

base current, I_B, modulates the collector current, I_C. This input current requirement makes the BJT less attractive than the MOSFET as a switching device. Furthermore, MOSFETs are more compact than BJTs. Since dense circuits such as memories and microprocessors can require millions of transistors, MOSFETs dominate the advanced digital markets. However, bipolar transistors are faster than MOSFETs and do have their state-of-the-art niches—super-fast/low-density digital circuitry that MOSFETs cannot handle and high-frequency (GHzs) analog applications such as cell phones. As with MOSFETs, there are two varieties of BJTs: the npn and pnp, which are similar to the n-channel and p-channel MOSFETs, respectively.

MODELING SWITCHING APPLICATIONS
In switching applications, the MOSFET can be treated as a voltage-controlled switch with appropriate models for the opened/closed scenarios. When the MOSFET is cut off, the current is very nearly zero and an open-circuit model, shown in Fig. 16.20a, is quite accurate. When the transistor is on—that is, the switch is "closed"—the model in Fig. 16.20b is appropriate. From Fig. 16.17, R_{on} is nonzero and a function of V_{GS}, making a single value for R_{on} impossible to specify. Typical values range from a few hundred ohms for low-current (well under a milliamp) applications to as low as 2 mΩ (circa 2000) for switching currents in the tens of amperes.

To investigate the utility of the MOSFET as a switch, consider the network in Fig. 16.21a. For $V_{GS} = 0$ V, the MOSFET is cut off and the resulting circuit is shown in Fig. 16.21b. The output voltage can be determined by voltage division to be

$$V_o = V_S \frac{R}{R + R} = \frac{V_S}{2}$$

Figure 16.20
(a) Model circuits for the MOSFET in cutoff and (b) the linear region.

(a)   (b)   (c)

Figure 16.21
A simple MOSFET switching circuit demonstrating the affects of R_{on}.

which is the ideal value. However, for the "closed" switch position, the network changes to that shown in Fig. 16.21c and the output voltage becomes

$$V_o = V_S \frac{R_X}{R + R_X}$$

where $R_X = R_{on} // R$. To isolate the effect of R_{on}, we rearrange the expression such that

$$V_o = \frac{V_S}{2 + \dfrac{R}{R_{on}}}$$ **16.3**

If the switch were perfect, then R_{on} and V_o would be zero. Obviously, a realistic value for R_{on} causes nonideal switching performance. However, the resistor ratio in Eq. (16.3) indicates that the critical issue is not the absolute value of R_{on} but its relative value with respect to the resistors around it. Since the ideal output voltage value for the "closed" scenario is 0 V, Eq. (16.3) is also the error caused by the switch. A plot of normalized output voltage versus R_{on}/R is shown in Fig. 16.22 with markers for specific error values.

Figure 16.22
Normalized output voltage and error for circuit in Fig. 16.21.

LEARNING Example 16.3

The MOSFET in Fig. 16.23 acts as a switch to energize and de-energize an inductor. Inductance is modeled by the ideal inductor and wiring resistance by R_L. Given $V_S = 10$ V, $R_1 = R_L = 1$ Ω, and $R_{on} = 0.25$ Ω, use Fig. 16.22 to determine the output voltage when the "switch is closed." Find the value of R_{on} required for an output voltage of no more than 0.1 V.

SOLUTION When the FET is conducting and all transients caused by the inductor have subsided, the network is accurately

modeled by the network in Fig. 16.21c. Since $R_1 = R_L = R$, the output voltage plot in Fig. 16.22 is applicable and at $R_{on}/R = 0.25$, the normalized output voltage is $V_o/V_S = 0.17$. Since $V_S = 10$ V, V_o is 1.7 V.

If we desire V_o to be less than 0.1 V, then the normalized output must be less than 0.01, or 1%. From the inset in Fig. 16.22, R_{on}/R must be less than 0.01. Given $R = 1$ Ω, R_{on} must be less than 10 mΩ.

Figure 16.23
Switching network for Example 16.3.

LEARNING Example 16.4

A power MOSFET is used to switch 10 A of current. The circuit, shown in Fig. 16.24, is supplied by a 40-V dc source. The voltage drop across the FET when "on" should be less than 4% of the supply voltage. What is the maximum value of R_{on}?

SOLUTION When "on" the MOSFET conducts 10 A, and the voltage across it must be less than $(40)(0.04) = 1.6$ V. Thus,

$$I_D R_{on} = 10 R_{on} < 1.6$$

and $R_{on} < 0.16$ Ω.

Figure 16.24 *The network for Example 16.4.*

MODELING AMPLIFIER APPLICATIONS Amplification is produced by operating the MOSFET in the saturation region of Fig. 16.18. To illustrate how amplification occurs, consider the MOSFET to be operating at point C. If V_{GS} increases slightly by v_{gs} then I_D will also increase by an amount i_d. If the changes are small enough, the $v_{gs} - i_d$ relationship can be approximated by the linear expression

$$i_d = g_m v_{gs} \qquad\qquad \textbf{16.4}$$

where g_m is the transconductance of the transistor. If the current i_d flows through a resistor, R, then a voltage is produced that is proportional to v_{gs}.

$$v_o = i_d R = (g_m R) v_{gs}$$

Note that v_o is larger than v_{gs} by a factor of $g_m R$, which can be as large as a few hundred. Should we apply the voltage we wish to amplify at v_{gs}, the transistor creates a scaled replica, v_o. This is amplification.

Linearization of nonlinear devices by restricting signal amplitude is very common in electrical engineering because it permits the use of linear circuit analysis techniques. Figure 16.25 shows the resulting MOSFET model, also called the small-signal model. Transconductance, defined in Eq. (16.4), is modeled by the dependent current source and the resistance r_{ds} models the nonzero slope of the I_D-V_{DS} curves in saturation. It is important to note that the model only predicts how changes in V_{GS} produce changes in I_D. Actual values for V_{GS} and I_D cannot be determined from our work here.

Figure 16.25
The MOSFET small-signal model equivalent circuit.

A popular transistor amplifier, the common source configuration, is shown in Fig. 16.26a. The gain of the circuit, $A = v_o/v_{gs}$ can be determined from the small-signal model. Recall that the small-signal model applies to changes or deviations about some operating point. Thus, any node voltage that cannot change must be a ground point in the small-signal circuit. These are sometimes called ac grounds in the ac circuit. Once the small-signal equivalent circuit is determined, any of the linear circuit analysis theories and methods discussed in earlier chapters can be employed. The resulting ac circuit is shown in Fig. 16.26b. The circuit equations are

$$\frac{v_o}{R_D} + i_d = 0 \qquad \text{where} \qquad i_d = \frac{v_o}{r_{ds}} + g_m v_{gs}$$

These equations yield

$$v_o = -g_m v_{gs}(r_{ds}//R_D)$$

The resulting gain is

$$A = -g_m(r_{ds}//R_D) \qquad\qquad\qquad \textbf{16.5}$$

Two interesting features of the gain expression are the negative sign— that is, phase inversion, a characteristic of the common source amplifier—and the role of r_{ds}. Note that if R_D is much less than r_{ds}, then R_D dominates the parallel combination. This is the desired scenario since r_{ds} is a characteristic of the transistor and difficult to control.

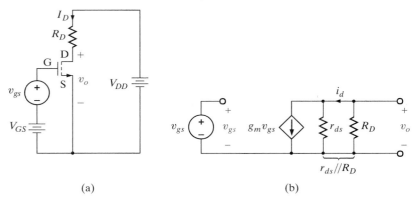

Figure 16.26
(a) The common source amplifier circuit diagram and (b) ac equivalent circuit.

LEARNING Example 16.5

A particular MOSFET has $g_m = 50$ mA/V and $r_{ds} = 2$ kΩ. Find the value of R_D that will produce a common source voltage gain of −50.

SOLUTION From Eq. (16.5), the gain expression is

$$A = -g_m(r_{ds}//R_D) = -(0.05)\left[\frac{(2k)R_D}{2k + R_D}\right] = -50$$

Solving for R_D yields $R_D = 2$ kΩ.

LEARNING Example 16.6

Thousands of the common source amplifiers described in Example 16.5 are to be manufactured. The MOSFET supplier guarantees r_{ds} to 2 kΩ ± 25%. What are the expected maximum and minimum gains?

SOLUTION The maximum and minimum r_{ds} values are

$$r_{ds\ min} = (2k)(1 - 0.25) = 1.5\ kΩ$$

$$r_{ds\ max} = (2k)(1 + 0.25) = 2.5\ kΩ$$

Minimum and maximum gains can be determined from (16.5).

$$A_{min} = -g_m(r_{ds\ min}//R_D) = -(0.05)\left[\frac{(1.5k)(2k)}{1.5k + 2k}\right] = -42.9$$

$$A_{max} = -g_m(r_{ds\ max}//R_D) = -(0.05)\left[\frac{(2.5k)(2k)}{2.5k + 2k}\right] = -55.5$$

LEARNING Example 16.7

Consider the circuit in Fig. 16.27a—a common source amplifier driving an inductive load. The transistor's small-signal characteristics are $g_m = 1.5$ mS and $r_{ds} = 110$ kΩ. We wish to find the value of R_D such that maximum average power operating at 30 krad/s is transferred to the load.

SOLUTION For maximum average power transfer, the output impedance of the amplifier, $\mathbf{Z}_O$, must be the complex conjugate of the load impedance.

$$\mathbf{Z}_O = \mathbf{Z}_L^* = R_L - j\omega L = 1 - j3\ kΩ$$

(continued)

(a)

(b)

(c)

(d)

Figure 16.27
(a) A common source amplifier with inductive load and (b) its small-signal equivalent circuit. (c) and (d) Circuits used to determine $\mathbf{V}_{oc}$ and $\mathbf{I}_{sc}$, respectively.

The output impedance is the Thévenin equivalent impedance defined in Fig. 16.27b—that is, $\mathbf{Z}_O = \mathbf{Z}_{Th}$, and is determined by $\mathbf{V}_{oc}$ and $\mathbf{I}_{sc}$ as defined in Figs. 16.27c and d. To find $\mathbf{V}_{oc}$, we apply KCL at the output node in the network in Fig. 16.27c.

$$(\mathbf{V}_{in} - \mathbf{V}_{oc})(j\omega C_1) = g_m\mathbf{V}_{in} + \frac{\mathbf{V}_{oc}}{R}$$

where $R = r_{ds}//R_D$. Solving for $\mathbf{V}_{oc}$, we have

$$\mathbf{V}_{oc} = \mathbf{V}_{in}\frac{(j\omega C_1 - g_m)R}{j\omega C_1 R + 1}$$

Similarly, for $\mathbf{I}_{sc}$ we employ KCL to the network in Fig. 16.27d and obtain

$$\mathbf{V}_{in}(j\omega C_1) = g_m\mathbf{V}_{in} + \mathbf{I}_{sc}$$

or

$$\mathbf{I}_{sc} = \mathbf{V}_{in}(j\omega C_1 - g_m)$$

The output impedance is

$$\mathbf{Z}_O = \frac{\mathbf{V}_{oc}}{\mathbf{I}_{sc}} = \frac{R}{j\omega C_1 R + 1}$$

$$= \frac{R(1 - j\omega C_1 R)}{\omega^2 C_1^2 R^2 + 1} = \frac{R}{\omega^2 C_1^2 R^2 + 1} - \frac{j\omega C_1 R^2}{\omega^2 C_1^2 R^2 + 1}$$

The requirements for maximum average power transfer are

$$\frac{R}{\omega^2 C_1^2 R^2 + 1} = 1000 \qquad \frac{\omega C_1 R^2}{\omega^2 C_1^2 R^2 + 1} = 3000 \qquad \textbf{16.6}$$

Dividing the two equations in Eq. (16.6) into one another yields

$$\omega C_1 R = 3$$

Given the values for C_1 and ω, the required value of R is 10 kΩ. Since $r_{ds} = 110$ kΩ, the required value of R_D is determined by

$$R = r_{ds} // R_D$$

which yields $R_D = 11$ kΩ.

To achieve large voltage gains, simple amplifiers can be combined as shown in Fig. 16.28a where an n-channel and a p-channel common source amplifier are cascaded. From the small-signal circuit model in Fig. 16.28b, we can determine the gain of each "stage" and the overall gain.

For the first stage, the gain is

$$A_1 = \frac{v_{o1}}{v_{in}} = -g_{m1}(r_{ds1} // R_1)$$

$$= -(2.5\,\text{m})\frac{(80\text{k})(8\text{k})}{80\text{k} + 8\text{k}} = -18.2$$

(a)

$$g_{m1} = 2.5 \text{ mS} \qquad g_{m2} = 1.5 \text{ mS}$$
$$r_{ds1} = 80 \text{ k}\Omega \qquad r_{ds2} = 80 \text{ k}\Omega$$
$$R_1 = 8 \text{ k}\Omega \qquad R_2 = 40 \text{ k}\Omega$$

(b)

Figure 16.28
(a) A cascaded amplifier with n- and p-channel MOSFETS and (b) its small-signal equivalent circuit.

and for the second stage

$$A_2 = \frac{v_o}{v_{o1}} = -g_{m2}(r_{ds2}//R_2)$$

$$= (1.5\,\text{m})\frac{(80\text{k})(40\text{k})}{80\text{k} + 40\text{k}} = -40.0$$

The overall gain is

$$A = \frac{v_o}{v_{in}} = \left(\frac{v_o}{v_{o1}}\right)\left(\frac{v_{o1}}{v_{in}}\right)$$

$$= A_1 A_2 = (-18.2)(-40.0) = 728$$

Note that the overall gain is the product of the individual stage gains; this is a feature of all cascaded amplifiers.

As a highly practical example of employing transistors as switches, we revisit the boost dc–dc converter discussed in Chapter 6 and redrawn in Fig. 16.29a. Having treated transistors and diodes, we can replace the idealized switches with functional components. The result is shown in Fig. 16.29b. When the FET is on (a closed switch), current flows through the inductor, which stores energy. The diode is reverse-biased and the capacitor provides energy to the load. At the instant the FET turns off, the inductor current cannot change and flows instead

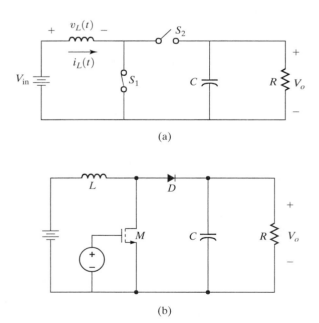

(a)

(b)

Figure 16.29
(a) A boost converter with idealized switching elements and (b) the standard MOSFET–diode implementation.

through the diode to the capacitor and load, both charging the capacitor and driving the load. Of course, in every switching cycle some power is lost in the FET and diode due to R_{on} and V_{on}. However, boost converter efficiencies above 85% are common.

Due to the switching nature of the converter, it is difficult to write a single equation for the performance of the circuit. We can however use the concept of energy balance to model the start-up transient and steady-state operation. As stated in Chapter 5, the energy stored in the inductor in each cycle is

$$W_L = \frac{1}{2} L I_p^2 \qquad \textbf{16.7}$$

where I_p is the peak current and is given by $I_p = V_{in}t_{on}/L$. Substituting into Eq. (16.7) yields

$$W_L = \frac{V_{in}^2 t_{on}^2}{2L} = \frac{V_{in}^2 D^2 T_S^2}{2L} \qquad \textbf{16.8}$$

where T_S is the switching period and D is the duty cycle. As energy is transferred to the capacitor, its voltage and energy build according to the relationship

$$W_C = \frac{1}{2} C(V_o^2 - V_{o-}^2) \qquad \textbf{16.9}$$

where V_{o-} is the output voltage at the end of the previous switching cycle—an initial condition for the present cycle. Finally, we model the energy delivered to the load as

$$W_{load} = \left[\frac{V_{o-}^2}{R_{load}}\right] T_S = \frac{V_{o-}^2}{f_S R_{load}} \qquad \textbf{16.10}$$

where f_S is the switching frequency. We assume that during each switching cycle the energy stored in the inductor is completely released to the capacitor and load.

$$W_L = W_C + W_{load}$$

Substituting from Eqs. (16.9) and (16.10) and solving for V_o yields the expression

$$V_o = \sqrt{\frac{2W_L}{C} - \frac{2V_{o-}^2}{R_{load} C f_S} + V_{o-}^2} \qquad \textbf{16.11}$$

Using EXCEL, Eqs. (16.8) and (16.11) can be easily programmed and V_o plotted as a function of time. One possible worksheet and plot are shown in Figs. 16.30a and b, respectively. Circuit parameters are listed at the top of the sheet and can be edited at any time. The first two full-column variables are the switching cycle number, n, and time $(\text{time} = n/f_S)$. Columns 3 and 4 are simply Eqs. (16.8) and (16.11). The plot indicates that the capacitor voltage builds until all of the inductor energy is being delivered to the load, $W_L = W_{load}$.

Boost Converter Transient Simulation

```
Vin =  5
  L  =  1.00E-05
  C  =  1.00E-06
  D  =  0.6
  fs =  1.00E+05
Rload =  2.0

  n        time         WL          Vo
  0     0.00E+00     4.50E-05        0
  1     1.00E-02     4.50E-05     9.486833
  2     2.00E-02     4.50E-05    13.0767
  3     3.00E-02     4.50E-05    15.6173
  4     4.00E-02     4.50E-05    17.5929
  5     5.00E-02     4.50E-05    19.19789
  6     6.00E-02     4.50E-05    20.53541
  7     7.00E-02     4.50E-05    21.66871
  8     8.00E-02     4.50E-05    22.64022
  9     9.00E-02     4.50E-05    23.48024
 10     1.00E-01     4.50E-05    24.21135
 11     1.10E-01     4.50E-05    24.85097
```

(a)

(b)

Figure 16.30 (a) An EXCEL spreadsheet for simulating the start-up and steady-state output voltage of a boost converter and (b) the resulting transient waveform.

LEARNING Check

Summary

▶ Silicon can be doped to produce both p-type and n-type material. A p-n junction, called a diode, is formed when the two materials are joined.

▶ Four models are used to describe the operation of a diode: the ideal model, the diode equation, the constant voltage model, and the piecewise linear model.

▶ Two popular types of transistors are the MOSFET and BJT.

▶ Both the MOSFET and BJT can be used as voltage-controlled switches and amplifiers.

Problems

SECTION 16.2

16.1 A diode has $I_S = 10^{-15}$ A and $n = 1$. Using the diode equation, find I_D for $V_D = 0$ V, 0.25 V, 0.5 V, and 0.75 V.

16.2 Plot an I_D-V_D curve for the diode in Problem 16.1 over the V_D range 0 to $+0.65$ V. On the same graph, add curves for constant voltage and piecewise linear models for currents between 10 and 100 μA. What are your estimates for V_{on}, V_F, and r_d?

16.3 Repeat Example 16.1 using the constant voltage model with $V_{on} = 0.6$ V.

16.4 For the circuit in Fig. P16.4, find V_o using (a) the ideal diode model, (b) the constant voltage model with $V_{on} = 0.7$ V, and (c) the piecewise linear model with $V_F = 0.6$ V and $r_d = 10$ Ω. Congratulate yourself if you can use the mathematical diode equation with $I_S = 1e{-}15$ A, $n = 1$, and $R = 1$ kΩ.

Figure P16.4

16.5 The circuit in Fig. P16.5 is called a half-wave rectifier. Plot the input and output voltages versus time for three cycles of the input voltage using (a) the ideal diode model and (b) the constant voltage model with $V_{on} = 1.0$ V.

Figure P16.5

SECTION 16.3

16.6 The MOSFET in Fig. P16.6 has $g_m = 1$ mS and $r_{ds} = 100$ kΩ. Draw the small-signal circuit and find the voltage gain v_o/v_{in}.

Figure P16.6

16.8 The amplifier in Fig. P16.8 is a common drain configuration. Find v_o/v_{in} and the output resistance, R_O, in terms of g_m, r_{ds}, and R_S. If $g_m = 10$ mS and $r_{ds} = 1$ kΩ, what is the value of R_O? What possible use could this configuration have?

Figure P16.8

16.7 For the amplifier in Problem 16.6, plot the gain v_o/v_{in} versus R_D for R_D between 1 kΩ and 1 MΩ. Why does the gain not increase linearly with R_D?

16.9 For the MOSFET in Fig. P16.9, $g_m = 2$ mS and assume that r_{ds} is infinite. Using superposition, find $v_o/(v_2 - v_1)$.

Figure P16.9

16.10 The amplifier in Fig. P16.10 is called a differential pair, or differential amplifier. MOSFETs M_1 and M_2 are identical devices with identical g_m values, and v_1 and v_2 are the small-signal inputs. Find the gain $v_o/(v_2 - v_1)$ in terms of g_m and R_D. Assume that r_{ds} is infinite.

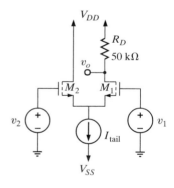

Figure P16.10

16.11 Find an expression for the complex gain $\mathbf{A}(j\omega) = \mathbf{V}_o/\mathbf{V}_{in}$ for the amplifier in Fig. P16.11 in terms of g_m, r_{ds}, R_D, and C. Find expressions for the gain at dc and the pole and zero locations. For $g_m = 500$ μS and $r_{ds} = 400$ kΩ, graph the Bode plot over the frequency range 1 Hz to 100 kHz.

Figure P16.11

16.12 The MOSFET in Fig. P16.12 acts as a switch to turn a red LED on or off. On-resistance for the MOSFET is given by the expression, $R_{on} = 2/(V_{GS} - V_T)$, where $V_T = 1$ V. A LED is nothing more than a diode that glows when forward-biased. The desired operating point for the LED is 2.0 V at 20 mA. Find R_{on} and the required value of R_1.

Figure P16.12

16.13 The MOSFETs in Fig. P16.13 are used to turn on a green or a red LED. V_{GS} is switched to either 0 or 3 V. M_1 is an NMOS device and M_2 is a PMOS device. Which voltage level turns on which LED? Use the constant-voltage model for diodes with $V_{on} = 2.1$ V and model the on-resistance of the MOSFET as $R_{on} = 17/(V_{GS} - 0.75)$. Find the value of R_1 such that the diode currents never exceed 10 mA.

Figure P16.13

16.14 The MOSFET in Fig. P16.14 is used as a switch to turn on the motor. It is desired that the circuit operate at an efficiency of at least 95% where efficiency is defined as

$$\eta = \frac{\text{motor power (W)}}{\text{total power (W)}}$$

Find the maximum allowable value of R_{on}.

Figure P16.14

16.15 Figure P16.15 shows an electric car motor controlled by a MOSFET. For the conditions shown, find the efficiency of the circuit and the MOSFET power when $R_{on} = 5$ mΩ. Efficiency is defined here as

$$\eta = \frac{\text{motor power (W)}}{\text{total power (W)}}$$

Find the value of R_{on} required for 99% efficiency. As mentioned in the text, MOSFETs with R_{on} values as low as 2 mΩ are commercially available (circa 2000). How can you produce the required R_{on} value?

Figure P16.15

Typical Problems Found on the FE Exam

16FE-1 For the circuit in Fig. P16FE-1, sketch the voltage V_o versus V_{in} over the range -10 to $+10$ V. What are the bias conditions for the diodes at $V_{in} = -10, 0,$ and $+10$ V? Assume that the diodes are ideal.

Figure 16PFE-1

16FE-2 The MOSFET in the circuit in Fig. P16FE-2 acts as a switch with an on-resistance of 1 Ω, and the diode is ideal.

(a) Assume that the MOSFET is on and the network has reached steady state. What is the value of the inductor current?

(b) With the circuit operating as described in (a), the MOSFET is turned off at time $t = 0$. Find expressions for $i_L(t)$ and $v_L(t)$ for $t \geq 0$.

Figure 16PFE-2

Complex numbers are typically represented in three forms: exponential, polar, or rectangular. In the exponential form a complex number **A** is written as

$$\mathbf{A} = ze^{j\theta} \qquad\qquad 1$$

The real quantity z is known as the amplitude or magnitude, the real quantity θ is called the *angle* as shown in Fig. 1, and j is the imaginary operator $j = \sqrt{-1}$. θ, which is the angle between the real axis and **A**, may be expressed in either radians or degrees.

The polar form of a complex number **A**, which is symbolically equivalent to the exponential form, is written as

$$\mathbf{A} = z\underline{/\theta} \qquad\qquad 2$$

and the rectangular representation of a complex number is written as

$$\mathbf{A} = x + jy \qquad\qquad 3$$

where x is the real part of **A** and y is the imaginary part of **A**.

The connection between the various representations of **A** can be seen via Euler's identity, which is

$$e^{j\theta} = \cos\theta + j\sin\theta \qquad\qquad 4$$

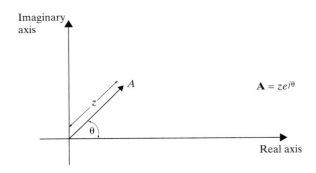

$\mathbf{A} = ze^{j\theta}$

Figure 1
The exponential form of a complex number.

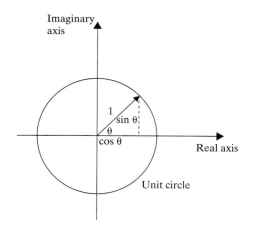

Figure 2
A graphical interpretation of Euler's identity.

Figure 2 illustrates that this function in rectangular form is a complex number with a unit amplitude.

Using this identity, the complex number **A** can be written as

$$\mathbf{A} = ze^{j\theta} = z\cos\theta + jz\sin\theta \tag{5}$$

which, as shown in Fig. 3, can be written as

$$\mathbf{A} = x + jy$$

Equating the real and imaginary parts of these two equations yields

$$x = z\cos\theta$$
$$y = z\sin\theta \tag{6}$$

From these equations we obtain

$$x^2 + y^2 = z^2\cos^2\theta + z^2\sin^2\theta = z^2 \tag{7}$$

Therefore,

$$z = \sqrt{x^2 + y^2} \tag{8}$$

Additionally,

$$\frac{z\sin\theta}{z\cos\theta} = \tan\theta = \frac{y}{x}$$

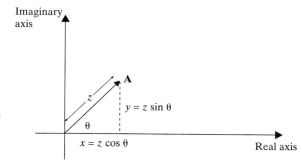

Figure 3
The relationship between the exponential and rectangular representation of a complex number.

and hence

$$\theta = \tan^{-1}\frac{y}{x} \qquad\qquad 9$$

The interrelationships among the three representations of a complex number are as follows.

Exponential	Polar	Rectangular
$ze^{j\theta}$	$z\,\underline{/\theta}$	$x + jy$
$\theta = \tan^{-1}y/x$	$\theta = \tan^{-1}y/x$	$x = z\cos\theta$
$z = \sqrt{x^2 + y^2}$	$z = \sqrt{x^2 + y^2}$	$y = z\sin\theta$

We will now show that the operations of addition, subtraction, multiplication, and division apply to complex numbers in the same manner that they apply to real numbers.

The *sum* of two complex numbers $\mathbf{A} = x_1 + jy_1$ and $\mathbf{B} = x_2 + jy_2$ is

$$\begin{aligned}\mathbf{A} + \mathbf{B} &= x_1 + jy_1 + x_2 + jy_2 \\ &= (x_1 + x_2) + j(y_1 + y_2)\end{aligned} \qquad\qquad 10$$

That is, we simply add the individual real parts, and we add the individual imaginary parts to obtain the components of the resultant complex number.

LEARNING Example 1

Suppose we wish to calculate the sum $\mathbf{A} + \mathbf{B}$ if $\mathbf{A} = 5\,\underline{/36.9°}$ and $\mathbf{B} = 5\,\underline{/53.1°}$.

SOLUTION We must first convert from polar to rectangular form.

$$\mathbf{A} = 5\,\underline{/36.9°} = 4 + j3$$
$$\mathbf{B} = 5\,\underline{/53.1°} = 3 + j4$$

Therefore,

$$\mathbf{A} + \mathbf{B} = 4 + j3 + 3 + j4 = 7 + j7$$
$$= 9.9\,\underline{/45°}$$

LEARNING Hint

Addition and subtraction of complex numbers are most easily performed when the numbers are in rectangular form.

The *difference* of two complex numbers $\mathbf{A} = x_1 + jy_1$ and $\mathbf{B} = x_2 + jy_2$ is

$$\begin{aligned}\mathbf{A} - \mathbf{B} &= (x_1 + jy_1) - (x_2 + jy_2) \\ &= (x_1 - x_2) + j(y_1 - y_2)\end{aligned} \qquad\qquad 11$$

That is, we simply subtract the individual real parts and we subtract the individual imaginary parts to obtain the components of the resultant complex number.

LEARNING Example 2

Let us calculate the difference $\mathbf{A} - \mathbf{B}$ if $\mathbf{A} = 5\,\underline{/36.9°}$ and $\mathbf{B} = 5\,\underline{/53.1°}$.

SOLUTION Converting both numbers from polar to rectangular form,

$$\mathbf{A} = 5\,\underline{/36.9°} = 4 + j3$$
$$\mathbf{B} = 5\,\underline{/53.1°} = 3 + j4$$

Then

$$\mathbf{A} - \mathbf{B} = (4 + j3) - (3 + j4) = 1 - j1 = \sqrt{2}\,\underline{/-45°}$$

The *product* of two complex numbers $\mathbf{A} = z_1\,\underline{/\theta_1} = x_1 + jy_1$ and $\mathbf{B} = z_2\,\underline{/\theta_2} = x_2 + jy_2$ is

$$\mathbf{AB} = \left(z_1 e^{j\theta_1}\right)\left(z_2 e^{j(\theta_2)}\right) = z_1 z_2\,\underline{/\theta_1 + \theta_2} \qquad \mathbf{12}$$

LEARNING Example 3

Given $\mathbf{A} = 5\,\underline{/36.9^\circ}$ and $\mathbf{B} = 5\,\underline{/53.1^\circ}$, we wish to calculate the product in both polar and rectangular forms.

SOLUTION

$$\mathbf{AB} = (5\,\underline{/36.9^\circ})(5\,\underline{/53.1^\circ}) = 25\,\underline{/90^\circ}$$
$$= (4 + j3)(3 + j4)$$
$$= 12 + j16 + j9 + j^2 12$$

$$= 25j$$
$$= 25\,\underline{/90^\circ}$$

LEARNING Hint

Multiplication and division of complex numbers are most easily performed when the numbers are in exponential or polar form.

LEARNING Example 4

Given $\mathbf{A} = 2 + j2$ and $\mathbf{B} = 3 + j4$, we wish to calculate the product $\mathbf{AB}$.

SOLUTION

$$\mathbf{A} = 2 + j2 = 2.828\,\underline{/45^\circ}$$
$$\mathbf{B} = 3 + j4 = 5\,\underline{/53.1^\circ}$$

and

$$\mathbf{AB} = (2.828\,\underline{/45^\circ})(5\,\underline{/53.1^\circ}) = 14.14\,\underline{/98.1^\circ}$$

The *quotient* of two complex numbers $\mathbf{A} = z_1\,\underline{/\theta_1} = x_1 + jy_1$ and $\mathbf{B} = z_2\,\underline{/\theta_2} = x_2 + jy_2$ is

$$\frac{\mathbf{A}}{\mathbf{B}} = \frac{z_1 e^{j\theta_1}}{z_2 e^{j\theta_2}} = \frac{z_1}{z_2} e^{j(\theta_1 - \theta_2)} = \frac{z_1}{z_2}\,\underline{/\theta_1 - \theta_2} \qquad \mathbf{13}$$

LEARNING Example 5

Given $\mathbf{A} = 10\,\underline{/30^\circ}$ and $\mathbf{B} = 5\,\underline{/53.1^\circ}$, we wish to determine the quotient $\mathbf{A}/\mathbf{B}$ in both polar and rectangular forms.

SOLUTION

$$\frac{\mathbf{A}}{\mathbf{B}} = \frac{10\,\underline{/30^\circ}}{5\,\underline{/53.1^\circ}}$$
$$= 2\,\underline{/-23.1^\circ}$$
$$= 1.84 - j0.79$$

LEARNING Example 6

Given $\mathbf{A} = 3 + j4$ and $\mathbf{B} = 1 + j2$, we wish to calculate the quotient $\mathbf{A}/\mathbf{B}$.

SOLUTION

$$\mathbf{A} = 3 + j4 = 5\,\underline{/53.1°}$$

and

$$\mathbf{B} = 1 + j2 = 2.236\,\underline{/63°}$$

$$\frac{\mathbf{A}}{\mathbf{B}} = \frac{5\,\underline{/53.1°}}{2.236\,\underline{/63°}} = 2.236\,\underline{/-9.9°}$$

LEARNING Example 7

If $\mathbf{A} = 3 + j4$, let us compute $1/\mathbf{A}$.

SOLUTION

$$\mathbf{A} = 3 + j4 = 5\,\underline{/53.1°}$$

and

$$1/\mathbf{A} = \frac{1\,\underline{/0°}}{5\,\underline{/53.1°}} = 0.2\,\underline{/-53.1°}$$

or

$$\frac{1}{\mathbf{A}} = \frac{1}{3 + j4} = \frac{3 - j4}{(3 + j4)(3 - j4)}$$

$$= \frac{3 - j4}{25} = 0.12 - j0.16$$

Index